Selected Titles in This Series

19 **Hermann Grassmann,** Extension theory, 2000

18 **Joe Albree, David C. Arney, and V. Frederick Rickey,** A station favorable to the pursuits of science: Primary materials in the history of mathematics at the United States Military Academy, 2000

17 **Jacques Hadamard (Jeremy J. Gray and Abe Shenitzer, Editors),** Non-Euclidean geometry in the theory of automorphic functions, 1999

16 **P. G. L. Dirichlet (with Supplements by R. Dedekind),** Lectures on number theory, 1999

15 **Charles W. Curtis,** Pioneers of representation theory: Frobenius, Burnside, Schur, and Brauer, 1999

14 **Vladimir Maz′ya and Tatyana Shaposhnikova,** Jacques Hadamard, a universal mathematician, 1998

13 **Lars Gårding,** Mathematics and mathematicians: Mathematics in Sweden before 1950, 1998

12 **Walter Rudin,** The way I remember it, 1997

11 **June Barrow-Green,** Poincaré and the three body problem, 1997

10 **John Stillwell,** Sources of hyperbolic geometry, 1996

9 **Bruce C. Berndt and Robert A. Rankin,** Ramanujan: Letters and commentary, 1995

8 **Karen Hunger Parshall and David E. Rowe,** The emergence of the American mathematical research community, 1876–1900: J. J. Sylvester, Felix Klein, and E. H. Moore, 1994

7 **Henk J. M. Bos,** Lectures in the history of mathematics, 1993

6 **Smilka Zdravkovska and Peter L. Duren, Editors,** Golden years of Moscow mathematics, 1993

5 **George W. Mackey,** The scope and history of commutative and noncommutative harmonic analysis, 1992

4 **Charles W. McArthur,** Operations analysis in the U.S. Army Eighth Air Force in World War II, 1990

3 **Peter L. Duren et al., Editors,** A century of mathematics in America, part III, 1989

2 **Peter L. Duren et al., Editors,** A century of mathematics in America, part II, 1989

1 **Peter L. Duren et al., Editors,** A century of mathematics in America, part I, 1988

Extension Theory

Hermann Grassmann

Die

Ausdehnungslehre.

Vollständig und in strenger Form

bearbeitet

von

Hermann Grassmann,
Professor am Gymnasium zu Stettin.

BERLIN, 1862.
VERLAG VON TH. CHR. FR. ENSLIN.
(ADOLPH ENSLIN.)

HISTORY OF MATHEMATICS SOURCES — VOL 19

Extension Theory

Hermann Grassmann

AMERICAN MATHEMATICAL SOCIETY | LONDON MATHEMATICAL SOCIETY

Die Ausdehnungslehre von 1862

by Hermann Grassmann

Published in German by Druck und Verlag von B. G. Teubner, Leipzig, 1896

Translated from the German by Lloyd C. Kannenberg

2000 *Mathematics Subject Classification*. Primary 01A55, 15A75.

Library of Congress Cataloging-in-Publication Data

Grassmann, Hermann, 1809–1877.
[Ausdehnungslehre. English.]
Extension theory / Hermann Grassmann ; translated by Lloyd Kannenberg.
p. cm. — (History of mathematics ; v. 19)
Includes bibliographical references and index.
ISBN 0-8218-2031-1 (alk. paper)
1.Ausdehnungslehre. I.Title. II. History of mathematics ; 19.
QA205.G7413 2000
512′.5—dc21 99-056134

11 10 9 8 7 6 5 4 3 2 29 28 27 26 25 24

Contents

Translator's Note

GRASSMANN's *Ausdehnungslehre* of 1862 is his most mature presentation of his system, and is unique in capturing the full sweep of his mathematical achievement. Stripped of the philosophical drapery of his earlier *Lineale Ausdehnungslehre* of 1844, the "Second Ausdehnungslehre" also contains an enormous amount of material not included (or only suggested) in the earlier book (which, as its title suggests, was deliberately restricted to the "lineale" aspects of the theory, by which he meant that part associated with *linear* constructions, or, in its two-dimensional realization, with constructions using the straightedge of elementary geometry). Among this new material is a detailed development of the inner product and its relation to the concept of angle, the "theory of functions" from the point of view of extension theory, and GRASSMANN's fundamental contributions to the problem of PFAFF. Curiously, one topic mentioned in the Foreword of the "First Ausdehnungslehre" that does *not* appear in the Second is the geometric exponential magnitude and the related geometric logarithm. In many ways, not least of which is its "rigorous Euclidean form", the Second Ausdehnungslehre is the version of GRASSMANN's system most accessible to contemporary readers.

This translation is based on the version of the *Ausdehnungslehre* ("A_2") appearing in *Hermann Grassmanns gesammelte mathematische und physikalische Werke*, published by B. G. TEUBNER under the general editorship of F. ENGEL (which volumes will be referred to here as the "Teubner Edition"). Since the editors of this volume (ENGEL and one of GRASSMANN's sons, Hermann Junior) hoped that it would become a working reference for active mathematicians rather than remain primarily of historical interest, they allowed themselves liberties with the A_2 that were not taken with the other publications in the collection (aside from the *Lehrbuch der Mathematik*, only excerpts of which were published in the Teubner Edition). Thus, in addition to the usual editorial work of correcting misprints, altering a word or construction here and there for clarity, and appending Editorial Comments as end notes, they *rearranged* some of GRASSMANN's theorems to smooth the flow of the argument (e.g. No. 132 was moved and renumbered 116b), *interpolated* new theorems of their own to flesh out and complete the discussion (e.g. No. 230a), *replaced* some of GRASSMANN's Proofs with versions of their own, where they regarded the originals as unsatisfactory (e.g. Part 1 of the Proof of No. 172), and in at least one instance *deleted* one of GRASSMANN's Remarks as being unclear (the Remark following No. 213). In fact, at the time of publication (1896) of this Volume their hope was not unreasonable. As Michael CROWE has documented in his magisterial *History of Vector Analysis* (Notre Dame, 1967; Dover reprint, 1985), there was at that time a "struggle for existence" between several vectorial systems, and such an "improved" A_2 might have been expected to promote the cause of the GRASSMANNian system. In the event, of course, the GRASSMANNian system was not generally adopted, and thus today interest in the A_2 is indeed more as a historical document (albeit an extraordinarily interesting one) than as a tool for mathematical research.

These circumstances left me with decisions to make about how to handle such editorial "improvements". Since my purpose has been to make GRASSMANN available to an English-speaking audience, I had no qualms about letting stand corrections to misprints and minor stylistic improvements; similarly, the rearrangements and even interpolations remained in place, although they are identified as such, either in Footnotes or by being enclosed {between braces}. Similarly, the Editorial Notes, excepting one or two lengthy digressions, have been included almost intact (more on the Editorial Notes below). On the other hand, the replacements and deletions could not pass muster as they stood. GRASSMANN's originals are restored to their proper places in this translation, but *they* are marked {with braces}, and the materials that replaced them in the Teubner Edition now appear in the Editorial Notes. Naturally there was no question of excluding the helpful figures that the editors of the Teubner Edition added to the Work.

GRASSMANN's nomenclature presents problems of an altogether different order. While a translator ought not inflict his technical problems on the patient reader, it is necessary to say at least a word or two here about my treatment of GRASSMANN's novel technical vocabulary. There is, unfortunately, no consistency among the experts about the proper English equivalents for GRASSMANN's technical terms. In fact GRASSMANN himself wavered somewhat in his native German; thus in the *Ausdehnungslehre* of 1844 (the "A_1") he used "Abschattung" for what, in the A_2, he called "Zurückleitung". Here and in my translation I have used "shadow" for this concept, which seems to have been the term preferred by WHITEHEAD in his *Universal Algebra*. Two other contentious terms are "Strecke" and "Verknüpfung", which I have translated as "displacement" and "conjunction", respectively. English-speaking GRASSMANN enthusiasts can and do argue about the best equivalents, and I am afraid my choices may not please them much; but if one were to wait until all such controversies were settled, at least another century would surely pass before an Englished GRASSMANN appeared. In general I have tried to be consistent with the choices used in my translation of the A_1. Finally, it is necessary to mention explicitly two expressions that might possibly lead to some confusion. First, the word "Hauptgebiet" is rendered throughout as "principal domain", and is so indicated in the Index of Technical Terms; it refers to the *underlying vector space* of the system, and must *not* be confused with the expression "principal ideal domain": Second, the word "Stufe", consistently translated as "order", is used in two distinct ways, the first as the "order of a (principal) domain", roughly the *dimension* of the underlying space, and second as the "order of an elementary magnitude", approximately its *rank* in tensor-like terms.

The Editorial Notes have their own interest, both as elaborations and elucidations of GRASSMANN's text, and as illustrating turn-of-the-century attitudes toward his work by competent and reasonably sympathetic mathematicians.

Following the Editorial Notes is a set of Supplementary Notes keyed to the Chapters in the two Parts of the book. These Notes are intended as a guide to help elucidate, in modern terms, what GRASSMANN actually did.

It is a pleasure to acknowledge the deep debt I owe to Prof. Alvin Swimmer and Prof. George B. Seligman, both of whom very kindly read through the entire manuscript and provided extensive lists of corrigenda together with invaluable suggestions and comments. The addition of Supplementary Notes was inspired by Prof. Seligman. With their help this will be a much more useful work. Whatever defects remain are entirely my responsibility.

To conclude: This completes the second installment of my project of translating GRASSMANN's principal works into English. It is pleasant to contemplate collecting the ten pounds (plus accrued interest?) that R. W. GENESE offered to subscribe (*Nature* **48**, 517 (1893)) for the translation of GRASSMANN's *Ausdehnungslehre*; but that is, I suppose, rather too much to hope for. On the other hand, if these translations produce even a few new GRASSMANN enthusiasts in the English-speaking community, the effort will have been worth the trouble. This translation I dedicate to the memory of my paternal grandfather, my first German teacher.

L. Kannenberg
Weston, 1999

Foreword

The present work comprises the complete theory of extension, a mathematical science, the first volume of which I already published seventeen years ago under the title *Linear Extension Theory, a New Branch of Mathematics*; Otto WIGAND Co., Leipzig 1844. In the Foreword to that work I had also indicated the principal topics which, according to my plan, were to be included in the contents of the second volume. Now, instead of publishing the second volume as a continuation of the first, and thus concluding the work in conformity with that plan, I have preferred to take up those topics previously treated in this new work as well, thus to provide a [single] coherent whole.

The principal reason that moved me to this decision is the difficulty which, according to all the mathematicians whose opinions I had the opportunity to hear, the study of that work caused the reader on account of what they believed to be its more philosophical than mathematical form. And this difficulty must in fact have been considerable, since the geometric articles that I have written on the interpretation of that work (*Crelles Journal* **31**, 111–137 (1846), **36**, 177–184 (1848), **42**, 187–212 (1851), **44**, 1–25 (1852), **49**, 1–65, 123–141 (1855), **52**, 254–275 (1856); *Geometrische Analyse*, Leipzig 1847)[1] are often cited and applied by other mathematicians, but the domain treated in that work itself is nowhere touched upon or applied, if I except an interesting short article by KYSAEUS (*Bedeutung und Anwendung der Zahlen in Geometrie*, Siegen 1850). As a consequence of this, no review of the book, or statement of its contents, has ever appeared, indeed not a single mention of it outside of the publisher's list, except for one I wrote myself (*Grunerts Archiv* **6**, 237–250 (1845)).[2]

To remove that difficulty thus became an essential task for me if I wanted the book to be read and understood by others as well as myself. This difficulty could not however be removed without substantially changing the plan of the whole. For it is implicit not in some arbitrarily chosen form but in the plan I had envisaged: To formulate the whole from the ground up, independent of other branches of mathematics. The direct implementation of this plan, although it must be the most expeditious for science as such, must, unless it has also become subjective, offer considerable difficulty in that type of presentation, particularly in a science such as extension theory, which extends and intellectualizes the sensual intuitions of geometry into general, logical concepts, and, with regard to abstract generality, is not simply one among the other branches of mathematics, such as algebra, combination theory, and function theory, but rather far surpasses them, in that all fundamental elements are unified under this branch, which thus as it were forms the keystone of the entire structure of mathematics.

Thus I had to abandon this plan, and in the present work I have presupposed the other branches of mathematics, at least in their elementary development. In addition I have adopted exactly the opposite method in the form of presentation, as I have applied the most rigorous mathematical form we know, the Euclidean, to the present work, and have relegated to the Remarks everything that serves to illustrate or motivate the method chosen.

A necessary consequence of the plan as altered was that all the results of the first part, insofar as they did not include applications to physics, had to be taken up and derived anew according to the altered plan of the new edition (as is seen in Nos. 1–136, 216–329). Yet because of the differences in method of the two editions the same topic becomes so dissimilar between them that one scarcely finds any duplication, with the exception of the results obtained themselves, from which in the nature of the subject no deviation can appear. Thus the previous edition is by no means made superfluous by the new one. Indeed, even the new method is by no means to be preferred to the old; on the contrary, the methods of the first edition that derive from the original idea, and proceed thence completely independently, probe more deeply into the essence of the subject, and thus from a purely scientific viewpoint have a decided advantage relative to the latter. On the other hand this latter method is more acceptable to, and in any case more easily understood by mathematicians, who do not willingly see lying fallow the wealth of mathematical knowledge obtained otherwise by their studies. Thus the two presentations are mutually complementary and illustrative.

The presentation chosen here very closely follows arithmetic, in the sense that it assumes the numerical magnitude as a continuous quantity. Now just as arithmetic develops all other magnitudes from a single magnitude, arbitrarily chosen from the rest, which is defined as the unit and may be symbolized by e (see my *Lehrbuch der Arithmetik*; ENSLIN, Berlin 1861),[3] so, within the framework given here, extension theory first defines several unit magnitudes, e_1, e_2, ... , none of which is derivable from the others (thus for example e_2 cannot be developed from e_1 by multiplication of e_1 with some numerical magnitude), and then considers the magnitudes, which I have called extensive magnitudes, resulting from these units by multiplication with numerical magnitudes and addition of these products. In this way there easily follow the laws of addition, subtraction, multiples, and fractions, discussed in Chapter 1.

It may seem surprising that such a simple idea, which basically consists in no more than regarding multiple sums of different magnitudes (which is how extensive magnitudes appear) as autonomous magnitudes, can in fact be developed into a new science, and indeed in this connection it has been objected that all of extension theory is merely an abbreviated method of notation, and that it is erroneous to regard as magnitudes expressions that are not magnitudes at all. But this objection is based on a complete misunderstanding of the essence of mathematics and of magnitudes. According to this argument all of arithmetic, indeed, one may say all of pure mathematics, is merely an abbreviated method of notation; for number is just an abbreviated expression for a sum of units, the product for a sum of equal numbers, the power for a product of them, and so on; however, without this abbreviated method of notation, or better expressed, without this {formal} unification, no progress toward a unification of concepts is imaginable. Without this unification, for example, it would not be possible to attain the concepts of reductive types of calculation (subtraction, division, roots, logarithms) and the new numerical forms obtained by means of them: negative, fractional, irrational, and imaginary. Thus it is important above all that one actually unify that which in its essence forms a unity, and which therefore must also lead to new results to which one would not be led without that unification.

Now in fact extension theory leads to an inexhaustible wealth of such relations which, without the formation of that conceptual unity appearing in the extensive magnitude, would in no way be conceived or derived. Whether one regards the name "magnitude" as appropriate for this concept is in itself of very subordinate importance, since little depends on names here. The question is only whether this new concept unites with the general concept of magnitude in such a way that in their essence they amalgamate into a general concept, and that a dividing line drawn between the two domains would separate related matters arbitrarily, and divide the subject inconsistently. Thus in this latter case it would even be erroneous *not* to attribute the name "magnitude" to this new concept.

Now I believe that between that which I have called an extensive magnitude and the general numerical magnitude, and particularly the imaginary magnitude $(a + bi)$, there exists a relation so intimate that it would be absurd to treat one and not the other as a magnitude, since the imaginary magnitude is derivable from two units, 1 and $i = \sqrt{-1}$, by real numerical coefficients, just as are extensive magnitudes from two or more units (cf. No. 413, Remark). Thus it seems to me perfectly justified if I designate the extensive magnitude as a *magnitude*. But I go even further, in that I designate them not just as magnitudes generally but as *elementary* magnitudes. Thus they contrast with other magnitudes that likewise bear as decided a character of unified magnitudes as such, as do those elementary ones, and which only enter upon addition of higher structures (cf. Nos. 77, 364, and 377).

I now proceed to clarify the course of development of the present work.

To addition, subtraction, multiples and fractions is adjoined (Chapter 2) the general concept of the multiplication of extensive magnitudes, which is based on the relation of multiplication to addition (that is, that one may multiply the summands in place of the sum). The multiplication of these magnitudes thereby reduces to that of their units, $(e_1, e_2, \dots)$, and from consideration of the products of these units there then follow the different species of multiplication. Now it is possible to separate from these species two, to which all the others can be reduced.

The laws of one of these coincide precisely with ordinary multiplication in algebra, and therefore I will call it algebraic. However it is by far the more complicated with respect to the magnitudes generated and can only be brought to full clarity by consideration of functions, as a consequence of which I have relegated it to the second part of the work. The symbol for this algebraic multiplication must in the nature of the subject coincide with the usual symbol, since it would be absurd to symbolize differently conjunctions that are subject to the same laws in all relations.

The second of those multiplications, which is treated in Chapter Three, is shown to be characteristic of extension theory, and as essential in developing it, as it yields the different orders of elementary magnitudes that enter into extension theory. Thus it is characterized by the condition that two elementary factors of the product can only be interchanged if one simultaneously changes the algebraic sign $(+-)$ of the product. While the relation of this product to addition is indeed the same as with standard multiplication, its other laws deviate substantially from that of standard multiplication, whence it is necessary to distinguish it by its symbol. In this work I have therefore chosen as its symbol brackets enclosing the product, whence $[ab] = -[ba]$ if a and b are elementary factors of this product. This product leads to an extraordinary manifold of apparent forms {Erscheinungsformen}, and permits an enormous abundance of relations to appear which throw a new and unsuspected light on all branches of mathematics, so that this product forms the true core of the new science. Once the concept of the supplementary magnitude is introduced, that product appears in a completely new aspect as the inner product (Chapter 4), so that the range of the subject presented in the first edition (cf. the Foreword to that work) emerges fully in this form. The first part of this volume concludes with applications to geometry (Chapter 5).

Now in the second part there appear combined magnitudes that we can characterize collectively as functions of elementary magnitudes. The first chapter of this part treats functions in general linked by multiplication and division, the second differential calculus, the third the theory of series, and finally the fourth integral calculus, all of these only to the extent that they relate to extensive magnitudes, to be sure. Even so I believe that the corresponding branches of ordinary mathematics (relating to numerical magnitudes), and in particular integral calculus, are not only essentially simplified by this presentation, but are also variously supplemented and developed further.

Because the material is considerably increased from the first edition I have found it necessary to drop all applications to physics; but I hope, if my time and strength permit, to be able to pursue a mathematical treatment of the most important branches of physics in an independent work, in which I will make use of the science reported upon here.[4]

I have earnestly endeavored to avoid superfluous technical words and to limit myself to the smallest possible number of new technical expressions; but since one cannot speak without any words, and thus must either employ new word formations or word combinations for new concepts, or else bestow new meanings on old words, there unavoidably still remains a considerable collection of technical expressions. To aid in comprehensibility I have above all chosen the technical expressions so that, I hope, they immediately suggest the concepts they represent by their own forms; and at the end I have provided an alphabetical index, giving references to the places where they are defined.

It still remains for me to indicate related efforts of other mathematicians. Almost without exception these refer to those topics that I have designated as applications of extension theory to geometry (thus to §§ 24, 28–30, 37–40, 56, 74–79, 91, 92, 101, 102, 114–119, 144–148, 159–170 of the *Ausdehnungslehre* of 1844 and Nos. 216–347 of the present work). In the first edition (1844) the only investigation included there that was known to me was the celebrated work *Barycentrische Kalkül* by the founder of geometric analysis, MÖBIUS, in which he treats the addition of points. Investigations on the geometric addition of displacements (of given length and direction) and of the significance of imaginary magnitudes were unknown to me.

This latter was first presented fully in an article by GAUSS (*Göttingische gelehrte Anzeigen*, 1831), to which GAUSS called my attention in a letter[5] because of a point handled in the same way in the Foreword to the *Ausdehnungslehre*. The concept of the geometric addition of displacements in the plane was already implicit in this presentation of imaginaries. The first to have dealt with the geometric addition of displacements in complete generality appears to have been BELLAVITIS, who had already formulated the relevant calculus in 1835 (*Annali delle scienze del regno Lombardo-Veneto* **5**, 244–259 (1835));[6] see No. 227, Remark.

MÖBIUS independently developed the laws for the geometric addition of displacements and applied them to problems of celestial mechanics in his *Mechanik des Himmels* (1843). After the appearance of my *Ausdehnungslehre* (1844), investigations in the domain of geometric analysis proliferated. In particular, it was again MÖBIUS and BELLAVITIS who materially advanced the science, and contributed significantly to its understanding and the wider circulation of the geometric methods of calculation on which I reported. There then appeared my own work on this subject, partly developed in my paper "*Geometrische Analyse*, geknüpft an die von Leibnitz erfundene geometrische Charakteristik, gekrönte Preisschrift, Leipzig 1847", which MÖBIUS sought to make more accessible to mathematicians by a lucid presentation annexed thereto, and partly in *Crelle's Journal* (in the articles cited above). In addition, a year after the appearance of my *Lineale Ausdehnungslehre*, SAINT-VENANT published on the geometric addition of lines (*Comptes rendus* **21**, 620–625 (1845)),[7] which was identical to the outer multiplication of displacements presented by me in my book (§ 28–40). Obviously he was not familiar with this work, and I sent two copies of it to CAUCHY with the request that one be delivered to SAINT-VENANT, whose address was unknown to me.

Later on, in several articles printed in the *Comptes rendus* of 1858*, CAUCHY published a method for solving algebraic equations and related problems by means of certain symbolic magnitudes that he called *clefs algébriques*, a method that agreed precisely with that presented in my *Ausdehnungslehre* of 1844 (§ 45, 46 and 93). I was far from accusing the celebrated mathematician of plagiarism, but I did believe it was owed me and the subject that I raise a priority claim to the Paris Academy.[9] But the Commission to which this claim was referred in April 1854 for examination and report (*Comptes rendus* **38**, 743f) has never been heard from, and in addition CAUCHY never afterward published anything on the subject.

These articles by CAUCHY are the only ones that had any point of contact with my *Ausdehnungslehre* (of 1844) outside the domain of geometry. Since however these articles claim an independent origin, it appears as if the characteristic core of my book, excepting the geometric incidentals themselves, has stimulated no related efforts. And therefore with renewed courage I have undertaken this new book, which encompasses the old one and brings it to a conclusion.

For I have every confidence that the effort I have applied to the science reported upon here, which has occupied a considerable span of my lifetime and demanded the most intense exertions of my powers, is not to be lost. Indeed I well know that the form I have given the science is, and *must* be, imperfect. But I also know and must declare, even at the risk of sounding presumptuous, — I know, that even if this work as well should lie idle yet another seventeen years or more without influencing the living development of science, a time will come when it will be drawn forth from the dust of oblivion and the ideas laid down here will bear fruit. I know that if I do not succeed to a position, as yet yearned for in vain, where I can gather a circle of students about me upon whom I can impress these ideas and who I can stimulate to develop and enrich them further, yet some day these ideas, even if in an altered form, will reappear and with the passage of time will participate in a lively intellectual exchange. For truth is eternal, it is divine; and no phase in

---

***36,** 70–75, 129–135, 161–169.[8]

the development of truth, however small the domain it embraces, can pass away without a trace. It remains even if the garments in which feeble men clothe it fall into dust.

Stettin, 29 August 1861

Part 1

The Elementary Conjunctions of Extensive Magnitudes

CHAPTER 1

Addition, Subtraction, Multiples and Fractions of Extensive Magnitudes

1. Concepts and laws of calculation

1. DEFINITION. I say that a magnitude a is *derived* from the magnitudes $b, c, \ldots$ by the numbers β, γ, $\ldots$ if

$$a = \beta b + \gamma c + \cdots,$$

where β, γ, $\ldots$ are real numbers, rational or irrational, different from zero or not. I also say that in this case a is *numerically derived* from b, c, $\ldots$.

2. DEFINITION. In addition I say that two or more magnitudes a, b, c, $\ldots$ stand in a *numerical relation* to one another, or that the assembly of magnitudes a, b, c, $\ldots$ is subject to a numerical relation, if any one of them is numerically derivable from the rest, thus if for example one can set

$$a = \beta b + \gamma c + \cdots,$$

where $\beta, \gamma, \ldots$ are real numbers. If the assembly consists of only a *single* magnitude a, then it will be said that the assembly is subject to a numerical relation only in case $a = 0$.

If *two* magnitudes a and b, neither of which is zero, stand in a numerical relation, then I symbolize this by

$$a \equiv b,$$

and say that a is *congruent* to b.

REMARK. Thus two real numbers always, two differently symbolized magnitudes never, stand in a numerical relation to one another. Zero is numerically derivable from any series of magnitudes, that is by the numbers 0, 0, $\ldots$. Thus several magnitudes, one of which is zero, always stand in a numerical relation to one another.

The symbol ($\equiv$) is used here in a sense similar to that of MÖBIUS (in his *Barycentrische Kalkül*).[10] The name (congruent) is based on geometric considerations. It was introduced by GAUSS to denote abstract relations.

3. DEFINITION. I define as a *unit* any magnitude that can serve for the numerical derivation of a series of magnitudes, and in particular I call such a unit an *original unit* if it is not derivable from another unit. The unit of numbers, that is *one*, I call the *absolute* unit, all others *relative*. Zero can never be a unit.

4. DEFINITION. I define as a *system of units* any assembly of magnitudes, standing in no numerical relation to one another, that can serve for the derivation of other magnitudes by means of arbitrary numbers.

REMARK. Here too belongs the case in which the assembly consists of only a single unit (which however according to No. 3 cannot be zero).

5. DEFINITION. I define as an *extensive magnitude* any expression that is derived from a system of units (none of which need be the absolute unit) by numbers, and I call the numbers that belong to the units the *derivation numbers* of that magnitude; for example the polynomial

$$\alpha_1 e_1 + \alpha_2 e_2 + \cdots$$

or

$$\sum \alpha e \quad \text{or} \quad \sum \alpha_r e_r,$$

where $\alpha_1, \alpha_2, \ldots$ are real numbers and $e_1, e_2, \ldots$ form a system of units, is an extensive magnitude, specifically the one derived from the units $e_1, e_2, \ldots$ by the numbers $\alpha_1, \alpha_2, \ldots$ belonging to them.

In general I will reserve the expression *magnitude* for these two species of them only. If the extensive magnitude can be derived from the *original units*, then I call that magnitude an extensive magnitude of *first order.*

REMARK. From elementary mathematics we assume the laws of calculation for numbers, and also for the so-called "named numbers" {benannten Zahlen}, that is, for the extensive magnitudes derived from a single unit; but only for the case that that unit is an original unit.

6. DEFINITION. To *add* two extensive magnitudes derived from the same system of units means to add their derivation numbers belonging to the same units, that is

$$\sum \alpha e + \sum \beta e = \sum (\alpha + \beta) e.$$

7. DEFINITION. To *subtract* an extensive magnitude from another derived from the same system of units means to subtract the derivation numbers of the first from the derivation numbers of the second belonging to the same units, that is

$$\sum \alpha e - \sum \beta e = \sum (\alpha - \beta) e.$$

REMARK. With respect to parentheses I use the convention that a polynomial or product of several factors written without parentheses is equivalent to that expression written with parentheses in which all parentheses commence at the beginning, thus

$$a + b + c = (a + b) + c, \quad abc = (ab)c,$$

and so on.

8. *For extensive magnitudes a, b, c the following fundamental formulas hold:*

1) $$a + b = b + a,$$
2) $$a + (b + c) = a + b + c,$$
3) $$a + b - b = a,$$
4) $$a - b + b = a.$$

PROOF. Suppose

$$a = \sum \alpha e, \qquad b = \sum \beta e, \qquad c = \sum \gamma e;$$

then

1) $$\begin{aligned} a + b &= \sum \alpha e + \sum \beta e = \sum (\alpha + \beta) e && \text{according to [6]} \\ &= \sum (\beta + \alpha) e = \sum \beta e + \sum \alpha e && [6] \\ &= b + a. \end{aligned}$$

2)
$$\begin{aligned} a+(b+c) &= \sum \alpha e + \Big(\sum \beta e + \sum \gamma e\Big) \\ &= \sum \alpha e + \sum (\beta+\gamma)e && [6] \\ &= \sum (\alpha + (\beta+\gamma))e && [6] \\ &= \sum (\alpha+\beta+\gamma)e \\ &= \sum (\alpha+\beta)e + \sum \gamma e && [6] \\ &= \sum \alpha e + \sum \beta e + \sum \gamma e && [6] \\ &= a+b+c. \end{aligned}$$

3)
$$\begin{aligned} a+b-b &= \sum \alpha e + \sum \beta e - \sum \beta e \\ &= \sum (\alpha+\beta)e - \sum \beta e && [6] \\ &= \sum (\alpha+\beta-\beta)e && [7] \\ &= \sum \alpha e = a. \end{aligned}$$

4)
$$\begin{aligned} a-b+b &= \sum \alpha e - \sum \beta e + \sum \beta e \\ &= \sum (\alpha-\beta)e + \sum \beta e && [7] \\ &= \sum (\alpha-\beta+\beta)e && [6] \\ &= \sum \alpha e = a. \end{aligned}$$

9. *All laws of algebraic addition and subtraction hold for extensive magnitudes.*

PROOF. As is known, these laws can be derived from the four fundamental formulas of No. 8.

10. DEFINITION. To *multiply* an extensive magnitude by a number means to multiply all its derivation numbers by that number, that is

$$\sum \alpha e \cdot \beta = \beta \cdot \sum \alpha e = \sum (\alpha\beta)e.$$

11. DEFINITION. To *divide* an extensive magnitude by a number not equal to zero means to divide its derivation numbers by this number, that is

$$\sum \alpha e : \beta = \sum \frac{\alpha}{\beta} e.$$

12. *For the multiplication and division of extensive magnitudes* (a, b) *by numbers* (β, γ) *the following fundamental formulas hold:*

1) $a\beta = \beta a$,
2) $a\beta\gamma = a(\beta\gamma)$,
3) $(a+b)\gamma = a\gamma + b\gamma$,
4) $a(\beta+\gamma) = a\beta + a\gamma$,
5) $a \cdot 1 = a$,
6) $a\beta = 0$ *if and only if either* $a = 0$ *or* $\beta = 0$,

7) $$a : \beta = a\frac{1}{\beta} \text{ if } \beta \neq 0.$$

PROOF. If $a = \sum \alpha e$, $b = \sum \beta e$, where the sum is referred to the system of units $e_1, \dots, e_n$, then

1) $$a\beta = \beta a \qquad \text{by definition [cf. the formula in No. 10]}$$

2) $$\begin{aligned} a\beta\gamma &= \left(\sum \alpha e\right)\beta\gamma = \left(\sum(\alpha\beta)e\right)\gamma && [10] \\ &= \sum(\alpha\beta\gamma)e && [10] \\ &= \sum \alpha(\beta\gamma)e = \left(\sum \alpha e\right)(\beta\gamma) && [10] \\ &= a(\beta\gamma). && [10] \end{aligned}$$

3) $$\begin{aligned} (a+b)\gamma &= \left(\sum \alpha e + \sum \beta e\right)\gamma = \sum(\alpha+\beta)e\cdot\gamma && [6] \\ &= \sum(\alpha+\beta)\gamma e && [10] \\ &= \sum(\alpha\gamma+\beta\gamma)e = \sum(\alpha\gamma)e + \sum(\beta\gamma)e && [6] \\ &= \sum \alpha e\cdot\gamma + \sum \beta e\cdot\gamma && [10] \\ &= a\gamma + b\gamma. \end{aligned}$$

4) $$\begin{aligned} a(\beta+\gamma) &= \sum \alpha e\cdot(\beta+\gamma) = \sum \alpha(\beta+\gamma)e && [10] \\ &= \sum(\alpha\beta+\alpha\gamma)e = \sum \alpha\beta e + \sum \alpha\gamma e && [6] \\ &= \sum \alpha e\cdot\beta + \sum \alpha e\cdot\gamma && [10] \\ &= a\beta + a\gamma. \end{aligned}$$

5) $$\begin{aligned} a\cdot 1 &= \sum \alpha e\cdot 1 = \sum \alpha e && [10] \\ &= a. \end{aligned}$$

6) a) If $a = 0$ then
$$a\beta = 0\cdot\beta = 0;$$

b) if $\beta = 0$ then
$$\begin{aligned} a\beta &= a\cdot 0 = \sum \alpha e\cdot 0 = \sum 0e && [10] \\ &= \sum 0 && [5, \text{Remark}] \\ &= 0; \end{aligned}$$

c) if $a\beta = 0$ then one has
$$0 = a\beta = \sum \alpha e\cdot\beta = \sum \alpha\beta e. \qquad [10]$$

From this it follows that all products $\alpha\beta$, that is $\alpha_1\beta$, $\alpha_2\beta$, ..., $\alpha_n\beta$, must be zero. For suppose that one of them, for example $\alpha_1\beta$, were not zero; then by multiplication of the equation

$$0 = \alpha_1\beta e_1 + \alpha_2\beta e_2 + \cdots + \alpha_n\beta e_n,$$

by $\frac{1}{\alpha_1\beta}$ one obtains the equation

$$0 = e_1 + \frac{\alpha_2\beta}{\alpha_1\beta}e_2 + \cdots + \frac{\alpha_n\beta}{\alpha_1\beta}e_n$$

or

$$e_1 = \left(-\frac{\alpha_2\beta}{\alpha_1\beta}\right)e_2 + \cdots + \left(-\frac{\alpha_n\beta}{\alpha_1\beta}\right)e_n,$$

that is, e_1 would be numerically derivable from e_2, ..., e_n, or in other words a numerical relation would obtain between the units e_1, ..., e_n, which is contrary to assumption since e_1, ..., e_n form a system of units. Consequently

$$0 = \alpha_1\beta = \alpha_2\beta = \cdots = \alpha_n\beta,$$

whence either $\beta = 0$ or, if $\beta \neq 0$,

$$0 = \alpha_1 = \alpha_2 = \cdots = \alpha_n,$$

and therefore

$$a = \sum \alpha e = \sum 0e = \sum 0 \qquad \text{[5, Remark]}$$
$$= 0,$$

that is, if $a\beta = 0$, then either β or a must equal zero.

7) $$a : \beta = \sum \alpha e : \beta = \sum \frac{\alpha}{\beta}e, \qquad [11]$$

since β is not zero,

$$= \sum\left(\alpha\frac{1}{\beta}\right)e = \sum \alpha e \cdot \frac{1}{\beta} \qquad [10]$$
$$= a\frac{1}{\beta}.$$

13. *All algebraic rules of multiplication and division hold for the multiplication and division of extensive magnitudes by numbers.*

PROOF. For from the fundamental formulas (1 to 6) of the previous theorem there follow in the well-known way all the laws of multiplication, and by formula (7) of the same theorem, division is reduced to multiplication, just as in algebra. Thus the algebraic law for division also holds for the division of extensive magnitudes by numbers.

2. Connection between the magnitudes derived from a system of units

14. DEFINITION. The collection of magnitudes numerically derivable from a series of magnitudes $a_1, a_2, \ldots, a_n$ I call the *domain* derivable from those magnitudes (the domain of the magnitudes $a_1, \ldots, a_n$), and in particular a domain of nth *order* if those magnitudes are of first order (that is, numerically derivable from n original units), and *if the domain is not derivable from less than n such magnitudes.* A domain that includes no magnitude but zero is called a domain of *zeroth* order.

REMARK. The domain of first order is therefore the collection of the multiples of a magnitude of first order, that is if one means by the *multiple of a magnitude* each product of the magnitude with a real numerical magnitude.

15. DEFINITION. Two domains are called *identical* if every magnitude of the first domain is simultaneously a magnitude of the second and conversely. If every magnitude of a domain (A) is simultaneously a magnitude of another (B) (without the converse obtaining), then I call the two domains *incident* upon one another, and say that the first domain (A) is *subordinate* to the second, the second *superordinate* to the first. The collection of magnitudes that simultaneously belong to two or more domains are called their *common* domain, and the collection of magnitudes that can be numerically derived from the magnitudes of two or more domains their *covering* domain.

REMARK. If for example the domain A is derivable from the units e_1, e_2, e_3 and the domain B from the units e_2, e_3, e_4, then the common domain of the domains A and B is the domain derived from the units e_2, e_3, and the covering domain of A and B that derived from the units e_1, e_2, e_3, e_4.

16. *Between n magnitudes $a_1, \ldots, a_n$ a numerical relation exists if and only if an equation*

$$\alpha_1 a_1 + \cdots + \alpha_n a_n = 0$$

can be found in which the numbers $\alpha_1, \ldots, \alpha_n$ are not all simultaneously zero.

PROOF. For if, in the equation

$$\alpha_1 a_1 + \cdots + \alpha_n a_n = 0,$$

only one of the numbers $\alpha_1, \ldots, \alpha_n$ is different from zero, for example α_1, then the magnitude bound to this number is numerically derivable from the rest; for then

$$a = -\frac{\alpha_2}{\alpha_1}a_2 - \frac{\alpha_3}{\alpha_1}a_3 - \cdots - \frac{\alpha_n}{\alpha_1}a_n.$$

Conversely, if any numerical relation between the magnitudes $a_1, \ldots, a_n$ exists, for example

$$a_1 = \beta_2 a_2 + \beta_3 a_3 + \cdots + \beta_n a_n,$$

then

$$-a_1 + \beta_2 a_2 + \beta_3 a_3 + \cdots + \beta_n a_n = 0,$$

an equation in which at least the coefficient of a_1 is not equal to zero.

17. *If n magnitudes stand in a numerical relation to one another, and they are not all zero, it must be possible to separate from them an assembly of less than n magnitudes subject to no numerical relation, from which the rest of the magnitudes are numerically derivable.*

PROOF. If $a_1, \dots, a_n$ are the magnitudes standing in a numerical relation to one another, then (according to No. 2) one of them must be numerically derivable from the rest; let this one be a_n and say

$$a_n = \alpha_1 a_1 + \cdots + \alpha_{n-1} a_{n-1}.$$

Now if there is again a numerical relation between the magnitudes $a_1, \dots, a_{n-1}$, then again one of them, say a_{n-1}, must be numerically derivable from the remaining $a_1, \dots, a_{n-2}$. Thus let

$$a_{n-1} = \beta_1 a_1 + \cdots + \beta_{n-2} a_{n-2}.$$

If one introduces this expression for a_{n-1} into the first equation, one obtains

$$a_n = (\alpha_1 + \alpha_{n-1}\beta_1)a_1 + \cdots + (\alpha_{n-2} + \alpha_{n-1}\beta_{n-1})a_{n-2},$$

whence a_n is also numerically derivable from $a_1, \dots, a_{n-2}$.

One can continue this procedure so long as a numerical relation still obtains between the remaining magnitudes. In this way either one will finally arrive at a group of *several* magnitudes that no longer stand in a numerical relation to one another, and from which the rest are numerically derivable, or there will remain at last only a single magnitude, say a_1, from which all the rest are numerically derivable. In the latter case this single magnitude a_1 may not be zero, since then all the other magnitudes, as numerically derivable from it, would also be zero, contrary to assumption. In both cases one thus (No. 2) arrives at an assembly no longer subject to a numerical relation, and from which all the rest of the n magnitudes $a_1, \dots, a_n$ are numerically derivable.

18. *If in an assembly of magnitudes $a_1, \dots, a_n$ the first, a_1, is not zero, and none of the subsequent members is numerically derivable from the preceding ones, then the assembly is subject to no numerical relation.*

PROOF. For if the assembly is subject to a numerical relation, then (according to 16) an equation

$$\alpha_1 a_1 + \alpha_2 a_2 + \cdots + \alpha_n a_n = 0$$

between the magnitudes $a_1, a_2, \dots, a_n$ can be found, in which not all the coefficients $\alpha_1, \alpha_2, \dots, \alpha_n$ are simultaneously zero. If α_r is the last of these coefficients that is different from zero, then one obtains

$$\alpha_1 a_1 + \alpha_2 a_2 + \cdots + \alpha_r a_r = 0,$$

whence, if r is greater than 1,

$$a_r = -\frac{\alpha_1}{\alpha_r}a_1 - \frac{\alpha_2}{\alpha_r}a_2 - \cdots - \frac{\alpha_{r-1}}{\alpha_r}a_{r-1},$$

that is, a_r is numerically derivable from the remaining magnitudes $a_1, \dots, a_{r-1}$, contrary to assumption. If however $r = 1$, then one has

$$\alpha_1 a_1 = 0;$$

thus, since α_1 is assumed to be different from zero,

$$a_1 = 0;$$

which is likewise contrary to assumption. Thus no numerical relation can exist between the magnitudes $a_1, \dots, a_n$.

19. *If a magnitude* a_1 *is numerically derivable from* n *magnitudes* $b_1, b_2, \dots, b_n$, *and the derivation number belonging to* b_1 *is not equal to zero, then the domain belonging to the* n *magnitudes* $b_1, b_2, \dots, b_n$ *is identical to that derivable from the* n *magnitudes* $a_1, b_2, \dots, b_n$.

PROOF. If $a_1 = \beta_1 b_1 + \beta_2 b_2 + \cdots + \beta_n b_n$, where β_1 is different from zero, then

$$b_1 = \frac{1}{\beta_1} a_1 - \frac{\beta_2}{\beta_1} b_2 - \cdots - \frac{\beta_n}{\beta_1} b_n.$$

Now if c is numerically derivable from $b_1, b_2, \dots, b_n$, for example

$$c = \gamma_1 b_1 + \gamma_2 b_2 + \cdots + \gamma_n b_n,$$

then one finds c derived from $a_1, b_2, \dots, b_n$ by replacing b_1 by the value found, thus

$$c = \frac{\gamma_1}{\beta_1} a_1 + \left(\gamma_2 - \frac{\gamma_1\beta_2}{\beta_1}\right) b_2 + \cdots + \left(\gamma_n - \frac{\gamma_1\beta_n}{\beta_1}\right) b_n.$$

Conversely, if c is numerically derivable from $a_1, b_2, \dots, b_n$, say

$$c = \alpha_1 a_1 + \alpha_2 b_2 + \cdots + \alpha_n b_n,$$

then one finds c derived from $b_1, b_2, \dots, b_n$ by replacing a_1 by its value $\beta_1 b_1 + \beta_2 b_2 + \cdots + \beta_n b_n$, that is

$$c = \alpha_1\beta_1 b_1 + (\alpha_2 + \alpha_1\beta_2) b_2 + \cdots + (\alpha_n + \alpha_1\beta_n) b_n.$$

Thus every magnitude belonging to one of the two domains also belongs to the other, that is, the two domains are identical.

20. *If* m *magnitudes* $a_1, \dots, a_m$, *that stand in no numerical relation to one another, are numerically derivable from* n *magnitudes* $b_1, \dots, b_n$, *then one can always adjoin to the* m *magnitudes* $a_1, \dots, a_m$ $(n-m)$ *further magnitudes* $a_{m+1}, \dots, a_n$ *of such nature that the magnitudes* $b_1, \dots, b_n$ *are also derivable from* $a_1, \dots, a_n$, *and thus the domain of the magnitudes* $a_1, \dots, a_n$ *is identical to the domain of the magnitudes* $b_1, \dots, b_n$; *in addition one can take those* $(n-m)$ *magnitudes from the magnitudes* $b_1, \dots, b_n$ *themselves.*

PROOF. By assumption a_1 is derivable from $b_1, \dots, b_n$. At least one of the numbers by which this derivation results must be different from zero, since otherwise a_1 itself would be zero, whence the assembly of m magnitudes would (according to 2) be subject to a numerical relation.

Let the number belonging to b_1 be different from zero, which one may always assume since one can choose the indices arbitrarily. Then (according to 19) the domain derivable from $b_1, b_2, \dots, b_n$ is identical to that derivable from $a_1, b_2, \dots, b_n$. One has now found for any $r < m$ that the domain of the magnitudes $b_1, b_2, \dots, b_n$ is identical to the domain of the magnitudes $a_1, a_2, \dots, a_r, b_{r+1}, \dots, b_n$, whence, since by hypothesis a_{r+1} is derivable from $b_1, b_2, \dots, b_n$, it is also (by the identity of domains) derivable from $a_1, a_2, \dots, a_r, b_{r+1}, \dots, b_n$. In the expression for this derivation

$$a_{r+1} = \alpha_1 a_1 + \cdots + \alpha_r a_r + \beta_{r+1} b_{r+1} + \cdots + \beta_n b_n$$

one of the coefficients belonging to $b_{r+1}, \dots, b_n$ must be different from zero, since otherwise a numerical relation would obtain between the magnitudes $a_1, \dots, a_{r+1}$, contrary to hypothesis. If this number is β_{r+1}, then (according to 19) the domain derivable from $a_1, \dots, a_r, b_{r+1}, \dots, b_n$ is identical to that derivable from $a_1, \dots,$

$a_{r+1}, b_{r+2}, \ldots, b_n$; thus this latter domain is also identical to the domain derivable from the magnitudes $b_1, \ldots, b_n$. One can therefore continue this procedure from $r = 1$ to $r = m$; that is, the domain of $a_1, \ldots, a_m, b_{m+1}, \ldots, b_n$ is identical to that of $b_1, \ldots, b_n$; and thus if one identifies the remaining arbitrary magnitudes $b_{m+1}, \ldots, b_n$ respectively with $a_{m+1}, \ldots, a_n$, then the domain of the magnitudes $a_1, \ldots, a_n$ is identical to the domain of the magnitudes $b_1, \ldots, b_n$.

21. *If n magnitudes $(a_1, \ldots, a_n)$ standing in no numerical relation are numerically derivable from n other magnitudes $(b_1, \ldots, b_n)$, then the domain of the former series of magnitudes is identical to that of the latter.*

PROOF. One has only to set $m = n$ in the previous theorem, and this theorem follows.

22. *If n magnitudes $(a_1, \ldots, a_n)$ are numerically derivable from less than n magnitudes $(b_1, \ldots, b_m)$, then those n magnitudes always stand in a numerical relation to one another.*

PROOF. Suppose the $a_1, \ldots, a_n$ are derivable from $b_1, \ldots, b_m$, where $m < n$. Now either $a_1, \ldots, a_m$ stand in a numerical relation to one another or not. In the first case the $a_1, \ldots, a_n$ also stand in a numerical relation to one another, since the magnitudes $a_1, \ldots, a_m$ appear among them. In the second case the domain of the magnitudes $a_1, \ldots, a_m$ is (according to 21) identical to the domain of the magnitudes $b_1, \ldots, b_m$, whence every magnitude numerically derivable from the $b_1, \ldots, b_m$ is also numerically derivable from the $a_1, \ldots, a_m$, thus in particular the magnitudes $a_{m+1}, \ldots, a_n$, which by hypothesis are derivable from the $b_1, \ldots, b_m$, are also derivable from the $a_1, \ldots, a_m$, that is, in this second case as well a numerical relation exists between the $a_1, \ldots, a_n$.

23. *If a domain of nth order is derivable from n magnitudes of first order, then these magnitudes stand in no numerical relation to one another, and conversely: If n magnitudes of first order stand in no numerical relation to one another, then the domain derivable from them is a domain of nth order.*

PROOF. Suppose A is the domain derivable from the n magnitudes of first order $a_1, \ldots, a_n$. First, if A is a domain of nth order, then the $a_1, \ldots, a_n$ can stand in no numerical relation to one another; for otherwise one of these magnitudes could be numerically derived from the rest, whence the domain would also be derivable from these $n - 1$, which (according to 14) contradicts the concept of a domain of nth order. Second, and conversely, if the $a_1, \ldots, a_n$ stand in no numerical relation to one another, then (according to 22) they cannot be numerically derived from less than n magnitudes, whence in particular not from the domain derivable from the $a_1, \ldots, a_n$, as a consequence of which (according to 14) this domain is of nth order.

24. *Every domain of nth order can be derived from n magnitudes of first order {belonging to it} that stand in no numerical relation to one another, and indeed from any n such magnitudes of the domain.*

PROOF. Let $a_1, \ldots, a_n$ be the magnitudes from which the domain considered was originally derived, and let $b_1, \ldots, b_n$ be n magnitudes of this domain standing in no numerical relation to one another. Since the $b_1, \ldots, b_n$ belong to the domain derived from the $a_1, \dots, a_n$, it follows (according to 14) that the magnitudes $b_1, \dots, b_n$ can be numerically derived from the $a_1, \dots, a_n$, and since at the same time those magnitudes $b_1, \dots, b_n$ stand in no numerical relation to one another (by assumption), we see that (according to 21) the domain derivable from the $b_1, \dots, b_n$ is identical to that derivable from the $a_1, \dots, a_n$.

REMARK. By this theorem any particular distinction between the original units and the magnitudes numerically derivable from them is removed, since one can derive every domain from n arbitrary magnitudes of this domain that stand in no numerical relation to one another instead of from the n original units taken as fundamental, and one can take the former magnitudes instead of the latter as units. This important theorem can also be derived directly from the theory of elimination. In fact our theorem is only a transformation of the theorem: If n magnitudes $y_1, \dots, y_n$ are completely homogeneous functions of first degree of n others $x_1, \dots, x_n$, and the former can all simultaneously be zero only in case the latter are also all zero, then the latter $(x_1, \dots, x_n)$ can also be represented as completely homogeneous functions of first degree of the former $(y_1, \dots, y_n)$.

However, the proof presented here is not only simple, but in addition has the advantage that thereby the essential, elementary relation between the extensive magnitudes appears more clearly.

25. *The order of two domains taken together is as great as the orders of their common domain and their covering domain, that is, if m and n are the orders of the given domains, r that of their common domain, and v that of their covering domain, then*

$$m + n = r + v.$$

PROOF. Let A and B be the two given domains, of mth and nth orders, and let A be derivable from the magnitudes $a_1, \dots, a_m$, B from the magnitudes $b_1, \dots, b_n$. Then (according to 23) no numerical relation can exist between the $a_1, \dots, a_m$ or between the $b_1, \dots, b_n$. Furthermore one can find an assembly of r, but no more, magnitudes that stand in no numerical relation to one another and simultaneously belong to both domains. This assumption will always be possible, since r may also be zero. Let $c_1, \dots, c_r$ be these magnitudes. Then (according to 20) in the series of magnitudes $a_1, \dots, a_m$ one can replace r of them, say $a_1, \dots, a_r$, by the magnitudes $c_1, \dots, c_r$ in such a way that the domain derivable from this new series of magnitudes is identical to the domain A. Likewise in the series of magnitudes $b_1, \dots, b_n$ one can replace r of them, say $b_1, \dots, b_r$, by the magnitudes $c_1, \dots, c_r$ in such a way that the domain derivable from this new series of magnitudes is identical to the domain B. Thus A is derivable from the m magnitudes $c_1, \dots, c_r, a_{r+1}, \dots, a_m$, and B from the n magnitudes $c_1, \dots, c_r, b_{r+1}, \dots, b_n$. None of these series of magnitudes is subject to a numerical relation (according to 23). Then it is clear that all the magnitudes derivable from the $c_1, \dots, c_r$ are common to the domains A and B; but no others are, since otherwise, contrary to assumption, there would be more than r magnitudes standing in no numerical relation to one another that were common to the two domains A and B. The domain common to the domains A and B is therefore the domain derived from the $c_1, \dots, c_r$, whence (according to 23) a domain of rth order.

Now in addition all the magnitudes $c_1, \dots, c_r, a_{r+1}, \dots, a_m, b_{r+1}, \dots, b_n$ form a series of magnitudes standing in no numerical relation to one another. For if there existed between them a numerical relation, then this must be in the form

$$a + b + c = 0,$$

where a is derivable from the $a_{r+1}, \ldots, a_m$, b from the $b_{r+1}, \ldots, b_n$, c from the $c_1, \ldots, c_r$. Neither a nor b can be zero here. For if a were zero, then $b + c = 0$, and there would then exist between the magnitudes $b_{r+1}, \ldots, b_n, c_1, \ldots, c_r$ a numerical relation, which as has been shown is impossible; and were b zero there would exist a numerical relation between the $a_{r+1}, \ldots, a_m, c_1, \ldots, c_r$, which likewise was already shown to be impossible.

If we present the above equation in the form

$$a = -b - c,$$

then the left side is derived from $a_{r+1}, \ldots, a_m$, and therefore belongs to the domain A, the right side is derived from $b_{r+1}, \ldots, b_n, c_1, \ldots, c_r$, and thus belongs to the domain B, whence a simultaneously belongs to both domains. Since however a is numerically derived from $a_{r+1}, \ldots, a_m$, and no numerical relation exists between $a_{r+1}, \ldots, a_m, c_1, \ldots, c_r$ (as shown above), a is not derivable from the $c_1, \ldots, c_r$. Thus the domains A and B would have more than r magnitudes standing in no numerical relation to one another, contrary to assumption. It therefore follows that the whole assembly of magnitudes $c_1, \ldots, c_r, a_{r+1}, \ldots, a_m, b_{r+1}, \ldots, b_n$ is subject to no numerical relation. But the domain derivable from these magnitudes consists of all the magnitudes that can be derived from the magnitudes of the domains A and B, that is, it is their covering domain. If its order is v, then (according to 23) v is equal to the number of the magnitudes $c_1, \ldots, c_r, a_{r+1}, \ldots, a_m, b_{r+1}, \ldots, b_n$, that is

$$v = m + n - r,$$

or

$$m + n = r + v.$$

26. *Two domains (A and B), which are respectively of αth and βth orders and lie in a domain of nth order, have at least a domain of $(\alpha + \beta - n)$th order in common if $\alpha + \beta > n$.*

PROOF. If the covering domain of A and B is of vth order, their common domain of rth order, then (according to 25)

$$\alpha + \beta = r + v \quad \text{or} \quad r = \alpha + \beta - v.$$

Since {however} A and B lie in a domain of nth order, their covering domain also lies in this domain of nth order, whence v is either as great as or smaller than n, and $r = \alpha + \beta - v$ is thus as great as $\alpha + \beta - n$ or greater.

REMARK. The theorems developed above are already, although mostly in other forms, in the first edition of my *Ausdehnungslehre*, of 1844, and in fact theorems 19 and 24 are precisely in the same form in § 20 of that work, theorem 25 is included in § 126, and in addition the idea of the proofs of these theorems is the same here and there.

3. Number as the quotient of extensive magnitudes and the replacement of the equations between extensive magnitudes by numerical equations

27. DEFINITION. I call two assemblies of equations mutually *replaceable* if each of the two sets can be derived from the other.

REMARK. This includes the case in which one or both of the two assemblies consists of only a *single* equation.

28. *A magnitude x, which is derived from n other magnitudes $a_1, \dots, a_n$ standing in no numerical relation to one another, is zero if and only if its n derivation numbers are zero, that is, the equation*

$$(a) \qquad x_1 a_1 + x_2 a_2 + \cdots + x_n a_n = 0$$

can be replaced by the n equations

$$(b) \qquad 0 = x_1 = x_2 = \cdots = x_n.$$

PROOF. Were any of the derivation numbers different from zero, then (according to 16) by means of equation (a) there would exist a numerical relation between the $a_1, \dots, a_n$, contrary to assumption. Thus if equation (a) is valid, so also is the assembly of equations (b). Conversely, if the latter assembly is valid, then equation (a) follows. Therefore this equation is replaceable by that assembly.

29. *Two magnitudes of a domain of nth order are equal if and only if their n derivation numbers belonging to the same units are equal, that is, the equation*

$$(a) \qquad \alpha_1 e_1 + \alpha_2 e_2 + \cdots + \alpha_n e_n = \beta_1 e_1 + \beta_2 e_2 + \cdots + \beta_n e_n$$

can be replaced by the n equations

$$(b) \qquad \alpha_1 = \beta_1, \quad \alpha_2 = \beta_2, \quad \dots, \quad \alpha_n = \beta_n.$$

PROOF. Equation (a) can be replaced by the equation

$$(\alpha_1 - \beta_1)e_1 + (\alpha_2 - \beta_2)e_2 + \cdots + (\alpha_n - \beta_n)e_n = 0,$$

and this (according to 28) by the n equations

$$0 = \alpha_1 - \beta_1 = \alpha_2 - \beta_2 = \cdots = \alpha_n - \beta_n,$$

that is, by the n equations

$$\alpha_1 = \beta_1, \quad \alpha_2 = \beta_2, \quad \dots, \quad \alpha_n = \beta_n.$$

30. DEFINITION. If an extensive magnitude b can be numerically derived from another a that is not zero, then by $\frac{b}{a}$ I mean the number by which b can be derived from a, that is

$$\frac{\alpha a}{a} = \alpha,$$

if $a \neq 0$.

31. *If two magnitudes (a and b) are numerically derived from the same magnitude (c), and the second (b) is not zero, then instead of dividing the first by the*

second one can divide the corresponding derivation numbers, that is

$$\frac{\alpha c}{\beta c} = \frac{\alpha}{\beta},$$

if $\beta c \neq 0$.

PROOF. If $\beta c \neq 0$, then (according to 12, 6) $\beta \neq 0$ and $c \neq 0$. Consequently (according to 13) $\alpha c = \frac{\alpha}{\beta}(\beta c)$, whence

$$\frac{\alpha c}{\beta c} = \frac{\frac{\alpha}{\beta}(\beta c)}{\beta c} = \frac{\alpha}{\beta}. \qquad [30]$$

32. *An equation, all the terms of which are numerically derivable from the same magnitude* (a), *is replaceable by an equation that one obtains by dividing all the terms of the original equation by an arbitrary nonzero magnitude numerically derivable from that magnitude* (a), *that is, the equation*

$$(a) \qquad \alpha a + \beta a + \cdots = \alpha_1 a + \beta_1 a + \cdots$$

can be replaced by the equation

$$(b) \qquad \frac{\alpha a}{\rho a} + \frac{\beta a}{\rho a} + \cdots = \frac{\alpha_1 a}{\rho a} + \frac{\beta_1 a}{\rho a} + \cdots$$

if $\rho a \neq 0$.

PROOF. If $\rho a \neq 0$ then (according to 12, 6) both $\rho \neq 0$ and $a \neq 0$. Consequently, since a is different from zero, one can regard it as a unit. Then (according to 29) equation (a) can be replaced by

$$\alpha + \beta + \cdots = \alpha_1 + \beta_1 + \cdots,$$

or, if one divides by $\rho \neq 0$, by the equation

$$\frac{\alpha}{\rho} + \frac{\beta}{\rho} + \cdots = \frac{\alpha_1}{\rho} + \frac{\beta_1}{\rho} + \cdots,$$

that is (according to 31), by the equation

$$\frac{\alpha a}{\rho a} + \frac{\beta a}{\rho a} + \cdots = \frac{\alpha_1 a}{\rho a} + \frac{\beta_1 a}{\rho a} + \cdots.$$

33. DEFINITION. If the magnitudes $a_1, \ldots, a_n$ stand in no numerical relation to one another, and the magnitude a is given by

$$a = \alpha_1 a_1 + \alpha_2 a_2 + \cdots + \alpha_n a_n,$$

then when m is smaller than n we call the magnitude

$$\alpha_1 a_1 + \alpha_2 a_2 + \cdots + \alpha_m a_m,$$

the "*shadow* of the magnitude a on the domain of $a_1, a_2, \ldots, a_m$, excluding the domain of $a_{m+1}, a_{m+2}, \ldots, a_n$." We say that the shadows of several magnitudes are taken *in the same sense* if the magnitudes are shadowed on the same domain and exclude the same domain.

REMARK. In particular, if

$$a = \alpha_1 a_1 + \alpha_2 a_2 + \cdots + \alpha_n a_n,$$

then for example $\alpha_1 a_1$ is the shadow of a on the domain of a_1, excluding the domain of $a_2, \ldots, a_n$; furthermore, if $a = \alpha_1 a_1 + \alpha_2 a_2$, then $\alpha_1 a_1$ is the shadow on the domain of a_1, excluding the domain of a_2.

34. *Every equation, each term of which is a product of an extensive magnitude with a number, is, provided the extensive magnitudes belong to a domain of nth order, replaceable by n numerical equations that one obtains by replacing all the extensive magnitudes in the given equation with their derivation numbers belonging to the same unit; and in fact this follows, whichever n magnitudes of the domain standing in no numerical relation to one another one may choose, that is, the equation*

$$\alpha a + \beta b + \cdots = \kappa k + \lambda l + \cdots, \tag{a}$$

in which

$$a = \alpha_1 e_1 + \cdots + \alpha_n e_n, \qquad b = \beta_1 e_1 + \cdots + \beta_n e_n, \qquad \ldots$$
$$k = \kappa_1 e_1 + \cdots + \kappa_n e_n, \qquad l = \lambda_1 e_1 + \cdots + \lambda_n e_n, \qquad \ldots$$

is replaceable by the equations

$$\begin{cases} \alpha\alpha_1 + \beta\beta_1 + \cdots = \kappa\kappa_1 + \lambda\lambda_1 + \cdots \\ \alpha\alpha_2 + \beta\beta_2 + \cdots = \kappa\kappa_2 + \lambda\lambda_2 + \cdots \\ \cdots\cdots\cdots\cdots\cdots\cdots \\ \alpha\alpha_n + \beta\beta_n + \cdots = \kappa\kappa_n + \lambda\lambda_n + \cdots \end{cases} \tag{b}$$

provided that $a_1, \ldots, a_n$ stand in no numerical relation to one another.

PROOF. If in equation (a) one replaces the expressions for $a, b, \ldots, k, l, \ldots$, one obtains

$$(\alpha\alpha_1 + \beta\beta_1 + \cdots)e_1 + (\alpha\alpha_2 + \beta\beta_2 + \cdots)e_2 + \cdots + (\alpha\alpha_n + \beta\beta_n + \cdots)e_n$$
$$= (\kappa\kappa_1 + \lambda\lambda_1 + \cdots)e_1 + \cdots + (\kappa\kappa_n + \lambda\lambda_n + \cdots)e_n.$$

This equation is (according to 29) replaceable by the n equations

$$\begin{aligned} \alpha\alpha_1 + \beta\beta_1 + \cdots &= \kappa\kappa_1 + \lambda\lambda_1 + \cdots \\ \alpha\alpha_2 + \beta\beta_2 + \cdots &= \kappa\kappa_2 + \lambda\lambda_2 + \cdots \\ &\cdots\cdots\cdots\cdots \\ \alpha\alpha_n + \beta\beta_n + \cdots &= \kappa\kappa_n + \lambda\lambda_n + \cdots. \end{aligned}$$

35. *Every equation, each term of which is the product of an extensive magnitude with a number, remains valid if one replaces all the extensive magnitudes with their shadows in the same sense.*

PROOF. If one assumes that the given equation is equation (a) of the previous theorem, in which $a, b, \ldots, k, l, \ldots$ have the same significance as before, then it must be shown that these equations continue to apply if one replaces the magnitudes $a, b, \ldots, k, l, \ldots$ with their shadows on the domain of $c_1, \ldots, c_m$, excluding the domain of $c_{m+1}, \ldots, c_n$, that is, that

$$\alpha a' + \beta b' + \cdots = \kappa k' + \lambda l' + \cdots, \tag{c}$$

where

$$a' = \alpha_1 e_1 + \cdots + \alpha_m e_m, \qquad b' = \beta_1 e_1 + \cdots + \beta_m e_m, \qquad \ldots$$
$$k' = \kappa_1 e_1 + \cdots + \kappa_m e_m, \qquad l' = \lambda_1 e_1 + \cdots + \lambda_m e_m, \qquad \ldots$$

Now equation (a) is replaceable by the n equations (b) of the previous number. If one multiplies the first m of these n equations respectively by $e_1, e_2, \ldots, e_m$ and adds the equations thus obtained, one obtains the desired equation (c).

REMARK. Simultaneously implicit in this is the more specialized theorem that equal magnitudes shadowed in the same sense yield equal shadows, or in other words that the shadow of a given magnitude is defined once the domain upon which it is shadowed, and that which is excluded, are given.

36. *If the numbers $x_1, \ldots, x_n$ by which an extensive magnitude x is derived from a system of n units $e_1, e_2, \ldots, e_n$ satisfy an equation of mth degree, then the numbers $y_1, \ldots, y_n$, by which the same magnitude is derived from a system of n other units $a_1, a_2, \ldots, a_n$ provided by the same domain, also satisfy an equation of mth degree, and in fact this latter equation is homogeneous if the former is.*

PROOF. By assumption

$$x = x_1 e_1 + x_2 e_2 + \cdots + x_n e_n,$$

and between these derivation numbers exists the equation

$$f(x_1, \ldots, x_n) = 0,$$

in which f symbolizes a function of mth degree. Now the new units $a_1, \ldots, a_n$ must be derivable from the units $e_1, \ldots, e_n$, since they belong to the domain of $e_1, \ldots, e_n$. Thus suppose

$$a_1 = \sum \alpha_{1,r} e_r = \alpha_{1,1} e_1 + \alpha_{1,2} e_2 + \cdots + \alpha_{1,n} e_n,$$
$$a_2 = \sum \alpha_{2,r} e_r = \alpha_{2,1} e_1 + \alpha_{2,2} e_2 + \cdots + \alpha_{2,n} e_n,$$
$$\cdots\cdots\cdots\cdots\cdots\cdots\cdots\cdots$$
$$a_n = \sum \alpha_{n,r} e_r = \alpha_{n,1} e_1 + \alpha_{n,2} e_2 + \cdots + \alpha_{n,n} e_n.$$

Further, since $y_1, \ldots, y_n$ are the derivation numbers with respect to these new units, one also has

$$x = y_1 a_1 + y_2 a_2 + \cdots + y_n a_n,$$

whence

$$\begin{aligned} x_1 e_1 &+ x_2 e_2 + \cdots + x_n e_n \\ &= y_1 a_1 + y_2 a_2 + \cdots + y_n a_n \\ &= y_1 \sum \alpha_{1,r} e_r + y_2 \sum \alpha_{2,r} e_r + \cdots + y_n \sum \alpha_{n,r} e_r \\ &= \sum y_r \alpha_{r,1} \cdot e_1 + \sum y_r \alpha_{r,2} \cdot e_2 + \cdots + \sum y_r \alpha_{r,n} \cdot e_n. \end{aligned}$$

Consequently (according to 29)

$$x_1 = \sum \alpha_{r,1} y_r = \alpha_{1,1} y_1 + \alpha_{2,1} y_2 + \cdots + \alpha_{n,1} y_n,$$
$$x_2 = \sum \alpha_{r,2} y_r = \alpha_{1,2} y_1 + \alpha_{2,2} y_2 + \cdots + \alpha_{n,2} y_n,$$
$$\cdots\cdots\cdots\cdots\cdots\cdots\cdots\cdots\cdots$$
$$x_n = \sum \alpha_{r,n} y_r = \alpha_{1,n} y_1 + \alpha_{2,n} y_2 + \cdots + \alpha_{n,n} y_n,$$

that is, $x_1, \ldots, x_n$ are all homogeneous functions of first degree of $y_1, \ldots, y_n$; consequently, if in

$$f(x_1, \ldots, x_n) = 0$$

one replaces $x_1, \ldots, x_n$ by these values, one obtains a function of mth degree of $y_1, \ldots, y_n$ and in fact it will be homogeneous if the first was.

CHAPTER 2

The Product Structure in General

1. Product of two magnitudes

37. DEFINITION. By the *product* $[ab]$ of an extensive magnitude a with another b I mean that extensive magnitude (or numerical magnitude) that one obtains when one first conjoins each of the units from which the first magnitude a is numerically derived with each of the units from which the second b is numerically derived in a product, the first factor of which is the unit of the first magnitude and the second factor of which is the unit of the second magnitude, then multiplies this product by the derivation numbers with which those units were conjoined, and {finally} adds the collection of products thus produced, that is,

$$\left[\sum \alpha_r e_r \cdot \sum \beta_s e_s\right] = \sum \alpha_r \beta_s [e_r e_s],$$

where e_r, e_s are the units from which the magnitudes are numerically derived, α_r, β_s denote the derivation numbers belonging to them, and the sum refers to the different values of the indices r and s.

REMARK. Since by definition the product of extensive magnitudes is again either an extensive magnitude or a numerical magnitude, it must (according to 5) be numerically derivable from a system of units. The definition states nothing about what this system of units is, or how from them the products $[e_r e_s]$, of which that product is composed, are numerically derived. Thus, should the concept of a particular product be precisely established, then the necessary definitions must be taken into account in that system of units and in these derivations. As soon as these definitions are taken into account a particular type of product structure results from the general product structure established above.

If for example one had the product

$$P = [(x_1 e_1 + x_2 e_2)(y_1 e_1 + y_2 e_2)],$$

then (according to 37) it is equal to

$$x_1 y_1 [e_1 e_1] + x_1 y_2 [e_1 e_2] + x_2 y_1 [e_2 e_1] + x_2 y_2 [e_2 e_2].$$

Particular types of product structure would now result if the units from which this product is numerically derived were established, and the way the four products

$$[e_1 e_1],\ [e_1 e_2],\ [e_2 e_1],\ [e_2 e_2]$$

are numerically derived from those units. Thus, for example, could it be established that these four products themselves form the system of units from which P is numerically derived, then $x_1 y_1$, $x_1 y_2$, $x_2 y_1$, $x_2 y_2$ are the derivation numbers of P; and we would have a particular type of product structure distinguished in that to establish it no equations are necessary. Or one could establish three of them, say $[e_1 e_1]$, $[e_1 e_2]$, $[e_2 e_2]$, as units, and introduce the definition that $[e_2 e_1] = [e_1 e_2]$; then the derivation numbers of P would be

$$x_1 y_1,\ (x_1 y_2 + x_2 y_1),\ x_2 y_2;$$

a type of product structure distinguished in that its law is identical with that of algebraic multiplication. Or one could establish one of them, for example $[e_1e_2]$, as the unit from which the product P is to be numerically derived, and for the rest perhaps use the definitions $[e_1e_1] = 0$, $[e_2e_1] = -[e_1e_2]$, $[e_2e_2] = 0$. Then the product P would have only *one* derivation number, $x_1y_2 - x_2y_1$; a product structure that I have called combinatorial. Indeed one could even take as fundamental another system of units in which $[e_1e_1]$, $[e_1e_2]$, $[e_2e_1]$, $[e_2e_2]$ do not appear, and then define how these four products can be derived from them; for example one could take the absolute unit as fundamental, and require that

$$[e_1e_1] = 1,\ [e_1e_2] = 0,\ [e_2e_1] = 0,\ [e_2e_2] = 1,$$

in which case P would be a single number, $x_1y_1+x_2y_2$. I have called this product structure *inner*.

For the present I will only consider those laws that follow from the general definition of the product in 37, and which is therefore valid for all types of product. I have enclosed the product in brackets in order to distinguish it from the ordinary product in algebra.

38. *Instead of adjoining a factor b to a magnitude a, one can adjoin it in the corresponding way to each unit in the derivation expression of the latter, that is*

$$\left[\sum \alpha_r e_r \cdot b\right] = \sum \alpha_r [e_r b].$$

PROOF. Suppose $b = \sum \beta_r e_r$; then

$$\begin{aligned}\left[\sum \alpha_r e_r \cdot b\right] &= \left[\sum \alpha_r e_r \cdot \sum \beta_s e_s\right] = \sum \alpha_r \beta_s [e_r e_s] && [37]\\ &= \sum \alpha_1 \beta_s [e_1 e_s] + \sum \alpha_2 \beta_s [e_2 e_s] + \cdots && [9]\\ \{ &= \alpha_1 \sum \beta_s [e_1 e_s] + \alpha_2 \sum \beta_s [e_2 e_s] + \cdots \} && [10]\\ &= \alpha_1 [e_1 \cdot \sum \beta_s e_s] + \alpha_2 [e_2 \cdot \sum \beta_s e_s] + \cdots && [37]\\ &= \alpha_1 [e_1 b] + \alpha_2 [e_2 b] + \cdots = \sum \alpha_r [e_r b].\end{aligned}$$

39. *A product of two factors, one factor of which is a sum, is equal to the sum of the products that one obtains by replacing the terms in the expansion of the original product by each term one by one, that is*

$$[(a+b+\cdots)p] = [ap] + [bp] + \cdots,$$
$$[p(a+b+\cdots)] = [pa] + [pb] + \cdots.$$

In particular

$$[(a+b)p] = [ap] + [bp],$$
$$[p(a+b)] = [pa] + [pb].$$

PROOF. Suppose

$$a = \sum \alpha_r e_r,\ b = \sum \beta_r e_r, \ldots;$$

then

$$\begin{aligned}
[(a+b+\cdots)p] &= \left[\left(\sum \alpha_r e_r + \sum \beta_r e_r + \cdots\right) p\right] \\
&= \left[\sum (\alpha_r + \beta_r + \cdots) e_r \cdot p\right] && [9] \\
&= \sum (\alpha_r + \beta_r + \cdots)[e_r p] && [38] \\
&= \sum \alpha_r [e_r p] + \sum \beta_r [e_r p] + \cdots && [9] \\
&= \left[\sum \alpha_r e_r \cdot p\right] + \left[\sum \beta_r e_r \cdot p\right] + \cdots && [38] \\
&= [ap] + [bp] + \cdots .
\end{aligned}$$

Thus the first formula is proven. One obtains the proof of the second formula from the first if one inserts p as the first factor everywhere.

40. *Instead of multiplying one factor of a product (of two factors) by a number $\{\alpha\}$, one can multiply the whole product by this number, that is*

$$[(\alpha a)b] = \alpha[ab],$$
$$[b(\alpha a)] = \alpha[ba].$$

PROOF. If $a = \sum \alpha_r e_r$, then

$$\begin{aligned}
[(\alpha a)b] &= \left[\left(\alpha \cdot \sum \alpha_r e_r\right) b\right] = \left[\left(\sum \alpha\alpha_r e_r\right) b\right] && [10] \\
&= \sum \alpha\alpha_r [e_r b] && [38] \\
&= \alpha \sum \alpha_r [e_r b] && [13] \\
&= \alpha \left[\sum \alpha_r e_r \cdot b\right] && [38] \\
&= \alpha[ab].
\end{aligned}$$

The proof of the second formula follows if one sets b as the first of the two factors everywhere.

41. *Instead of adjoining a factor p to a magnitude numerically derived from arbitrary magnitudes a, b, ... , one can adjoin it in the corresponding way to each of the magnitudes a, b, ... in the expression of this derivation, that is*

$$[(\alpha a + \beta b + \cdots)p] = \alpha[ap] + \beta[bp] + \cdots$$

and

$$[p(\alpha a + \beta b + \cdots)] = \alpha[pa] + \beta[pb] + \cdots .$$

PROOF.

$$\begin{aligned}
[(\alpha a + \beta b + \cdots)p] &= [(\alpha a)p] + [(\beta b)p] + \cdots && [39] \\
&= \alpha[ap] + \beta[bp] + \cdots . && [40]
\end{aligned}$$

42. *To find the product of two factors numerically derived from arbitrary magnitudes one first replaces each factor with one of the magnitudes from which it is*

derived, multiplies the product thus obtained with the product of the derivation numbers belonging to the substituted magnitudes, and adds the collection of magnitudes produced in this way, that is

$$\left[\sum \alpha_r a_r \cdot \sum \beta_s b_s\right] = \sum \alpha_r \beta_s [a_r b_s],$$

where a_r, b_s *are arbitrary magnitudes,* α_r, β_s *arbitrary numbers.*

PROOF.

$$\begin{aligned}\left[\sum \alpha_r a_r \cdot \sum \beta_s b_s\right] &= \sum \alpha_r \left[a_r \cdot \sum \beta_s b_s\right] && [41]\\ &= \sum \alpha_r \left(\sum \beta_s [a_r b_s]\right) && [41]\\ &= \sum \alpha_r \beta_s [a_r b_s]. && [13]\end{aligned}$$

2. Product of several magnitudes

43. DEFINITION. If one product can be derived from another by replacing each factor appearing in the first product by one of the second (equal to or different from that of the first), I say that the two products *correspond* to one another, and that each factor of the first product corresponds to that of the other substituted for it. To adjoin two factors in a corresponding way to two magnitudes or two corresponding products respectively means to adjoin them so that the resulting products again correspond to one another, and in particular so that the factor adjoined in the one and in the other product remain corresponding factors, and the previously corresponding factors also continue to correspond.

A product in which the factors a, b, ... are included in any way I will, where it seems appropriate, denote by $P_{a,b,\ldots}$; then, within the same expression, $P_{h,i,\ldots}$ denotes the corresponding product in which the factors h, i, ... correspond {respectively} to a, b, ...

REMARK. These definitions are indispensable if one is to avoid all ambiguities. For, since the factors of a product of extensive magnitudes can neither be interchanged nor unified into particular products under all circumstances, the way a factor enters a product must be defined. For example, let $a(bc)$ and $d(ef)$ be two corresponding products, where the factors correspond in the order they appear. Should the factors g and h be adjoined to each of them correspondingly, this can occur in various ways, for example so that the products $a(bc)g$ and $d(ef)h$ result, or $a(bgc)$ and $d(ehf)$, and so forth. For the significance of the omitted parentheses I refer to the Remark following No. 7.

44. *If a given product includes a factor p, derivable from arbitrary magnitudes* a, b, c, ... *by the numbers* q, r, s, ... , *and in the product one replaces the factor* p *one by one by the magnitudes* a, b, c, ... , *multiplies the resulting products by* q, r, s, ... *respectively, and adds these expressions, their sum is equal to the given product, that is*

$$P_{qa+rb+\cdots} = qP_a + rP_b + \cdots .$$

PROOF. However the product may have been composed, one can always regard it as resulting when the other factors are progressively adjoined to the factor p in their corresponding ways, that is, to p there is first adjoined another factor (either as first or second factor of the product), to this product another, and so forth. But now instead of adjoining a factor to a numerically derived magnitude one can (according to 41) adjoin it, in the expression of that derivation, in a corresponding way to each of the magnitudes from which the first was derived. Consequently, instead of progressively adjoining the other factors in the way specified to $p = qa + rb + \cdots$, one can adjoin them in the corresponding way to each of the magnitudes a, b, ... in the derivation expression, that is

$$P_p = qP_a + rP_b + \cdots, \quad \text{if} \quad p = qa + rb + \cdots.$$

45. *Theorem 42 is also valid for more than two factors, that is*

$$\left[\sum \alpha_r a_r \cdot \sum \beta_s b_s \cdots\right] = \sum \alpha_r \beta_s \cdots [a_r b_s \cdots].$$

PROOF. Suppose the theorem holds for any given number of factors, for example for $[\sum \alpha_r a_r \cdot \sum \beta_s b_s \cdots \sum \kappa_m k_m]$, so that

$$\text{(a)} \qquad \left[\sum \alpha_r a_r \cdot \sum \beta_s b_s \cdots \sum \kappa_m k_m\right] = \sum \alpha_r \beta_s \cdots \sum \kappa_m [a_r b_s \cdots k_m].$$

Then it also holds if another factor, say $\sum \lambda_n l_n$, is adjoined; thus

$$\begin{aligned} &\left[\sum \alpha_r a_r \cdot \sum \beta_s b_s \cdots \sum \kappa_m k_m \cdot \sum \lambda_n l_n\right] \\ &\quad = \left[\sum \alpha_r \beta_s \cdots \kappa_m [a_r b_s \cdots k_m] \cdot \sum \lambda_n l_n\right] && \text{[according to (a)]} \\ &\quad = \sum \alpha_r \beta_s \cdots \kappa_m \lambda_n [a_r b_s \cdots k_m l_n]. && [42] \end{aligned}$$

Thus, if formula 45 holds for any given number of factors, it also holds if another factor is adjoined. But (according to 42) it holds for two factors, and therefore also for three, four, and so on, and thus for arbitrarily many.

46. *Instead of multiplying the factors of a product by a series of numbers $\{q, r, \ldots\}$, one can multiply the whole product by the product of these numbers, that is*

$$P_{qa,rs,\ldots} = qr \cdots P_{a,b,\ldots}.$$

PROOF. According to 44, $P_{qa} = qP_a$, whence also

$$\begin{aligned} P_{qa,rb,sc,\ldots} &= qP_{a,rb,sc,\ldots} && [44] \\ &= qrP_{a,b,sc,\ldots} && [44] \end{aligned}$$

and so forth, so that finally

$$= qrs \cdots P_{a,b,c,\ldots}. \qquad [44]$$

47. *If there appear in a product two factors that stand in a numerical relation to one another, then one can interchange them without changing the value of the product, that is*

$$P_{qa,ra} = P_{ra,qa}.$$

PROOF. For

$$\begin{aligned} P_{qa,ra} &= qrP_{a,a} && [46] \\ &= rqP_{a,a} = P_{ra,qa}. && [46] \end{aligned}$$

3. The various types of product structure

48. DEFINITION. If the product structure is more precisely defined, so that numerical relations exist between the products of the units, then I call every equation that expresses such a numerical relation a *defining equation* of that type of product structure. An assembly of p defining equations, no one of which can be produced from the rest, I call a *system of defining equations* belonging to that product structure, provided there exists no numerical relation between the products other than those that can be produced from those equations.

49. *Every system of m defining equations between the n unit products $E_1, \ldots, E_n$ can be cast into a form such that each of the equations expresses how from $n-m$ of those unit products, for example from $E_1, \ldots, E_{n-m}$, the remaining m are numerically derivable. Then $E_1, \ldots, E_{n-m}$ form a system of units from which all products belonging to this product structure are derivable.*

PROOF. According to 48 each equation of the system of m defining equations expresses a numerical relation between the $E_1, \ldots, E_n$. Each such numerical relation can (according to 16) be cast into the form

$$\alpha_1 E_1 + \cdots + \alpha_n E_n = 0,$$

in which the numbers $\alpha_1, \ldots, \alpha_n$ are not all simultaneously zero. Consider one of them, in which α_n, say, is not zero; then one can express E_n by $E_1, \ldots, E_{n-1}$. If one substitutes this expression into the remaining $(m-1)$ equations, they will be of the form

$$\alpha_1 E_1 + \cdots + \alpha_{n-1} E_{n-1} = 0.$$

In none of the resulting equations may the numbers $\alpha_1, \ldots, \alpha_{n-1}$ all simultaneously be zero, since this defining equation could then have been found from the former set, which would contradict the concept of a system of defining equations (according to 48). Consider one of these resulting equations, and suppose that in it the coefficient of E_{n-1} is not zero; then E_{n-1} can be expressed by $E_1, \ldots, E_{n-2}$, and if this expression is substituted into the remaining $(m-2)$ equations, they will be of the form

$$\alpha_1 E_1 + \cdots + \alpha_{n-2} E_{n-2} = 0.$$

Since in this way, by the application of each new equation, another of the magnitudes $E_1, \ldots, E_n$ disappears from the remaining equations, we will always assume it is the last of those still remaining, whence one is finally left with just the magnitudes $E_1, \ldots, E_{n-m}$, in terms of which all the other $E_{n-m+1}, \ldots, E_n$ can be expressed.

50. DEFINITION. Every product structure, the defining equations of which remain valid when one replaces the units appearing therein by arbitrary magnitudes numerically derived from them, is a *linear* product structure (multiplication).

51. *For product structures of two factors there are, besides that product structure having no defining equation and that in which all products are zero, only* ***two*** *species of linear product structure, and the system of defining equations for the one is*

(1) $$[e_r e_s] + [e_s e_r] = 0,$$

while that for the other is

$$(2) \qquad [e_r e_s] = [e_s e_r],$$

where, if $e_1, e_2, \ldots, e_n$ *are the units,* r *and* s *are to be taken, one by one, as every pair of numbers from among* $1, \ldots, n$.

PROOF. Each defining equation, composed of pairs of factors derived from the units $e_1, \ldots, e_n$, will have the form

$$(a) \qquad \sum \alpha_{r,s}[e_r e_s] = 0,$$

where the coefficients $\alpha_{r,s}$ are arbitrary numbers, which however cannot all simultaneously be zero, and where for r and s every pair of values from $1, \ldots, n$ are to be substituted one by one into the sum. We assume that the product structure is to be linear; that is (according to 50), every defining equation is to remain valid when one replaces the units with arbitrary magnitudes numerically derived from them.

Suppose that in (a) one replaces e_r with $\sum x_{r,u}e_u$ and e_s with $\sum x_{s,v}e_v$, where the sum refers only to the indices u and v, and $x_{r,u}$ and $x_{s,v}$ denote any chosen numbers. Then we obtain

$$\begin{aligned} 0 &= \sum \alpha_{r,s}\left[\sum x_{r,u}e_u \cdot \sum x_{s,v}e_v\right] \\ &= \sum \alpha_{r,s} \sum x_{r,u}x_{v,s}[e_u e_v], \qquad [42] \end{aligned}$$

whence

$$0 = \sum \alpha_{r,s}x_{r,u}x_{s,v}[e_u e_v], \qquad [13]$$

where the sum now refers to all four values r, s, u, v. If one now interchanges r with s and u with v, as one can since in each term r, s, u, v are completely arbitrary numbers from $1, \ldots, n$, then one obtains

$$0 = \sum \alpha_{s,r}x_{s,v}x_{r,u}[e_v e_u],$$

and upon adding this equation to the previous one, one obtains

$$(b) \qquad 0 = \sum x_{r,u}x_{s,v}(\alpha_{r,s}[e_u e_v] + \alpha_{s,r}[e_v e_u]),$$

an equation which is more convenient for applications than the former two. It must hold for all values that one may give the magnitudes $x_{r,u}$, $x_{s,v}$.

Now if in (b) one sets any one of the magnitudes $x_{r,u}$, say $x_{a,c}$, first to $+1$, then to -1, subtracts one of the resulting equations from the other, and divides by 2, then all terms drop out that contain $x_{a,c}$ either not at all or twice, and there remains only

$$\sum x_{s,v}(\alpha_{a,s}[e_c e_v] + \alpha_{s,a}[e_v e_c]) = 0,$$

in which however those pairs of values of s and v are omitted for which simultaneously $s = a$ and $v = c$. Now if in this expression one again sets any one of the magnitudes $x_{s,v}$, say $x_{b,d}$, first equal to 1 and then equal to -1, subtracts the resulting equations, and divides by 2, again all terms that do not contain $x_{b,d}$ drop out, and there remains

$$(c) \qquad a_{a,b}[e_c e_d] + \alpha_{b,a}[e_d e_c] = 0$$

at first only for every four indices a, b, c, d for which it is not simultaneously the case that the first is equal to the second, the third equal to the fourth.

Thus equation (b) is reduced to

$$0 = \sum x_{r,u} x_{r,u} \alpha_{r,r} [e_u e_u].$$

Now if in this equation one inserts for one of the magnitudes $x_{r,u}$, say $x_{a,c}$, two values among the numbers that are not mutually opposed, for example 1 and 2, subtracts one of the equations from the other, and divides the residual equation in this case by 3, there remains

$$(d) \qquad 0 = \alpha_{a,a} [e_c e_c],$$

that is, equation (c) also holds for the previously excluded case that $a = b$, $c = d$.

Thus from equation (a) necessarily follows the group of equations (c), provided (a) is a linear defining equation, that is if it continues to hold, with whatever magnitudes numerically derived from the units one may replace them; and also conversely, given the group of equations (c), equation (b) follows, which expresses that equation (a) is linear.

If in (c) we set the indices c and d equal it goes over into

$$(e) \qquad (\alpha_{a,b} + \alpha_{b,a}) [e_c e_c] = 0,$$

while if we set $a = b$ it becomes

$$(f) \qquad \alpha_{a,a} ([e_c e_d] + [e_d e_c]) = 0.$$

In such products equated to zero one or the other factor must always (according to 12, 6) be zero.

If *first* we assume that $[e_c e_c]$ is different from zero, then for each pair of indices a and b it necessarily follows that

$$\alpha_{a,b} + \alpha_{b,a} = 0, \quad \text{that is} \quad -\alpha_{a,b} = \alpha_{b,a}.$$

Inserting this into (c), we obtain

$$\alpha_{a,b} ([e_c e_d] - [e_d e_c]) = 0.$$

Should $[e_c e_d] - [e_d e_c]$ be different from zero here, then the other factor, $\alpha_{a,b}$, must be zero for every pair of indices a and b, that is equation (a) would be identically zero, contrary to assumption. Consequently in this case, when $[e_c e_c]$ differs from zero, it necessarily follows that

$$[e_c e_d] - [e_d e_c] = 0, \quad \text{that is} \quad [e_c e_d] = [e_d e_c],$$

which is to say, the group of equations (51, 2) applies.

Now *in the second case*, if $[e_c e_c] = 0$, or replacing c by a, $[e_a e_a] = 0$, we can use this equation as defining equation in place of equation (a); then $\alpha_{a,a} = 1$ while all the other coefficients are zero, and it then follows upon inserting $\alpha_{a,a} = 1$ in (f) that

$$[e_c e_d] + [e_d e_c] = 0,$$

that is, the group (51, 1) applies.

Now it is still possible that both groups (51,1) and (51,2) hold simultaneously. But it would then follow that $[e_c e_d] = 0$, whence all products would be zero, a case that we have already excluded.

Thus no other product structures are possible but those stated in the theorem. That these are in fact linear follows at once from equation (c) compared with (a). Thus if

$$(g) \qquad [e_c e_d] \pm [e_d e_c] = 0$$

are the defining equations, and one applies any one of them as equation (a), then for it $\alpha_{a,b} = 1$ and $\alpha_{b,a} = \pm 1$, all other coefficients being zero. Then the group of equations (c) transforms into

$$[e_c e_d] \pm [e_d e_c] = 0,$$

which is already included in the group (g) given. Thus those two species of product structure are linear and in fact are the only ones possible except for that without defining equations and that with all products zero.

REMARK. Thus if the previously self-evident principle, that every law expressible by an equation remains valid when one replaces the units with any magnitudes derivable from them, is to continue to hold in the next step of development, then no other procedure is possible except that in the two product structures cited.[11] If for simplicity we assume only two units, e_1 and e_2, then for the first species the system of defining equations is simultaneously

$$[e_1 e_1] = [e_2 e_2] = 0 \quad \text{and} \quad [e_1 e_2] = -[e_2 e_1],$$

and for the second

$$[e_1 e_2] = [e_2 e_1].$$

With respect to the operations the latter species is the simpler. Indeed, since its defining equations express nothing but the interchangeability of the factors, the multiplication equation that these operations concern is identical to ordinary multiplication in algebra, whence I have called it *algebraic.** It is obvious that along with it there is also an algebraic division, raising to a power, ... , and that one may immediately assume as valid the algebraic laws for all these conjunctions of extensive magnitudes. On the other hand this multiplication, as regards the magnitudes generated by it, is much more complicated than the first species, which I have called *combinatorial.*

Thus if for two units e_1 and e_2 we consider the product of two factors

$$[(q_1 e_1 + q_2 e_2)(r_1 e_1 + r_2 e_2)] = q_1 r_1 [e_1 e_1] + q_2 r_2 [e_2 e_2] + q_1 r_2 [e_1 e_2] + q_2 r_1 [e_2 e_1],$$

then this is reduced, in the first species, where $[e_1 e_1] = [e_2 e_2] = 0$, $[e_2 e_1] = -[e_1 e_2]$, to $(q_1 r_2 - q_2 r_1)[e_1 e_2]$, that is to only one unit, $[e_1 e_2]$; indeed, if no more than those two original units e_1 and e_2 appear in the course of a development, one can without loss of generality set $[e_1 e_2] = 1$, and one then obtains a number as the result of the multiplication. Matters stand completely otherwise with the other species, where that product is reduced to

$$q_1 r_1 [e_1 e_1] + q_2 r_2 [e_2 e_2] + (q_1 r_2 + q_2 r_1)[e_1 e_2],$$

that is, to not less than three units.

Since in the development of science it is above all important to grasp the gradually emerging magnitudes in their simplest concepts, the transition to the first species of multiplication is suggested forcefully here. Indeed, since the structures generated by algebraic multiplication of extensive magnitudes are no longer presented as elementary magnitudes, but rather appear much more in parallel with the functions of algebra, we will only encounter this multiplication again in the second part of this work, which will treat of functions.

For the development of the various types of multiplication I refer to the article in *Crelles Journal* mentioned above. There I have given a much more detailed but also more elementary proof of the above theorem. The general idea of multiplication, as it is developed in the first paragraph of this chapter, I have already taken as fundamental in the first edition of my *Ausdehnungslehre* (§§10–12).

---

* *Crelles Journal* **49**, 123–141 (1855): "Sur les differents genres de multiplication."

CHAPTER 3

Combinatorial Product

1. General laws of combinatorial multiplication

52. DEFINITION. If the factors of a product P are derived from a system of units, and every pair of products of the units that result from the interchange of the last two factors yield zero upon summation, but every product that includes only different units as factors is different from zero, then I call that product P *combinatorial*, and those factors of it its *elementary factors*; that is, if b and c are *units*, and A is an arbitrary *series of units*, then the above definition is expressed by the formula

$$[Abc] + [Acb] = 0.$$

REMARK. The reason for beginning directly with this particular species of multiplication is explained in the Remark to No. 51.

53. *In every combinatorial product one can interchange the last two (elementary) factors, only if one changes the algebraic sign (±) into its opposite at the same time, that is, it is also true that*

$$[Abc] + [Acb] = 0,$$

when A is an ***arbitrary series of factors*** *and b and c are* ***elementary factors****.*

PROOF. 1. Suppose first that b and c are units. Since A is an arbitrary series of factors, and the factors are numerically derivable from the units, upon replacing the factors of A by their derivation expressions and dropping the parentheses, one obtains (according to 45) an expression numerically derivable from the units, and thus has the form

$$A = \sum \alpha_r E_r,$$

where E_r are products of the units. If one substitutes this in, then

$$\begin{aligned}
[Abc] + [Acb] &= \left[\sum \alpha_r E_r \cdot bc\right] + \left[\sum \alpha_r E_r \cdot cb\right] \\
&= \sum \alpha_r [E_r bc] + \sum \alpha_r [E_r cb] && [44] \\
&= \sum \alpha_r ([E_r bc] + [E_r cb]) && [12, 3] \\
&= \sum \alpha_r 0 && [52] \\
&= 0.
\end{aligned}$$

2. Suppose b and c are numerically derived from the units $e_1, e_2, \dots$, and let

$$b = \sum \beta_r e_r, \qquad c = \sum \gamma_r e_r;$$

then

$$\begin{aligned}
[Abc] + [Acb] &= \left[A \cdot \sum \beta_r e_r \cdot \sum \gamma_r e_r\right] + \left[A \cdot \gamma_r e_r \cdot \sum \beta_r e_r\right] \\
&= \sum \beta_r \gamma_s [A e_r e_s] + \sum \gamma_s \beta_r [A e_s e_r] && [45] \\
&= \sum \beta_r \gamma_s ([A e_r e_s] + [A e_s e_r]) && [12] \\
&= \sum \beta_r \gamma_s 0 && [\text{Proof, } 1] \\
&= 0.
\end{aligned}$$

54. *In a combinatorial product one can interchange any two sequential elementary factors provided one simultaneously reverses the sign* ($\pm$), *that is*

$$[AbcD] + [AcbD] = 0,$$

where A and D are arbitrary series of factors, and b and c are elementary factors.

PROOF.

$$\begin{aligned}
[AbcD] + [AcbD] &= [[Abc]D] + [[Acb]D] && [7, \text{Remark}] \\
&= [([Abc] + [Acb])D] && [39] \\
&= 0 \cdot D && [53] \\
&= 0.
\end{aligned}$$

55. *In a combinatorial product one can interchange any two elementary factors, provided one simultaneously reverses the sign* ($\pm$), *that is*

$$P_{a,b} = -P_{b,a},$$

or

$$P_{a,b} + P_{b,a} = 0.$$

PROOF. Assume that n elementary factors stand between a and b in $P_{a,b}$. Now if one interchanges b with the next adjacent factor, that is, if one moves b one place to the left, then (according to 54) the sign is changed; thus if one moves b one by one past the n factors that originally stood between a and b, then the sign is changed n times. Now b immediately follows a, and if one interchanges a with b the sign changes one more time. Now b stands in the place where a originally stood; in order to bring a to the place where b originally stood one need only move a n places to the right, whence the sign changes another n times. In all it has been changed $(2n+1)$ times; but the sign is returned to its original value by the $2n$-fold change, and since the single further change is adjoined, the final sign is the opposite of the original, whence

$$P_{a,b} = -P_{b,a}, \quad \text{or} \quad P_{a,b} + P_{b,a} = 0.$$

56. DEFINITION. If two series of magnitudes contain each of the factors a and b, and each of them one time, and in both series a stands earlier than b, then I say that these two magnitudes are *equally ordered* in those series; alternatively they are

oppositely ordered in those series if in the one a stands earlier than b, in the other b earlier than a.

57. *Two combinatorial products that include the same elementary factors (but in different orders) are equal or opposite to one another according as the number of oppositely ordered factor pairs in the two products is even or odd, that is*

$$P = (-1)^r Q,$$

if P and Q are combinatorial products that include the same elementary factors, and if r is the number of factor pairs that are in P oppositely ordered compared to Q.

PROOF. First, if every immediately adjacent pair of factors in one of the products, say Q, is equally ordered in both products, then it is clear that the two products are identical, and therefore they can contain no oppositely ordered factor pairs. Thus so long as in Q there is still a factor pair oppositely ordered relative to P, then there are also at least two immediately adjacent factors in Q that are oppositely ordered in Q relative to P. Assume a and b are two such factors. If one interchanges them one obtains a product Q_1, which (according to 54) is signed opposite to Q, and in which all factor pairs with the exception of the factor pair a, b are ordered just as in Q, while the factor pair a, b is oppositely ordered relative to Q, and thus equally ordered relative to P. Consequently the number of factor pairs that are oppositely ordered in Q_1 and P is one less than the number that are oppositely ordered in Q and P. If this latter is r, then the former is $r - 1$.

Should $r-1$ not be zero, that is, if there are still factor pairs that are oppositely ordered in Q_1 and P, then one can proceed with Q_1 just as previously with Q; thus one obtains from Q_1 the product Q_2, where $Q_2 = -Q_1$, and thus $= (-1)^2 Q$, and the number of factor pairs oppositely ordered in Q_2 and P is $r - 2$. If one proceeds in this way until one arrives at Q_r, then $Q_r = (-1)^r Q$, and the number of factor pairs oppositely ordered in Q_r and P is $r - r$, that is zero, whence the factor pairs in Q_r and P are all equally ordered, so that $Q_r = P$, and consequently $P = Q_r = (-1)^r Q$.

58. *If in a combinatorial product one interchanges a series of r elementary factors with an* ***immediately adjacent*** *series of s elementary factors (without changing the order of factors otherwise), then the resulting product is equal or opposite to the original, according as rs is even or odd, that is*

$$[BC] = (-1)^{rs}[CB], \qquad [ABC] = (-1)^{rs}[ACB],$$

where B represents a series of r, C a series of s, elementary factors.

PROOF. Let $C = c_1 c_2 \cdots c_s$; then

$$[ABC] = [ABc_1 c_2 \cdots c_s].$$

Now if one interchanges c_1 with the last elementary factor of B, that is, advances c_1 by one place, then (according to 54) the sign of the product is changed; thus if one advances c_1 by r elementary factors, that is, advances it by the factors of B, the sign is changed r times, whence

$$\begin{aligned}[ABc_1c_2\cdots c_s] &= (-1)^r[Ac_1Bc_2c_3\cdots c_s]\\ &= (-1)^r(-1)^r[Ac_1c_2Bc_3\cdots c_s]\\ &= (-1)^{2r}[Ac_1c_2Bc_3\cdots c_s]\\ &= (-1)^{3r}[Ac_1c_2c_3Bc_4\cdots c_s]\\ &= \cdots\\ &= (-1)^{sr}[Ac_1c_2\cdots c_sB],\end{aligned}$$

or

$$[ABC] = (-1)^{rs}[ACB],$$

and if one sets $A = 1$ in this expression,

$$[BC] = (-1)^{rs}[CB].$$

59. *If in a combinatorial product one interchanges a series of q elementary factors with a series of s elementary factors separated from it by r elementary factors, the resulting product is equal or opposite to the first, according as $rs+sq+qr$ is even or odd, that is*

$$[ABC] = (-1)^{rs+sq+qr}[CBA],$$

where A, B, C represent series of q, r, s elementary factors, respectively.

PROOF.

$$\begin{aligned}[ABC] &= (-1)^{(q+r)s}[CAB] && [58]\\ &= (-1)^{qs+rs}(-1)^{qr}[CBA] && [58]\\ &= (-1)^{rs+sq+sr}[CBA].\end{aligned}$$

60. *If two elementary factors of a combinatorial product are equal, the product is zero, that is*

$$P_{a,a} = 0.$$

PROOF. Let $P_{a,b}$ be any combinatorial product containing the factors a and b; and $P_{b,a}$ be that resulting from their interchange; then (according to 55)

$$P_{a,b} + P_{b,a} = 0,$$

whence, if a equals b,

$$P_{a,a} + P_{a,a} = 0,$$

that is $2P_{a,a} = 0$, whence also $P_{a,a} = 0$.

61. *A combinatorial product is zero if there is a numerical relation between its elementary factors, that is*

$$[a_1a_2a_3\cdots a_m] = 0$$

if one of the magnitudes a_1, a_2, ... , a_m is numerically derived from the rest, for example

$$a_1 = \alpha_2a_2 + \alpha_3a_3 + \cdots + \alpha_ma_m.$$

PROOF. Upon inserting the value of a_1 in the product one obtains

$$[a_1a_2a_3\cdots a_m] = [(\alpha_2a_2 + \alpha_2a_2 + \cdots + \alpha_ma_m)a_2a_3\cdots a_m],$$

whence (according to 44)

$$\begin{aligned}[a_1a_2a_3\cdots a_m] &= \alpha_2[a_2a_2a_3\cdots a_m] + \alpha_3[a_3a_2a_3\cdots a_m] + \cdots + \alpha_m[a_ma_2a_3\cdots a_m]\\ &= \alpha_2\cdot 0 + \alpha_3\cdot 0 + \cdots + \alpha_m\cdot 0 \qquad [60]\\ &= 0.\end{aligned}$$

62. DEFINITION. By the *determinant* of n series of any n numbers one means, if one denotes the rth number of the sth series by $\alpha_r^{(s)}$, that polynomial one obtains from the product $\alpha_1^{(1)}\alpha_2^{(2)}\cdots\alpha_n^{(n)}$ by interchanging the lower indices in all possible ways one by one, leaving the upper indices unchanged, then furnishes these products with a $+$ or $-$ sign, according as the number of those pairs of indices that are oppositely ordered relative to those above is even or odd, and adds this collection of terms. One denotes this determinant by $\sum\pm\alpha_1^{(1)}\alpha_2^{(2)}\cdots\alpha_n^{(n)}$, that is, one sets

$$\sum\pm\alpha_1^{(1)}\alpha_2^{(2)}\cdots\alpha_n^{(n)} = \sum(-1)^u\alpha_r^{(1)}\alpha_s^{(2)}\cdots\alpha_w^{(n)},$$

where r, s, $\ldots$, w are equal to the numbers 1, 2, $\ldots$, n, taken in any order, the sum refers to all possible orderings of this type, and u denotes the number of index pairs below that are oppositely ordered relative to those above.

REMARK. For completeness I believe this concept of the determinant had to be presented here, especially since it seemed expedient to establish the sign convention in the simple form in which it appears here, while the usual form of the sign convention introduced by *Cauchy* would necessitate a return to the law of permutations. It is clear that one could also interchange the upper indices instead of the lower indices, but it is inappropriate to introduce such a twofold specification in a rigorous definition.[12]

63. *One obtains the combinatorial product of n elementary factors, each of which is derived from n magnitudes* a_1, a_2, $\ldots$, a_n, *by forming the determinant from the n series of numbers by which those factors are derived from the n magnitudes* a_1, a_2, $\ldots$, a_n, *and multiplying this by the combinatorial product of the magnitudes* a_1, a_2, $\ldots$, a_n, *where the numbers, by which the first of those factors is derived from* a_1, a_2, $\ldots$, a_n, *form the first row in the determinant, and so forth, that is*

$$\begin{aligned}&[(\alpha_1^{(1)}a_1 + \cdots + \alpha_n^{(1)}a_n)(\alpha_1^{(2)}a_1 + \cdots + \alpha_n^{(2)}a_n)\cdots(\alpha_1^{(n)}a_1 + \cdots + \alpha_n^{(n)}a_n)]\\ &\qquad = \sum\pm\alpha_1^{(1)}\alpha_2^{(2)}\cdots\alpha_n^{(n)}\cdot[a_1a_2\cdots a_n].\end{aligned}$$

PROOF. [13] The product on the left side is (according to 45)

$$= \sum\alpha_r^{(1)}\alpha_s^{(2)}\cdots\alpha_w^{(n)}[a_ra_s\cdots a_w],$$

where each of the indices r, s, $\ldots$, w assume each of the values 1, $\ldots$, n, one by one. If two or more of these values are equal, then the product $[a_ra_s\cdots a_w]$ includes equal factors, and is therefore (according to 60) zero. Thus, if we drop the terms that include these products, there remain only the rest, in which the n indices r, s, $\ldots$, w are equal to the values 1, 2, $\ldots$, n in any order whatever. Then (according to 57) the product $[a_ra_s\cdots a_w]$ is equal to $(-1)^u[a_1a_2\cdots a_n]$, if u is the number of factor pairs which in the product $[a_ra_s\cdots a_w]$ are oppositely ordered relative to those in $[a_1 a_2 \cdots a_n]$, that is, the number of those pairs of lower indices which in the product $\alpha_r^{(1)} \alpha_s^{(2)} \cdots \alpha_w^{(n)}$ are oppositely ordered relative to those above; consequently the given product is

$$= \sum (-1)^u \alpha_r^{(1)} \alpha_s^{(2)} \cdots \alpha_w^{(n)} \cdot [a_1 a_2 \cdots a_n],$$

that is (according to 62)

$$= \sum \pm \alpha_1^{(1)} \alpha_2^{(2)} \cdots \alpha_n^{(n)} \cdot [a_1 a_2 \cdots a_n].$$

64. Definition. By the *multiplicative combinations* of a series of magnitudes I mean the combinations without repetition of these magnitudes, and in particular each combination understood as a combinatorial product whose elementary factors are the elements of the combination; thus for example $[ab]$, $[ac]$, $[bc]$ are the multiplicative combinations of second class of the magnitudes a, b, c.

65. *Every combinatorial product of m elementary factors, each of which is numerically derived from n magnitudes $a_1, \ldots, a_n$ that stand in no numerical relation, is numerically derivable from the multiplicative combinations of mth class of these magnitudes, and in particular the derivation number belonging to any given one of these combinations is the determinant of those m^2 derivation numbers of those m factors that belong to the m elements of this combination, that is*

$$\left[\sum \alpha_{\mathfrak{a}} a_{\mathfrak{a}} \cdot \sum \beta_{\mathfrak{b}} a_{\mathfrak{b}} \cdots \right] = \sum \left\{ \sum \pm \alpha_r \beta_s \cdots \right\} [a_r a_s \cdots],$$

where $r < s < \cdots$.

Proof. Let

$$\left[\sum \alpha_{\mathfrak{a}} a_{\mathfrak{a}} \cdot \sum \beta_{\mathfrak{b}} a_{\mathfrak{b}} \cdots \right] = \sum (\alpha_{\mathfrak{a}} \beta_{\mathfrak{b}} \cdots)[a_{\mathfrak{a}} a_{\mathfrak{b}} \cdots]. \qquad [45]$$

Since (according to [60]) $[a_{\mathfrak{a}} a_{\mathfrak{b}} \cdots]$ is zero when two of the factors, that is, here, two of the indices $\mathfrak{a}$, $\mathfrak{b}$, ..., are equal, we can add the stipulation that $\mathfrak{a}$, $\mathfrak{b}$, ... are all to be different. Now let $\mathfrak{a}$, $\mathfrak{b}$, ..., in ascending order, $= r, s, t, \ldots$, whence $r < s < t < \ldots$, and let u be the number of pairs of magnitudes which, in the series $\mathfrak{a}$, $\mathfrak{b}$, $\mathfrak{c}$, ..., are ordered oppositely with respect to r, s, t, ... ; then $[a_{\mathfrak{a}} a_{\mathfrak{b}} \cdots] = (-1)^u [a_r a_s \cdots]$, and thus the given product

$$= \sum \{\alpha_{\mathfrak{a}} \beta_{\mathfrak{b}} \cdots (-1)^u [a_r a_s \cdots]\}.$$

But according to definition (62), if $\mathfrak{a}$, $\mathfrak{b}$, ..., taken in any order, are equal to r, s, ..., then $(-1)^u \alpha_{\mathfrak{a}} \beta_{\mathfrak{b}} \cdots$ is equal to the determinant $\sum \pm \alpha_r \beta_s \cdots$, whence

$$\left[\sum \alpha_{\mathfrak{a}} a_{\mathfrak{a}} \cdot \sum \beta_{\mathfrak{b}} a_{\mathfrak{b}} \cdots \right] = \sum \left\{ \sum \pm \alpha_r \beta_s \cdots \right\} [a_r a_s \cdots],$$

where $r < s < \cdots$.

66. [Converse of 61.] *If a combinatorial product is zero, then its elementary factors stand in a numerical relation to one another, that is, if*

$$[a_1 a_2 \cdots a_m] = 0,$$

then it must be possible to find an equation

$$\alpha_1 a_1 + \alpha_2 a_2 + \cdots + \alpha_m a_m = 0,$$

in which the numbers $\alpha_1, \alpha_2, \ldots, \alpha_m$ are not all simultaneously zero.

PROOF. Suppose the combinatorial product

$$[a_1 a_2 \cdots a_m] = 0.$$

It is to be shown that a_1, a_2, ... , a_m must stand in a numerical relation.

Suppose they stand in no numerical relation to one another. Then if e_1, ... , e_n form the system of units from which a_1, ... , a_n are numerically derived, one can (according to 20) append to the m magnitudes a_1, ... , a_m another $n - m$ magnitudes a_{m+1}, ... , a_n, so that the units e_1, ... , e_n can be numerically derived from the a_1, ... , a_n. If one introduces the expressions of these derivations into the combinatorial product $[e_1 e_2 \cdots e_n]$, and removes the parentheses, then (according to 63) one obtains an equation of the form

$$[e_1 e_2 \cdots e_n] = \alpha[a_1 a_2 \cdots a_n],$$

where α is a number.

But $[a_1 a_2 \cdots a_m] = 0$, whence also

$$[a_1 a_2 \cdots a_m a_{m+1} \cdots a_n] = 0,$$

and thus

$$[e_1 e_2 \cdots e_n] = \alpha \cdot 0 = 0.$$

But this contradicts the assumption in 52, according to which $[e_1 e_2 \cdots e_n]$ is different from zero. The assumption that a_1, ... , a_m stand in no numerical relation to one another is therefore impossible; they must stand in a numerical relation to one another.

67. *A combinatorial product does not change its value if one adds to one of its elementary factors an arbitrary multiple of another of its elementary factors, that is*

$$P_{a,b+qa} = P_{a,b},$$

if q is a number and P denotes a combinatorial product.

PROOF. Observe that

$$\begin{aligned} P_{a,b+qa} &= P_{a,b} + qP_{a,a} && [44] \\ &= P_{a,b}. && [60] \end{aligned}$$

68. *All the theorems on combinatorial multiplication remain valid if one replaces the n original units with any n magnitudes derived from them that stand in no numerical relation to one another.*

PROOF. First, all conditions given in the definition of the combinatorial product are also valid if one replaces the system of the n original units by n such magnitudes as satisfy the conditions of the theorem. Thus, for this case as well (according to 58), one has

$$[Abc] + [Acb] = 0,$$

and the product of the collection of n magnitudes is different from zero, since if it were zero, then (according to 66) there must be a numerical relation between the n factors, contrary to assumption. Now these two conditions were the only ones included in the definition. But in addition all the laws established in the first two chapters remain valid for the case of that substitution; and all the laws of combinatorial multiplication were derived from that definition and these laws. Thus these laws are also valid after that substitution.

2. The combinatorial product as magnitude

PRELIMINARY REMARK. If a conjunction of magnitudes is to be recognized as again a single magnitude, then the following questions must be answered: When are two such conjunctions equal to or different from one another? When do they stand in a numerical relation to one another, and when they do, in what relation? For the completion of the concept it is then still necessary to be able to derive all the different series of magnitudes, each of which produce the same magnitude as the others when subjected to the conjunction in question.

These questions will be answered here for the combinatorial product, whereby we take as fundamental the concept of the multiplicative combinations.

69. If the magnitudes $a_1, a_2, \ldots, a_n$ stand in no numerical relation to one another, then their multiplicative combinations of any given class also stand in no numerical relation to one another, that is, the equation

$$(a) \qquad \alpha A + \beta B + \cdots = 0,$$

in which $A, B, \ldots$ denote the multiplicative combinations of $a_1, \ldots, a_n$ of any given class, and $\alpha, \beta, \ldots$ denote numbers, can be replaced by the group of equations

$$(b) \qquad \alpha = 0, \quad \beta = 0, \ldots.$$

PROOF. Assume equation (a) is valid. Multiply the whole equation combinatorially by those of the magnitudes $a_1, \ldots, a_n$ that do not appear in the product A; let A_1 be this series of factors, so that the combinatorial product $[AA_1]$ includes all the magnitudes $a_1, \ldots, a_n$ as factors. Then one obtains

$$\alpha[AA_1] + \beta[BA_1] + \cdots = 0.$$

Now since A and B are different combinations, B must contain at least one factor not included in A. If a_r is such a factor, then a_r must be included in A_1, since A_1 includes all of the factors $a_1, \ldots, a_n$ that do not appear in A. Thus a_r appears in B as well as A_1, whence the combinatorial product $[BA_1]$ is zero (according to 60). On the same grounds so is $[CA_1]$ and so forth. The equation is thus reduced to

$$\alpha[AA_1] = 0.$$

Consequently (according to 12, 6) either α or $[AA_1]$ must be zero. Now since $[AA_1]$ is a combinatorial product of n magnitudes that stand in no numerical relation to one another, it is not zero (according to 66). Consequently the other factor, α, must be zero. On the same grounds $\beta, \ldots$ are zero, that is, there exists no numerical relation between the combinations $A, B, \ldots$.

70. *Two combinatorial products (A and B) that are not zero stand in a numerical relation to one another if and only if the domains derived from their elementary factors are identical; that is,*

$$(a) \qquad A \equiv B$$

if and only if the elementary factors of A yield the same domain as do those of B; or

$$(b) \qquad [a_1a_2\cdots a_m] \equiv [b_1b_2\cdots b_m]$$

if and only if each of the magnitudes numerically derivable from $a_1, \ldots, a_m$ *can also be derived from* $b_1, \ldots, b_m$*, and thus if one can always set*

$$(c) \qquad x_1a_1 + x_2a_2 + \cdots + x_ma_m = y_1b_1 + y_2b_2 + \cdots + y_mb_m$$

whatever values either $x_1, \ldots, x_m$ *or* $y_1, \ldots, y_m$ *may have.*

PROOF. 1. Assume first that the domain $a_1, \ldots, a_m$ is identical to the domain $b_1, \ldots, b_m$, so that the magnitudes $a_1, \ldots, a_m$ are numerically derivable from the $b_1, \ldots, b_m$. Then (according to 63)

$$[a_1a_2\cdots a_m] = \alpha[b_1b_2\cdots b_m],$$

where α is a number (the determinant described there). This equation expresses that the two combinatorial products stand in a numerical relation, and since in fact neither of them is zero, one has (according to 2) the congruence

$$[a_1a_2\cdots a_m] \equiv [b_1b_2\cdots b_m].$$

2. Conversely, suppose the congruence holds, whence the two combinatorial products stand in a numerical relation to one another without being zero, and let

$$[a_1a_2\cdots a_m] = \alpha[b_1b_2\cdots b_m].$$

If one appends the combinatorial factor b_1 to both sides, one obtains

$$[a_1a_2\cdots a_mb_1] = \alpha[b_1b_2\cdots b_mb_1].$$

But $[b_1b_2\cdots b_mb_1]$ is zero (according to 60), since it contains two equal factors (b_1); thus also

$$[a_1a_2\cdots a_mb_1] = 0.$$

Consequently (according to 66) the elementary factors of this product, that is a_1, $a_2, \ldots, a_m, b_1$, stand in a numerical relation to one another. Thus (according to 16) it must be possible to find an equation of the form

$$\alpha_1a_1 + \alpha_2a_2 + \cdots + \alpha_ma_m + \beta_1b_1 = 0,$$

in which the numbers $\alpha_1, \alpha_2, \ldots, \alpha_m, \beta_1$ are not all simultaneously zero. In addition β_1 cannot be zero, since then (according to 16) a numerical relation between the magnitudes $a_1, a_2, \ldots, a_m$ would hold, whence the combinatorial product $[a_1a_2\cdots a_m]$ would have to be zero (according to 61), contrary to assumption. If however β_1 is not equal to zero, one can divide the above equation by β_1, and obtain

$$b_1 = -\frac{\alpha_1}{\beta_1}a_1 - \frac{\alpha_2}{\beta_1}a_2 - \cdots - \frac{\alpha_m}{\beta_1}a_m,$$

that is, b_1 is numerically derivable from $a_1, \ldots, a_m$. On the same grounds $b_2, \ldots, b_m$ are numerically derivable from $a_1, \ldots, a_m$.

But $b_1, \ldots, b_m$ also stand in no numerical relation to one another, since otherwise the combinatorial product $[b_1b_2\cdots b_m]$ would have to vanish (according to 61), contrary to assumption; thus the m magnitudes $b_1, \ldots, b_m$, which stand in no numerical relation to one another, are numerically derivable from the m magnitudes $a_1, \ldots, a_m$, whence (according to 21) the domain derivable from the first series of magnitudes identical to that derivable from the second.

REMARK. Since two equal magnitudes always stand in a numerical relation to one another, it follows at once from the preceding theorem that two equal combinatorial products always have one and the same domain that belongs to their elementary factors, and therefore that, aside from this domain, only the metric value, represented by a number, needs to be given in order to specify completely the whole value of the combinatorial product. Thus if, in the domain, any combinatorial product is given, from whose elementary factors the domain is derivable, then every other combinatorial product, from whose elementary factors the same domain is derivable, is defined by a simple number, which represents the ratio of this product to that.

71. DEFINITION. If, from one series of magnitudes, one derives a second by adding to any magnitude of the series a multiple of the neighboring magnitude of the series, while leaving all the other magnitudes of the series unchanged, then I say that the first series is transformed into the second by an *elementary linear evolution*; if one derives from this second series a third by an elementary linear evolution, and so forth, then I say that the first series is transformed into the last by a *multiple* linear evolution. Thus in both cases I say that the first series is transformed into the last by a linear evolution.

Thus if p and q are any two sequential magnitudes of the series, then by an elementary linear evolution the series

$$\ldots, \quad p, \quad q, \quad \ldots$$

is transformed into

$$\ldots, \quad p+\alpha q, \quad q, \quad \ldots$$

or into

$$\ldots, \quad p, \quad q+\alpha p, \quad \ldots,$$

where α is an arbitrary number.[14]

REMARK. The choice of the expression is related to the contrast with an evolution to be considered further below, which I call circular evolution. Both expressions refer back to geometry and indeed to the two fundamental structures of geometry, the straight line and the circle, or rather to the compass and straightedge, since, as I will show later, linear evolution in geometry can be accomplished simply by means of the straightedge, the circular by means of the compass.[15]

72. *The combinatorial product of a series of magnitudes remains unchanged under the linear evolution of the series.*

PROOF. According to 67 a combinatorial product does not change its value if one adds to one factor an arbitrary multiple of one of the other factors, whence it does not change its value under elementary linear evolutions of its factors, and thus also not under multiples.

73. *One can change, by linear evolutions, any two magnitudes of a series by reciprocal ratios, that is, by linear evolutions change the series*

$$\ldots, p, \ldots, q, \ldots$$

into

$$\ldots, \alpha p, \ldots, \frac{q}{\alpha}, \ldots .$$

PROOF. *First*, if p and q are two sequential magnitudes of the series, then (according to 71) they can gradually be transformed by linear evolutions:

$$p, \qquad q$$

into

$$p, \qquad q + (\alpha - 1)p,$$

and this again into

$$p + q + (\alpha - 1)p, \qquad q + (\alpha - 1)p,$$

that is into

$$\alpha p + q, \qquad q + (\alpha - 1)p;$$

this into

$$\alpha p + q, \qquad q + (\alpha - 1)p - \frac{\alpha - 1}{\alpha}(\alpha p + q),$$

that is, into

$$\alpha p + q, \qquad \frac{q}{\alpha},$$

and finally this into

$$\alpha p + q - q, \qquad \frac{q}{\alpha},$$

that is, into

$$\alpha p, \qquad \frac{q}{\alpha}.$$

Second: If p and q are separated by the magnitudes $p_1, p_2, \ldots, p_n$, then they are transformed by linear evolutions by applying the transformation shown to be admissible in the previous part, thus

$$p, \quad p_1, \quad p_2, \quad \ldots, \quad p_n, \quad q$$

into

$$\alpha p, \quad \frac{p_1}{\alpha}, \quad p_2, \quad \ldots, \quad p_n, \quad q,$$

this into

$$\alpha p, \quad p_1, \quad \frac{p_2}{\alpha}, \quad \ldots, \quad p_n, \quad q,$$

and, continuing in this way, finally into

$$\alpha p, \quad p_1, \quad p_2, \quad \ldots, \quad p_n, \quad \frac{q}{\alpha},$$

that is

$$\ldots, p, \qquad \ldots, \qquad q, \ldots$$

goes over into

$$\ldots, \alpha p, \qquad \ldots, \qquad \frac{q}{\alpha}, \ldots .$$

74. *From an arbitrary series of magnitudes one can by linear evolutions derive every other sequence of the same magnitudes, assuming, for the case that the combinatorial product of the derived series of magnitudes is the opposite of that of the original, the algebraic sign of one of the magnitudes in the new series changes; that is, if* a', b', c', ... *are the same magnitudes as* a, b, c, ... , *but in another sequence, then by linear transformations it is possible to change*

$$a, b, c, \ldots \quad \textit{into} \quad a', b', c', \ldots$$

if

$$[abc\cdots] = [a'b'c'\cdots],$$

but

$$a, b, c, \ldots \quad \textit{into} \quad -a', b', c', \ldots$$

if

$$[abc\cdots] = -[a'b'c'\cdots].$$

PROOF. 1. If p and q are two arbitrary {sequential} magnitudes of that series, then they are interchanged by linear evolutions upon adding and subtracting the second and first factors alternately to the first and second factors; step by step

$$p, \qquad q,$$

transforms into

$$p+q, \qquad q,$$

this into

$$p+q, \quad q-(p+q),$$

that is, into

$$p+q, \qquad -p,$$

and this into

$$q, \qquad -p.$$

2. Thus by linear evolutions one can bring two sequential magnitudes of the series into the reverse order, provided one changes the algebraic sign of one of them. One can therefore also bring every magnitude of the series to any place by linear evolutions, with the appropriate sign change.

Now if a', b', c', ... are the same magnitudes as a, b, c, ... , but in another sequence, then by linear evolutions one can transform the series a, b, c, ... into a series, the sequence of whose magnitudes is either equal or opposite to that of a', b', c', ... Now (according to 73) one can by a linear evolution evolve two arbitrary magnitudes p, q of a series reciprocally, that is, evolve them so that if one of the magnitudes, p, transforms into αp, the other, q, transforms into $\frac{1}{\alpha}q$; thus in particular one can evolve the series just found so that any arbitrary magnitude p of it opposed to one of the magnitudes a, b, c, ... transforms into $(-1)p$, that is into $-p$, while the first magnitude of that series, $\pm a'$, transforms into $\pm\frac{a'}{-1}$, that is, into $\mp a'$. If one applies this evolution, one by one, to each magnitude of the series that is opposed to one of the magnitudes a, b, c, ... , except not to the first magnitude $\pm a'$ of that series, then at last one obtains either the series

$$a', b', c', \dots \quad \text{or} \quad -a', b', c', \dots,$$

where the sign of a' is yet to be determined. Now since this series results from a, b, c, ... by linear evolutions, it follows (according to 72) that in the first case it must be that

$$[abc\cdots] = [a'b'c'\cdots]$$

and in the second,

$$[abc\cdots] = [-a'b'c'\cdots] = -[a'b'c'\cdots].$$

75. *If one adds to any given magnitude* (p) *of a series of magnitudes a multiple of another magnitude* (q) *of that series, thus replacing* p *with* $p+\alpha q$, *while leaving all the other magnitudes of that series unaltered, then the resulting series can be derived from the original series by a linear evolution, that is, by a linear evolution*

$$p, \quad \dots, \quad q$$

can be transformed into

$$p+\alpha q, \quad \dots, \quad q$$

or alternatively into

$$p, \quad \dots, \quad q+\alpha p.$$

PROOF. If no magnitudes stand between p and q in the given series, then the proof follows immediately from the definition [71]. If p_1, p_2, ... , p_n stand between p and q, then (according to 74) by linear evolutions

$$p, \quad p_1, \quad p_2, \quad \dots, \quad p_n, \quad q$$

can be transformed into

$$p, \quad q, \quad p_1, \quad p_2, \quad \dots, \quad \pm p_n$$

and this (according to 71) into

$$p+\alpha q, \quad q, \quad p_1, \quad p_2, \quad \dots, \quad \pm p_n;$$

and this again (according to 74) into

$$p+\alpha q, \quad p_1, \quad p_2, \quad \dots, \quad p_n, \quad \pm q,$$

where the sign of q is yet to be determined.

Since the last series is produced from the first by linear evolutions, the combinatorial product of the first series is (according to 72) equal to that of the last; thus

$$\begin{aligned}[pp_1p_2\cdots p_nq] &= [(p+\alpha q)p_1p_2\cdots p_n(\pm q)]\\ &= \pm[(p+\alpha q)p_1p_2\cdots p_nq].\end{aligned}$$

But (according to 67)

$$[pp_1p_2\cdots p_nq] = +[(p+\alpha q)p_1p_2\cdots p_nq],$$

that is, the + sign is appropriate in the previous formula, whence we are to replace $\pm q$ with q, that is, the series attained is $p+\alpha q$, p_1, p_2, … , p_n, q. Consequently it is possible to transform

$$p, \ldots, q \text{ into } p+\alpha q, \ldots, q,$$

by linear evolutions, and in the same way it follows that it is also possible to transform

$$p, \ldots, q \text{ into } p, \ldots, q+\alpha p$$

by linear evolutions.

76. *If two nonzero combinatorial products are equal, then the elementary factors of the one can be derived from those of the other by linear evolutions, that is, if*

(a) $$[abc\cdots] = [ABC\cdots] \neq 0,$$

then it is possible to transform the series of magnitudes

(b) $$a, b, c, \ldots \text{ into } A, B, C, \ldots$$

by linear evolutions [Converse of 72].

PROOF. Since the nonzero products $[abc\cdots]$ and $[ABC\cdots]$ are equal, and consequently they stand in a numerical relation to one another, the domains derivable from their elementary factors must (according to 70) be identical; that is, the magnitudes numerically derivable from the series of magnitudes a, b, c, … must also be numerically derivable from A, B, C, … and conversely. In particular the A, B, C, … must themselves be numerically derivable from a, b, c, … . In the expressions for these derivations the coefficients of any one of the magnitudes a, b, c, … , for example those of a, cannot all simultaneously be zero; for then the magnitudes A, B, C, … , whose number is n, would be derivable from the $n-1$ magnitudes b, c, … . Consequently (according to 22) there would be a numerical relation between them and their product would therefore (according to 61) be zero, contrary to assumption.

Let a' be one of the magnitudes A, B, C, … , and in fact one in whose derivation expression the coefficient of a is not zero, and let

$$a' = \alpha_1 a + \beta_1 b + \cdots$$

where $\alpha_1 \neq 0$; then by linear evolutions

$$a, \qquad b,\ c,\ \ldots,\ l$$

can be transformed step by step into

$$\alpha_1 a, \qquad b,\ c,\ \ldots,\ \frac{l}{\alpha_1},$$

and this into

$$\alpha_1 a + \beta_1 b + \cdots, \quad b,\ c,\ \ldots,\ \frac{l}{\alpha_1},$$

that is, into

$$a', \qquad b,\ c,\ \ldots,\ \frac{l}{\alpha_1}.$$

But now (according to 19) the domain derivable from a', b, c, ... must be identical with that of a, b, c, ... , whence in particular the magnitudes $A, B, C, \ldots$ must be derivable from a', b, c, ... Now again it is impossible, on the same grounds as before, that the coefficients of b in all the expressions of this derivation are simultaneously zero. Let b' be one of the magnitudes A, B, C, ... , and in fact one in which that coefficient is not zero, and let

$$b' = \alpha_2 a' + \beta_2 b + \gamma_2 c + \cdots ,$$

then by linear evolutions, in the same way as before,

$$a', \ b, \ c, \ \ldots, \ \frac{l}{\alpha_1}$$

can be transformed into

$$a', \ b', \ c, \ \ldots, \ \frac{l}{\alpha_1 \beta_2}.$$

Likewise let {c' be one of the magnitudes A, B, C, ... , such that}

$$c' = \alpha_3 a' + \beta_3 b' + \gamma_3 c + \delta_3 d + \cdots ,$$

where γ_3 is not zero; then again by linear evolutions

$$a', \ b', \ c, \ \ldots, \ \frac{l}{\alpha_1 \beta_2}$$

can be transformed into

$$a', \ b', \ c', \ d, \ldots, \ \frac{l}{\alpha_1 \beta_2 \gamma_3}.$$

In this way one proceeds through to the next to last magnitude. If this is k, then one finally obtains the series of magnitudes

$$a', b', c', \ldots, k', \frac{l}{\alpha_1 \beta_2 \gamma_3 \cdots \kappa_{n-1}}.$$

Now since this series of magnitudes was produced from the original a, b, c, ... , its combinatorial product must (according to 72) be equal to that of that series of magnitudes; thus

$$\left[a'b'c' \cdots k' \frac{l}{\alpha_1 \beta_2 \gamma_3 \cdots \kappa_{n-1}}\right] = [abc \cdots kl].$$

Furthermore, the $n-1$ magnitudes a', b', c', ... , k' are all taken from A, B, C, ... , K, L, and since a', b', c', ... , k' must all be different from one another, for otherwise (according to 60) their combinatorial product would be zero (which on account of the equation just developed would be contrary to assumption), there can only be *one* of the magnitudes A, B, C, ... , K, L remaining which is not included among the magnitudes a', b', c', ... , k'. Let this be l', and let $l = \alpha_n a' + \beta_n b' + \cdots + \kappa_n k' + \lambda_n l'$; then the last series obtained is transformed by linear evolutions into

$$a', b', c', \ldots, k', \frac{\lambda_n l'}{\alpha_1 \beta_2 \gamma_3 \cdots \kappa_{n-1}}.$$

But the series of magnitudes a', b', c', ... , k', l' includes the same magnitudes as does the series A, B, C, ... , K, L, only in another order; thus (according to 74)

a', b', c', ... , k', l' can by linear evolutions be transformed into A, B, C, ... , K, $\pm L$, whence also

$$a', b', c', \ldots, k', \frac{\lambda_n l'}{\alpha_1 \beta_2 \gamma_3 \cdots \kappa_{n-1}}.$$

into

$$A, B, C, \ldots, K, \pm \frac{\lambda_n L}{\alpha_1 \beta_2 \gamma_3 \cdots \kappa_{n-1}},$$

since (according to 73) it is of no consequence to which magnitude one appends the numerical factor $\frac{\pm\lambda_n}{\alpha_1\beta_2\gamma_3\cdots\kappa_{n-1}}$. If this numerical factor is $= \varepsilon$, then by linear evolutions from a, b, c, ... , k, l there finally results A, B, C, ... , K, εL. Thus (according to 72)

$$[abc \cdots kl] = [ABC \cdots K \cdot \varepsilon L] = \varepsilon [ABC \cdots KL].$$

However, by hypothesis

$$[abc \cdots kl] = [ABC \cdots KL],$$

whence also

$$\varepsilon [ABC \cdots KL] = [ABC \cdots KL],$$

that is $\varepsilon = 1$ and so $\varepsilon L = L$, and therefore by linear evolutions one has transformed the series

$$a, b, c, \ldots, k, l \text{ into } A, B, C, \ldots, K, L,$$

a transformation whose possibility was to be proved.[16]

77. Definition. Multiplicative combinations of mth class of the original units I call *units of mth order*, a magnitude numerically derived from these units a *magnitude of mth order*, and in particular an *elementary* one if it can be represented as the combinatorial product of m magnitudes of first order, a *compound* if this is not possible.* {Numbers are therefore magnitudes of zeroth order.}[17]

I call the domain derivable from the elementary factors of an elementary magnitude the domain *belonging* to this magnitude, or briefly the domain of this magnitude. Finally I call an elementary magnitude A *superordinate*, *subordinate*, or *incident* on another, according as this applies to the domains of these magnitudes (cf. No. 15).

77b. Corollary. *A combinatorial product of m magnitudes of first order is an* ***elementary*** *magnitude of mth order* {77}*, and is numerically derivable from the units of mth order* {65}.

Remark. As an example of a compound magnitude I introduce here the sum $[ab] + [cd]$, where a, b, c, d are four magnitudes standing in no numerical relation to one another. Thus if $[ab] + [cd]$ were equal to an elementary magnitude, say $= [pq]$, then necessarily (according to 60)

$$[(ab + cd)(ab + cd)] = [pqpq] = 0;$$

but

$$[(ab + cd)(ab + cd)] = [abcd] + [cdab],$$

since $[abab]$ and $[cdcd]$ are zero. But (according to 58) $[abcd] = [cdab]$, whence

$$[(ab + cd)(ab + cd)] = 2[abcd].$$

---

*Grassmann's repeated use of "order" [Stufe] seems unfortunate, since the order of a *magnitude* defined here is quite distinct from the order of a *domain* defined in No. 14. *Tr.*

Thus if $[ab] + [cd]$ were an elementary magnitude, then necessarily $[abcd] = 0$, whence (according to 66) a, b, c, d would stand in a numerical relation, which conflicts with the assumption.[18]

3. Outer multiplication of magnitudes of higher order

78. DEFINITION. To *outer multiply* two units of higher order means to combinatorially multiply their elementary factors without changing their serial order, that is

$$[(e_1 e_2 \cdots e_m)(e_{m+1} \cdots e_n)] = [e_1 e_2 \cdots e_n].$$

REMARK. I have chosen the name outer multiplication in order to indicate that the product only has a nontrivial value if one factor lies completely outside the domain of the other. Outer multiplication is the opposite of inner multiplication (cf. Ch. 4).

79. *Instead of outer multiplying an elementary magnitude A by another B, one can combinatorially multiply the elementary factors of the first serially by those of the second, that is*

$$[(ab \cdots)(cd \cdots)] = [ab \cdots cd \cdots].$$

PROOF. Let $e_1, \ldots, e_n$ be the original units, and let

$$a = \sum \alpha_{\mathfrak{a}} e_{\mathfrak{a}}, \; b = \sum \beta_{\mathfrak{b}} e_{\mathfrak{b}}, \; \ldots, \; c = \sum \gamma_{\mathfrak{c}} e_{\mathfrak{c}}, \; d = \sum \delta_{\mathfrak{d}} e_{\mathfrak{d}}, \; \ldots;$$

then

$$\begin{aligned} [(ab \cdots)(cd \cdots)] &= \left[\left(\sum \alpha_{\mathfrak{a}} e_{\mathfrak{a}} \cdot \sum \beta_{\mathfrak{b}} e_{\mathfrak{b}} \cdots\right)\left(\sum \gamma_{\mathfrak{c}} e_{\mathfrak{c}} \cdot \sum \delta_{\mathfrak{d}} e_{\mathfrak{d}} \cdots\right)\right] \\ &= \left[\sum\{\alpha_{\mathfrak{a}} \beta \mathfrak{b} \cdots [e_{\mathfrak{a}} e_{\mathfrak{b}} \cdots]\} \cdot \sum\{\gamma_{\mathfrak{c}} \delta \mathfrak{d} \cdots [e_{\mathfrak{c}} e_{\mathfrak{d}} \cdots]\}\right] && [45] \\ &= \sum\{\alpha_{\mathfrak{a}} \beta_{\mathfrak{b}} \cdots \gamma_{\mathfrak{c}} \delta_{\mathfrak{d}} \cdots [(e_{\mathfrak{a}} e_{\mathfrak{b}} \cdots)(e_{\mathfrak{c}} e_{\mathfrak{d}} \cdots)]\} && [42] \\ &= \sum\{\alpha_{\mathfrak{a}} \beta_{\mathfrak{b}} \cdots \gamma_{\mathfrak{c}} \delta_{\mathfrak{d}} \cdots [e_{\mathfrak{a}} e_{\mathfrak{b}} \cdots e_{\mathfrak{c}} e_{\mathfrak{d}} \cdots]\} && [78] \\ &= \left[\sum \alpha_{\mathfrak{a}} e_{\mathfrak{a}} \cdot \sum \beta_{\mathfrak{b}} e_{\mathfrak{b}} \cdots \sum \gamma_{\mathfrak{c}} e_{\mathfrak{c}} \cdot \sum \delta_{\mathfrak{d}} e_{\mathfrak{d}} \cdots\right] && [45] \\ &= [ab \cdots cd \cdots]. \end{aligned}$$

79B. COROLLARY. *If a nonzero elementary magnitude A is subordinate to another, B, which is also nonzero, then the latter can be represented as an outer product, one factor of which is A and the other of which is an elementary magnitude C, that is, in the form*

$$B = [AC].$$

PROOF. According to 77, A is subordinate to B if the domain of A is subordinate to that of B, that is, if every magnitude of the first domain is simultaneously a magnitude of the second. If $A = [a_1 a_2 \cdots a_m]$, where $a_1, \ldots, a_m$ are magnitudes of first order, then since A cannot be zero, these stand in no numerical relation to one another (61). Suppose further that $B = [b_1 \cdots b_n]$. Now since the magnitudes $a_1, \ldots, a_m$ belong to the domain B, they must be numerically derivable from the $b_1, \ldots, b_n$. But then (according to 20) one can adjoin to the magnitudes $a_1, \ldots, a_m$ $(n - m)$ further magnitudes $a_{m+1}, \ldots, a_n$ such that the domains $a_1, \ldots, a_n$ and

$b_1, \ldots, b_n$ are identical. But in this case (according to 70) the products $[a_1 \cdots a_n]$ and $[b_1 \cdots b_n]$ stand in a numerical relation to one another. Thus let

$$\begin{aligned}[b_1 \cdots b_n] &= \alpha[a_1 \cdots a_n] = \alpha[a_1 \cdots a_m a_{m+1} \cdots a_n] \\ &= [(a_1 \cdots a_m) \cdot \alpha(a_{m+1} \cdots a_n)]. && [79, \{46\}]\end{aligned}$$

Thus, if one sets

$$\alpha[a_{m+1} \cdots a_n] = C,$$

then

$$B = [AC].$$

80. *The placement of parentheses in an outer product is of no importance for the result, that is*

$$[A(BC)] = [ABC].$$

PROOF. 1. If A, B, C are elementary magnitudes, $A = [a_1 \cdots a_q]$, $B = [b_1 \cdots b_r]$, $C = [c_1 \cdots c_s]$, then

$$\begin{aligned}[A(BC)] &= [a_1 \cdots a_q((b_1 \cdots b_r)(c_1 \cdots c_s))] \\ &= [a_1 \cdots a_q(b_1 \cdots b_r c_1 \cdots c_s)] && [79] \\ &= [a_1 \cdots a_q b_1 \cdots b_r c_1 \cdots c_s] && [79] \\ &= [(a_1 \cdots a_q)(b_1 \cdots b_r)(c_1 \cdots c_s)] && [79] \\ &= ABC.\end{aligned}$$

2. If A, B, C are sums of elementary magnitudes, $A = \sum A_\mathfrak{a}$, $B = \sum B_\mathfrak{b}$, $C = \sum C_\mathfrak{c}$, then

$$\begin{aligned}[A(BC)] &= \left[\sum A_\mathfrak{a} \cdot \left(\sum B_\mathfrak{b} \cdot \sum C_\mathfrak{c}\right)\right] = \sum [A_\mathfrak{a}(B_\mathfrak{b} C_\mathfrak{c})] && [42] \\ &= \sum [A_\mathfrak{a} B_\mathfrak{b} C_\mathfrak{c}] && [\text{Proof } 1] \\ &= \left[\sum A_\mathfrak{a} \sum B_\mathfrak{b} \sum C_\mathfrak{c}\right] && [45] \\ &= [ABC].\end{aligned}$$

81. *If $a_1, \ldots, a_m$, $b_1, \ldots, b_n$ are magnitudes of first order standing in no numerical relation to one another, and A results from $a_1, \ldots, a_m$, and B from $b_1, \ldots, b_n$, by addition and multiplication, and*

$$[AB] = 0,$$

then either $A = 0$ or $B = 0$.

PROOF. If A is of αth order, B of βth order, and if $A_1, A_2, \ldots$ are the multiplicative combinations of $a_1, \ldots, a_m$ of αth class, $B_1, B_2, \ldots$ those of βth class of $b_1, \ldots, b_n$, then (according to 77) A and B are representable in the forms

$$A = \sum \alpha_\mathfrak{a} A_\mathfrak{a}, \quad B = \sum \beta_\mathfrak{b} B_\mathfrak{b},$$

and thus

$$[AB] = \sum \alpha_\mathfrak{a} \beta_\mathfrak{b} [A_\mathfrak{a} B_\mathfrak{b}].$$

Here the $[A_{\mathfrak{a}}B_{\mathfrak{b}}]$ are regarded as multiplicative combinations of $a_1, \ldots, a_m, b_1, \ldots, b_n$. Thus (according to 69) they stand in no numerical relation to one another. Consequently (according to 28)

$$\alpha_r\beta_s = 0$$

for each r and s; thus if $B \neq 0$, that is, if any one of the magnitudes β_s is different from zero, it follows that $\alpha_r = 0$ for each r, that is $A = 0$.

82. *If a sum S of elementary magnitudes gives zero when outer multiplied by a magnitude of first order, a, different from zero, then the former (S) is represented as an outer product in which a is a factor, that is in the form*

$$S = [aP], \text{ if } [aS] = 0.$$

PROOF. If S is a sum of magnitudes of mth order and $e_1, e_2, \ldots, e_n$ are the original units, then one can (according to 20) introduce $(n-1)$ magnitudes b, $c, \ldots$ in addition to a, such that the $e_1, \ldots, e_n$ can be numerically derived from the $a, b, c, \ldots$. Then every elementary factor in each of the magnitudes of mth order, whose sum is S, can also be numerically derived from the $a, b, c, \ldots$. Thus each of these magnitudes, and therefore also their sum S, can be numerically derived from the multiplicative combinations of mth class of the $a, b, c, \ldots$.

Now let $[aB_1], [aB_2], \ldots$ be those of these combinations that include a, and $C_1, C_2, \ldots$ be those that do not include a, and let

$$S = \beta_1[aB_1] + \beta_2[aB_2] + \cdots + \gamma_1 C_1 + \gamma_2 C_2 + \cdots.$$

Now since by assumption $[aS] = 0$, one has

$$0 = [aS] = \gamma_1[aC_1] + \gamma_2[aC_2] + \cdots,$$

since $[aaB_1], [aaB_2], \ldots$ are zero. Because a is not included in $C_1, C_2, \ldots, [aC_1], [aC_2], \ldots$ are {clearly different} multiplicative combinations that stand in no numerical relation to one another. It therefore follows at once from the above equation $0 = \gamma_1[aC_1] + \gamma_2[aC_2] + \cdots$ that the $\gamma_1, \gamma_2, \ldots$ are all zero (according to 28). Consequently

$$\begin{aligned} S &= \beta_1[aB_1] + \beta_2[aB_2] + \cdots \\ &= [a(\beta_1 B_1 + \beta_2 B_2 + \cdots)] = [aP], \end{aligned}$$

if one sets

$$P = \beta_1 B_1 + \beta_2 B_2 + \cdots.$$

83. *If a sum S of elementary magnitudes gives zero upon outer multiplication by each of m magnitudes of first order $a_1, a_2, \ldots$ standing in no numerical relation to one another, then S can be expressed as an outer product in which the $a_1, a_2, \ldots$ are factors, that is in the form*

$$S = [a_1 a_2 \cdots a_m S_m],$$

if

$$0 = [a_1 S] = [a_2 S] = \cdots = [a_m S].$$

PROOF. If $e_1, \ldots, e_n$ are the original units, then (according to 20) to the $a_1, \ldots, a_m$ there can be adjoined $n-m$ other magnitudes $a_{m+1}, \ldots, a_n$ such that the $e_1, \ldots, e_n$ can be numerically derived from the $a_1, \ldots, a_n$. Consequently all the magnitudes of first order appearing in S can be numerically derived from $a_1, \ldots, a_n$.

Now if it is assumed that $[a_1S] = 0$, then (according to 82) S can be represented in the form $S = [a_1S_1]$. If one represents the elementary factors of this magnitude as multiple sums of $a_1, \ldots, a_n$, then (according to 67) one can, without changing the value of the product $[a_1S_1]$, drop the terms containing a_1 in all these multiple sums. This done, S_1 is transformed into P_1, and thus

$$S = [a_1S_1] = [a_1P_1],$$

where P_1 is produced only from the magnitudes $a_2, \ldots, a_n$ (no a_1 included).

Now if in addition $[a_2S] = 0$, that means

$$0 = [a_2a_1P_1], \text{ or (according to 55) } 0 = [a_1a_2P_1].$$

Now $[a_2P_1]$ is generated only from the magnitudes $a_2, \ldots, a_n$, and thus (according to 81) either a_1 or $[a_2P_1]$ must be zero. The former is contrary to assumption, so $[a_2P_1] = 0$. Consequently P_1 must be representable in the form $[a_2S_2]$; here again the magnitude a_2 can be removed from S_2 without changing the value of the product $[a_2S_2]$; if $P_1 = [a_2S_2] = [a_2P_2]$, where P_2 is generated only from $a_3, \ldots, a_n$ (without a_1 and a_2), then

$$S = [a_1a_2S_2] = [a_1a_2P_2].$$

Next, if $[a_3S] = 0$, then

$$[a_3a_1a_2P_2] = 0, \text{ or } [a_1a_2a_3P_2] = 0.$$

Now $[a_3P_2]$ is only generated from $a_3, \ldots, a_n$, and thus (according to 81) either $[a_1a_2]$ or $[a_3P_2]$ must be zero. The first is not possible, since if it were, then (according to 66) there would be a numerical relation between a_1 and a_2, contrary to assumption. Consequently it must be that $[a_3P_2] = 0$, whence P_2 is representable in the form $P_2 = [a_3P_3]$, where again P_3 is generated only from $a_4, \ldots, a_n$, and so forth, until finally

$$S = [a_1a_2 \cdots a_mS_m].$$

84. *If a sum S of magnitudes of mth order gives zero upon outer multiplication by each of m magnitudes of first order $a_1, \ldots, a_m$, then S is congruent to the outer product of these m magnitudes, that is, if*

$$0 = [a_1S] = \cdots = [a_mS], \text{ then } S \equiv [a_1 \cdots a_m].$$

PROOF. First, S is (according to 83) representable in the form $[a_1 \cdots a_mS_m]$; thus, since S is an expression of mth order here, S_m must be of zeroth order, and therefore a number [77], and therefore (according to 2) instead of $S = [a_1 \cdots a_mS_m]$ we can write

$$S \equiv [a_1 \cdots a_m].$$

85. *If there are $m + 1$ magnitudes* {*of first order*} $a_1, \dots, a_{m+1}$, *each of which gives zero upon outer multiplication by a sum S of magnitudes of mth order, then either $S = 0$ or $[a_1 \cdots a_{m+1}] = 0$.*

PROOF. Assuming $[a_1 \cdots a_{m+1}]$ is not zero, then also $[a_1 \cdots a_m]$ is not zero, whence $a_1, \dots, a_m$ stand in no numerical relation to one another; then since at the same time $[a_1 S] = [a_2 S] = \cdots = [a_m S] = 0$, it follows that $S = \alpha[a_1 \cdots a_m]$ (according to 84). But if in addition $[a_{m+1} S] = 0$, that is $\alpha[a_1 \cdots a_{m+1}] = 0$, then since by assumption $[a_1 \cdots a_{m+1}]$ is different from zero, necessarily $\alpha = 0$, whence also $S = \alpha[a_1 \cdots a_m] = 0$, that is, either $[a_1 \cdots a_{m+1}]$ or S is zero.

4. Supplement of magnitudes with respect to a principal domain

86. DEFINITION. By a *principal domain* I mean the domain of the original units from which all the magnitudes under consideration are produced.*

87. *Two elementary magnitudes A and B, the sum of whose orders exceed that of the principal domain by γ, can be expressed in the forms*

$$A = [CA_1], \quad B = [CB_1],$$

where {*A_1 and B_1 are elementary magnitudes, and*} *C is an elementary magnitude of γth order.*

PROOF. If α is the order of A, β that of B, n that of the principal domain, then

$$\alpha + \beta = n + \gamma.$$

Now (according to 26) the domains of A and B have at least a domain of $(\alpha + \beta - n)$th, that is of γth order, in common. If C is an elementary magnitude of γth order of this domain, then C is subordinate to the magnitude A as well as the magnitude B, whence (according to 79B) A is representable in the form $[CA_1]$ and B in the form $[CB_1]$ {where A_1 and B_1 are also elementary magnitudes}.

88. *The sum of elementary magnitudes of $(n-1)$th order in a principal domain of nth order give again an elementary magnitude of $(n-1)$th order.***

PROOF. Suppose A and B are the two magnitudes of $(n-1)$th order that lie in a principal domain of nth order; then (according to 87) since the sum $(2n-2)$ of their orders exceeds that of the principal domain by $n-2$, they must be representable in the form

$$A = [Ca], \quad B = [Cb],$$

where C represents a series of $n-2$ elementary factors of first order, a and b factors of first order, and thus

$$A + B = [Ca] + [Cb] = [C(a + b)]. \qquad [39]$$

Here $a + b$, as the sum of two magnitudes of first order, is again a magnitude of first order, whence $A + B$ is representable as a combinatorial product of $n-1$ magnitudes of first order, and therefore itself an elementary magnitude of $(n-1)$th order.

---

*A "principal domain" is to be distinguished from a "principal ideal domain".

**For the different meanings of "order", see the footnote on p. 44.

REMARK. If in a principal domain of nth order one has two elementary magnitudes A and B, whose orders are greater than 1 but less than $n-1$, then their sum is in general no longer an elementary magnitude. Thus, for example, if a, b, c, d are four magnitudes of first order standing in no numerical relation to one another, the sum $S = [ab] + [cd]$ cannot be represented in the form of a combinatorial product of factors of first order. For if it could, then (according to 60)

$$[SS] = 0,$$

whence

$$0 = [SS] = [(ab + cd)(ab + cd)] = [abcd] + [cdab],$$

since $[abab]$ and $[cdcd]$ are zero (according to 60). But since (according to 58) $[cdab] = [abcd]$, one has

$$0 = 2[abcd],$$

that is $[abcd]$ must be zero, whence (according to 66) a, b, c, d must stand in a numerical relation to one another, contrary to assumption. Thus S is not representable in the form of a combinatorial product of magnitudes of first order, and is therefore a compound magnitude.[19]

89. DEFINITION. If in a principal domain of nth order the combinatorial product of the original units e_1, e_2, ... , e_n is set equal to 1, and E is a unit of arbitrary order, that is, either one of the original units or a combinatorial product of several of them, then by the "supplement of E" I mean that magnitude that is equal or opposite to the combinatorial product E' of all the units not appearing in E, according as $[EE']$ is equal or opposite to the absolute unit; I symbolize the supplement of a magnitude by a vertical line in front of the symbol for the magnitude, thus that of E by $|E$. I set the supplement of a number equal to the number. Thus:

$$|E = [EE']E',$$

if E and E' include the elementary factors e_1, ... , e_n and

$$[e_1 e_2 \cdots e_n] = 1;$$

and

$$|\alpha = \alpha,$$

if α is a number.[20]

REMARK. In the definition it is assumed that $[EE']$ can only equal either $+1$ or -1. But in fact, since E and E' are combinatorial products of the original units, and E' includes all the units missing from E, it follows that $[EE']$ is distinguished from $[e_1 e_2 \cdots e_n]$ only by the sequence of its factors, and therefore the two expressions are (according to 57) either equal or opposite, whence since $[e_1 e_2 \cdots e_n] = 1$, $[EE'] = \pm 1$.

90. DEFINITION. By the supplement of an arbitrary magnitude A I mean that magnitude $|A$ that one obtains if, in the expression representing the numerical derivation of that magnitude in terms of the units, one replaces these units by their supplements, that is

$$|(\alpha_1 E_1 + \alpha_2 E_2 + \cdots) = \alpha_1 |E_1 + \alpha_2 |E_2 + \cdots,$$

where E_1, E_2, ... are units of arbitrary order.[21]

COROLLARY. If n is the order of the principal domain and α that of the magnitude A, then $n - \alpha$ is that of the supplement.

REMARK. According to these definitions the vertical line appears as one of the attributes of a factor. This factor, as will be shown below, is a striking analogue of the imaginary expression $\sqrt{-1}$, so that one can under certain circumstances replace the one by the other. I have chosen the vertical line to indicate as well that, as I will show below, this supplement is represented geometrically by the structure standing orthogonal to a given structure.

91. *The outer product of a unit and its supplement is* 1, *that is*

$$[E|E] = 1.$$

PROOF. If E' is the combinatorial product of all the original units not included in E, then (according to 89)

$$|E = \pm E',$$

according as

$$[EE'] = \pm 1.$$

Thus, if the upper sign applies, then

$$[E|E] = [EE'] = 1,$$

and, if the lower applies,

$$[E|E] = -[EE'] = -(-1) = 1.$$

92. *The supplement of the supplement of a magnitude A is the equal or opposite of this magnitude A, according as the product of the order of this magnitude on the one hand and that of its supplement on the other is even or odd, that is*

$$||A = (-1)^{qr}A,$$

where q is the order of A and r that of $|A$.

PROOF. Supposing *first* that A is a combinatorial product of the original units, and $B = |A$ is its supplement, then by definition B includes all the units missing from A, such that $[A|A] = 1$, whence

$$[AB] = 1.$$

Again, since A includes all the units missing from B, the supplement of B is (according to 89) equal or opposite to A, according as $[BA]$ is equal or opposite to the absolute unit. Now (according to 58) $[BA] = (-1)^{qr}[AB]$, where q and r are the orders of A and B; thus, since $[AB] = 1$,

$$[BA] = (-1)^{qr},$$

whence also the supplement of B is equal to $+A$ or $-A$ according as $(-1)^{qr}$ is equal to $+1$ or -1, that is

$$|B = (-1)^{qr}A.$$

But B was assumed to equal $|A$, whence

$$||A = (-1)^{qr},$$

if A is a combinatorial product of the original units.

Second, suppose A is an arbitrary magnitude of qth order and its supplement of rth order, and let

$$A = \alpha_1 E_1 + \alpha_2 E_2 + \cdots,$$

where E_1, E_2, ... are combinatorial products of the original units, α_1, α_2, ... numbers; then (according to 90)

$$|A = \alpha_1 |E_1 + \alpha_2 |E_2 + \cdots.$$

Consequently, since $|E_1$, $|E_2$, ... are again products of units,

$$||A = \alpha_1||E_1 + \alpha_2||E_2 + \cdots .$$

Now E_1, E_2, ... are of the same order as A, that is of qth order, and their supplements of rth order; thus according to the first part of the Proof $||E_1 = (-1)^{qr}|E_1$, $||E_2 = (-1)^{qr}E_2, \ldots$, and therefore

$$\begin{aligned} ||A &= (-1)^{qr}(\alpha_1 E_1 + \alpha_2 E_2 + \cdots) \\ &= (-1)^{qr} A. \end{aligned}$$

93. *If the order of the principal domain* (n) *is odd, then*

$$||A = A.$$

If n *is even, then*

$$||A = (-1)^q A,$$

where q *is the order of* A.

PROOF. The order of $|A$ is equal to $(n-q)$, and therefore (according to 92) $||A = (-1)^{q(n-q)}A$. Now if n is odd, then either q or $n-q$ is even, and thus $(-1)^{q(n-q)} = 1$, and thus $||A = A$. If n is even, then $q(n-q)$ is even or odd according as q is, whence $(-1)^{q(n-q)} = (-1)^q$, and $||A = (-1)^q A$.

REMARK. If q and r are both odd, as for example if one considers the supplement of magnitudes of first order in a domain of second order, then $||A = -A$, so that in this case the symbol $|$ is subject to the same laws as $i = \sqrt{-1}$, and we therefore obtain here a real interpretation of the imaginaries.

In the applications to geometry it will be shown that displacements, that is lines of definite direction and length, are regarded as magnitudes of first order, and that with respect to them the plane appears as a domain of second order, so that the above-mentioned case $||A = -A$ arises here. I will show that, if one takes two perpendicular displacements of equal length as the original units, the supplement of a displacement is the displacement perpendicular to it, and thus one already sees here that the real significance that we attribute to the imaginaries corresponds precisely to the geometric significance itself which GAUSS first perceived, except that this significance enters here in a more general form.

5. Product with respect to a principal domain

94. DEFINITION. If the sum of the orders of two units is less than or equal to the order n of the principal domain, then by their *progressive product* I mean their outer product, with the condition that the progressive product of the n original units is 1. On the other hand, if the sum of the orders of two units is greater than the order (n) of the principal domain, then by their *regressive product* I mean that magnitude whose supplement is the progressive product of the supplements of those units.

I combine the progressive and regressive products under the name products *relative to* a principal domain. For all these products the symbol is the same, that is brackets enclosing the elements of the product. Thus

$$|[EF] = [|E|F],$$

if the sum of the orders of E and F is greater than the order (n) of the principal domain, and

$$[e_1 e_2 \cdots e_n] = 1,$$

if $e_1, e_2, \ldots, e_n$ is the series of the original units.[22]

REMARK. All multiplications that are of interest for science can be reduced to the multiplication considered here and to algebraic multiplication. It is therefore only a question of unambiguously distinguishing these two species of multiplication from one another by means of their symbols.

If in the algebraic multiplication one eliminates all superfluous parentheses and thus never encloses an *algebraic* product, which is combined with another magnitude by addition, subtraction, multiplication, division, then parentheses applied in these cases are always an unambiguous sign of *relative* multiplication. In those cases where the algebraic product is so conjoined that parentheses are necessary, for example if the product is to be raised to a power or its logarithm to be taken, or if it is to be inserted into a function symbol (among which summation symbols, differential symbols and so forth may be reckoned), then again all ambiguity is removed if in this case one either uses *two* parentheses for relative multiplication or, as seems more convenient, if for this case one assigns parentheses to the algebraic product, brackets to the relative, while in the first case one has both at one's disposal for convenience.

It is yet to be mentioned that the symbolism chosen for the different types of relative multiplication, once the principal domain is known, is completely sufficient and leaves no room for error, and that the calculational laws are the same throughout, and depend only on the orders of the magnitudes to be conjoined and that of their principal domain.[23]

95. *If q and r are the orders of two magnitudes A and B, and n that of the principal domain, then the order of the product $[AB]$ is* ***first*** *equal to $q+r$ if $q+r$ is smaller than n, and* **second** *equal to $q+r-n$ if $q+r$ is greater than or equal to n, whence in both cases congruent to the sum of the orders relative to the modulus n, that is, if*

$$C = [AB],$$

then

$$s \equiv q + r \ (Mod.\ n),$$

where q, r, s, n are respectively the orders of A, B, C, and the principal domain.

PROOF. If $q + r < n$ and

$$A = [a_1 a_2 \cdots a_q], \quad B = [b_1 b_2 \cdots b_r],$$

where $a_1, \ldots, a_q, b_1, \ldots, b_r$ are magnitudes of first order, then $[AB]$ is to be treated as a progressive product, whence

$$C = [AB] = [(a_1 a_2 \cdots a_q)(b_1 b_2 \cdots b_r)]$$
$$= [a_1 a_2 \cdots a_q b_1 b_2 \cdots b_r], \qquad [79]$$

and therefore (according to 77) the order of the product $= q + r$.

If $q + r = n$, then since (according to 94) the product of the n units is set $= 1$ so that the product of n magnitudes of first order is a number, but (according to 77) the numbers are interpreted as magnitudes of zeroth order, the order of the product is zero, and therefore $= q + r - n$.

Finally, if $q + r$ is greater than n, then (according to 94) if A and B are units of arbitrary order,

$$|C = [|A|B].$$

But (according to the Corollary to 90) the orders of $|A$, $|B$, $|C$ are equal to $n-q$, $n-r$, $n-s$. Now $n-q+n-r=n-(q+r-n)$, that is smaller than n, whence (according to the first part of the Proof) the order of the product $[|A|B]$ is equal to the sum of the orders of its factors; thus

$$n-s=n-q+n-r,$$

that is

$$s=q+r-n.$$

The theorem is therefore valid for the case that A and B are units of arbitrary order. Since however every magnitude numerically derived from the units is of the same order as the units, the theorem is valid for arbitrary magnitudes.[24]

96. *If n is the order of the principal domain, then the order of an arbitrary relative product of this domain is congruent to the sum of the orders of its factors with respect to the modulus n, or, the order of the product is equal to the remainder when one divides the sum of the orders of all the factors by the order of the principal domain; thus if*

$$R=[ABC\cdots],$$

then

$$\rho \equiv \alpha+\beta+\gamma+\cdots \ (Mod.\ n),$$

where ρ, α, β, γ, ... are the orders of R, A, B, C, ... and n is that of the principal domain.

Proof. In 95 it was shown that the order of the product of two magnitudes is congruent with the sum of the orders of these magnitudes, with respect to the modulus n. Now if another factor is adjoined to the product the law remains in effect on the same grounds, and so forth, whence it is valid for arbitrarily many factors. Since the order is always smaller than n and is never negative, the theorem is also correct in its alternative form.

97. *The product of the supplements of two magnitudes is the supplement of the product of these magnitudes, that is*

$$[|A|B]=|[AB].$$

Proof. 1. Suppose the orders α and β of the magnitudes A and B are together greater than the order n of the principal domain, that is $\alpha+\beta>n$.

Let

$$A=\sum \alpha_r E_r, \quad B=\sum \beta_s F_s,$$

where E_r, F_s are units; then (according to 90)

$$|A=\sum \alpha_r|E_r \text{ and } |B=\sum \beta_s|F_s.$$

Thus

$$\begin{aligned}
[|A|B] &= \left[\sum \alpha_r|E_r \cdot \sum \beta_s|F_s\right] = \sum \alpha_r\beta_s[|E_r|F_s] && [42]\\
&= \sum \alpha_r\beta_s|[E_rF_s] && [94]\\
&= \left|\sum \alpha_r\beta_s[E_rF_s]\right. && [90]\\
&= \left|\left[\sum \alpha_rE_r \cdot \sum \beta_sF_s\right]\right. && [42]\\
&= |[AB].
\end{aligned}$$

2. If $\alpha + \beta = n$, then the theorem applies first of all to the units.

Let E and F be units, that is, combinatorial products of the original units e_1, $\dots$, e_n.

First, if E includes one original unit that also appears in F, then not all the original units can appear in E and F, since together they contain only n elementary factors. Consequently at least one of these units, say e_1, must be missing from both magnitudes E and F; now $|E$ includes all original units missing from E, and thus in particular e_1, and $|F$ all those missing from F, and thus also e_1, and since both $|E$ and $|F$ include the original unit e_1, it follows (according to 60) that $[|E|F] = 0$, but also $[EF] = 0$ since E and F both include one and the same original unit by assumption, and therefore

$$[|E|F] = [EF].$$

Second, if E includes no original unit that appears in F, then since E and F include n factors in all, each of which is one of the original units, $[EF]$ must be a product of all n original units. But then (according to 89)

$$|E = [EF]F \quad\text{and}\quad |F = [FE]E,$$

where $[EF]$ and $[FE]$ can only be either $+1$ or -1. Consequently

$$[|E|F] = [EF][FE][FE].$$

But $[FE][FE]$ is either $1 \cdot 1$ or $(-1) \cdot (-1)$, or 1 in both cases. Consequently

$$[|E|F] = [EF],$$

as in the previous case. Now since $[EF]$ is a number, $[EF] = |[EF]$ (according to 89); thus in both cases

$$[|E|F] = |[EF].$$

Since the law is true for units, it follows in exactly the same way as in Part 1 of the Proof that it is also true for arbitrary magnitudes, provided the sum of their orders is n.

3. If $\alpha + \beta < n$, then let $A = |A'$ and $B = |B'$. Now if, first of all, n is odd, then (according to 93)

$$|A = ||A' = A' \quad\text{and likewise}\quad |B = ||B' = B'.$$

Thus

$$[|A|B] = [A'B'] = ||[A'B']. \qquad \text{[according to 93]}$$

But since the orders of A' and B' are respectively $= n - \alpha$, $n - \beta$ (90, Corollary), the orders of A' and B' together are $= 2n - \alpha - \beta = n + (n - \alpha - \beta)$. Now $n - \alpha - \beta$ is positive since by assumption $\alpha + \beta$ is less than n, whence $n + (n - \alpha - \beta) > n$, and thus the sum of the orders of A' and B' is greater than n. Therefore, according to Part 1 of the Proof,

$$|[A'B'] = [|A'|B'] = [AB].$$

Consequently

$$||[A'B'] = |[AB],$$

but also

$$||[A'B'] = [|A|B]$$

as shown above, whence

$$[|A|B] = |[AB].$$

{Finally, if n is *even*, then (according to 93)

$$|A = ||A' = (-1)^{n-\alpha}A' \quad\text{and}\quad |B = ||B' = (-1)^{n-\beta}B',$$

whence

$$[|A|B] = (-1)^{2n-\alpha-\beta}[A'B'].$$

Now the sum of the orders of A' and B' is greater than n, as already shown previously in the case of an odd n. Thus (according to Part 1 of the Proof)

$$|[A'B'] = [|A'|B'] = [AB],$$

and therefore

$$|[AB] = ||[A'B'].$$

But since, as just mentioned, the sum of the orders of the factors of the product $[A'B']$ is greater than n, the order of the product $[A'B']$ itself has (according to 95) the value

$$(n - \alpha) + (n - \beta) - n = n - \alpha - \beta;$$

thus (according to 93)

$$|[AB] = ||[A'B'] = (-1)^{n-\alpha-\beta}[A'B'],$$

or, since by assumption n is an even number,

$$|[AB] = (-1)^{2n-\alpha-\beta}[A'B'].$$

One therefore obtains the same value for $|[AB]$ as above for $[|A|B]$, and thus the formula

$$[|A|B] = |[AB]$$

is also valid in this last case.}[25]

COROLLARY. *If the product of two magnitudes is progressive, then that of their supplements is regressive, provided that one regards the product of zeroth order as both progressive and regressive.*

PROOF. If $[AB]$ is a progressive product, then (according to 94) $\alpha + \beta \leq n$, if α and β are the orders of A and B. Then (according to the Corollary to 90) those of the supplements are $n - \alpha$ and $n - \beta$, but $n - \alpha - \beta \geq n$, whence also $n - \alpha + n - \beta \geq n$, that is (according to 94), the product of the supplements is regressive.

98. *The product of the supplements of several magnitudes is the supplement of the product of these magnitudes, that is*

$$[|A|B|C\cdots] = |[ABC\cdots].$$

PROOF. If the theorem holds for m factors, that is, if

$$[|A_1|A_2\cdots|A_m] = |[A_1A_2\cdots A_m],$$

then it also holds for $m+1$ factors. For if a factor A_{m+1} is appended to both sides of the above equation, then

$$\begin{aligned}[|A_1|A_2\cdots|A_m|A_{m+1}] &= [|(A_1A_2\cdots A_m)|A_{m+1}] \\ &= |[A_1A_2\cdots A_mA_{m+1}]. \qquad [97]\end{aligned}$$

Thus if the theorem holds for any given number of factors, it holds for the next higher number of factors. Since (according to 97) it holds for two factors, it therefore holds for arbitrarily many.

99. *In particular,*

$$[|a|b\cdots] = |[ab\cdots],$$

if a, b, ... are magnitudes of first order.

COROLLARY. It follows from this that a regressive product, whose factors are the supplements of magnitudes of first order, can be regarded as a combinatorial product, whose elementary factors are of $(n-1)$th order.

100. *One obtains the supplement of a polynomial by taking the supplement of each of its terms without changing their algebraic signs, that is*

$$|(A \pm B \pm \cdots) = |A \pm |B \pm \cdots .$$

PROOF. If

$$A = \sum \alpha_r A_r, \ B = \sum \beta_r B_r, \ \ldots,$$

then

$$\begin{aligned}|(A \pm B \pm \cdots) &= \Big|\Big(\sum \alpha_r E_r \pm \sum \beta_r E_r \pm \cdots\Big) = \Big|\sum(\alpha_r \pm \beta_r \pm \cdots)E_r \\ &= \sum(\alpha_r \pm \beta_r \pm \cdots)|E_r \qquad [90] \\ &= \sum \alpha_r|E_r \pm \sum \beta_r|E_r \pm \cdots \\ &= \Big|\sum \alpha_r E_r \pm \Big|\sum \beta_r E_r \pm \cdots \qquad [90] \\ &= |A \pm |B \pm \cdots .\end{aligned}$$

101. *An equation in which no conjunctions other than those treated in Chapters 1 and 3 appear, remain valid if one replaces the magnitudes appearing therein by their supplements, that is, if*

$$f(A, B, \ldots) = \phi(A', B', \ldots),$$

where f and ϕ symbolize conjunctions that belong to the chapters cited, then

$$f(|A, |B, \ldots) = \phi(|A', |B', \ldots).$$

{And conversely the latter equation follows from the former.}[26]

PROOF. Since equal magnitudes subject to the same conjunction yield equals, it follows that if

$$f(A, B, \dots) = \phi(A', B', \dots),$$

then

$$|f(A, B, \dots) = |\phi(A', B', \dots)$$

as well. Now no other conjunctions can appear but addition, subtraction, and relative multiplication, under which multiplication by numbers may be reckoned. For addition and subtraction it was shown in theorem 100 that one can replace the supplements of the conjunctions by the conjunctions of the supplements of the conjoined terms, and the same holds (according to 98) for relative multiplication, and thus for all conjunctions appearing on both sides of the equation.

{On the other hand, from 92 it follows at once that the equation

$$||||A = (-1)^{q(n-q)}(-1)^{q(n-q)}A = A$$

holds for every magnitude A of qth order in a principal domain of nth order. Thus if one knows that an equation of the form

$$f(|A, |B, \dots) = \phi(|A', |B', \dots)$$

obtains, then by applying the first part of the theorem thrice to this equation one recognizes that the equation

$$f(A, B, \dots) = \phi(A', B', \dots)$$

holds as well.}

REMARK. From this there follows the complete reciprocity between arbitrary magnitudes and their supplements, and thus generally between magnitudes of mth and $(n-m)$th order, if the principal domain is of nth order; of particular interest is the reciprocity between magnitudes of first and $(n-1)$th order. This reciprocity will become even clearer as we proceed.[27]

102. *If E, F, G are units whose orders taken together amount to n (the order of the principal domain), then*

$$[EF \cdot EG] = [EFG]E.$$

PROOF. We can distinguish two cases, either $[EFG]$ includes equal factors or not. If it includes equal factors, then since by assumption the number of its elementary factors equals n, and is therefore equal to the number of original units, one of these, say e_1, is missing from the factors of $[EFG]$.

Now if $[EF] = |Q$, then (according to 89) Q must include this factor e_1 also missing from $[EF]$; likewise if $[EG] = |R$, then R must include this factor as well. Thus (according to 60) $[QR]$ is equal to zero. Now

$$[EF \cdot EG] = [|Q|R] = |[QR], \qquad [97]$$

whence the supplement of $[QR] = 0$, and thus (according to 99) it is itself zero. But $[EFG]$ is also zero, since by assumption it contains equal factors, whence both sides of the original equation are zero.

If on the other hand $[EFG]$ includes no equal factors, then since it contains n factors and indeed none other than original units, it must be a product of the n original units. Thus (according to 89 {and 80})

$$|G = [GEF][EF], \quad |F = [FEG][EG].$$

Since here $[GEF]$ and $[FEG]$, as products of all the units $e_1, \ldots, e_n$ are equal to $\pm[e_1 \cdots e_n]$ and hence (according to 94) to ± 1, and multiplication by ± 1 yields the same result as division by ± 1, we can also write the above equations as

$$[EF] = [GEF]|G, \quad [EG] = [FEG]|F.$$

Then, since one may arbitrarily order the numerical factors,

$$\begin{aligned} [EF \cdot EG] &= [GEF][FEG][|G|F] = [GEF][FEG]|[GF] && [97] \\ &= [GEF][FEG][GFE]E. && [89] \end{aligned}$$

Now $[EF] = \pm[FE]$ (according to 58). Thus if in the equation just obtained one interchanges E and F *twice*, then its value remains unchanged, and thus it

$$= [GEF][EFG][GEF]E.$$

But since $[GEF] = \pm 1$, $[GEF][GEF] = 1$, and consequently one obtains

$$[EF \cdot EG] = [EFG]E.$$[28]

103. *If A, B, C are elementary magnitudes, and the sum of their orders is equal to the order of the principal domain, then*

$$[AB \cdot AC] = [ABC]A.$$

PROOF. Assuming that formula 103 holds for the case that A, B, C include no other factors than those belonging to a given series of n magnitudes of first order $a_1, a_2, a_3, \ldots$, then I will show that it still holds if one replaces one of these magnitudes, say a_1, by one numerically derived from them, say

$$a' = \alpha_1 a_1 + \alpha_2 a_2 + \cdots = \sum \alpha_r a_r.$$

Now a_1 can occur in A or B or C. If a_1 is included in B, then let $B = [a_1 D]$, and under the above substitution B is transformed into $B' = [a'D]$. Then

$$\begin{aligned} [AB' \cdot AC] &= \left[A\left\{\sum \alpha_r a_r)D\right\} \cdot AC\right] = \sum \alpha_r [A(a_r D) \cdot AC] \\ &= \sum \alpha_r [A(a_r D)C]A, \end{aligned}$$

since by assumption formula 103 holds for the case that the magnitudes considered only include $a_1, a_2, \ldots$. Thus

$$\begin{aligned} [AB' \cdot AC] &= \sum \alpha_r [A(a_r D)C]A = \left[A\left\{\left(\sum \alpha_r a_r\right) D\right\} C\right] A \\ &= [A(a'D)C]A = [AB'C]A. \end{aligned}$$

Precisely the same conclusion holds if a_1 is included in C. Thus it remains only to consider the case that a_1 is included in A.

In this case suppose first that a_1 is transformed into $a' = \alpha_1 a_1 + \alpha_2 a_2$, and let $A = [a'D]$, whence

$$A' = [a'D] = \alpha_1 [a_1 D] + \alpha_2 [a_2 D],$$

that is

$$(*) \qquad A' = \alpha_1 A + \alpha_2 [a_2 D].$$

Should a_2 be included in D, then this last summand is zero (according to 60); then A is transformed only into its multiple, whence

$$[A'B \cdot A'C] = \alpha_1^2[AB \cdot AC] = \alpha_1^2[ABC]A = [\alpha_1 ABC]\alpha_1 A = [A'BC]A'.$$

Thus it only remains to consider the case that a_2 appears in B or C, for example in B. In this case let $B = [a_2 E]$. Now as shown in $(*)$, $A' = \alpha_1 A + \alpha_2[a_2 D]$, and since a_2 is included as a factor in B so that $[a_2 DB] = 0$, it follows that

$$(**) \qquad [A'B] = \alpha_1[AB];$$

furthermore,

$$[A'C] = [(\alpha_1 A + \alpha_2[\alpha_2 D])C] = \alpha_1[AC] + \alpha_2[a_2 DC],$$

whence

$$[A'B \cdot A'C] = \alpha_1^2[AB \cdot AC] + \alpha_1\alpha_2[AB \cdot a_2 DC] = \alpha_1^2[ABC]A + \alpha_1\alpha_2[AB \cdot a_2 DC].$$

But

$$\begin{aligned} [AB \cdot a_2 DC] &= [a_1 D(a_2 E) \cdot a_2 DC] = -[a_2 D(a_1 E) \cdot a_2 DC] && [\{80\}, 55] \\ &= -[a_2 D(a_1 E)C][a_2 D]. \end{aligned}$$

Finally there is the case that the three magnitudes $[a_2 D]$, $[a_1 E]$, and C include no factors other than those belonging to the series of magnitudes $a_1, a_2, a_3, \ldots$, and the sum of whose orders is n, and therefore satisfy all the conditions under which the validity of the formula

$$[AB \cdot AC] = [ABC]A$$

is assumed. Consequently

$$\begin{aligned} [A'B \cdot A'C] &= \alpha_1^2[ABC]A - \alpha_1\alpha_2[a_2 D(a_1 E)C][a_2 D] \\ &= \alpha_1^2[ABC]A + \alpha_1\alpha_2[a_1 D(a_2 E)C][a_2 D] && [\{80\}, 55] \\ &= \alpha_1^2[ABC]A + \alpha_1\alpha_2[ABC][a_2 D] \\ &= \alpha_1[ABC](\alpha_1 A + \alpha_2[a_2 D]) \\ &= [a_1 ABC]A' && [*] \\ &= [A'BC]A'. && (**) \end{aligned}$$

This also holds if a_2 is included in C instead of B. It has thus been shown that the formula always applies if one transforms a factor a_1 into $\alpha_1 a_1 + \alpha_2 a_2$, and therefore also if one transforms this into $\alpha_1 a_1 + \alpha_2 a_2 + \alpha_3 a_3$ and so forth.

It has therefore been completely proved that, if formula 103 holds for any series of n magnitudes of first order $a_1, a_2, \ldots, a_n$ that form the elementary factors of the magnitudes A, B, C, it also holds if one replaces *any one* of these factors by a magnitude numerically derived from those magnitudes $a_1, \ldots, a_n$. Since it can again be applied to the resulting series of magnitudes, it follows that the formula still holds *if one replaces the elementary factors* of the magnitudes A, B, C *by arbitrary magnitudes numerically derived from those magnitudes* $a_1, \ldots, a_n$.

Thus suppose for example that $b_1, \ldots, b_n$ are such magnitudes numerically derived from the $a_1, \ldots, a_n$. However they are supplied, among them (according to 17) an assembly of m magnitudes can be given, standing in no numerical relation to one another, from which, if m is less than n, the rest can be numerically derived.

Suppose $b_1, \ldots, b_m$ are these magnitudes, from which $b_{m+1}, \ldots, b_n$ are numerically derived; then (according to 20) one can replace m of the magnitudes $a_1, \ldots, a_n$ with the magnitudes $b_1, \ldots, b_m$ in such a way that the domain of the magnitudes so obtained is identical to the domain of the magnitudes $a_1, \ldots, a_n$. If $a_1, \ldots, a_m$ are the magnitudes that one can replace in this way with $b_1, \ldots, b_m$, then the domain of the magnitudes $a_1, \ldots, a_n$ is identical to the domain of the magnitudes $b_1, \ldots, b_m, a_{m+1}, \ldots, a_n$, or, applying the same conclusion step by step, the domain

$a_1, a_2, \ldots, a_n$ is identical to the domain $b_1, a_2, \ldots, a_n$,
this is identical to that of $b_1, b_2, \ldots, a_n$,
. . . .
and finally, is identical to that of $b_1, b_2, \ldots, b_m, a_{m+1}, \ldots, a_n$.

Now if formula 103 is valid for $a_1, \ldots, a_n$, it is also valid if one replaces a_1 by the magnitude b_1 derived from $a_1, \ldots, a_n$, and thus for $b_1, a_2, \ldots, a_n$. Since the domain of $a_1, \ldots, a_n$ is identical to the domain of $b_1, a_2, \ldots, a_n$, and by assumption b_2 is derivable from the $a_1, \ldots, a_n$, it is also derivable from $b_1, a_2, \ldots, a_n$, whence formula 103 still holds if one replaces a_2 with b_2, that is for the series $b_1, b_2, a_3, \ldots, a_n$ and so on, and finally for the series $b_1, b_2, \ldots, b_m, a_{m+1}, \ldots, a_n$. Furthermore, since by assumption $b_{m+1}, \ldots, b_n$ are numerically derivable from the $b_1, b_2, \ldots, b_m$, 103 still holds if in the series $b_1, \ldots, b_m, a_{m+1}, \ldots, a_n$ one replaces $a_{m+1}, \ldots, a_n$ one by one with $b_{m+1}, \ldots, b_n$, and therefore it holds as well for the series $b_1, \ldots, b_n$, that is, for any arbitrary series of magnitudes numerically derived from the $a_1, \ldots, a_n$. But 103 holds for the original units (according to 102), and therefore for an arbitrary series of magnitudes of first order that form the elementary factors of A, B, C, that is, for any elementary magnitudes A, B, C.[29]

104. If B and C are compound magnitudes, but A is an elementary magnitude, and the sum of the orders of A, B, C is equal to the order of the principal domain, then

$$[AB \cdot AC] = [ABC]A.$$

PROOF. If

$$B = \sum \beta_r E_r, \quad C = \sum \gamma_s F_s,$$

where E_r and F_s are products of units, then

$$\begin{aligned} [AB \cdot AC] &= \sum \beta_r \gamma_s [AE_r \cdot AF_s] && [42] \\ &= \sum \beta_r \gamma_s [AE_r F_s] A && [103] \\ &= [A \sum \beta_r E_r \cdot \sum \gamma_s F_s] A && [45] \\ &= [ABC]A. \end{aligned}$$

REMARK. This theorem no longer holds in general if A is a compound magnitude. If for example $A = [ab] + [cd]$, where a, b, c, d stand in no numerical relation to one another, and if $B = c$, $C = d$, then with respect to a domain of fourth order

$$[AB \cdot AC] = [(ab + cd)c \cdot (ab + cd)d] = [abc \cdot abd],$$

since $[cdc]$ and $[cdd]$ vanish; but $[abc \cdot abd] = [abcd][ab]$. Thus

$$[AB \cdot AD] = [abcd][ab].$$

On the other hand

$$\begin{aligned}[ABC]A &= [(ab + cd)cd]([ab] + [cd]) = [abcd]([ab] + [cd]) \\ &= [abcd][ab] + [abcd][cd].\end{aligned}$$

The two expressions therefore differ from one another by $[abcd][cd]$.

105–107. *If A, B, C are elementary magnitudes, and their product is of zeroth order, then*

105. $[AB \cdot AC] = [ABC]A,$
106. $[AB \cdot BC] = [ABC]B,$
107. $[AC \cdot BC] = [ABC]C,$

that is, if two products (P and Q) of elementary magnitudes having a common factor are to be multiplied, and this common factor appears either as the first factor in the second product (Q) or as the second factor in the first product (P), then one can multiply this factor by the product of the remaining factors, provided that this last product is of zeroth order. In both cases the common product has the same value.

PROOF. 1. If α, β, γ are the orders of A, B, C, and n is that of the principal domain, then since $[ABC]$ is of zeroth order, it follows (according to 96) that $\alpha + \beta + \gamma$ is divisible by n, whence, since α, β, γ are smaller than n, it must equal n or $2n$. If $\alpha + \beta + \gamma = n$, then the validity of formula 105 is already proven in 103. On the other hand, if $\alpha + \beta + \gamma = 2n$, then if

$$A = |A', \quad B = |B', \quad C = |C'$$

and α', β', γ' are the orders of A', B', C', then

$$\alpha' = n - \alpha, \quad \beta' = n - \beta, \quad \gamma' = n - \gamma, \qquad \text{[Corollary to 90]}$$

whence

$$\alpha' + \beta' + \gamma' = 3n - (\alpha + \beta + \gamma) = 3n - 2n = n.$$

Consequently

$$\begin{aligned}[AB \cdot AC] &= ([|A'|B'][|A'|C']) = |[A'B' \cdot A'C'] &&[97] \\ &= |[A'B'C' \cdot A'], &&[103]\end{aligned}$$

since $\alpha' + \beta' + \gamma' = n$. But (according to 98) this is

$$= [|A'|B'|C']|A' = [ABC]A.$$

2. If $[AB] = [BD]$, then D is of the same order as A, whence

$$\begin{aligned}[AB \cdot BC] &= [BD \cdot BC] = [BDC]B &&[105] \\ &= [ABC]B,\end{aligned}$$

since $[BD] = [AB]$.

3. If $[BC] = [CD]$, then D is of the same order as B, whence

$$[AC \cdot BC] = [AC \cdot CD] = [ACD]C \qquad [106]$$
$$= [ABC]C,$$

since $[CD] = [BC]$.[30]

REMARK. The laws established in 105–107 can be so extended that they also hold for cases where $[ABC]$ is not of zeroth order, provided one represents them in the forms

$$[AB \cdot AC] = [A \cdot ABC],$$
$$[AB \cdot BC] = [B \cdot ABC],$$
$$[AC \cdot BC] = [C \cdot ABC].$$

The proof of these formulas, which I will not use in this more general form in the following, I leave to the reader.[31]

108. *If A, B, C are elementary magnitudes, and the sum of the orders of A and C is equal to that of their principal domain, and B is subordinate to A, then*

$$[A \cdot BC] = [AC]B,$$
$$[CB \cdot A] = [CA]B.$$

PROOF. Since (according to 79B) A is representable in the form $[BD]$, it follows that

$$[A \cdot BC] = [BD \cdot BC] = [BDC]B$$
$$= [AC]B \qquad [105]$$

and

$$[CB \cdot A] = [CB \cdot BD] = [CBD]B$$
$$= [CA]B. \qquad [106]^{32}$$

109. *A relative product of two nonzero elementary magnitudes is zero if and only if the order of their common domain has the smallest, or, what is the same thing, if the order of the covering domain has the greatest value that it can have with the given orders of the two factors and that of the principal domain, that is, if α and β are the orders of the factors A and B, n that of the principal domain, γ that of the common, and δ that of the covering domain, then*

*if $\alpha + \beta \le n$, that is, the product is **progressive**, then*

$$[AB] \ne 0$$

if and only if

$$\gamma = 0,$$

or, what is the same thing,

$$\alpha + \beta = \delta;$$

*and if $\alpha + \beta > n$, that is, the product is **regressive**, then*

$$[AB] \ne 0$$

if and only if

$$\gamma = \alpha + \beta - n,$$

or, what is the same thing,

$$\delta = n.$$

PROOF. 1. If $\alpha + \beta \leq n$, so that the product is progressive (according to 94), then (according to 61, 66) it is zero if and only if its elementary factors stand in a numerical relation to one another. Thus if $[AB] = 0$, one of the elementary factors of the product $[AB]$ can be numerically derived from the other $\alpha+\beta-1$ (according to 2). Therefore the collection of elementary factors of that product are included in a domain of lower than $(\alpha + \beta)$th order, that is, $\delta < \alpha + \beta$. On the other hand, if $[AB] \neq 0$, then (according to 61) the elementary factors of this product stand in no numerical relation to one another, and their covering domain is therefore of $(\alpha + \beta)$th order, that is $\alpha + \beta = \delta$. Consequently, if $\alpha + \beta \leq n$, $[AB]$ is different from zero if and only if $\alpha + \beta = \delta$.

2. Suppose $\alpha + \beta > n$ and $\alpha + \beta - n = \gamma$. Then (according to 26) the domains of A and B, being of orders α and β respectively, have at least a domain of order $(\alpha + \beta - n)$th, that is, γth order, in common. Let C be a magnitude of γth order in this domain; then (according to 79B) A and B are representable in the form $A = [CA_1]$, $B = [CB_1]$. Consequently

$$[AB] = [CA_1 \cdot CB_1] = [CA_1B_1]C, \qquad [103]$$

since the sum of the orders of C, A_1, and B_1

$$= \gamma + (\alpha - \gamma) + (\beta - \gamma) = \alpha + \beta - \gamma = n.$$

However, since $[CA_1B_1]$ is of zeroth order, and thus a number, $[CA_1B_1]C$ is zero if and only if $[CA_1B_1]$, that is $[AB_1]$, is zero. But according to part 1 of the Proof $[AB_1]$ is zero if and only if A and B_1 are contained in a domain of lower than nth order; but since C lies in $A = [CA_1]$, it is also true that A and $[CB_1]$, that is A and B, are contained in a domain of lower than nth order, i.e. $\delta < n$. Consequently, if $\alpha + \beta > n$, $[AB]$ differs from zero if and only if $\delta = n$.

3. According to 25 the sum of the orders of two domains is equal to the sum of the orders of their common domain and their covering domain, that is

$$\alpha + \beta = \gamma + \delta.$$

The condition in Part 1 of the proof, that $\alpha + \beta = \delta$, is therefore identical with the condition that $\gamma = 0$, and the condition in Part 2, that

$$\alpha + \beta - n = \gamma.$$

Consequently the second form of our theorem is proved. But it is clear that, if $\alpha + \beta \leq n$, the smallest order that the common domain of the magnitudes A and B can have is zero, and the largest that the common domain can have is $\alpha + \beta$. On the other hand, if $\alpha + \beta > n$, then the greatest order that the common domain can have is n, whence (according to 26) the smallest order that the common domain can have is $\alpha + \beta - n$. Thus the first form of the theorem accords with the second, and the theorem is proved.

110. *All laws of relative multiplication in a principal domain are also valid if one* ***everywhere*** *replaces the original units by an arbitrary series of* n *magnitudes numerically derivable from them, and whose combinatorial product is* 1.

PROOF. Let $e_1, \ldots, e_n$ be the original units, and $a_1, \ldots, a_n$ be numerically derived from them, so that

$$[a_1 a_2 \cdots a_n] = 1.$$

According to 89, by the supplement $|E$ of a product of units is meant that magnitude equal or opposite to the combinatorial product E' of all the units not appearing in E, according as $[EE']$ is equal or opposite to the absolute unit, and thus that magnitude that satisfies the equation

$$|E = [EE']E'.$$

Now if for the moment we denote that magnitude formed in the same way from a product A in which only the magnitudes $a_1, \ldots, a_n$ appear, by IA; that is, if we denote by IA that magnitude which is equal or opposite to the combinatorial product A of all those magnitudes of that series $a_1, \ldots, a_n$ that do not appear in A, according as $[AA']$ is equal or opposite to the absolute unit, whence

$$(a) \qquad IA = [AA']A'$$

{and if in addition, when α denotes a number, we set

$$(b) \qquad I\alpha = \alpha$$

and finally

$$(c) \qquad I\sum \alpha_r A_r = \sum \alpha_r I A_r,$$

corresponding to definition 90}, then first we have to prove that the stipulation formulated (in 94) as a definition of the relative product remains valid with this substitution of the magnitudes $a_1, \ldots, a_n$ for the original units, that is, that

$$(d) \qquad I[AB] = [IA \cdot IB]$$

if A and B include only magnitudes from the series $a_1, \ldots, a_n$ as elementary factors, and the sum $(\alpha + \beta)$ of the orders of A and B is greater than the order n of their principal domain.

{First, if $[AB]$ is zero, then (according to 109) the domains of A and B must have a domain of higher order than zero in common. If they have a domain of γth order in common, and thus γ elementary factors, then since IA includes only those factors not appearing in A, these γ elementary factors are missing from IA, and on the same grounds from IB. Thus IA and IB are included in a domain of lower than nth order, so that (according to 109)

$$[IAIB] = 0,$$

whence, since $[AB]$ is also zero and the supplement of a number is equal to the number (according to 89), so that that of zero is also zero, it follows that

$$I[AB] = [IAIB].$$

Second, if $[AB]$ is different from zero, then it includes $\alpha+\beta$ different elementary factors of the series $a_1, \ldots, a_n$. If C is the product of the rest of them, then (according to 57) $[ABC]$ is either equal or opposite to the product $[a_1 \cdots a_n]$, and since the latter is 1, $[ABC] = \pm 1$. Now since $[BC]$ is the product of the factors not appearing in A, according to the notation used here one has

$$IA = [ABC] \cdot [BC],$$

and likewise

$$IB = [BAC] \cdot [AC] \quad \text{and} \quad I[AB] = [ABC]C. \qquad [*]$$

Thus, since $[ABC]$ and $[BAC]$ are numbers $(= \pm 1)$,

$$\begin{aligned}[IAIB] &= [ABC][BAC][BC \cdot AC] \\ &= [ABC][BAC][BAC] \cdot C. \qquad [107]\end{aligned}$$

Now since $[BAC] = \pm 1$, it follows that $[BAC][BAC] = 1$. Thus

$$[IAIB] = [ABC] \cdot C = I[AB]. \qquad [*]\}^{33}$$

Thus the requirement upon which the concept of relative multiplication was based is also valid if one replaces the original units by the magnitudes $a_1, \ldots, a_n$; but all earlier laws are valid if one replaces the n original units by n magnitudes that stand in no numerical relation to one another, that is, whose combinatorial product is not zero, and thus also, if one replaces them by the magnitudes $a_1, \ldots, a_n$. But all the following laws are derived from these earlier theorems and the requirement established in the definition, whence these are also still valid upon the given substitution.

REMARK. From the continued validity of the multiplication laws when one substitutes a series of linearly derivable units from the original ones the multiplication is defined as *linear*, and only in the following chapter will we proceed to a product structure in which the continuation of the laws valid for the original units only obtains to a very limited extent.

It is yet to be noted that the magnitude denoted above by IA does *not* in general coincide with the supplement of A, which we have denoted by $|A$. For example, in a domain of third order, where e_1, e_2, e_3 are the original units and $[e_1e_2e_3] = 1$, the supplement of $e_1 + e_2$ is equal to $[e_2e_3] + [e_3e_1]$ since $|e_1 = [e_2e_3]$, $|e_2 = [e_3e_1]$; for (according to 90)

$$|(e_1 + e_2) = |e_1 + |e_2 = [e_2e_3] + [e_3e_1].$$

On the other hand, if

$$a_1 = e_1 + e_2, \quad a_2 = e_2, \quad a_3 = e_3,$$

then indeed $[a_1a_2a_3] = 1$, but, applying the notation in the above theorem,

$$Ia_1 = [a_1a_2a_3][a_2a_3] = [a_2a_3] = [e_2e_3],$$

whence a_1 differs from $[e_2e_3]$. In the following chapter {cf. Nos. 167 and 168} will be given the relations that must obtain between $e_1, \ldots, e_n$ and $a_1, \ldots, a_n$ in order that $|A = IA$.

111. *If*

$$1 = [a_1 \cdots a_n] = [PP'] = [AA'] = [BB'] = [CC'] = \cdots,$$

and P, P', A, A', B, B', C, C', ... *include no other elementary factors than those belonging to the series* $a_1, \ldots, a_n$, *and*

$$P = [ABC \cdots],$$

then so also is

$$P' = [A'B'C' \cdots].$$

PROOF. If by IA is meant the same as in the preceding Proof, then

$$IP = [PP']P' = P', \quad IA = [AA']A' = A', \ \ldots .$$

Now since according to the preceding theorem all earlier theorems, and in particular theorem 98, remain valid if one replaces the sign | by the sign I everywhere,

$$I[ABC\cdots] = [IA \cdot IB \cdot IC \cdots],$$

whence

$$IP = [IA \cdot IB \cdot IC \cdots].$$

Thus, since $IP = P'$, $IA = A'$, $IB = B'$, $IC = C'$, ... ,

$$P' = [A'B'C'\cdots].$$

112. *If, from n magnitudes of first order whose combinatorial product is unity, one forms the multiplicative combinations of* $(n-1)$*th class, and orders the elements of each combination alphabetically, the combinations themselves lexicographically under the assumption that the series of n magnitudes is regarded as an alphabet, then the product of the first* $n - m$ *of these combinations is equal to the product of the first m of those n magnitudes, that is*

$$[A_n \cdots A_{m+1}] = [a_1 \cdots a_m],$$

if

$$A_r = [a_1 \cdots a_{r-1}a_{r+1} \cdots a_n]$$

and

$$[a_1 \cdots a_n] = 1.$$

PROOF. 1. First I prove that

$$[a_1 \cdots a_r A_r] = [a_1 \cdots a_{r-1}].$$

According to the notation chosen,

$$A_r = [a_1 \cdots a_{r-1}a_{r+1} \cdots a_n].$$

Thus

$$\begin{aligned}[a_1 \cdots a_r A_r] &= [a_1 \cdots a_r(a_1 \cdots a_{r-1}a_{r+1} \cdots a_n)] \\ &= [a_1 \cdots a_r a_{r+1} \cdots a_n][a_1 \cdots a_{r-1}] \qquad [105 \text{ \{and 80\}}] \\ &= [a_1 \cdots a_{r-1}],\end{aligned}$$

since $[a_1 \cdots a_n] = 1$.

2. Consequently

$$[A_n A_{n-1}] = [a_1 \cdots a_{n-1} A_{n-1}] = [a_1 \cdots a_{n-2}],$$
$$[A_n A_{n-1} A_{n-2}] = [a_1 \cdots a_{n-2} A_{n-2}] = [a_1 \cdots a_{n-3}],$$

and so forth. Thus

$$[A_n A_{n-1} \cdots A_{n-r}] = [a_1 \cdots a_{r-1}].$$

If therefore $n - r - 1 = m$, then

$$[A_n A_{n-1} \cdots A_{m+1}] = [a_1 \cdots a_m].$$ [34]

113. *If C_1, C_2, ... are the multiplicative combinations {of a given class} of the elementary factors {of first order} of a nonzero magnitude B, and D_r consists of those factors of B that are missing from C_r, and these factors are so ordered that in each case*

$$[C_r D_r] = B,$$

then for every magnitude A whose order supplements the order of D_r to that of the principal domain,

$$[AB] = \sum [AD_r]C_r = [AD_1]C_1 + [AD_2]C_2 + \cdots .$$

PROOF. Suppose m is the number of elementary factors of B and n the order of the principal domain, α the order of A, and let $B = [b_1 b_2 \cdots b_m]$.

Now since by assumption B is different from zero, $b_1, \ldots, b_m$ stand in no numerical relation to one another (according to 61); consequently (according to 20) to them can be added $n - m$ further magnitudes of first order $b_{m+1}, \ldots, b_n$ such that all magnitudes of first order, belonging to the principal domain under consideration, can be numerically derived from them. Then A can be numerically derived as a magnitude of αth order from the multiplicative combinations of the n magnitudes $b_1, \ldots, b_n$ of αth class, and therefore can be represented in the form

$$A = \alpha_1 A_1 + \alpha_2 A_2 + \cdots = \sum \alpha_r A_r,$$

if A_1, A_2, ... are the multiplicative combinations of $b_1, \ldots, b_n$ of αth class. Let these combinations be so chosen that in each case A_r consists of those of the n magnitudes that are missing in D_r. This is always possible, since by assumption D_r includes $n - \alpha$ of those magnitudes. Then

$$[AB] = \left[\sum \alpha_r A_r \cdot B\right] = \sum \alpha_r [A_r \cdot B] \qquad [41]$$
$$= \sum \alpha_r [A_r \cdot C_r D_r],$$

since by assumption $B = [C_r D_r]$. Now since C_r includes only such of those n magnitudes $b_1, \ldots, b_n$ as are missing in D_r, and A_r includes all the magnitudes of $b_1, \ldots, b_n$ that are missing from D_r, C_r is subordinate to A_r, whence (according to 108)

$$[A_r \cdot C_r D_r] = [A_r D_r]C_r,$$

and therefore

$$[AB] = \sum \alpha_r [A_r D_r] C_r.$$

But {according to 60} $[A_s D_r] = 0$ if s is different from r, since then A_s includes at least one factor that also appears in D_r, whence instead of $\alpha_r[A_r D_r]$ one can write $\sum \alpha_s [A_s D_r]$, where the sum relates only to the index s, that is,

$$\alpha_r [A_r D_r] = \sum \alpha_s [A_s D_r] = \left[\sum \alpha_s A_s \cdot D_r\right] = [AD_r],$$

and therefore

$$[AB] = \sum [AD_r] C_r.$$ [35]

6. Interchange of factors and removal of parentheses in a pure and in a mixed product

114. DEFINITION. If more than two magnitudes A, B, C, ... are so conjoined in a product that they are subject to nothing but progressive multiplication, then I call the product a *pure progressive* product of those magnitudes, while if they are subject to nothing but regressive multiplication, or, if the net product is of zeroth order, only the last, the ultimate multiplication forming the product, is progressive, then I call the product *purely regressive*, in both cases a *pure* product, and all other cases *mixed*. That is, if in the product $[ABCD\cdots JK]$ the product $[AB]$ is progressive, the product of the two magnitudes $[AB]$ and C is again progressive, likewise the product of the two magnitudes $[ABC]$ and D, and so forth, and finally the product of the two magnitudes $[ABCD\cdots J]$ and K is progressive, then $[ABCD\cdots JK]$ is a pure progressive product of A, B, C, D, ..., J, K. If on the other hand all those products are regressive, or at most only the last, namely the product of the two magnitudes $[ABCD\cdots J]$ and K is progressive, and that of zeroth order, then $[ABCD\cdots JK]$ is a pure regressive product of the magnitudes A, B, C, D, ..., J, K.

REMARK. That the product can also be regarded as purely regressive when the last multiplication, which makes the whole product of zeroth order, is progressive is because the progressive multiplication that forms a product of zeroth order can also simultaneously be regarded as regressive multiplication, insofar as all special laws of regressive multiplication are likewise valid for it, as well as all special laws of progressive multiplication. The product $[ab\cdot ac\cdot bc]$, where the principal domain is of third order, serves as an example of such a pure regressive multiplication.

115. *If a product of several magnitudes* $[ABC\cdots]$ *is purely progressive, then the product of the supplements* $[|A|B|C\cdots]$ *is purely regressive and conversely.*

PROOF. For (according to the first Corollary to 97) this is true for two factors, whence since $[AB]$ is progressive, $[|A|B]$ is regressive, and since $[(AB)C]$ is progressive, $[|(AB)|C]$ is regressive, that is $[|A|B|C]$ is regressive, and so forth.

116. *A product of* m *magnitudes* A, B, C, ..., J, K *is purely progressive if the sum of the orders of these magnitudes is less than or equal to the order* (n) *of the principal domain, but purely regressive if that sum is as great as or greater than* $n(m-1)$, *and a mixed product if that sum is greater than* n *and less than* $n(m-1)$.

PROOF. 1. Let α, β, γ, ... be the orders of the magnitudes A, B, C, ... If

$$\alpha+\beta+\gamma+\cdots+\iota+\kappa\leq n,$$

then $\alpha+\beta<n$, whence (according to 94) the product $[AB]$ is progressive; but also $\alpha+\beta+\gamma<n$, whence the product of the two magnitudes $[AB]$ and C is progressive, and so forth. Finally,

$$\alpha+\beta+\gamma+\cdots+\iota+\kappa\leq n,$$

and thus the product of the two magnitudes $[ABC\cdots J]$ and K is also progressive, since (according to 96) the order of $[ABC\cdots J]$ is $\alpha+\beta+\gamma+\cdots+\iota$. The converse follows in the same way, since if $[ABC\cdots JK]$ is a pure progressive product, then necessarily

$$\alpha+\beta+\gamma+\cdots+\iota+\kappa\leq n.$$

2. If

$$\alpha+\beta+\gamma+\cdots+\iota+\kappa>n(m-1),$$

we can also write this as

$$(n-\alpha)+(n-\beta)+(n-\gamma)+\cdots+(n-\iota)+(n-\kappa)\leq n,$$

since m is the number of magnitudes A, B, C, ... , J, K. Now since $n-\alpha$ is the order of the supplement of A, that is the order of $|A$, the product

$$[|A|B|C\cdots|J|K]$$

is purely progressive according to part 1 of the Proof, whence (according to 115) $[ABC\cdots JK]$ is purely regressive.

116B.* *The pure product of magnitudes of first order or of magnitudes of* $(n-1)$*th order in a principal domain of* n*th order is a combinatorial product of these magnitudes.*

PROOF. The pure product of units of first order is (according to 114, 94) progressive, thus (according to 94) outer, and thus (according to 78) a combinatorial product of units; thus (according to 52) the pure product of any magnitudes of first order is a combinatorial product of these magnitudes. But the pure product of magnitudes of $(n-1)$th order is (according to 101) subject to exactly the same laws as that of magnitudes of first order, and thus of the laws of combinatorial multiplication, that is that product is a combinatorial product of those magnitudes of $(n-1)$th order.

117. *The order of a pure progressive product is zero if the sum of the orders of its factors is equal to the order,* n*, of the principal domain, otherwise the order of that product is equal to the sum of the orders of its factors. The order of a pure regressive product* $=\sigma-(m-1)n$*, where* σ *is the sum of the orders of its factors and* m *the number of these factors.*

PROOF. The theorem is proved in 95 for two factors. Now if $[ABC\cdots JK]$ is purely progressive, and if α, β, γ, ... , ι, κ are the orders of A, B, C, ... , J, K, then that of $[AB]$ equals $\alpha+\beta$, so that of $[(AB)C]$ equals $\alpha+\beta+\gamma$, ... ; consequently that of $[ABC\cdots J]$ equals $\alpha+\beta+\gamma+\cdots+\iota$. If

$$\alpha+\beta+\gamma+\cdots+\iota+\kappa<n,$$

then according to the same theorem (95) the order of $[(ABC\cdots J)K]$ equals $\alpha+\beta+\gamma+\cdots+\iota+\kappa$, but if

$$\alpha+\beta+\gamma+\cdots+\iota+\kappa=n,$$

then according to the same theorem it is zero.

Second, if the product $[ABC\cdots JK]$ is purely regressive, then according to the theorem previously adduced the order of $[AB]$ equals $\alpha+\beta-n$, whence that of $[(AB)C]$ equals $\alpha+\beta+\gamma-2n$, ... , and thus, if m is the number of factors of $[ABC\cdots JK]$, the order of this product equals $\alpha+\beta+\gamma+\cdots+\iota+\kappa-(m-1)n$.

*{In the original edition this theorem was No. 132; but it belongs here after No. 116, and is placed here in the Teubner Edition.}

118. *The domain of a pure progressive product is equal to that of the covering domain of all its factors, and the domain of a pure regressive product is equal to that of the common domain of all its factors, assuming that in both cases the product is not zero.*

PROOF. 1. If $[AB\cdots]$ is a pure progressive product and

$$A = [a_1 \cdots a_q], \quad B = [b_1 \cdots b_r], \ \ldots,$$

where $a_1, \ldots, a_q, b_1, \ldots, b_r, \ldots$ are magnitudes of first order, one finds

$$[AB\cdots] = [(a_1 \cdots a_q)(b_1 \cdots b_r)\cdots].$$

Now since the progressive product is always also an outer product, one can (according to 80) drop the parentheses and thus the last expression is

$$= [a_1 \cdots a_q b_1 \cdots b_r \cdots].$$

The domain of the product is thus (according to 77) the domain numerically derivable from its elementary factors. Likewise the domain of A is that derivable from $a_1, \ldots, a_q$, and so on, and (according to 15) the domain derivable from the magnitudes of two or more domains $A, B, \ldots$ is the covering domain of these latter magnitudes, whence the domain of the progressive product $[AB\cdots]$ is the covering domain of the factors $A, B, \ldots$.

2. If $[AB]$ is a nonzero regressive product, so that the orders α and β of the factors A and B together are greater than n, then A and B have a domain of $(\alpha+\beta-n)$th order in common {26}; but none of higher order, since otherwise (according to 109) the product would be zero. Thus {according to 87} A and B can be expressed in terms of a common factor D of $(\alpha+\beta-n)$th order, so that $A = [DE]$, $B = [DF]$, and D, E, F are elementary factors; then (according to 105)

$$[AB] = [DE \cdot DF] = [DEF]D \equiv D,$$

since $[DEF]$ is a number different from zero, that is, the domain of $[AB]$ is equal to the common domain of the factors A and B. If one appends another factor, C, then

$$[ABC] \equiv [DC] \equiv G,$$

where G is the common domain of D and C, and thus the common domain of A, B, C, and so on.

119. *In a pure product one can arbitrarily insert and drop parentheses, that is*

$$[A(BC)] = [ABC]$$

if $[ABC]$ is a pure product.

PROOF. 1. If the product is purely progressive, then (according to 94) it is also an outer product, and thus (according to 80), placement of parentheses is of no importance for the resultant.

2. If the product $[ABC]$ is purely regressive, then (according to 115) the product $[|A|B|C]$ is purely progressive, and thus according to Part 1 of the Proof

$$[|A(|B|C)] = [|A|B|C],$$

that is (according to 101)

$$[A(BC)] = [ABC].$$

119B. *A pure product maintains its value if one resolves its factors into pure factors of first or* $(n-1)$*th order, according as the given product is progressive or regressive. In addition, with these new factors the product maintains its character as purely progressive or regressive, that is, if*

$$P = [AB \cdots E]$$

is a pure product of the factors A, B, $\ldots$, E, *and*

$$A = [a_1 \cdots a_q], \quad B = [a_{q+1} \cdots a_r], \quad \ldots, \quad E = [a_{r+1} \cdots a_u],$$

and $a_1, \ldots, a_u$ *are magnitudes of first or* $(n-1)$*th order, according as the product* $[AB \cdots E]$ *is progressive or regressive, then*

$$P = [a_1 a_2 \cdots a_u],$$

and this product too is purely progressive or regressive, according as was the original product $[AB \cdots E]$.

PROOF. If the product $[AB \cdots E]$ is purely progressive, then according to 116 the sum of the orders of $A, B, \ldots, E$, that is u, is $\leq n$, whence it remains {according to 116} purely progressive with respect to the factors $a_1, \ldots, a_u$ if one replaces A with $[a_1 \cdots a_q]$ and so forth, whence one can (according to 119) drop the parentheses and obtain $P = [a_1 a_2 \cdots a_u]$.

But if $[AB \cdots E]$ is a pure regressive product, then (according to 115) $[|A|B \cdots |E]$ is purely progressive; and if $A = [a_1 \cdots a_q]$, $\ldots$, and $a_1, \ldots, a_u$ are magnitudes of $(n-1)$th order, then $|A = [|a_1 \cdots |a_q]$ and so on, where $|a_1, \ldots, |a_u$ are magnitudes of first order, whence according to the first part of the Proof

$$[|A|B \cdots |E] = [|a_1 \cdots |a_u].$$

Thus also (according to 101)

$$[AB \cdots E] = [a_1 \cdots a_u],$$

and this is a pure progressive product (according to 115).

119C.* *If* a, b, $\ldots$ *are the orders of {the factors of} a pure product* P (114), n *that of the principal domain,* v *the order of the covering,* g *that of the common domain {of all the factors} and* p *that of the product, and* $n-a=a'$, $n-b=b'$, $\ldots$, $n-g=g'$, *then*

1) *if the product* P *is purely progressive,* P *is nonzero if and only if*

$$v = a + b + \cdots,$$

and in fact $p = v$,

2) *if the product* P *is regressive, then* P *is nonzero if and only if*

$$g = a' + b' + \cdots,$$

and in fact $p = g$.

PROOF. 1. If the product P is progressive one can (according to 119B) resolve its factors into pure factors of first order; the number of these factors is (according to 77) $a + b + \cdots$, their product (according to 61 and 66) is then nonzero if and only if the factors of first order stand in no numerical relation to one another, that is (according to 23) if the covering domain is of $(a+b+\cdots)$th order, and thus

*{In the original edition this theorem was No. 287, but it belongs here after No. 119B, and is placed here in the Teubner Edition.}

$v = a+b+\cdots$. Then (according to 77) the order of the product $= a+b+\cdots$, that is $= v$.

2. If the product P is regressive the theorem is true for two factors. For {in this case} according to 109 P is nonzero if and only if $g = a + b - n$, that is

$$n - g = n - a + n - b,$$

whence

$$g' = a' + b'.$$

Furthermore, according to 95 $p = a+b-n$, that is $= g$. Thus the theorem is true for two factors. But by continued application one obtains the theorem for arbitrarily many factors.

120. *A pure product remains congruent with itself if one arbitrarily changes the ordering of its factors, that is*

$$P_{A,B} \equiv P_{B,A}.$$ [36]

PROOF. 1. If q and r are the orders of A and B, and if the product $[AB]$ is progressive, then (according to 58)

$$[AB] = (-1)^{qr}[BA],$$

whence

$$[AB] \equiv [BA].$$

If the product $[AB]$ is regressive and the order of the principal domain n, then $[|A|B]$ is (according to 115) a progressive product, and since $n-q$ and $n-r$ are the orders of $|A$ and $|B$,

$$[|A|B] = (-1)^{(n-q)(n-r)}[|A|B],$$

whence (according to 101)

$$[AB] = (-1)^{(n-q)(n-r)}[AB],$$

and thus also

$$[AB] \equiv [BA].$$

2. Further, if $[PAB]$ is a pure product, then

$$[PAB] = [P \cdot AB] \qquad [119]$$
$$\equiv [P \cdot BA]$$

according to Part 1 of the Proof and according to 2, 40; again this

$$\equiv [PBA],$$

and thus

$$[PAB] \equiv [PBA],$$

that is, the product remains congruent to itself if one interchanges two sequential factors.

3. By interchanging two sequential factors one can now move any given factor to any position, and thus set the factors in any arbitrary ordering, while according to Part 2 of the Proof the product remains congruent with itself.

121. *If a pure product contains two incident factors, neither of which is of order zero, the product is zero, that is $P_{A,B} = 0$ if P is a pure product and A and B are the incident factors.*

PROOF. 1. If A and B are the incident factors, and thus one of them is subordinate to the other, say B to A, then B is the common domain and A the covering, whence (according to 109) the product $[AB]$ is zero, since the order of B is greater than zero and that of A less than n.

2. Now if P contains two incident factors A and B, then (according to 120) one can so order the factors that A and B are sequential, while the product remains congruent to itself, whence (according to 2) the first case remains zero if the second case is. Then one can (according to 119) enclose these two factors in parentheses. Their product is zero according to part 1 of the Proof, whence a factor of P is zero, and thus P is itself zero.

122. *A mixed product $[ABC]$ of three {elementary} magnitudes {the third of which, C, is not zero}, is zero if and only if either $[AB] = 0$, or all three magnitudes A, B, C are contained in a domain of less than nth order, or have a domain of higher than zeroth order in common.*

PROOF. If α, β, γ are the orders of A, B, C, then $\alpha + \beta + \gamma > n$ and $< 2n$ (according to 116). Suppose $[AB] \neq 0$; then

First, if $\alpha+\beta > n$, say $= n+\delta$, then (according to 87) A and B can be expressed in terms of a common factor of δth order D, so that $A = [DE]$, $B = [DF]$, whence

$$[AB] = [DEF]D, \qquad [103]$$

and thus, since $[AB] \neq 0$ by assumption, $[DEF] \neq 0$ as well. Then

$$[ABC] = [DEF][DC],$$

and therefore, since $[DEF]$ is a nonzero number, $[ABC]$ is zero if and only if $[DC]$ is. The order of $[DC]$ is $= \delta+\gamma = \alpha+\beta-n+\gamma$, and thus $< n$, since $\alpha+\beta+\gamma < 2n$. Therefore {since C and D are not zero} (according to 109) the product $[DC]$ is zero if and only if D and C have a domain of higher than zeroth order in common, that is (since D is the common domain of A and B), if A, B, and C have a domain of higher than zeroth order in common.

Second, if $\alpha + \beta \leq n$, then $[AB]$ is a progressive product, and thus, since by assumption $[AB] \neq 0$ {and C is not zero}, (according to 109) the product $[(AB)C]$ is zero if and only if $[AB]$ and C, that is A, B, and C, are contained in a domain of less than nth order. Thus the proof.

123. *The order in which one progressively multiplies two incident elementary magnitudes is of no importance for the resultant, that is*

$$[ABC] = [ACB]$$

if B and C are incident.

PROOF. 1. If B or C is of zeroth order, that is a number, then (according to 13) the equality of both sides certainly obtains. If the product is pure, then (according to 121) both sides are zero, whence the theorem is yet to be proved only for the case of mixed products, in which B and C are of higher than zeroth order; that is (if α, β, γ are the orders of A, B, C, and n that of the principal domain) for the case that $\alpha + \beta + \gamma > n$ and $< 2n$ {116}, and β and γ are not zero.

Since the formula to be proved is not changed if one interchanges B and C, we can assume that $\beta \leq \gamma$, that is, since B and C are incident, that B is subordinate to C. In addition we first assume that A is an elementary magnitude. Since the sum $\alpha+\beta$ is less than or equal to the sum $\alpha+\gamma$, only three cases are possible; either both sums are smaller than n, both greater than n, or $\alpha+\gamma \geq n$ and $\alpha+\beta \leq n$.

2. If both sums are smaller than n, and thus $\alpha+\gamma < n$, then the three magnitudes A, B, C, of which B is subordinate to the magnitude C, are contained in a domain of $(\alpha+\gamma)$th order, and thus in a domain of less than nth order, whence (according to 122) both $[ABC]$ and $[ACB]$ are zero, and thus

$$[ABC] = [ACB].$$

3. If $\alpha+\beta$ and $\alpha+\gamma > n$, then $(n-\alpha)+(n-\beta)$ and $(n-\alpha)+(n-\gamma) < n$, and thus, since $n-\alpha$, $n-\beta$, $n-\gamma$ are the orders of the supplements of A, B, C,

$$[|A|B|C] = [|A|C|B] \qquad \text{[Part 2 of Proof]}$$

and thus (according to 101)

$$[ABC] = [ACB].$$

4. If $\alpha+\gamma \geq n$ and $\alpha+\beta \leq n$, with, say, the former $= n+\delta$, where δ could be zero, then (according to 87) it must be possible to express A and C in terms of a common factor D of δth order in such a way that $C = [DE]$, where D and E are elementary magnitudes and D is subordinate to A. Then E is of $(\gamma-\delta)$th order, and thus the sum of the orders of A and E is $\alpha+\gamma-\delta = n$, whence (according to 108)

$$[AC] = [A \cdot DE] = [AE]D,$$

and therefore

$$[ACB] = [AE][DB].$$

Here the product $[DB]$ is progressive, because

$$\delta+\beta = \alpha+\beta+\gamma-n < n,$$

since the product is supposed to be mixed. Now if $[DB]$ is zero, then (according to 109) D and B have a domain of higher than zeroth order in common, whence however, since D is subordinate to A, A and B have this domain in common; but since $\alpha+\beta \leq n$, the product $[AB]$ is progressive, whence (according to 109) it is zero. But then also $[ABC] = 0$, just as $[ACB]$, and thus they are equal. But if $[DB] \neq 0$, then since D and B are both subordinate to C, their covering domain $[DB]$ is also subordinate to the domain of C, and thus (according to 79B) C can be represented in the form $[DBF]$, where again F is an elementary magnitude. Then, since we set $C = [DE]$ above, it is possible to set $E = [BF]$, and one obtains

$$[ACB] = [AE][DB] = [ABF][DB].$$

Furthermore,

$$[ABC] = [AB \cdot DBF] = [ABF][DB], \qquad \text{(according to 108)}$$

since D is subordinate to A, and thus $[DB]$ to $[AB]$, and since $[ABF] = [AE]$, as shown above, is of zeroth order. Thus one obtains $[ACB] = [ABC]$.

5. Since it was assumed that $\beta \leq \gamma$, this exhausts all cases insofar as A is an elementary magnitude. Now if A is a compound magnitude, then (according to 77) it is always numerically derivable from elementary magnitudes. If $A =$

$\sum \alpha_r[A_rBC]$, where all the A_r are elementary magnitudes, then

$$\begin{aligned}[ABC] &= \sum \alpha_r[A_rBC] && [44]\\ &= \sum \alpha_r[A_rCB] && \text{[parts 1–4 of Proof]}\\ &= [ACB]. && [44]\end{aligned}$$

124. *If q, r, s are the orders of three {nonzero} elementary magnitudes A, B, C and n that of the principal domain, the products $[ABC]$ and $[ACB]$ are congruent* ***only*** *in the following {six} cases:*

$$[ABC] \equiv [ACB],$$

a) if $q + r + s \leq n$, then

$$[ABC] = (-1)^{rs}[ACB],$$

b) if $q + r + s \geq 2n$, then

$$[ABC] = (-1)^{(n-r)(n-s)}[ACB],$$

c) if $[AB]$ and $[AC]$ are zero, then

$$[ABC] = [ACB] = 0,$$

d) if $[ABC]$ is a mixed product and A, B, C either have a domain of higher than zeroth order in common or are included in a domain of less than nth order, then

$$[ABC] = [ACB] = 0,$$

e) if {the product $[ABC]$ is mixed and} $q + r + s = n + t$, and B and C either have a domain of tth order in common or are included in a domain of tth order, then

$$[ABC] = (-1)^{(r-t)(s-t)}[ACB],$$

f) if B and C are incident, then

$$[ABC] = [ACB].$$

PROOF. Formula *a*) is proved in 58.

If $q + r + s \geq 2n$, then $(n - q) + (n - r) + (n - s) \leq n$, whence, since $n - q$, $n - r$, $n - s$ are the orders of the supplements of A, B, C, in this case

$$[|A|B|C] = (-1)^{(n-r)(n-s)}[|A|C|B]. \qquad \text{(Formula } a)$$

Thus (according to 101)

$$[ABC] = (-1)^{(n-r)(n-s)}[ACB],$$

whence formula *b*) is proved.

Since in these two cases $q + r + s$ is either $\leq n$ or $\geq 2n$, there remains only the case where $q + r + s > n$ and $< 2n$, that is the case of mixed products.

Assume first that $[ABC]$ is zero. A mixed product $[ABC]$ {whose third factor C is not zero} is (according to 122) zero if and only if either $[AB] = 0$, or all three magnitudes A, B, C are contained in a domain of less than nth order, or if they have a domain of higher than zeroth order in common. When either of the latter two cases eventuate, both $[ABC]$ and $[ACB]$ are zero, and thus $[ABC] = [ACB] = 0$, proving formula *d*). If neither of these latter cases obtain, then $[ABC]$ cannot be zero unless $[AB]$ is zero; should this be the case and should $[ABC]$ be congruent with $[ACB]$, then $[ACB]$ must also be zero, but since the two cases mentioned above are excluded {and $B \neq 0$}, this cannot occur unless $[AC]$ is also zero, and thus case c) obtains.

Thus there remains only the case of a nonzero mixed product. Since $q+r+s < 2n$ and $> n$, we can set $q+r+s = n+t$, where $t > 0$ and $< n$. Now here (precisely as in 123) three cases must be distinguished.

First, that in which the sums $q+s$ and $q+r$ are both *smaller* than n. Then, since $[AB]$ and $[AC]$ are nonzero progressive products, $q+r$ is the order of $[AB]$ and $q+s$ that of $[AC]$. Thus (according to 26, {87}) $[AB]$ and C have a factor of $(q+r+s-n)$th, that is tth order, in common. Let this be D, and let $C = [DE]$; then the sum of the orders of A, B, E equals n, and D is subordinate to $[AB]$. Consequently (according to 108)

$$[ABC] = [AB \cdot DE] = [ABE]D.$$

Thus, since $[ABE]$ is a number, $[ABC] \equiv D$; consequently, if $[ABC] \equiv [ACB]$, it must also be true that $[ACB] \equiv D$, that is, $[AC]$ and B must be expressible in terms of a common factor congruent with D, and thus also in terms of D itself. Thus D must be subordinate to B, but it was also subordinate to C, that is B and C can be expressed in terms of a common factor of tth order, D, which is to say that they have a domain of tth order in common, which is the first condition of formula e).

If $B = [DF]$, then

$$[ABC] = [AB \cdot DE] = [ABE]D \qquad [108]$$
$$= [A(DF)E]D = [ADFE]D, \qquad [119]$$
$$[ACB] = [AC \cdot DF] = [ACF]D \qquad [108]$$
$$= [A(DE)F]D = [ADEF]D. \qquad [119]$$

Now since $C = [DE]$, E is of $(s-t)$th order, and since $B = [DF]$, F is of $(r-t)$th order, whence

$$[ADFE]D = (-1)^{(r-t)(s-t)}[ADEF]D, \qquad [58]$$

and thus

$$[ABC] = (-1)^{(r-t)(s-t)}[ACB],$$

which is formula e).

Second, if the sums $q+r$ and $q+s$ are *greater* than n, then $(n-q)+(n-r)$ and $(n-q)+(n-s)$ are less than n, and $(n-q)+(n-r)+(n-s) = n+(n-t)$. Consequently in this case (according to Case e) the products $[|A|B|C]$ and $[|A|C|B]$ are only congruent if $|B$ and $|C$ can be expressed in terms of a common factor of $(n-t)$th order. Let this be $|D$ and let $|B = [|D|F]$, $|C = [|D|E]$; then (according to e)

$$[|A|B|C] = (-1)^{(t-r)(t-s)}[|A|C|B].$$

But (according to 98 {and 101})

$$B = [DF], \quad C = [DE]$$
$$[ABC] = (-1)^{(t-r)(t-s)}[ACB] = (-1)^{(r-t)(s-t)}[ACB],$$

that is, the second of the conditions of formula e) obtains, and this itself eventuates, since the product $B = [DF]$ appears as regressive because B is of lower order than

D, and likewise $[DE]$, and thus B and C are both subordinate to D, and are therefore contained in the domain of D.

There now remains only the *third* case, where one of the sums $q+r$ and $q+s$, say the former, is less than or equal to n, the other greater than or equal to n. Then (according to 26 {and 87}) $[AB]$ and C can be expressed in terms of a common factor of $(q+r+s-n)$th, that is tth, order. Let it be D and let $C = [DE]$;* then

$$[ABC] = [AB \cdot DE] = [ABE]D. \qquad [108]$$

Further, let $q+s = n+v$; then A and C have a factor of vth order in common. Let this factor be F, and let $C = [FG]$; then

$$[AC] = [A \cdot FG] = [AG]F. \qquad [108]$$

Thus

$$[ACB] = [AG][FB].$$

Consequently if $[ABC] \equiv [ACB]$, then since $[ABE]$ and $[AG]$ are not zero, it is necessary that $D \equiv [FB]$, that is, B is subordinate to D, but D is also subordinate to C, and thus B is subordinate to C, that is B and C are incident. This is the condition for formula f) and (according to 123) therefore

$$[ABC] = [ACB].$$

The proof of the theorem is therefore complete.

125. *In the same cases (as in* 124*), and no others,*

$$[BAC] \equiv [B \cdot AC].$$

PROOF. In the cases considered in 124,

$$\begin{aligned}[BAC] &\equiv [ABC] && [120]\\ &\equiv [ACB] && [124]\\ &\equiv [B \cdot AC] && [120]\end{aligned}$$

and conversely from the congruence of 125 there follows that of 124.

REMARK. In particular for cases c) and d) it follows that

$$[BAC] = [B \cdot AC] = 0,$$

for cases a) and b) {cf. 119}

$$[BAC] = [B \cdot AC].$$

On the other hand case e) splits into two cases; namely, if the sums $q+r$ and $q+s$ are smaller than n, then

$$[BAC] = (-1)^{qt}[B \cdot AC],$$

and if those sums are greater than n,

$$[BAC] = (-1)^{(n-q)(n-t)}[B \cdot AC].$$

Case f) splits into two cases, namely when $q+r$ is smaller and $q+s$ greater than n, then

$$[BAC] = (-1)^{r(n-s)}[B \cdot AC],$$

but if $q+r$ is greater, $q+s$ smaller than n, then

$$[BAC] = (-1)^{(n-s)r}[B \cdot AC].$$ [37]

---

*Should $q+r=n$, E would be of zeroth order, which is included in the above proof; the same goes in the following for F.

If one of the sums is equal to n, then both the formula in which the sum is assumed to be greater than n, and that in which the sum is assumed to be smaller, are valid, since both formulas then become identical. It should also be noted that if in f), that is in the case of B and C being incident, the condition obtains that both sums are greater than n, or both sums are less than n, both $[BAC]$ and $[B \cdot AC]$ are zero, and thus at the same time case c) or d) obtains.[38]

126. *A product of zeroth order remains congruent with itself if one reverses the ordering of all its factors, or if any number of its final factors are combined in reverse order in a product, that is*

$$[A_1 A_2 \cdots A_{n-1} A_n] \equiv [A_n A_{n-1} \cdots A_2 A_1] \equiv [A_1 A_2 \cdots A_{n-r-1} \cdot A_n A_{n-1} \cdots A_{n-r}].$$

PROOF. First let

$$[A_1 A_2 \cdots A_{n-2}] = P;$$

then

$$[A_1 A_2 \cdots A_{n-1} A_n] = [P A_{n-1} A_n].$$

Since the product is supposed to be of zeroth order, the sum of the orders of P, A_{n-1}, A_n (according to 96) must be divisible by n, and thus, since the individual orders are > 0 and $< n$, this sum must either equal n or $2n$; in the former case the product of the three magnitudes P, A_{n-1}, A_n is purely progressive, in the latter purely regressive, but in either case pure, whence (according to 119, 120)

$$[P A_{n-1} A_n] \equiv [P \cdot A_n A_{n-1}],$$

or

$$[A_1 A_2 \cdots A_{n-1} A_n] \equiv [A_1 A_2 \cdots A_{n-2} \cdot A_n A_{n-1}].$$

If we treat this expression as a product of the three magnitudes $[A_1 \cdots A_{n-3}]$, A_{n-2}, and $[A_n A_{n-1}]$, then in the same way as with the last expression found we obtain

$$\equiv [A_1 \cdots A_{n-3} \cdot A_n A_{n-1} A_{n-2}].$$

If one applies this procedure r times, one finds

$$[A_1 \cdots A_n] \equiv [A_1 A_2 \cdots A_{n-r-1} \cdot A_n A_{n-1} \cdots A_{n-r}],$$

that is, the product remains congruent with itself if one combines any number (r) of its final factors in reverse order in a product. Consequently also

$$[A_1 A_2 \cdots A_n] \equiv [A_1 \cdot A_n A_{n-1} \cdots A_2],$$

and thus (according to 58)

$$\equiv [A_n A_{n-1} \cdots A_2 A_1],$$

whence the first part of the theorem is also proved.

7. Shadow and replacement

127. DEFINITION. If n is the order of the principal domain, $A_1, \ldots, A_v$ the multiplicative combinations, of any given class, of the n magnitudes of first or $(n-1)$th order $a_1, \ldots, a_n$ standing in no numerical relation to one another, and $A_1, \ldots, A_u$ the multiplicative combinations of m of them, say of $a_1, \ldots, a_m$, of the same class, and

$$C = \alpha_1 A_1 + \cdots + \alpha_v A_v,$$
$$C_1 = \alpha_1 A_1 + \cdots + \alpha_u A_u,$$

then I call C_1 the *shadow* of C on the domain $[a_1 \cdots a_m]$, *excluding* the domain $[a_{m+1} \cdots a_n]$, and I call the shadow *progressive* if $a_1, \ldots, a_n$ are magnitudes of first order, *regressive* if $a_1, \ldots, a_n$ are magnitudes of $(n-1)$th order. The shadows of several magnitudes are said to be taken *in the same sense* if they are shadowed on the same domain and excluding the same domain (cf. 33).

REMARK. If for example the principal domain is of fourth order (as for example is space), and if a, b, c, d are four magnitudes of first order standing in no numerical relation to one another (for example four noncoplanar points), then $C_1 = [bc]+[ca]+[ab]$ (in space a line) is the (progressive) shadow of

$$C = [bc] + [ca] + [ab] + [ad] + [bd] + [cd]$$

on the domain $[abc]$ (thus on the plane abc), excluding the domain d.

Further, if we denote $[bcd]$ by a', $[cad]$ by b', $[abd]$ by c', and $[acb]$ by d', then a', b', c', d' are magnitudes of $(n-1)$th order, and if in addition we set $[abcd] = 1$, then

$$[b'c'] = [ad],\ [c'a'] = [bd],\ [a'b'] = [cd],$$
$$[a'd'] = [bc],\ [b'd'] = [ca],\ [c'd'] = [ab],$$

and

$$C = [a'd'] + [b'd'] + [c'd'] + [b'c'] + [c'a'] + [a'b'].$$

Then, if

$$C' = [b'c'] + [c'a'] + [a'b'],$$

C' is the (regressive) shadow of C on the domain $[a'b'c']$, excluding the domain d', that is, since $[a'b'c'] = [cd \cdot abd] \equiv d$, and $d' \equiv [abc]$, C' is the shadow of C on the domain d, excluding the domain $[abc]$.

Thus in geometry the shadow of a line on a plane appears as a progressive shadow, and that of a line on a point as a regressive shadow. In geometry the shadow itself is equivalent to projection in the broadest and most pregnant sense of the word (cf. below {Nos. 129, 164–166, 200}).

We have above (33) called the magnitude C_1 specified in the definition the shadow of the magnitude C on the domain of the magnitudes $A_1, A_2, \ldots, A_u$. We have here set aside this nomenclature for convenience in applications.

128. *According as the order of the domain on which the shadow is taken is greater or less than the order of the shadowed magnitude, the shadow appears as progressive or regressive. If the order of that domain is equal to the order of the shadowed magnitude, the shadow can be progressive as well as regressive.*

PROOF. If $a_1, \ldots, a_n$ are magnitudes of *first order* that stand in no numerical relation to one another, $A_1, \ldots, A_v$ are the multiplicative combinations of $a_1, \ldots, a_n$ of pth class, and $A_1, \ldots, A_u$ are the multiplicative combinations of $a_1, \ldots, a_m$ of pth class, and

$$C = \alpha_1 A_1 + \cdots + \alpha_v A_v, \quad C_1 = \alpha_1 A_1 + \cdots + \alpha_u A_u,$$

so that (according to 127) C_1 is the progressive shadow of the magnitude C on the domain $[a_1 \cdots a_m]$, excluding the domain $[a_{m+1} \cdots a_n]$, then if $m < p$, there are no combinations at all of $a_1, \ldots, a_m$ of pth class, and thus no progressive shadow. Thus for the progressive shadow it is necessary that $m \geq p$, but since m is the order of the domain $[a_1 \cdots a_m]$ and p that of the multiplicative combinations $A_1, \ldots, A_u$, and thus also of C, the order of the domain on which the shadow is taken must be greater than or equal to that of the shadowed magnitude.

Under the same assumptions as above, with the sole distinction that $a_1, \ldots, a_n$ are magnitudes of $(n-1)$th order (in the principal domain of nth order), the shadow is regressive; and here too, on the same grounds as above, it is necessary that $m \geq p$. But now if one interprets the magnitudes of $(n-1)$th order $a_1, \ldots, a_m$ as supplements of magnitudes of first order, then (according to 98) their product $[a_1 \cdots a_m]$ is the supplement of a product of m magnitudes of first order, its order is therefore (according to the Corollary to 90) equal to $n-m$; likewise that of C equals $n-p$. Since however $n-m \leq n-p$, the order of the domain on which the shadow is taken is less than or equal to that of the shadowed magnitude.

Thus if the order of that domain is greater or less than the order of the shadowed magnitude, then the shadow is progressive in the first case, regressive in the second. If on the other hand the orders are equal, then the shadow can be both progressive and regressive.

129. *The shadow A' of a magnitude A on a domain B, excluding the domain C, is*

$$A' = \frac{[B \cdot AC]}{BC},$$ [39]

whence

$$A' = [B \cdot AC],$$

if $[BC] = 1$.

PROOF. If $a_1, \ldots, a_n$ are n magnitudes of first or $(n-1)$th order standing in no numerical relation to one another, and $A_1, \ldots, A_v$ are the multiplicative combinations {of a given class} of $a_1, \ldots, a_n$, and $A_1, \ldots, A_u$ those of $a_1, \ldots, a_m$, and $A = \alpha_1 A_1 + \cdots + \alpha_v A_v$, whence $A' = \alpha_1 A_1 + \cdots + \alpha_u A_u$, and if

$$[a_1 \cdots a_m] = B, \quad [a_{m+1} \cdots a_n] = C,$$

then

$$[AC] = [(\alpha_1 A_1 + \cdots + \alpha_u A_u + \alpha_{u+1} A_{u+1} + \cdots + \alpha_v A_v)C].$$

But since $A_1, \ldots, A_u$ are the combinations of $a_1, \ldots, a_m$ and $A_{u+1}, \ldots, A_v$ are those combinations of $a_1, \ldots, a_n$ that are not at the same time combinations of $a_1, \ldots, a_m$, each of the magnitudes $A_{u+1}, \ldots, A_v$ must contain at least one of the factors $a_{m+1}, \ldots, a_n$, and thus must have at least one factor in common with $C = [a_{m+1} \cdots a_n]$. The products $[A_{u+1}C], \ldots, [A_v C]$ are however pure with respect to the factors $a_1, \ldots, a_n$ (according to 114 {127, 128, and 119B}), whence, since they contain equal factors, they are zero {according to 121}, and therefore

$$[AC] = [(\alpha_1 A_1 + \cdots + \alpha_u A_u)C] = \alpha_1 [A_1 C] + \cdots + \alpha_u [A_u C].$$

Consequently

$$[B \cdot AC] = \alpha_1 [B \cdot A_1 C] + \cdots + \alpha_u [B \cdot A_u C].$$ [40]

Now since each of the magnitudes $A_1, \ldots, A_u$ consists of factors included in B, each of them is incident on B, whence, since the orders of B and C together amount to n,

$$[B \cdot A_1C] = [BC]A_1, \;\ldots,\; [B \cdot A_uC] = [BC]A_u$$

(according to 108),[41] and thus

$$[B \cdot AC] = [BC](\alpha_1 A_1 + \cdots + \alpha_u A_u) = [BC]A'.$$

Consequently, since $[BC]$ is a number,

$$A' = \frac{[B \cdot AC]}{[BC]}.$$ [42]

130. *Every equation whose terms are multiples of any given magnitudes of mth order remains valid if one replaces all of these magnitudes with their shadows in the same sense; or, if*

$$P = \alpha A + \beta B + \cdots,$$

and P', A', B', ... are the shadows in the same sense of P, A, B, ... , then

$$P' = \alpha A' + \beta B' + \cdots.$$

PROOF. If Q is the domain on which the shadow is taken, and R the excluded domain and $[QR] = 1$, then one obtains from the equation

$$P = \alpha A + \beta B + \cdots$$

by multiplication the expressions

$$[PR] = \alpha[AR] + \beta[BR] + \cdots$$

and

$$[Q \cdot PR] = \alpha[Q \cdot AR] + \beta[Q \cdot BR] + \cdots,$$

that is (according to 129)

$$P' = \alpha A' + \beta B' + \cdots.$$

131. *The progressive shadow of a pure progressive product and the regressive shadow of a pure regressive product are equal to the product of the shadows, taken in the same sense, of the factors of that product, that is, if*

$$P = [AB \cdots E]$$

is the pure product P, and P', A', B', ... , E' are the shadows, taken in the same sense, of P, A, B, ... , E (and progressive or regressive according as the product is progressive or regressive), then

$$P' = [A'B' \cdots E'].$$

PROOF. 1. Let

$$A = [a_1 \cdots a_q], \; B = [a_{q+1} \cdots a_r], \; \ldots, \; E = [a_{t+1} \cdots a_v],$$

where $a_1, \ldots, a_v$ are magnitudes of first or $(n-1)$th order, according as the product $[AB \cdots E]$ is progressive or regressive. Then {according to 119B}

$$P = [a_1 \cdots a_v]$$

is a pure product of magnitudes of first or $(n-1)$th order. Further, let $Q = [u_1 \cdots u_m]$ be the domain on which the shadow is taken, $R = [u_{m+1} \cdots u_n]$ be the excluded domain, and $[QR] = 1$, whence $u_1, \ldots, u_n$ are magnitudes of first or $(n-1)$th order according as $a_1, \ldots, a_n$ are. Then (according to 129)

$$\begin{aligned} P' &= [Q \cdot PR] \\ &= [u_1 \cdots u_m(a_1 \cdots a_v \cdot u_{m+1} \cdots u_n)]. \end{aligned}$$

Now if the original product $[AB \cdots E]$ is *progressive*, then the shadow is also progressive, that is (according to 128), the order of Q is greater than or equal to that of P, that is $m \geq v$, whence $v+n-m \leq n$, that is the product $[a_1 \cdots a_v \cdot u_{m+1} \cdots u_n]$ is purely progressive, whence also

$$= [a_1 \cdots a_v u_{m+1} \cdots u_n],$$

and since all the factors are of first order, it is combinatorial (according to 94, 78). Now if

$$a_r = \alpha_1^{(r)} u_1 + \cdots + \alpha_n^{(r)} u_n,$$

then (according to 67) in the product $[a_1 \cdots a_v u_{m+1} \cdots u_n]$ we can replace $a_r = \alpha_1^{(r)} u_1 + \cdots + \alpha_n^{(r)} u_n$ by $\alpha_1^{(r)} u_1 + \cdots + \alpha_m^{(r)} u_m$ since $u_{m+1}, \ldots, u_n$ appear as factors in that product; but $\alpha_1^{(r)} u_1 + \cdots + \alpha_m^{(r)} u_m$ is the shadow of a_r on the domain $Q = [u_1 \cdots u_m]$, excluding the domain $R = [u_{m+1} \cdots u_n]$. Consequently, if we denote this shadow by a_r', then

$$P' = [u_1 \cdots u_m(a_1' \cdots a_v' \cdot u_{m+1} \cdots u_n)].$$

Here $[a_1' \cdots a_v']$ is subordinate to $[u_1 \cdots u_m]$, whence (according to 108)

$$P' = [u_1 \cdots u_n][a_1' \cdots a_v'] = [a_1' \cdots a_v'].$$

On the same grounds

$$A' = [a_1' \cdots a_q'], \; B' = [a_{q+1}' \cdots a_r'], \; \ldots, \; E' = [a_{t+1}' \cdots a_v'].$$

Consequently

$$P' = [A'B' \cdots E'].$$

2. If the original product $[AB \cdots E]$ is *regressive*, and thus also the shadow of P on Q, excluding R, is regressive, that is {according to 128} the order of P is greater than or equal to that of Q, then {one first replaces all the magnitudes by their supplements. Then} the regressive product is exchanged for the progressive product and likewise the regressive shadow for the progressive shadow (according to 115 and the Corollary to 90). Thus if P_1, Q_1, R_1, A_1, B_1, $\ldots$, E_1 are the supplements of P, Q, R, A, B, $\ldots$, E, and if P_1', A_1', B_1', $\ldots$, E_1' are the shadows of P_1, A_1, B_1, $\ldots$, E_1 on Q_1, excluding R_1, then (according to 101)

$$P_1 = [A_1 B_1 \cdots E_1],$$

and according to Part 1 of the Proof

(*) $$P_1' = [A_1' B_1' \cdots E_1'].$$

Furthermore (according to 129)

$$P' = [Q \cdot PR], \; P_1' = [Q_1 \cdot P_1 R_1] = |[Q \cdot PR], \qquad \text{[according to 97]}$$

whence $P_1' = |P'$ and likewise

$$A_1' = |A', \; B_1' = |B', \ldots$$

and therefore (according to $*$)

$$|P' = [|A'|B' \cdots |E']$$

$$P' = [A'B' \cdots E']. \qquad \text{[according to 101]}$$

3. Finally, if A, B, ... , E are compound magnitudes,

$$A = \sum \alpha_r A_r, \quad B = \sum \beta_s B_s, \quad \ldots, \quad E = \sum \varepsilon_v E_v,$$

where A_r, B_s, ... , E_v are elementary magnitudes, and A'_r, B'_s, ... , E'_v are the shadows of A_r, B_s, ... , E_v, then

$$P = \left[\sum \alpha_r A_r \cdot \sum \beta_s B_s \cdots \sum \varepsilon_v E_v\right] = \sum \alpha_r \beta_s \cdots \varepsilon_v [A_r B_s \cdots E_v] \qquad [45]$$

$$= \sum \alpha_r \beta_s \cdots \varepsilon_v P_{r,s,\ldots,v},$$

where

$$P_{r,s,\ldots,v} = [A_r B_s \cdots E_v].$$

Thus (according to 130)

$$P' = \sum \alpha_r \beta_s \cdots \varepsilon_v P'_{r,s,\ldots,v} = \sum \alpha_r \beta_s \cdots \varepsilon_v [A'_r B'_s \cdots E'_v]$$

according to Parts 1 and 2 of the Proof; and this

$$= \left[\sum \alpha_r A'r \cdot \sum \beta_s B'_s \cdots \sum \varepsilon_v E'_v\right] \qquad \text{[according to 45]}$$

$$= [A'B' \cdots E'].$$

{132 of the original edition has been moved and renumbered 116B.}

133. *An equation whose terms are magnitudes of mth order can, when the order of the principal domain is n, be replaced by as many numerical equations as there are combinations without repetition of n elements of mth class; and in particular one obtains a replacement assembly of equations by multiplying the given equation one by one with the multiplicative combinations of $(n - m)$th class of an arbitrary collection of n magnitudes of first order, whose product is* 1.

PROOF. If

$$(a) \qquad P = A + B + \cdots$$

is the original equation, in which P, A, B, ... are magnitudes of mth order, and if in addition e_1, ... , e_n are magnitudes of first order whose product $[e_1 \cdots e_n] = 1$, and if E_1, E_2, ... , E_v are the multiplicative combinations of mth class of e_1, ... , e_n, and F_1, F_2, ... , F_v are the supplementary combinations, that is, the combinations of the same elements of $(n - m)$th class, so ordered that

$$[E_1 F_1] = [E_2 F_2] = \cdots = [E_v F_v] = 1,$$

then it is to be shown that the above equation can be replaced with the assembly of equations resulting from

$$(b) \qquad [PF_r] = [AF_r] + [BF_r] + \cdots,$$

where one sets r equal to 1, 2, ... , v one by one.[43]

Now (according to 77) P, A, B, ... are numerically derivable from E_1, ... , E_v; so let

$$P = \pi_1 E_1 + \cdots + \pi_v E_v, \quad A = \alpha_1 E_1 + \cdots + \alpha_v E_v, \quad B = \beta_1 E_1 + \cdots + \beta_v E_v,$$

and so on; then

$$[PF_r] = \left[\sum \pi_s E_s \cdot F_r\right] = \sum \pi_s [E_s F_r].$$

But since E_s and F_r necessarily include equal elements $(e_1, \ldots, e_n)$ as factors if $s \neq r$, for this case {according to 60} $[E_s E_r] = 0$, whence

$$[PF_r] = \pi_r [E_r f_r] = \pi_r.$$

On the same grounds

$$[AF_r] = \alpha_r, \quad [BF_r] = \beta_r, \quad \ldots .$$

Now if equation (a) is satisfied, so also is the group of equations (b) resulting from it through multiplication. Conversely, if the latter are satisfied, then upon replacing $[PF_r]$, $[AF_r]$, $[BF_r]$, ... by the values found for them, one has for each r from $1, \ldots, v$ that

$$\pi_r = \alpha_r + \beta_r + \cdots ,$$

and thus also

$$\pi_r E_r = \alpha_r E_r + \beta_r E_r + \cdots$$

for each r from 1 to v. If one adds all these equations one obtains

$$\sum \pi_r E_r = \sum \alpha_r E_r + \sum \beta_r E_r + \cdots ,$$

that is

$$P = A + B + \cdots .$$

Thus equation (a) and the assembly of equations (b) can be replaced indifferently.

Corollary. In particular, the equation

$$A = \alpha_1 E_1 + \alpha_2 E_2 + \cdots$$

can be replaced by the assembly of equations

$$[AF_r] = \alpha_r,$$

that is by the equations

$$[AF_1] = \alpha_1, \quad [AF_2] = \alpha_2, \quad \ldots .$$

8. Elimination of the unknowns from algebraic equations by combinatorial multiplication

134. Problem. *To find the unknowns from n equations of first degree.*

Solution 1. Let the n equations be

$$(a) \qquad \begin{cases} \alpha_1^{(1)} x_1 + \alpha_2^{(1)} x_2 + \cdots + \alpha_n^{(1)} x_n = \beta^{(1)}, \\ \alpha_1^{(2)} x_1 + \alpha_2^{(2)} x_2 + \cdots + \alpha_n^{(2)} x_n = \beta^{(2)}, \\ \cdots\cdots\cdots\cdots\cdots\cdots\cdots\cdots \\ \alpha_1^{(n)} x_1 + \alpha_2^{(n)} x_2 + \cdots + \alpha_n^{(n)} x_n = \beta^{(n)}. \end{cases}$$

If one multiplies these equations respectively by n extensive magnitudes $e^{(1)}$, $e^{(2)}, \ldots, e^{(n)}$, whose combinatorial product is unity, and adds them, then upon setting

$$(b)\qquad \begin{cases} \alpha_1^{(1)}e^{(1)} + \alpha_1^{(2)}e^{(2)} + \cdots + \alpha_1^{(n)}e^{(n)} = a_1, \\ \alpha_2^{(1)}e^{(1)} + \alpha_2^{(2)}e^{(2)} + \cdots + \alpha_2^{(n)}e^{(n)} = a_2, \\ \dots\dots\dots\dots\dots\dots\dots\dots \\ \alpha_n^{(1)}e^{(1)} + \alpha_n^{(2)}e^{(2)} + \cdots + \alpha_n^{(n)}e^{(n)} = a_n, \\ \beta^{(1)}e^{(1)} + \beta^{(2)}e^{(2)} + \cdots + \beta^{(n)}e^{(n)} = b, \end{cases}$$

one obtains the equation

$$(c)\qquad x_1a_1 + x_2a_2 + \cdots + x_na_n = b,$$

an equation which {according to 29} replaces the given equation (a). In order to find x_1 from it one appends the factors $a_2, a_3, \dots, a_n$ to both sides of the equation, whence, since

$$[a_2a_2a_3\cdots a_n],\ [a_3a_2a_3\cdots a_n],\ \dots,\ [a_na_2a_3\cdots a_n]$$

are zero (according to 60), one finds

$$(d)\qquad x_1[a_1a_2\cdots a_n] = [ba_2a_3\cdots a_n].$$

And likewise

$$x_2[a_1a_2\cdots a_n] = [a_1ba_3\cdots a_n], \dots .$$

Now assuming, *first*, that the product $[a_1a_2\cdots a_n]$ {which as a magnitude of nth order is a number} is not zero, one obtains

$$(e)\qquad x_1 = \frac{[ba_2a_3\cdots a_n]}{[a_1a_2a_3\cdots a_n]}$$

and likewise

$$(e)\qquad \begin{cases} x_2 = \dfrac{[a_1ba_3\cdots a_n]}{[a_1a_2a_3\cdots a_n]}, \\ \dots\dots\dots\dots \\ x_n = \dfrac{[a_1a_2\cdots a_{n-1}b]}{[a_1a_2a_3\cdots a_{n-1}a_n]}. \end{cases}$$

For this case it is still to be shown that in fact these values for the unknowns satisfy equation (c).

Since for this case the product $[a_1a_2\cdots a_n]$ is not zero by assumption, the $a_1, \dots, a_n$ stand in no numerical relation to one another. Now since the $a_1, \dots, a_n$ are numerically derived from the $e_1, \dots, e_n$ and stand in no numerical relation to one another, it is necessarily true (according to 21) that every magnitude numerically derivable from the $e_1, \dots, e_n$, and thus in particular b, is also numerically derivable from $a_1, \dots, a_n$; thus let

$$b = y_1a_1 + y_2a_2 + \cdots + y_na_n.$$

If one substitutes this value into (e) one finds

$$x_1 = y_1,\ x_2 = y_2,\ \dots,\ x_n = y_n,$$

whence

$$x_1a_1 + x_2a_2 + \cdots + x_na_n = y_1a_1 + y_2a_2 + \cdots + y_na_n,$$

that is, $= b$. Thus equation (c) is satisfied by the values (e), and consequently so also are the original equations.

If, *second*, the combinatorial product $[a_1 a_2 \cdots a_n]$ is zero, then the $a_1, \ldots, a_n$ stand in a numerical relation to one another, whence there must (according to 17) be some among them that stand in no numerical relation to one another, and from these the rest are numerically derivable; let these be $a_1, \ldots, a_r$ and let $a_{r+1}, \ldots, a_n$ be numerically derivable from them. Then by virtue of equation (c), b must also be numerically derivable from the magnitudes $a_1, \ldots, a_r$. Should it also be the case that, by virtue of the nature of the given equations, b is not numerically derivable from the $a_1, \ldots, a_r$, even though $a_{r+1}, \ldots, a_n$ are, then those equations contain a contradiction. If on the other hand this condition is satisfied, let equation (c) be written

$$x_1 a_1 + x_2 a_2 + \cdots + x_r a_r = c,$$

where

$$c = b - x_{r+1} a_{r+1} - x_{r+2} a_{r+2} - \cdots - x_n a_n,$$

and one obtains

$$(f)\qquad \begin{cases} x_1 = \dfrac{[c a_2 a_3 \cdots a_r]}{[a_1 a_2 a_3 \cdots a_r]}, \\ x_2 = \dfrac{[a_1 c a_3 \cdots a_r]}{[a_1 a_2 a_3 \cdots a_r]}, \\ \cdots\cdots\cdots\cdots\cdots \\ x_r = \dfrac{[a_1 a_2 \cdots a_{r-1} c]}{[a_1 a_2 \cdots a_{r-1} a_r]}, \end{cases}$$

while x_{r+1} to x_n are completely arbitrary.[44]

REMARK. If for the magnitudes $a_1, \ldots, a_n$ and b one inserts their values from (b) into equation (e), then by virtue of 63 one finds the well-known equation

$$x_1 = \frac{\sum \pm \beta^{(1)} \alpha_2^{(2)} \alpha_3^{(3)} \cdots \alpha_n^{(n)}}{\sum \pm \alpha_1^{(1)} \alpha_2^{(2)} \alpha_3^{(3)} \cdots \alpha_n^{(n)}}, \quad \ldots .$$

Here I append a second method of solution, which it is true does not seem so simple at first glance, but which has great advantages, and whose particular characteristics will later appear in an even clearer light {cf. No 500–527}.

SOLUTION 2. One casts all the equations into a form in which their right sides are zero. Let the equations be

$$(\alpha)\qquad \begin{cases} \alpha_0^{(1)} + \alpha_1^{(1)} x_1 + \alpha_2^{(1)} x_2 + \cdots + \alpha_n^{(1)} x_n = 0, \\ \alpha_0^{(2)} + \alpha_1^{(2)} x_1 + \alpha_2^{(2)} x_2 + \cdots + \alpha_n^{(2)} x_n = 0, \\ \cdots\cdots\cdots\cdots\cdots\cdots\cdots\cdots \\ \alpha_0^{(n)} + \alpha_1^{(n)} x_1 + \alpha_2^{(n)} x_2 + \cdots + \alpha_n^{(n)} x_n = 0. \end{cases}$$

The equations are thus the same as in the previous solution, except that $\alpha_0(r) = -\beta_0(r)$. On account of the symmetry we can even incorporate into the first term of each equation the unknown x_0 as a factor, which we can then ultimately set equal to 1.

Now one assumes a system of $(n+1)$ units $e_0, e_1, \ldots, e_n$, whose product is unity. Then (according to 91)

$$(\beta)\qquad [e_0|e_0] = [e_1|e_1] = [e_2|e_2] = \cdots = [e_n|e_n] = [e_0 e_1 e_2 \cdots e_n] = 1.$$

Furthermore, since (according to 89) $|e_s$ contains all the units except e_s as factors,

$$[e_r|e_s] = 0, \text{ if } r \neq s. \qquad \text{[according to 60]}$$

Now if one sets

$$(\gamma) \qquad \begin{cases} x_0|e_0 + x_1|e_1 + \cdots + x_n e_n = X, \\ \alpha_0^{(1)} e_0 + \alpha_1^{(1)} e_1 + \cdots + \alpha_n^{(1)} e_n = a^{(1)}, \\ \alpha_0^{(2)} e_0 + \alpha_1^{(2)} e_1 + \cdots + \alpha_n^{(2)} e_n = a^{(2)}, \\ \cdots\cdots\cdots\cdots\cdots\cdots\cdots\cdots \\ \alpha_0^{(n)} e_0 + \alpha_1^{(n)} e_1 + \cdots + \alpha_n^{(n)} e_n = a^{(n)}, \end{cases}$$

then it easily follows that the given equations (α) are identical to the equations

$$(\delta) \qquad [a^{(1)}X] = 0, \; [a^{(2)}X] = 0, \; \ldots, \; [a^{(n)}X] = 0.$$

Thus if for example in the first of these equations one replaces $a^{(1)}$ and X by their values from (γ), then by virtue of the system of equations (β) this equation would be identical to the first equation in (α), and similarly with the rest.

Now assume, *first*, that $a^{(1)}, \ldots, a^{(n)}$ stand in no numerical relation to one another.

Since X is a magnitude of nth order, and yields zero when bound with each of the n magnitudes of first order $a^{(1)}, a^{(2)}, \ldots, a^{(n)}$, which stand in no numerical relation to one another, in a combinatorial product, X must (according to 84) stand in a numerical relation with the combinatorial product of those magnitudes, whence, if λ is a yet undetermined number,

$$(\varepsilon) \qquad X = \lambda A, \text{ where } A = [a^{(1)}a^{(2)} \cdots a^{(n)}];$$

and again since conversely equations (δ) follow from this equation, so also do the original equations (α). If one now annexes the factor e_0 to the above equation, then since

$$[e_0 X] = [e_0(x_0|e_0 + x_1|e_1 + \cdots)] = x_0[e_0|e_0] + x_1[e_0|e_1] + \cdots,$$

that is, by virtue of (β), $= x_0$, one obtains the equation

$$(\zeta) \qquad x_0 = \lambda[e_0 A],$$

and likewise

$$(\zeta) \qquad x_1 = \lambda[e_1 A], \; x_2 = \lambda[e_2 A], \ldots,$$

that is

$$(\eta) \qquad x_0 : x_1 : x_2 : \cdots = [e_0 A] : [e_1 A] : [e_2 A] : \cdots,$$

and, since x_0 is equal to 1, one has $1 = \lambda[e_0 A]$, {and consequently}

$$(\theta) \qquad x_1 = \frac{e_1 A}{e_0 A}, \; x_2 = \frac{e_2 A}{e_0 A}, \; \ldots .$$

The solution is thus only possible if $[e_0 A]$ is different from zero; if on the other hand $[e_0 A] = 0$, while A is different from zero, then the equation $1 = \lambda[e_0 A]$ shows that the given equations contain a contradiction.

Furthermore it can be shown that, in the case assumed ($A \neq 0$ and $[e_0 A] \neq 0$), the values (θ) satisfy the given equations (α). For if

$$\lambda = \frac{1}{e_0 A}$$

then equations (ζ) are satisfied, as we can also write them thus:

$$[e_0 X] = \lambda[e_0 A],\ [e_1 X] = \lambda[e_1 A],\ \ldots,$$

or

$$0 = [e_0(X - \lambda A)] = [e_1(X - \lambda A)] = \cdots .$$

Thus the magnitude of nth order $X - \lambda A$ multiplied combinatorially with each of the system of $n+1$ units $e_0, \ldots, e_n$ gives zero; thus (according to 85) that magnitude is itself zero, that is

$$X - \lambda A = 0$$

or

$$X = \lambda A,$$

which, according to the above, replaces the given equations (α).

Assuming, *second*, that $a^{(1)}, a^{(2)}, \ldots, a^{(n)}$ stand in a numerical relation to one another, without all of them being zero, then (according to 17) there is among them a collection of magnitudes that stand in no numerical relation to one another, and from them the rest are numerically derivable. Suppose $a^{(1)}, a^{(2)}, \ldots, a^{(r)}$ form such a collection, and that the rest, $a^{(r+1)}, \ldots, a^{(n)}$, are numerically derivable from them, and for example

$$a^{(n)} = \alpha_1 a^{(1)} + \alpha_2 a^{(2)} + \cdots + \alpha_r a^{(r)},$$

then it follows for any X that

$$[a^{(n)} X] = \alpha_1 [a^{(1)} X] + \alpha_2 [a^{(2)} X] + \cdots + \alpha_r [a^{(r)} X],$$

that is, the nth of the given equations is found from the first r, when one multiplies these respectively with $\alpha_1, \alpha_2, \ldots, \alpha_r$; that is, the $n - r$ last equations are numerically derivable from the first r, each value of X that satisfies these also satisfies the last $n - r$. There thus remain only r equations to satisfy, and consequently the last $(n - r)$ unknowns can be chosen arbitrarily, and the rest are then specified by the above procedure.[45]

REMARK. The second method of solution has the advantage that it shows how to replace the collection of n unknowns with a single unknown of nth order and how most easily to find it.

135. PROBLEM. *To eliminate the unknowns from $n+1$ equations which are of first degree with respect to n unknowns.*

SOLUTION. Let the equations be

$$(a)\quad \begin{cases} \alpha_0^{(0)} + \alpha_1^{(0)} x_1 + \alpha_2^{(0)} x_2 + \cdots + \alpha_n^{(0)} x_n = 0, \\ \alpha_0^{(1)} + \alpha_1^{(1)} x_1 + \alpha_2^{(1)} x_2 + \cdots + \alpha_n^{(1)} x_n = 0, \\ \cdots\cdots\cdots\cdots\cdots\cdots\cdots\cdots \\ \alpha_0^{(n)} + \alpha_1^{(n)} x_1 + \alpha_2^{(n)} x_2 + \cdots + \alpha_n^{(n)} x_n = 0. \end{cases}$$

Upon multiplying these by the $n+1$ magnitudes $e^{(0)}, e^{(1)}, e^{(3)}, \ldots, e^{(n)}$, which stand in no numerical relation to one another, respectively, adding the equations so obtained, and setting

$$(b)\qquad \begin{cases} \alpha_0^{(0)}e^{(0)} + \alpha_0^{(1)}e^{(1)} + \alpha_0^{(2)}e^{(2)} + \cdots + \alpha_0^{(n)}e^{(n)} = a_0, \\ \alpha_1^{(0)}e^{(0)} + \alpha_1^{(1)}e^{(1)} + \alpha_1^{(2)}e^{(2)} + \cdots + \alpha_1^{(n)}e^{(n)} = a_1, \\ \cdots\cdots\cdots\cdots\cdots\cdots \\ \alpha_n^{(0)}e^{(0)} + \alpha_n^{(1)}e^{(1)} + \alpha_n^{(2)}e^{(2)} + \cdots + \alpha_n^{(n)}e^{(n)} = a_n, \end{cases}$$

one obtains

$$(c)\qquad a_0 + x_1a_1 + x_2a_2 + \cdots + x_na_n = 0.$$

If one annexes the combinatorial factors $a_1, a_2, \ldots, a_n$ to this equation, one finds, since $[a_1a_1a_2\cdots a_n], \ldots$ are zero, that

$$(d)\qquad [a_0a_1a_2\cdots a_n] = 0,$$

which is the equation sought.

136. PROBLEM. *To eliminate an unknown from two equations, each of which is algebraic and of arbitrary degree with respect to that unknown.*

SOLUTION 1. Let the first equation be of mth degree with respect to the unknown y, the other be of nth degree, and let the two equations be

$$(a)\qquad \begin{cases} a_0 + a_1y + \cdots + a_my^m = 0, \\ b_0 + b_1y + \cdots + b_ny^n = 0, \end{cases}$$

where $a_0, a_1, \ldots, a_m$ and $b_0, b_1, \ldots, b_n$ are arbitrary functions of the other unknowns. If one multiplies the first of these by $1, y, y^2, \ldots, y^{n-1}$ one by one, and the second by $1, y, y^2, \ldots, y^{m-1}$ one by one, one obtains the equations

$$(b)\qquad \begin{aligned} a_0 + a_1y + \cdots + a_my^m &= 0, \\ a_0y + \cdots\cdots + a_my^{m+1} &= 0, \\ \cdots\cdots\cdots\cdots & \\ a_0y^{n-1} + \cdots\cdots + a_my^{m+n-1} &= 0, \\ b_0 + b_1y + \cdots + b_ny^n &= 0, \\ b_0y + \cdots\cdots + b_ny^{n+1} &= 0, \\ \cdots\cdots\cdots\cdots & \\ b_0y^{m-1} + \cdots\cdots + b_ny^{m+n-1} &= 0. \end{aligned}$$

If one multiplies these sequentially by $n+m$ units standing in no numerical relation to one another, namely $e_1, e_2, \ldots, e_{n+m}$, adds the resulting equations and sets

$$(c)\qquad \begin{cases} a_0e_1 + b_0e_{n+1} = u_1, \\ a_1e_1 + a_0e_2 + b_1e_{n+1} + b_0e_{n+2} = u_2, \\ a_2e_1 + a_1e_2 + a_2e_3 + b_2e_{n+1} + b_1e_{n+2} + b_0e_{n+3} = u_3, \\ \cdots\cdots\cdots\cdots\cdots\cdots \\ a_me_n + b_ne_{m+n} = u_{m+n}, \end{cases}$$

one obtains the equation

$$(d)\qquad u_1 + u_2y + u_3y^2 + \cdots + u_{m+n}y^{m+n-1} = 0;$$

and if one annexes to it the combinatorial factors $u_2, u_3, \ldots, u_{m+n}$, one obtains

$$(e) \qquad [u_1 u_2 \cdots u_{m+n}] = 0,$$

which is the equation sought.

SOLUTION 2. Let the given equations be the same as in Solution 1 (a), and let system (b) be derived from them. One takes $m+n$ units $e_0, e_1, \ldots, e_{m+n-1}$, whose product is $=1$.

Now if

$$(\alpha) \qquad \begin{aligned} |e_0 + y|e_1 + \cdots + y^{m+n-1}e_{m+n-1} &= Y, \\ a_0e_0 + a_1e_1 + \cdots + a_me_m &= c_0, \\ a_0e_1 + a_1e_2 + \cdots + a_me_{m+1} &= c_1, \\ \cdots\cdots\cdots\cdots\cdots\cdots\cdots\cdots & \\ a_0e_{n-1} + a_1e_n + \cdots + a_me_{m+n-1} &= c_{n-1}, \\ b_0e_0 + b_1e_1 + \cdots + b_ne_n &= d_0, \\ \cdots\cdots\cdots\cdots\cdots\cdots\cdots\cdots & \\ b_0e_{m-1} + b_1e_m + \cdots + b_ne_{m+n-1} &= d_{n-1}, \end{aligned}$$

then equations (b) are equivalent to the equations

$$(\beta) \qquad \begin{cases} [c_0Y] = [c_1Y] = \cdots = [c_{n-1}Y] = 0, \\ [d_0Y] = [d_1Y] = \cdots = [d_{m-1}Y] = 0. \end{cases}$$

Now since Y is a nonzero magnitude of $(n+m-1)$th order, the $n+m$ magnitudes $c_0, c_1, \ldots, c_{n-1}, d_0, d_1, \ldots, d_{m-1}$ stand in a numerical relation to one another (according to 85). Thus one has

$$[c_0c_1 \cdots c_{n-1}d_0d_1 \cdots d_{m-1}] = 0,$$

which is the equation sought.

REMARK 1. With this last method even the unknown y can be expressed in a simple way, if it is assumed that there are, among the $n+m$ magnitudes $c_0, c_1, \ldots, c_{n-1}, d_0, d_1, \ldots, d_{m-1}$, $n+m-1$ magnitudes that stand in no numerical relation to one another; if one supposes that these are $c_1, \ldots, c_{n-1}, d_0, d_1, \ldots, d_{m-1}$, and lets their combinatorial product be symbolized briefly as A, then from the equations

$$0 = [c_1Y] = [c_2Y] = \cdots = [c_{n-1}Y] = [d_0Y] = [d_1Y] = \cdots = [d_{m-1}Y],$$

it follows (according to 84) that

$$Y = pA,$$

where p represents an unknown number. But now $Y = |e_0 + y|e_1 + \cdots$, whence $[e_0Y] = [e_0|e_0] = 1$ and $[e_1Y] = y$, whence one has

$$\begin{aligned} 1 &= [e_0Y] = p[e_0A], \\ y &= [e_1Y] = p[e_1A], \end{aligned}$$

whence, if one divides the second by the first,

$$y = \frac{[e_1A]}{[e_0A]},$$

whereby y is found, while the equation eliminating the unknowns appears as

$$[c_0A] = 0.$$

REMARK 2. These methods of solution (in the first of the forms given here) I already presented in the first edition of the *Ausdehnungslehre* (1844), and gave an abstract of them in *Grunerts Archiv* (1845). Later CAUCHY, in a series of articles published in the *Comptes Rendus* for 1853, presented the same method, without however having mentioned me or my works, which I had previously sent him in 1847. In consequence of a priority claim which I submitted to the Paris Academy of Science regarding this matter, a review committee was named by it, without however its ever having returned a report upon the claim; which indeed scarcely seems necessary, as the circumstances themselves permit no room for doubt. I should also mention that I was induced by CAUCHY's articles to use brackets to symbolize combinatorial multiplication and multiplication relative to a principal domain in general.[46]

CHAPTER 4

Inner Product

1. Fundamental laws of inner multiplication

137. DEFINITION. By the *inner product* of two units of arbitrary order I mean the relative product of the first with the supplement of the second; that is, if E and F are units of arbitrary order, then

$$[E|F]$$

is the inner product of the units E and F.

138. *The inner product of two arbitrary magnitudes equals the relative product of the first with the supplement of the second, that is,*

$$[A|B]$$

is the inner product of the magnitudes A and B.

PROOF. Let $A_1, \dots, A_n$ be the units from which A is numerically derived, $B_1, \dots, B_m$ the units for B, and let

$$A = \alpha_1 A_1 + \cdots + \alpha_n A_n, \quad B = \beta_1 B_1 + \cdots + \beta_m B_m;$$

further, for the moment let $\times$ be chosen as the symbol for inner multiplication; then

$$\begin{aligned} [A \times B] &= [(\alpha_1 A_1 + \cdots + \alpha_n A_n) \times (\beta_1 B_1 + \cdots + \beta_m B_m)] \\ &= \sum \alpha_r \beta_s [A_r \times B_s], \end{aligned} \qquad [42]$$

where the sum refers to the values $1, \dots, n$ for r and $1, \dots, m$ for s. Now since A_r and B_s are units, (according to 137) $[A_r \times B_s]$ equals $[A_r|B_s]$, whence

$$\begin{aligned} [A \times B] &= \sum \alpha_r \beta_s [A_r \times |B_s] = \left[\alpha_r A_r \cdot \sum \beta_s |B_s\right] && [42] \\ &= \left[A \sum \beta_s |B_s\right] = \left[A \middle| \sum \beta_s B_s\right] && [90] \\ &= [A|B]. \end{aligned}$$

REMARK. Thus a special symbol for the inner product now seems superfluous, since the symbol for the supplement fully replaces the symbol for inner multiplication. And it only remains to be noted that this symbol can also be treated as a symbol for multiplication.

In my earlier work (*Geometrische Analyse*, gekronte Preisschrift, Leipzig 1847) I had introduced the symbol $\times$ for the inner product, a symbol which is now unnecessary.[47]

139. *The order of the inner product whose two factors have the orders* α *and* β*, respectively, while the principal domain has order* n*, equals either* $n+\alpha-\beta$ *or* $\alpha-\beta$*, according as* β *is greater than* α *or not.*

PROOF. If A and B are the two factors, whose orders are respectively α and β, then {according to the Corollary to 90} the order of $|B$ is $n-\beta$. Now if, first, β is greater than α, then n is also greater than $\alpha+n-\beta$; that is, the sum of the orders of A and $|B$ is less than that of the principal domain, and thus (according to 95) the order of the product $[A|B]$ is equal to that sum, that is, equal to $\alpha+n-\beta$. If however β is less than or equal to α, then n is less than or equal to $\alpha+n-\beta$, that is, the sum of the orders of A and $|B$ is greater than or equal to n, whence (according to 95) the order of the product $[A|B]$ is smaller by n than that sum, that is, it is equal to $\alpha-\beta$.

140. *The number of units from which an inner product can be numerically derived is equal to the number of combinations of as many elements as the order of the principal domain, and of the same class as the positive difference of the orders of the two factors.*

PROOF. According to 139 the order of the product is either equal to $n+\alpha-\beta$ or to $\alpha-\beta$, according as β is greater than α or not. The units of the same order are, in the first case, the multiplicative combinations of the n original units of $(n+\alpha-\beta)$th class, and in the second of $(\alpha-\beta)$th class. But the number of combinations of the n elements of $(n+\alpha-\beta)$th class is, according to a well-known theorem of combination theory, equal to the number of combinations of n elements of $(\alpha-\beta)$th class. The number of units from which the product can be derived is therefore in the *first* case equal to the number of combinations of n elements of $(\beta-\alpha)$th class, in the *second* of $(\alpha-\beta)$th class. Thus in *both* cases the class of these combinations is equal to the positive difference of α and β.

141. *The inner product of two magnitudes of the same order is a number.*

PROOF. For the difference of the orders is then zero, and thus {according to 139} the product is of zeroth order, that is a number.

142. *The inner product of two equal units is one, that of two different units of the same order is zero, that is*

$$[E_r|E_r]=1,\ \ [E_r|E_s]=0.$$

PROOF. $[E_r|E_r]=1$ (according to {137 and} 91). Furthermore (according to 89), $|E_s$ is equal to the combinatorial product of all the units of first order not appearing in the product E_s; now since E_r is different from E_s but both are products of an equal number of original units, E_r necessarily contains as factors such units as are lacking in E_s, and which therefore appear in $|E_s$; thus (according to 60) $[E_r|E_s]$ is equal to zero.

143. *If* $E_1, \dots, E_m$ *are all units of arbitrary but equal order, then*

$$[(\alpha_1E_1+\cdots+\alpha_mA_m)|(\beta_1E_1+\cdots+\beta_nE_n)]=\alpha_1\beta_1+\cdots+\alpha_m\beta_m.$$

PROOF. Denoting

$$\alpha_1 E_1 + \cdots + \alpha_m A_m \text{ by } \sum \alpha_r E_r \quad \text{and} \quad \beta_1 E_1 + \cdots + \beta_n E_n \text{ by } \sum \beta_s E_s,$$

one has

$$\left[\sum \alpha_r E_r \middle| \sum \beta_s E_s\right] = \sum \alpha_r \beta_s [E_r | E_s]. \qquad [42]$$

Now (according to 142) the product $[E_r|E_s]$ is zero if E_r and E_s are different units, and unity if r equals s, whence the resulting expression is

$$= \sum \alpha_r \beta_r = \alpha_1 \beta_1 + \cdots + \alpha_m \beta_m.$$

144. *The two factors of an inner product are interchangeable if they are of the same order, that is*

$$[A|B] = [B|A]$$

if A and B are of the same order.

PROOF. If $E_1, \ldots, E_m$ represent the units of the same order as A and B, and $A = \sum \alpha_r A_r$, $B = \beta_s B_s$, then (according to 143)

$$[A|B] = \sum \alpha_r \beta_r = \sum \beta_r \alpha_r = [B|A].$$

145. DEFINITION. For short we write

$$[A|A] = A^2,$$

and call it the *inner square* of A.

146. *One has*

$$(\alpha_1 E_1 + \cdots + \alpha_m E_m)^2 = \alpha_1^2 + \cdots + \alpha_m^2.$$

PROOF.

$$\begin{aligned}(\alpha_1 E_1 + \cdots + \alpha_m E_m)^2 &= [(\alpha_1 E_1 + \cdots + \alpha_m E_m)|(\alpha_1 E_1 + \cdots + \alpha_m E_m)] \qquad [145] \\ &= \alpha_1 \alpha_1 + \cdots + \alpha_m \alpha_m.\end{aligned}$$

147. The inner product of two units E and F is different from zero if and only if one is incident on the other, that is

$$[E|F] = 0$$

if E and F are not incident on one another,

$$[E|F] \neq 0$$

if E and F are incident on one another.

PROOF. The theorem was proved in 142 for units of the same order. Now suppose E and F are two units of different order, and {*first*, that F is *of higher order than* E, so that the sum of the orders of E and $|F$ is smaller than n (cf. the Proof of 139), and thus the product $[E|F]$ is *progressive.* Let

$$E = [F_1G],$$

where F_1 is subordinate to F and the domain of G has no magnitude of first order in common with F. Then E is incident on F or not according as G is of zeroth order (a number) or not. Further, let

$$F = [F_1F_2]$$

and let $[F_1F_2GH]$ equal the product of all the original units, and thus equal the absolute unit. Then (according to 89) $[GH]$ is the supplement of $[F_1F_2]$, that is of F, and thus

$$[E|F] = [F_1G \cdot GH].$$

But the product on the right side here is (according to 109) different from or equal to zero according as the order of the common domain G is equal to or greater than zero, that is, according as E and F are incident on one another or not.

Now suppose, *second,*} that E is *of higher order than* F {so that the sum of the orders of E and $|F$ is greater than n (cf. the Proof of 139) and therefore the product $[E|F]$ is *regressive*}. Let

$$F = [E_1G],$$

where E_1 is subordinate to E and the domain of G has no magnitude of first order in common with E. Then F is incident on E or not, according as G is of zeroth order (a number) or not. Further, let

$$E = [E_1E_2]$$

and let $[E_1GE_2H]$ equal the product of all n original units and thus equal the absolute unit. Then (according to 89) $[E_2H]$ is the supplement of $[E_1G]$, that is $|[E_1G] = [E_2H]$, whence

$$[E|F] = [E_1E_2|E_1G] = [E_1E_2 \cdot E_2H].$$

If G is of *zeroth order*, that is, if E is incident on F, then $[E_1E_2H]$ is of zeroth order, whence (according to 106) the expression

$$[E_1E_2 \cdot E_2H] = [E_1E_2H]E_2$$

is different from zero, since E_2 and $[E_1E_2H]$ are different from zero. If however G is of *higher than zeroth order*, then {the order of the covering domain of the magnitudes $[E_1E_2]$ and $[E_2H]$, namely} the sum of the orders of E_1, E_2, and H is less than the sum of the orders of E_1, G, E_2, H, that is, smaller than n, whence (according to 109) $[E_1E_2 \cdot E_2H] = 0$, that is, if E and F are not incident on one another, then $[E|F] = 0$.[48]

148. *If E and F are units, and $[E|F]$ is not zero, then*

$$[EF|E] = F \quad \textit{and} \quad [F|EF] = |E.$$

PROOF. If the product $[EF]$ is *progressive* and $[EFG]$ is equal to the product of all original units and thus equal to one, then $|E = [FG]$, whence

$$\begin{aligned}[EF|E] &= [EF \cdot FG] = [EFG]F && [106]\\ &= F.\end{aligned}$$

Furthermore, it then follows that $|[EF] = G$, and thus $[F|EF] = [FG] = |E$.

{If the product is *regressive* the theorem still applies if one replaces E and F with their supplements (according to 115), and thus, according to 101, it also holds for E and F themselves.}

{149. *If E, F, G are units, and neither $[EF]$ nor $[EG]$ are zero, then either*

$$[EF|EG] = [F|G] \quad \textit{or} \quad [FE|GE],$$

the former if F is of higher order than G, the latter if G is of higher order than F. If they are of the same order, then both formulas apply.

PROOF. 1. If F and G are not incident on one another, then $[EF]$ and $[EG]$ are also not incident on one another, whence (according to 147) both sides of the equation to be verified are zero.

2. If G is subordinate to F, then $F = [GH]$. Consequently

$$\begin{aligned}[EF|EG] &= [EGH|EG] = H && [148]\\ &= [GH|G] && [148]\\ &= [F|G].\end{aligned}$$

3. If F is subordinate to G, then $G = [HF]$. Consequently

$$\begin{aligned}[FE|GE] &= [FE|HFE] = |H && [148]\\ &= [F|HF] && [148]\\ &= [F|G].\end{aligned}$$

4. If F and G are of the same order, and therefore, except for the case in Part 1 of the Proof, coincide, then (according to 70) G is subordinate to F and F to G, and consequently, according to Parts 2 and 3 of the Proof, both formulas apply.}[49]

150. *If q and r are the orders of A and B and q is smaller than r, then*

$$[A|B] = (-1)^{q(r-1)}|[B|A],^{50}$$

that is, $[A|B]$ is opposite to the supplement of $[B|A]$ if the order of A is odd and at the same time that of B is even; otherwise $[A|B]$ is equal to the supplement of $[B|A]$.

PROOF. One has

$$\begin{aligned}|[B|A] &= [|B||A] && [97]\\ &= (-1)^{q(n-q)}[|B \cdot A] && [92]\\ &= (-1)^{q(n-q)}(-1)^{q(n-r)}[A|B], && [58]\end{aligned}$$

{for the product $[|B \cdot A]$ is progressive, as the sum of the orders $n - r + q = n - (r - q) < n$. The expression $|[B|A]$ is therefore}

$$= (-1)^{q(2n-q-r)}[A|B].$$

Now with respect to the modulus 2 the magnitude $q(2n - r - q)$ is congruent to $q(r - q)$, or to $q(r - 1)$, since q^2 is even or odd together with q, whence

$$|[B|A] = (-1)^{q(r-1)}[A|B],$$

or equivalently

$$[A|B] = (-1)^{q(r-1)}|[B|A].$$

REMARK. With the theorem just proved one can always reduce the case in which the second factor of an inner product is of higher order than the first to the other case in which the first factor is of higher order than the second. From now on I will mostly consider this latter case, which in the formulas developed above[51] turns out to be the simpler.

2. Concept of the normal and its correlates

151. DEFINITION. The *numerical value* of a magnitude A means the positive square root of the inner square of this magnitude. Two *numerically equal* magnitudes have the same numerical value, that is, they are magnitudes whose inner squares are equal.

REMARK. For numbers, real or imaginary, this nomenclature is already in use in the same way, since the numerical value of a positive number is itself, while that of a negative $-a$ is the corresponding positive number a, that is, in both cases it is the positive square root of its square. If one has an imaginary number $a + b\sqrt{-1}$, then its units are 1 and $\sqrt{-1}$. Denoting one of the two roots of -1 by i, and taking 1 and i as units, whence $i^2 = -1$, the numerical value of $a + bi$ is, by definition, equal to $\sqrt{a^2 + b^2}$, which is also already interpreted as the numerical value of the imaginary magnitude $a+bi$. In geometry the numerical value of a line is its length measured by the unit of length, and so forth.

152. DEFINITION. Two nonzero magnitudes are *normal to one another* if their inner product is zero. Two domains are normal to one another if their elements are.[52] Two domains are *completely* normal to one another if every magnitude of first order belonging to one of the domains is normal to every one belonging to the other; and two magnitudes are called completely normal to one another if their domains are.[53]

REMARK. The basis of this nomenclature lies in geometry. There, if one takes the original units as equally long displacements orthogonal to one another, as must always be the case,[54] then it follows easily that the inner product of two displacements is zero if and only if these displacements are orthogonal to one another. Instead of the expression "orthogonal" {senkrecht} I have chosen "normal" as the more abstract, and also as permitting applications to nonspatial relations.

153. DEFINITION. A *normal system of nth order* is an assembly of n (nonzero) numerically equal magnitudes of first order, each normal to the others; and if at the same time n is the order of the principal domain, it is called a *complete normal system.* The numerical value of those n magnitudes is also the numerical value of the normal system. Every normal system whose numerical value is one is an *elementary normal system.*

REMARK. In space, for example, three equally long and mutually orthogonal displacements form a normal system.

154. DEFINITION. A *circular evolution* is any transformation of an assembly in which two magnitudes a and b of the assembly are transformed into $xa + yb$ and $\pm(xb - ya)$ respectively, provided $x^2 + y^2 = 1$. I call the circular evolution *positive*

or *negative* according as a and b are transformed into $xa+yb$ and $(xb-ya)$ or into $xa+yb$ and $-(xb-ya)$. Herewith, if $x=\cos\alpha$ and $y=\sin\alpha$, and a and b are numerically equal and normal to one another, I say that *the assembly has evolved from a toward b by the angle α.*[55]

REMARK. If one represents a and b as two equally long and mutually orthogonal displacements, then one easily sees that by the circular evolution through which a is transformed into $a_1=a\cos\alpha+b\sin\alpha$, b into $b_1=b\cos\alpha-a\sin\alpha$, there result two displacements a_1 and b_1 of the same length as a and b, which are again mutually orthogonal. Thus by this evolution a and b remain conjugate radii of a fixed circle, thus justifying the name *circular evolution*. One also perceives that the angle from a to a_1 is equal to α.

Otherwise, if a and b are arbitrary displacements, then a_1 and b_1 are conjugate radii of a constant ellipse, in which a and b are also conjugate radii. From this consideration the name elliptical evolution would seem to recommend itself. But the ellipse is always reducible to the circle, and the circle is a simpler curve, so I have preferred the former name as the simpler. See also *Crelle's Journal* **49**, 134 {in the article "Sur les differents genres de multiplication"}.[56]

155. *Under a circular evolution every normal system transforms into a numerically equal normal system.*

PROOF. Let a, b, c, ... be the magnitudes of a normal system, that is $a^{\underline{2}}=b^{\underline{2}}=c^{\underline{2}}=\cdots$ and $0=[a|b]=[a|c]=[b|c]=\cdots$, and let a be transformed into $a_1=xa+yb$ and b into $b_1=\pm(xb-ya)$, where $x^2+y^2=1$; then it is to be shown that a_1, b_1, c, ... form a normal system in which $a_1^{\underline{2}}=a^{\underline{2}}$.

Since $[a|b]=0$ one has

$$\begin{aligned} a_1^{\underline{2}} &= (xa+yb)^{\underline{2}} = x^2a^{\underline{2}}+y^2b^{\underline{2}} \\ &= (x^2+y^2)a^{\underline{2}} && (\text{since } b^{\underline{2}}=a^{\underline{2}}) \\ &= a^{\underline{2}} && (\text{since } x^2+y^2=1). \end{aligned}$$

On the same grounds $b_1^{\underline{2}}=a^{\underline{2}}$. Furthermore,

$$\begin{aligned} [a_1|b_1] &= \pm[(xa+yb)|(xb-ya)] = \pm xy(b^{\underline{2}}-a^{\underline{2}}) && (\text{since } [a|b]=0) \\ &= 0 && (\text{since } b^{\underline{2}}=a^{\underline{2}}). \end{aligned}$$

Finally

$$[a_1|c]=[(xa+yb)|c]=x[a|c]+y[b|c]=0,$$

since $[a|c]$ and $[b|c]=0$. On the same grounds $[b_1|c]=0$, ... Consequently the system a_1, b_1, c, ... is a normal system whose numerical value is the same as that of the given system.

156. *Under a* ***positive*** *circular evolution the combinatorial product of the magnitudes of a normal system remains unchanged, and under a* ***negative*** *it is transformed into its opposite.*

PROOF. Let a be transformed into $a_1=xa+yb$, b into $b_1=xb-ya$, where $x^2+y^2=1$; then

$$\begin{aligned} [a_1b_1] &= [(xa+yb)(xb-ya)] \\ &= x^2[ab]-y^2[ba] && (\text{since } [aa] \text{ and } [bb] \text{ are zero according to 60}) \\ &= (x^2+y^2)[ab] && (\text{since } [ba]=-[ab] \text{ according to 55}) \\ &= [ab] && (\text{since } x^2+y^2=1). \end{aligned}$$

Thus $[a_1 b_1] = [ab]$. Now if the same combinatorial factors are appended in the corresponding places to the equal products $[ab]$ and $[a_1 b_1]$, the products remain equal. {The corresponding holds for negative circular evolutions.} Thus the proof.

157. *The magnitudes of a normal system stand in no numerical relation to one another, and every magnitude of first order can be numerically derived from an arbitrary complete normal system.*

PROOF. 1. Let a, b, c, ... be the magnitudes of a normal system. Suppose then that they stand in a numerical relation to one another, such that, say,

$$a = \beta b + \gamma c + \cdots ;$$

then if one inner multiplies both sides with a, one finds

$$a^2 = \beta[b|a] + \gamma[c|a] + \cdots = 0,$$

since $[b|a]$, $[c|a]$, ... are zero (according to 53 {and 152}). Thus $a^2 = 0$, contrary to 153. Thus none of the magnitudes a, b, c, ... can be numerically derived from the rest, that is (according to 2) they stand in no numerical relation to one another.

2. A complete normal system in a principal domain of nth order consists {according to 153} of n magnitudes, and since, according to Part 1 of the Proof, these stand in no numerical relation to one another, any magnitude of first order, which must always belong to the principal domain, can (according to 24) be numerically derived from them.[57]

158. *If a magnitude A is normal to several magnitudes B, C, ... , then it is also normal to every magnitude numerically derivable from them.*

PROOF. If A is normal to B, C, ... , then (according to 152)

$$0 = [A|B] = [A|C] = \cdots .$$

Consequently

$$[A|(\beta B + \gamma C + \cdots)] = \beta[A|B] + \gamma[A|C] + \cdots \qquad [41]$$
$$= 0,$$

since $[A|B]$, $[A|C]$, ... are zero.

159. *All the magnitudes of first order that are normal to m magnitudes of first order belonging to a complete normal system of nth order belong to the domain of the $n - m$ remaining magnitudes of the system.*

PROOF. Let $a_1, \ldots , a_n$ be a complete normal system, and let m of its magnitudes, say $a_1, \ldots , a_m$, be normal to some given magnitude of first order a; then it is to be shown that a belongs to the domain $a_{m+1}, \ldots , a_n$.

According to 157, a can be numerically derived from the complete normal system $a_1, \ldots , a_n$. Let the expression for this derivation be

$$a = \alpha_1 a_1 + \cdots + \alpha_n a_n .$$

Now since a is normal to $a_1, \ldots , a_m$, upon inner multiplying by a_1 one obtains

$$0 = [a_1|a] = \alpha_1 a_1^2 + \alpha_2[a_1|a_2] + \cdots + \alpha_n[a_1|a_n] = \alpha_1 a_1^2,$$

since $[a_1|a_2]$ up to $[a_1|a_n]$ are zero as inner products of the magnitudes of a normal system. Now since (according to 153) a_1^2 is not zero, it follows from the equation $\alpha_1 a_1^2 = 0$ that $\alpha_1 = 0$. In the same way it follows, upon multiplying one by one with $a_2, \ldots, a_m$, that in addition $\alpha_2, \ldots, \alpha_m$ are zero. Consequently

$$a = \alpha_{m+1}a_{m+1} + \cdots + \alpha_n a_n,$$

that is, a belongs to the domain $a_{m+1}, \ldots, a_n$.

160. *Every normal system can be so transformed by repeated circular evolutions that one of its magnitudes is identical to any other given magnitude of first order whose numerical value is equal to that of the normal system, and which belongs to its domain.*

PROOF. Let $a_1, \ldots, a_n$ be the magnitudes of the given normal system and k be the given magnitude, numerically equal to a_1, and let

$$k = \alpha_1 a_1 + \cdots + \alpha_n a_n.$$

Now circularly evolve a_1 and a_2 so that a_1 is transformed into

$$c_2 = \frac{\alpha_1 a_1 + \alpha_2 a_2}{\sqrt{\alpha_1^2 + \alpha_2^2}}$$

which is always possible (according to 154). Then

$$\alpha_1 a_1 + \alpha_2 a_2 = \sqrt{\alpha_1^2 + \alpha_2^2}\, c_2,$$

whence

$$k = \sqrt{\alpha_1^2 + \alpha_2^2}\, c_2 + \alpha_3 a_3 + \cdots + \alpha_n a_n.$$

Next one circularly evolves c_2 and a_3 so that c_2 is transformed into

$$c_3 = \frac{\sqrt{\alpha_1^2 + \alpha_2^2}\, c_2 + \alpha_3 a_3}{\sqrt{\alpha_1^2 + \alpha_2^2 + \alpha_3^2}}.$$

If one proceeds in this way until

$$k = \sqrt{\alpha_1^2 + \alpha_2^2 + \cdots + \alpha_{n-1}^2}\, c_{n-1} + \alpha_n a_n,$$

and at last circularly evolves c_{n-1} and a_n so that c_{n-1} is transformed into

$$c_n = \frac{\sqrt{\alpha_1^2 + \cdots + \alpha_{v-1}^2}\, c_{n-1} + \alpha_n a_n}{\sqrt{\alpha_1^2 + \cdots + \alpha_n^2}},$$

then

$$(*) \qquad k = \sqrt{\alpha_1^2 + \cdots + \alpha_n^2}\, c_n.$$

Now by hypothesis

$$a_1^2 = k^2 = (\alpha_1 a_1 + \cdots + \alpha_n a_n)^2 = \alpha_1^2 a_1^2 + \cdots + \alpha_n^2 a_n^2,$$

since $[a_1|a_2], \ldots$ are zero. And since in addition $a_1^2 = a_2^2 = \cdots = a_n^2$, it follows that

$$a_1^2 = (\alpha_1^2 + \cdots + \alpha_n^2)a_1^2,$$

that is, $\alpha_1^2 + \cdots + \alpha_n^2 = 1$. If this is substituted into equation $(*)$, then, upon taking the positive value of the root, one has

$$k = c_n,$$

that is, in the ultimate normal system one of the magnitudes, c_n, is equal to k, as required.

161. *If two normal systems have the same numerical value, and their domains are incident, then by repeated circular evolutions each can be derived from the other if they are both of the same order; but if they are of different order, the higher order is so transformed that it includes the magnitudes of the other.*[58]

PROOF. Let a, b, c, ... and a_1, b_1, c_1, ... be two normal systems of the same numerical value, and let their domains be incident, and let the latter be of order equal to or higher than the former; then (according to 15) all the magnitudes a, b, c, ... must belong to the domain a_1, b_1, c_1, ... Consequently one can (according to 160) circularly transform the normal system a_1, b_1, c_1, ... so that one of its magnitudes $= a$. The resulting normal system consists of the magnitudes a, b_2, c_2, ... Now since b, c, ..., as magnitudes of the normal system a, b, c, ..., are normal to a, and thus to a magnitude of the normal system a, b_2, c_2, ..., they must (according to 159) belong to the domain of the remaining magnitudes of this system, that is to the domain b_2, c_2, ... Consequently one can again circularly transform the system b_2, c_2, ... so that one of its magnitudes $= b$. The resulting normal system consists of the magnitudes b, c_3, d_3 ..., and thus, on the same grounds as before, c, d, ... must belong to the domain c_3, d_3, ... The normal system a_1, b_1, c_1, d_1, ... is thus transformed by circular evolutions into a, b, c_3, d_3, ...

Thus, if the system a_1, b_1, ... is of higher order than a, b, ..., one can proceed until the ultimately resulting normal system includes all the magnitudes of the given system a, b, c, ..., or, if both systems are of the same order, until it includes all the magnitudes of the system a, b, c, ... except the last. If this last is q, and its predecessor p, and the resulting normal system a, b, ..., p, q_n, then according to the procedure applied q must belong to the domain q_n, that is, the two of them must stand in a numerical relation to one another. Now if $q_n = xq$, where x is a number, then since they are numerically equal, $q_n^2 = q^2$, whence $x^2 = 1$, and consequently $q_n = \pm q$. If $q_n = -q$ one has only to replace the last circular evolution with its opposite,[59] and the last magnitude of the resulting normal system will coincide with q, whence one of the given normal systems is then circularly derived from the other, as required.

162. *The system of original units is a (complete) normal system, whose numerical value is one.*

PROOF. If e_1, ..., e_n are the original units, then (according to 142)

$$1 = e_1^2 = \cdots = e_n^2,$$
$$0 = [e_1|e_2] = \cdots .$$

163. *In every domain of mth order one can find a normal system of the same order and arbitrary numerical value, and in fact such that this system is part of a complete normal system.*

PROOF. Let a_1 be a magnitude of first order in the given system of mth order A, and let its numerical value be 1. Now since (according to 162) the system of original units $e_1, \ldots, e_n$ is a complete normal system whose numerical value is 1, this normal system can (according to 160) be circularly transformed so that a_1 is one of the magnitudes of the resulting normal system. Then a_1 is normal to the $n-1$ remaining magnitudes of this normal system, and thus also (according to 158) to every magnitude of their domain A_1. This domain is of $(n-1)$th order and thus (according to 26) has in common with the domain of mth order A a domain whose order is $n-1+m-n=m-1$.

In this common domain let a_2 be a magnitude of first order whose numerical value is 1. Since a_2 also belongs to the domain A_1, it is according to the above normal to a_1, but numerically equal to a_1, namely $=1$, whence a_1 and a_2 form a normal system with numerical value 1. Thus (according to 161) the complete normal system $e_1, \ldots, e_n$ is transformed into another normal system that includes a_1 and a_2. The domain A_2 of the remaining $n-2$ magnitudes of this normal system is of $(n-2)$th order, and all magnitudes of first order belonging to this domain are normal to a_1 and a_2. Now A and A_2 have a domain of $(m-2)$th order in common; in it let a_3 be an arbitrary magnitude of numerical value 1. Then one already has a normal system of three magnitudes a_1, a_2, a_3 in A, and one can continue thus.

If one has in this way obtained in A a normal system of $(m-1)$ magnitudes $a_1, \ldots, a_{m-1}$, then the complete normal system to which it belongs includes in addition $n-m+1$ magnitudes; its domain, A_{m-1}, is of $(n-m+1)$th order, and thus still has in common with the domain of mth order A yet another domain, whose order is $n-m+1+m-n=1$. Let a_m be a magnitude of this domain, of numerical value 1; then a_m, lying in A_{m-1}, is normal to $a_1, \ldots, a_{m-1}$, and thus $a_1, \ldots, a_m$ form a normal system of mth order in the domain of mth order A. To this normal system one can give any numerical value by multiplying all its magnitudes with the same arbitrary number.

3. Laws of the inner product associated with the concept of the normal

164. DEFINITION. By the *normal shadow* A' of a magnitude A on a domain B I mean the shadow of the magnitude A on the domain B, excluding the domain supplementary to B (cf. 127 and 33).

REMARK. If for example a, b, c are a complete normal system and $p = qa + rb + sc$ is an arbitrary magnitude of the principal domain, then the normal shadow of the magnitude p on the domain $[bc]$ equals $rb+sc$. For geometry it is identical to the orthogonal projection.[60]

165. *The normal shadow A' of a magnitude A on a domain B is*

$$A' = \frac{[B(A|B)]}{B^{\underline{2}}}, \quad or \ = [B(A|B)],$$

the latter, if the numerical value of B equals 1.[61]

PROOF. According to 164, A' is the shadow of A on B, excluding the supplementary domain to B, that is the domain $|B$. If we denote $|B$ by C, then (according to 129)

$$A' = \frac{[B \cdot AC]}{[BC]}, \text{ and thus } = \frac{[B(A|B)]}{[B|B]} = \frac{[B(A|B)]}{B^{\underline{2}}}.$$

166. COROLLARY. *In particular, if A and B are of the same order, then the shadow is*

$$A' = \frac{[A|B]B}{B^{\underline{2}}}, \quad or \ = [A|B]B, \ \ if\ B^{\underline{2}} = 1.$$

PROOF. For then (according to 141) $[A|B]$ is a number and therefore $[B(A|B)]$ can be replaced by $[A|B]B$.

167. *The supplement of the combinatorial product A of m magnitudes of a complete normal system of numerical value one is the equal or opposite of the combinatorial product $\{B\}$ of the $(n-m)$ other magnitudes of the system, according as $[AB] = +1$ or $= -1$, that is*

$$(*) \qquad |A = [A|B]B,$$

if the n elementary factors of $[A|B]$ are the n magnitudes of the normal system.[62]

PROOF. 1. For the system of the original units this relation is presented in 89 as a definition.

2. I now show that, if this relation (represented by equation $(*)$) holds for any given normal system a, b, c, ... , it also holds for any normal system produced from it by circular evolution.

Let a and b be transformed into $a_1 = xa + yb$ and $b_1 = xb - ya$ by a {positive} circular evolution. Under this evolution A is transformed into A_1, B into B_1; thus it is to be shown that $|A_1 = [A_1B_1]B_1$. Since A and B together include all the magnitudes a, b, ... of the normal system, and in fact each of these magnitudes only once, *a and b appear either both in A, both in B, or one in A and the other in B.*

We have already shown in 156 that the product $[a_1b_1]$ remains equal to $[ab]$ under this evolution; consequently *in the first two cases* both A and B remain unchanged, and thus the above equation, which only contains A and B, also remains as before.

In the *third case*, let a be included in A, b in B, and let A' be the magnitude produced from A if one replaces a by b, and B' be the magnitude produced from B if one replaces b by a. Then the combinatorial products $[A'B']$ and $[AB]$ are distinguished only by the interchange of the elementary factors a and b, whence (according to 55) $[A'B'] = -[AB]$. Furthermore, one has

$$A_1 = xA + yA', \ B_1 = xB - yB',$$

and consequently

$$A_1 = x|A + y|A'. \qquad [101]$$

Now since A and A' only include magnitudes of the normal system a, b, c, ... as elementary factors, and B and B' the rest of them in each case {that is, B includes those not in A, B' those not in A'}, equation $(*)$ above holds (by assumption) for them, that is,

$$|A = [AB]B, \ |A' = [A'B']B' = -[AB]B',$$

the latter since $[A'B'] = -[AB]$; consequently

$$|A_1 = x[AB]B - y[AB]B' = [AB](xB - yB') = [AB]B_1.$$

Finally (according to 156), $[A_1B_1] = [AB]$, since the elementary factors of $[A_1B_1]$ result from those of $[AB]$ by a positive circular evolution. Thus

$$|A_1 = [A_1B_1]B_1,$$

that is, if equation (∗) holds for any given normal system, it holds as well for any produced from it by a *positive* circular evolution, and likewise for one produced by a *negative* evolution. For the positive circular evolution, which we assumed above, would (according to 154) be changed to a negative if one changed the sign of b_1, which would change the sign of B_1 as well, leaving the equation unchanged.

Thus in general equation (∗) continues to hold under any circular evolution of the normal system, if it holds for any given system. But according to Part 1 of the Proof, it holds for the normal system of the original units, and thus also for any circularly derived from it. Now (according to 161) any normal system whose numerical value is 1 can be derived from that one, and therefore the equation holds for any normal system whose numerical value is 1.

168. *All theorems previously established still hold if one replaces the system of the original units with an arbitrary complete normal system whose numerical value is one.*

PROOF. All the laws of calculation developed in the first three chapters continue to hold (according to 110) if one replaces the n original units with n arbitrary magnitudes standing in no numerical relation to one another, and thus also if one uses a complete normal system. Furthermore (according to 167) the concept of the supplement, as it is established in 89 with respect to the system of the original units, continues to apply with respect to every normal system whose numerical value is one. But all the theorems on the inner product developed so far depend on this concept of the supplement and the laws of calculation developed in the first three chapters. Consequently these theorems also apply if one replaces the system of the original units with a normal system whose numerical value is one.

REMARK. By virtue of the theorem just proved the concept of the inner product is no longer associated with the system of the original units, insofar as one can replace this system with an arbitrary {complete} normal system, whose numerical value is one, without any of the previously established theorems suffering a change. Thus the concept of the inner product only appears to be associated with the concept of the normal system, and this therefore replaces in prominence the system of the original units in the following developments.

169. *The inner product of two {elementary} magnitudes does not change its value if one replaces one of the factors by its {progressive} normal shadow on the domain of the other, that is*

$$[A|B] = [A|B']$$

and

$$[B|A] = [B'|A],$$

if B' is the progressive shadow of B on the domain A (and thus A is of order equal to or higher than B).[63]

{PROOF.} Let A be of mth order, B be of pth, and the principal domain of nth; then (according to 163) one can find a complete normal system $a_1, \ldots, a_n$ such that m of its magnitudes, say $a_1, \ldots, a_m$, lie in A, and its numerical value is 1. The p factors of B are then (according to 157) numerically derivable from the a_1,

$\dots, a_n$, whence B is numerically derivable from the multiplicative combinations of $a_1, \dots, a_n$ of pth class. Let these combinations be $B_1, B_2, \dots, B_q, B_{q+1}, \dots, B_r$, where $B_1, \dots, B_q$ are the combinations of $a_1, \dots, a_m$, and let

$$B = \beta_1 B_1 + \dots + \beta_q B_q + \beta_{q+1} B_{q+1} + \dots + \beta_r B_r;$$

then (according to 147 and 168) $[A|B_{q+1}], \dots, [A|B_r]$ all equal zero, since each of the magnitudes B_{q+1} to B_r includes such factors as do not appear in A, and these magnitudes are therefore not incident on A, whence

$$\begin{aligned}[A|B] &= \beta_1[A|B_1] + \dots + \beta_q[A|B_q] \\ &= [A|(\beta_1 B_1 + \dots + \beta_q B_q)].\end{aligned}$$

But (according to 127) $\beta_1 B_1 + \dots + \beta_q B_q$ is the shadow of B on the domain $[a_1 \cdots a_m]$, excluding the domain $[a_{m+1} \cdots a_n]$, but the latter domain is (according to 167) the supplement of the former; thus {according to 164} $\beta_1 B_1 + \dots + \beta_q B_q$ is the normal shadow of B on the domain $[a_1 \cdots a_m]$, that is, on the domain of A. Thus $\beta_1 B_1 + \dots + \beta_q B_q$ equals B', and consequently

$$[A|B] = [A|B'].$$

On the same grounds $[B|A] = [B'|A]$.

170. *If, in an inner product of two {elementary} magnitudes* ***of the same order****, one normally shadows one magnitude on the domain of the other, and measures this shadow and the magnitude on whose domain the shadow is taken by the same standard, whose numerical value is one, then the product of the two measure quotients equals the given inner product, that is*

$$[A|B] = \alpha\beta',$$

if $A = \alpha E$, and the normal shadow B' of B on A equals $\beta' E$, and the numerical value of E is one.

PROOF. According to 169,

$$[A|B] = [A|B'].$$

Let E be an element of the domain of A whose numerical value is one, and let $A = \alpha E$, $B' = \beta' E$; then $[A|B'] = \alpha\beta'[E|E] = \alpha\beta' E^{\underline{2}} = \alpha\beta'$, since $E^{\underline{2}} = 1$.

171. *If the domains of {two elementary magnitudes} A and B are perfectly normal to one another, and C is an arbitrary magnitude of order lower than or equal to B, then*

$$[AB|AC] = A^{\underline{2}}[B|C]$$

and

$$[CA|BA] = A^{\underline{2}}[C|B].$$

PROOF. Let a normal system be given whose magnitudes are divided between the domains A and B, and whose numerical value is one, and let it be extended to a complete normal system;[64] then C is numerically derivable from the multiplicative combinations of the magnitudes of this normal system (69, 77). Let $C = \gamma_1 C_1 + \gamma_2 C_2 + \cdots$, and further $A = \alpha A_1$, $B = \beta B_1$, where A_1, B_1 are combinatorial products of magnitudes of the normal system; then

$$\begin{aligned}[AB|AC] &= \alpha^2\beta[A_1 B_1|A_1(\gamma_1 C_1 + \gamma_2 C_2 + \cdots)] \\ &= \alpha^2\beta\gamma_1[A_1 B_1|A_1 C_1] + \alpha^2\beta\gamma_1[A_1 B_1|A_1 C_2] + \cdots.\end{aligned}$$

{But since A_1, B_1, C_r are units of higher order, that is combinatorial products of the original units, $[A_1B_1|A_1C_r]$ is (according to 149) equal to $[B_1|C_r]$. Now (according to 168) this still obtains if those magnitudes are combinatorial products of the magnitudes of an elementary normal system, and thus in our case.}[65] Consequently

$$\begin{aligned}[AB|AC] &= \alpha^2\beta\gamma_1[B_1|C_1] + \alpha^2\beta\gamma_2[B_1|C_2] + \cdots \\ &= \alpha^2\beta[B_1|(\gamma_1C_1 + \gamma_2C_2 + \cdots)] \\ &= \alpha^2\beta[B_1|C].\end{aligned}$$

Now as {according to 142 and 168} A_1^2 equals 1, since A_1 is a combinatorial product of magnitudes of an elementary normal system, the resulting expression

$$\begin{aligned}&= \alpha^2\beta A_1^2[B_1|C] = (\alpha A_1)^2[\beta B_1|C] \\ &= A^2[B|C].\end{aligned}$$

The second formula of the theorem follows in the same way.

172. *If {the elementary magnitude} A is of the same order as* A, *{the elementary magnitude} B of order greater than or equal to that of* B, *and $[AB]$ does not vanish, {and finally, if $[AB]$ and* [AB] *are progressive products} then*

(*a*) $$[AB|\mathrm{AB}] = [A|\mathrm{A}][B|\mathrm{B}] + [A_1|\mathrm{A}][B_1|\mathrm{B}] + \cdots$$

and

(*b*) $$[\mathrm{BA}|BA] = [\mathrm{B}|B][\mathrm{A}|A] + [\mathrm{B}|B_1][\mathrm{A}|A_1] + \cdots,$$

where A, A_1, ... are {all} the multiplicative combinations of the elementary factors {of first order} of $[AB]$, and B, B_1, ... are the combinations supplementary to A, A_1, ... (so that $[AB] = [A_1B_1] = \cdots$).

REMARK. Thus if A is one of the multiplicative combinations of $a_1, a_2, \ldots, a_m$, then I call the multiplicative combination B that includes all the elements {of the series a_1, a_2, ..., a_m} not included in A, and is furnished with such a sign ($\pm$) that $[AB] = [a_1a_2\cdots a_m]$, the *supplementary combination* to A.

PROOF. 1. {Let the elementary factors of $[AB]$ all be normal to one another. Since A is of the same order as A, it is numerically derivable from the multiplicative combinations A, A_1, ... Let $\mathrm{A} = \alpha A + \alpha_1 A_1 + \cdots$; then

$$\begin{aligned}[AB|\mathrm{AB}] &= [AB|(\alpha A + \alpha_1A_1 + \cdots)\mathrm{B}] \\ &= \alpha[AB|A\mathrm{B}] + \alpha_1[AB|A_1\mathrm{B}] + \cdots.\end{aligned}$$

Now since (by assumption) $[AB] = [A_1B_1] = \cdots$, we obtain the expression

$$= \alpha[AB|A\mathrm{B}] + \alpha_1[A_1B_1|A_1\mathrm{B}] + \cdots.$$

Because the elementary factors of $[AB]$ are all normal to one another, and identical to those of $\pm[A_1B_1]$, ... (by assumption), A is completely normal to B, and likewise A_1 to B_1, ... Consequently (according to 171) the previous expression

$$= \alpha A^2[B|\mathrm{B}] + \alpha_1A_1^2[B_1|\mathrm{B}] + \cdots.$$

But now

$$[A_r|\mathrm{A}] = [A_r|(\alpha A + \alpha_1A_1 + \cdots)] = \alpha_rA_r^2,$$

since (according to 147, 168) A_r inner multiplied by the magnitudes A, A_1, ... gives zero, except for A_r. Thus in the equation previously established one can replace $\alpha_r A_r^2$ by $[A_r|\mathrm{A}]$, and that expression becomes

$$= [A|\mathrm{A}][B|\mathrm{B}] + [A_1|\mathrm{A}][B_1|\mathrm{B}] + \cdots ,$$

that is, formula (a) holds under our assumption.}[66]

2. Now I show that, if formula (a) holds for any series of elementary factors that $[AB]$ consists of, it continues to hold if one linearly evolves this series of factors (see 71), that is, if one replaces any given factor a by $a + \beta b$, where b is one of the other factors and β is a number.

If this is done, then (according to 72) the product $[AB]$, and thus the left side of our formula, maintains the same value. If one now considers any given term of the right side, for example $[A_r|\mathrm{A}][B_r|\mathrm{B}]$, then a and b can either both appear in A_r, or both in B_r, or one in A_r, the other in B_r. In the first two cases the values of both A_r and B_r remain unchanged, and therefore also that of the term under consideration. In the last case there appears {in the above development} another term $[A_s|\mathrm{A}][B_s|\mathrm{B}]$ such that A_r and A_s include all the same factors except that, where one of these products contains the factor a, the other contains the factor b. And then B_r and B_s, as respectively including the factors missing from A_r and A_s, stand in the same reciprocal relation to one another. Thus a appears in one of the magnitudes A_r and A_s; let it appear in A_r.

Now let A' be the magnitude that results from A_r when one replaces a by b, and B' be the magnitude that results from B_r when one replaces b by a. Then A' includes the same factors as A_s and B' the same as B_s; consequently (according to 57) the magnitudes A' and B' are either equal or opposite to the magnitudes A_s and B_s. Since $[A'B']$ results from $[A_r B_r]$ by interchange of the two elementary factors a and b, it follows (according to 55) that $[A'B'] = -[A_r B_r]$, and this $= -[A_s B_s]$ (by assumption). Thus if $A' = \mp A_s$, then $B' = \pm B_s$. If one now introduces the linear substitution of $a + \beta b$ for a, then

$$[A_r|\mathrm{A}][B_r|\mathrm{B}] + [A_s|\mathrm{A}][B_s|\mathrm{B}] = [A_r|\mathrm{A}][B_r|\mathrm{B}] - [A'|\mathrm{A}][B'|\mathrm{B}]$$

is transformed into

$$[(A_r + \beta A')|\mathrm{A}][B_r|\mathrm{B}] - [A'|\mathrm{A}][(B' + \beta B_r)|\mathrm{B}],$$

since B_r and A' include no a and therefore remain unchanged, while A_r and B' are transformed into $A_r + \beta A'$ and $B' + \beta B_r$. Consequently that sum is transformed into

$$[A_r|\mathrm{A}][B_r|\mathrm{B}] - [A'|\mathrm{A}][B'|\mathrm{B}] + \beta[A'|\alpha][B_r|\mathrm{B}] - \beta[A'|\mathrm{A}][B_r|\mathrm{B}],$$

that is, since the last two terms cancel out, the value of the sum remains unchanged. Consequently the whole right side of our formula remains unchanged under a linear evolution, since either the terms remain unchanged individually, or, if they do change, they are grouped in pairs of terms whose sum remains unchanged. Since both sides of the formula remain unchanged under a linear evolution, the formula continues to hold under a linear evolution if it holds for any given series of factors.

3. Finally, let the series of factors a, b, ... be completely arbitrary, but such that their combinatorial product $[AB]$ is not zero; then (according to 163) a series of mutually normal magnitudes of first order a_1, a_2, ... can always be found for which $[ab\cdots] = [a_1 a_2 \cdots]$. But then (according to 76) the series of magnitudes a, b, ... can be derived from a_1, a_2, ... by linear evolutions. Now according to Part 1

of the Proof our formula (a) holds for the series of mutually normal factors a_1, a_2, ... , and thus also, according to Part 2 of the Proof, for the series of factors a, b, ... resulting from it by progressive linear evolutions, and thus generally.

{Formula (b) of theorem 172 is demonstrated in the same way.}

173. *If A and A are of the same order, and likewise B and B, ... , and finally L and Λ, but M is of order greater than or equal to M, and $[AB\cdots LM]$ is a nonvanishing combinatorial product of magnitudes of first order, then*

$$[AB\cdots LM|\mathrm{AB}\cdots\Lambda\mathrm{M}] = \sum[A_a|\mathrm{A}][B_b|\mathrm{B}]\cdots[L_l|\Lambda][M_m|\mathrm{M}],$$

where $[A_aB_b\cdots L_lM_m]$ includes the same elementary factors as $[AB\cdots LM]$, but in another order, but such that the two products are equal, where further A_a includes as many factors as A, B_b as B, ... , and finally where the sum is taken over all possible different expressions of this type, with $A_aB_b\cdots L_lM_m$ and $A_{a'}B_{b'}\cdots L_{l'}M_{m'}$ being regarded as different expressions if at least one of the pairs of magnitudes A_a and $A_{a'}$, B_b and $B_{b'}$, ... consists of two magnitudes that stand in no numerical relation to one another.

PROOF. 1. The theorem is proven in 172 for two factors; thus we can also write the formula in 172 in the following way:

$$(*) \qquad [AB|\mathrm{AB}] = \sum[A_a|\mathrm{A}][B_b|\mathrm{B}],$$

where the A_a, B_b have the meaning given in the theorem, which coincides with the meaning of the pairs of magnitudes A, B, A_1, B_1, ... in 172.

2. By repeated application of the theorem for two factors one is led to the theorem for arbitrarily many factors. Thus one can first regard the product $[AB\cdots LM]$ as consisting of the two factors A and $[BC\cdots LM]$. Then

$$\begin{aligned}[AB\cdots LM|\mathrm{AB}\cdots\Lambda\mathrm{M}] &= [A(BC\cdots LM)|\mathrm{A}(\mathrm{B}\Gamma\cdots\Lambda\mathrm{M})]\\ &= \sum[A_a|\mathrm{A}][(BC\cdots LM)_b|\mathrm{B}\Gamma\cdots\Lambda\mathrm{M}],\end{aligned}$$

where the index following the parentheses indicates that the expression in parentheses is to be regarded as a single magnitude in formula ($*$). On the same grounds the formula just obtained is again

$$= \sum[A_a|\mathrm{A}][B_b|\mathrm{B}][(CD\cdots LM)_c|\Gamma\Delta\cdots\Lambda\mathrm{M}],$$

and if one continues thus, one obtains at last

$$= \sum[A_a|\mathrm{A}][B_b|\mathrm{B}]\cdots[L_l|\Lambda][M_m|\mathrm{M}].$$

REMARK. One can combinatorially develop all the members of the series of magnitudes A_a, B_b, ... , M_m in the following way: If one regards the elementary factors of the product $[AB\cdots]$ as combinatorial elements and develops from them the multiplicative combinations of such a class as amounts to the order of A, one obtains the magnitudes A_a; if for each of these one develops the multiplicative combinations of those elements not appearing in it, of such a class as amounts to the order of B, then for each A_a one obtains all the magnitudes B_b belonging to it, and so forth; finally the last of these multiplicative combinations, which belong to the magnitude M, one sets equal to $\pm M_m$, where the sign is chosen so that

$$[A_aB_b\cdots M_m] = [AB\cdots M].$$

For example, if $A = [ab]$, $B = [cd]$, $C = M = e$, one obtains the following collection of three magnitudes, the first of which in each case represents a magnitude A_a, the second a magnitude B_b belonging to it, the third a magnitude C_c belonging to both of them:

$ab,\ cd,\ e$	$ad,\ bc,\ e$	$bc,\ ad,\ e$	$be,\ ac,\ d$	$ce,\ ab,\ -d$
$ab,\ ce,\ -d$	$ad,\ be,\ -c$	$bc,\ ae,\ -d$	$bc,\ ad,\ -c$	$ce,\ ad,\ b$
$ab,\ de,\ c$	$ad,\ ce,\ b$	$bc,\ de,\ a$	$be,\ cd,\ a$	$ce,\ bd,\ -a$
$ac,\ bd,\ -e$	$ae,\ bc,\ -d$	$bd,\ ac,\ -e$	$cd,\ ab,\ e$	$de,\ ab,\ c$
$ac,\ be,\ d$	$ae,\ bd,\ c$	$bd,\ ae,\ c$	$cd,\ ae,\ -b$	$de,\ ac,\ -b$
$ac,\ de,\ -b$	$ae,\ cd,\ -b$	$bd,\ ce,\ -a$	$cd,\ be,\ a$	$de,\ bc,\ a.$

174. COROLLARY. *If in the inner product*

$$[AB\cdots|\mathrm{AB}\cdots]$$

the magnitudes A *and* A *are of the same order, and likewise* B *and* B, *and so forth, {and if further* A, B, … *are elementary magnitudes and* $[AB\cdots]$ *is an outer product,} then*

$$[AB\cdots|\mathrm{AB}\cdots] = \frac{[A'B'\cdots]}{[AB\cdots]},$$

where

$$A' = \sum [A_r|\mathrm{A}]A_r,\ \ B' = \sum [B_r|\mathrm{B}]B_r,\ \ldots$$

and where the A_r *are the multiplicative combinations of the elementary factors of the outer product* $[AB\cdots]$ *of such a class as amounts to the order of* A, *and correspondingly the* B_r, *and so forth.*

PROOF. According to 173,

$$[AB\cdots|\mathrm{AB}\cdots] = \sum [A_a|\mathrm{A}][B_b|\mathrm{B}]\cdots,$$

where

$$[A_aB_b\cdots] = [AB\cdots],$$

with the further conditions given in 173.

Now since A is of the same order as A, and thus also A_a is of the same order as A, (according to 141) $[A_a|\mathrm{A}]$ is a number, and on the same grounds so is $[B_b|\mathrm{B}]$, …. Consequently, instead of $\sum[A_a|\mathrm{A}][B_b|\mathrm{B}]\cdots$ we can write

$$= \sum \frac{[A_a|\mathrm{A}][B_b|\mathrm{B}]\cdots[A_aB_b\cdots]}{[A_aB_b\cdots]}.$$

Thus, since $[A_aB_b\cdots]$ is equal to $[AB\cdots]$,

$$= \sum [A_a|\mathrm{A}][B_b|\mathrm{B}]\cdots[A_aB_b\cdots] : [AB\cdots].$$

Or, since (according to 46) the numerical factors can be annexed to any factors of a product,

$$= \sum ([A_a|\mathrm{A}]A_a \cdot [B_b|\mathrm{B}]B_b\cdots) : [AB\cdots].$$

Here (according to 173) each product $[A_aB_b\cdots]$ includes the same factors of first order as $[AB\cdots]$, whence in each of them A_a includes factors other than those in B_b and so forth. But since those products, in which {any two of the magnitudes} A_a, B_b, … include the same factors of first order, are zero, we can add those products to the above expression, and thus (according to 45) obtain for it

$$= \left[\sum [A_r|\mathrm{A}]A_r \cdot \sum [B_r|\mathrm{B}]B_r \cdots\right] : [AB\cdots],$$

that is

$$= \frac{[A'B'\cdots]}{[AB\cdots]}.$$

175. *The inner product of two magnitudes of mth order A and B, each of which consists of m elementary factors, equals the determinant of m series, each of m terms, that one obtains when one conjoins each elementary factor of A, in order, with each of B in an inner product, that is*

$$[abc\cdots|a'b'c'\cdots] = Determ. \begin{cases} [a|a'] & [a|b'] & [a|c'] & \ldots \\ [b|a'] & [b|b'] & [b|c'] & \ldots \\ [c|a'] & [c|b'] & [c|c'] & \ldots \\ \ldots & \ldots & \ldots & \ldots \end{cases}$$

$$= \sum \pm(\alpha\beta_1\gamma_2\cdots),$$

where

$$\begin{array}{llll} \alpha = [a|a'], & \alpha_1 = [a|b'], & \alpha_2 = [a|c'], & \ldots \\ \beta = [b|a'], & \beta_1 = [b|b'], & \beta_2 = [b|c'], & \ldots \\ \gamma = [c|a'], & \gamma_1 = [c|b'], & \gamma_2 = [c|c'], & \ldots \\ \ldots & \ldots & \ldots & \ldots \end{array}$$

PROOF. According to 174,

$$[abc\cdots|a'b'c'\cdots] = \frac{[a_1b_1c_1\cdots]}{[abc\cdots]},$$

where

$$\begin{aligned} a_1 &= [a|a']a + [b|a']b + [c|a']c + \cdots = \alpha a + \beta b + \gamma c + \cdots, \\ b_1 &= [a|b']a + [b|b']b + [c|b']c + \cdots = \alpha_1 a + \beta_1 b + \gamma_1 c + \cdots, \\ c_1 &= [a|c']a + [b|c']b + [c|c']c + \cdots = \alpha_2 a + \beta_2 b + \gamma_2 c + \cdots, \\ &\ldots \quad \ldots \quad \ldots \quad \ldots \quad \ldots \quad \ldots \quad \ldots \end{aligned}$$

But (according to 63),

$$[(\alpha a + \beta b + \gamma c + \cdots)(\alpha_1 a + \beta_1 b + \gamma_1 c + \cdots)(\alpha_2 a + \beta_2 b + \gamma_2 c + \cdots)\cdots]$$
$$= \sum \pm(\alpha\beta_1\gamma_2\cdots) \cdot [abc\cdots].$$

Consequently

$$[abc\cdots|a'b'c'\cdots] = \frac{[a_1b_1c_1\cdots]}{[abc\cdots]} = \frac{\sum \pm(\alpha\beta_1\gamma_2)\cdot[abc\cdots]}{[abc\cdots]}$$
$$= \sum \pm(\alpha\beta_1\gamma_2\cdots).$$

176–179. COROLLARIES. *In particular,*

176. $[ab|a'b'] = [a|a'][b|b'] - [a|b'][b|a']$,

177. $[ab]^{\underline{2}} = a^{\underline{2}}b^{\underline{2}} - [a|b]^2$,

178. $[abc]^{\underline{2}} = a^{\underline{2}}b^{\underline{2}}c^{\underline{2}} - a^{\underline{2}}[b|c]^2 - b^{\underline{2}}[c|a]^2 - c^{\underline{2}}[a|b]^2 + 2[a|b][b|c][c|a]$.

179. $[abcd] = \text{Determ.} \begin{cases} a^{\underline{2}} & [a|b] & [a|c] & [a|d] \\ [b|a] & b^{\underline{2}} & [b|c] & [b|d] \\ [c|a] & [c|b] & c^{\underline{2}} & [c|d] \\ [d|a] & [d|b] & [d|c] & d^{\underline{2}} \end{cases}$

180. $[ab|c] = [a|c]b - [b|c]a$,

181. $[abc|d] = [a|d][bc] + [b|d][ca] + [c|d][ab]$,

182. $[abcd|e] = [a|e][bcd] + [b|e][cad] + [c|e][abd] + [d|e][cba]$.

For in 180 to 182 one can regard the second factor of the inner product (c, d, or e) as a product whose second factor is 1 (thus $c \cdot 1$, $d \cdot 1$, or $e \cdot 1$), and can then apply No. 173; whereby one has to observe that, according to the laws of combinatory multiplication,

$$[ab] = -[ba], \quad [a \cdot bc] = [b \cdot ca] = [c \cdot ab]$$

and

$$[a \cdot bcd] = [b \cdot cad] = [c \cdot abd] = [d \cdot cba].$$

183. *If, from a series of* (n) *magnitudes of first order, one forms the multiplicative combinations of any given class, and conjoins each of them in an inner product with its supplementary combination, then the sum of these products is zero, that is*

$$[A|B] + [A_1|B_1] + \cdots = 0,$$

if A, A_1, ... are the multiplicative combinations of the n magnitudes of first order a_1, a_2, ... , a_n of any given {say the mth} class, and B, B_1, ... are the supplementary combinations.

PROOF. 1. First let it be assumed that $m \geq n - m$. Now since A is one of the multiplicative combinations of $a_1, \ldots, a_n$, it has the form

$$A = [a_r a_s \cdots a_z],$$

where $r, s, \ldots, z$ are any m different numbers from among $1, \ldots, n$. Since furthermore B is the supplementary combination to A, it must include as factors those $n - m$ of the magnitudes $a_1, a_2, \ldots, a_n$ that do not appear among the magnitudes $a_r, a_s, \ldots, a_z$. Let these be $a_{r'}, a_{s'}, \ldots$, so that

$$B = (-1)^p [a_{r'} a_{s'} \cdots a_{n'}].$$

Further, the sign indicated by $(-1)^p$ must (according to the Remark following 172) be chosen so that $[AB] = [a_1 a_2 \cdots a_n]$, that is so that

$$(*) \qquad (-1)^p [a_r a_s \cdots a_u a_v \cdots a_z a_{r'} a_{s'} \cdots a_{u'}] = [a_1 a_2 \cdots a_n].$$

{Thus

$$[A|B] = (-1)^p [a_r a_s \cdots a_u a_v \cdots a_z | a_{r'} a_{s'} \cdots a_{u'}].\}$$

All the rest of the products $[A_1|B_1]$, ... are of the same form. For the combinations A, B, A_1, B_1, ... to be well ordered, one must append the conditions that $r < s < \cdots < u < v < \cdots < z$ and $r' < s' < \cdots < u'$. If we append these conditions, then

$$[A|B] + [A_1|B_1] + \cdots = \sum (-1)^p [a_r a_s \cdots a_u a_v \cdots a_z | a_{r'} a_{s'} \cdots a_{u'}].$$

If we interpret $a_v \cdots a_z$ as one factor and annex to the last place of the second factor of the inner product the factor 1, the conditions of No. 173 will be satisfied, whence the above expression will be

$$(**) \qquad [A|B] + [A_1|B_1] + \cdots = \sum (-1)^p [a_r|a_{r'}][a_s|a_{s'}] \cdots [a_u|a_{u'}][a_v a_w \cdots a_z],$$

whereby equation $(*)$ is still maintained, and the conditions $r' < s' < \cdots < u'$ and $v < w < \cdots < z$ continue to hold, whereas the condition that $r < s < \cdots < u$ is lost, and the sum refers to all terms possible under those conditions.

I now show that in this sum all the terms are opposite to one another pairwise, and therefore that they cancel.

Consider any one of these terms, say

$$(-1)^p [a_r|a_{r'}][a_s|a_{s'}] \cdots [a_u|a_{u'}][a_v a_w \cdots a_z],$$

where the indices have fixed (different) values satisfying the above conditions, and where according to the above p has a value such that equation $(*)$ is satisfied. Since the indices r, r', s, s', ... , u, u' are all different, one among them must be the smallest, and among the products $[a_r|a_{r'}]$, $[a_s|a_{s'}]$, ... , $[a_u|a_{u'}]$ one of them must include this smallest index; let this be the product $[a_r|a_{r'}]$ for example. This being assumed, if one interchanges r and r' and changes the sign, one obtains the expression

$$(***) \qquad (-1)^{p+1} [a_{r'}|a_r][a_s|a_{s'}] \cdots [a_u|a_{u'}][a_u a_v \cdots a_z],$$

which I will show likewise appears as a term in the above sum $(**)$. Should the index r be greater than s', then one is to give the factor $[a_{r'}|a_r]$ such a position among the remaining factors $[a_s|a_{s'}] \cdots [a_u|a_{u'}]$ as will satisfy the condition that the second index in each of these factors will be smaller than the second index in the next sequential factor. I will assume that this condition is satisfied if one moves the factor by m positions to the right, which is permitted since all these factors are numbers.

It must still be shown that the condition expressed by equation $(*)$ holds for the term resulting in this way, that is, that it continues to apply if in it one replaces p by $p+1$, interchanges a_r with $a_{r'}$ on the left side, and moves these two factors by q positions to the right.

If one denotes by P the product that results in this way from

$$(-1)^p [a_r a_s \cdots a_u a_v \cdots a_z a_{r'} a_{s'} \cdots a_{u'}],$$

then

$$P = (-1)^p [a_{r'} a_s \cdots a_u a_v \cdots a_z a_r a_{s'} \cdots a_{u'}].$$

For one can in this product P simultaneously move the factors a_r and $a_{r'}$ back by q places, without thereby changing the value of the product (according to 58).

Furthermore, (according to 55) upon interchanging a_r and $a_{r'}$, the last expression

$$= -(-1)^{p+1}[a_r a_s \cdots a_u a_v \cdots a_z a_{r'} a_{s'} \cdots a_{u'}]$$
$$= (-1)^p [a_r a_s \cdots a_u a_v \cdots a_z a_{r'} a_{s'} \cdots a_{u'}]$$
$$= [a_1 a_2 \cdots a_n] \qquad \text{[according to } *].$$

Thus $P = [a_1 a_2 \cdots a_n]$. Consequently the expression $(***)$ is subject to all the conditions that the terms of the sum $(**)$ are subject to, and is therefore a term of that sum, since the sum includes all terms that satisfy those conditions. Now this term cancels with the term first considered; for

$$(-1)^p [a_r|a_{r'}][a_s|a_{s'}] \cdots [a_u|a_{u'}][a_v a_w \cdots a_z] + \\ + (-1)^{p+1}[a_{r'}|a_r][a_s|a_{s'}] \cdots [a_u|a_{u'}][a_v a_w \cdots a_z]$$

is zero, since $[a_{r'}|a_r] = [a_r|a_{r'}]$ (according to 144) and $(-1)^{p+1} = -(-1)^p$. But in the same way as the latter of these two terms results from the former, the former results from the latter. And in the same way one finds for each term of that sum another paired with it, with which it cancels; consequently the sum is zero, whence also

$$[A|B] + [A_1|B_1] + \cdots = 0.$$

2. If $m < n - m$, then (according to 150) upon setting $m(n - m - 1) = c$,

$$[A|B] + [A_1|B_1] + \cdots = (-1)^c|[B|A] + (-1)^c|[B_1|A_1] + \cdots$$
$$= (-1)^c|([B|A] + [B_1|A_1] + \cdots). \qquad [100]$$

Here (according to Part 1 of the Proof) the sum enclosed in parentheses is zero, whence

$$[A|B] + [A_1|B_1] + \cdots = (-1)^p|(0) = 0. \qquad [89]$$

184. Corollary. *If from a series of $4m$ magnitudes of first order $a_1, \ldots, a_{4m}$ one forms all the multiplicative combinations A, B, C, … of $2m$th class that include one of the magnitudes, say a_1, and conjoins each of them in an inner product with the supplementary combination, then the sum of these products is zero, that is*

$$[A|A'] + [B|B'] + \cdots = 0,$$

where A, B, … are the multiplicative combinations of $a_1, \ldots, a_{4m}$ of $2m$th class that include a_1, and A', B', … are the supplementary combinations.

Proof. Since A, B, … are all the multiplicative combinations of $4m$ elements of $2m$th class that include a_1, their supplementary combinations A', B', … are all the combinations of the same elements of the same class that do *not* contain a_1. Further, since the orders of $A, B, \ldots, A', B', \ldots$ are even, (according to 58) $[AA'] = [A'A]$, and thus (according to the Remark following 172) if A' is the supplementary combination to A, A is also the supplementary combination to A', and likewise B to B', whence (according to 183)

$$[A|A'] + [B|B'] + \cdots + [A'|A] + [B'|B] + \cdots = 0.$$

But (according to 144)

$$[A|A'] = [A'|A], \ [B|B'] = [B'|B], \ \ldots .$$

Consequently

$$2[A|A'] + 2[B|B'] + \cdots = 0,$$

that is

$$[A|A'] + [B|B'] + \cdots = 0.$$

185–187. COROLLARIES. *In particular,*

185. $[ab|cd] + [ac|db] + [ad|bc] = 0.$

186. $[ab|c] + [bc|a] + [ca|b] = 0.$

187. $[abc|d] - [bcd|a] + [cda|b] - [dab|c] = 0.$

4. Special theorems on the inner multiplication of two magnitudes of first order

188. *The defining equations for the inner multiplication of two magnitudes of first order are*

(*a*) $$[e_r|e_s] = 0$$

if $r \neq s$,

(*b*) $$[e_r|e_r] = [e_s|e_s] = \cdots,$$

and these continue to apply if one replaces the units $e_1, e_2, \ldots, e_n$ *with an arbitrary complete normal system.*

PROOF. The validity of the two groups of equations is proven for the units in 142. Thus (according to 168) they are also valid for every complete elementary normal system. But they also hold for any *arbitrary* complete normal system. For if a, b are two of its magnitudes and λ the numerical value of the normal system, so that one can set $a = \lambda a'$, $b = \lambda b'$, where a' and b' have the numerical value 1, then $[a'|b'] = 0$, and thus also $[\lambda a'|\lambda b'] = 0$, that is $[a|b] = 0$ and $[a'|a'] = 1$, whence $[a|a] = [\lambda a'|\lambda a'] = \lambda^2[a'|a'] = \lambda^2$. Likewise $[b|b] = \lambda^2$, and thus $[a|a] = [b|b]$.

It is still to be shown that the above two groups include all of the defining equations, that is (according to 84), there is no other numerical relation between the products $[e_r|e_s]$ than those derivable from the two groups.

By virtue of the two groups all products $[e_r|e_s]$ for which r equals s can be set equal to $[e_1|e_1]$, whereas they vanish whenever r is different from s. Thus if one had a defining equation

$$\sum \alpha_{r,s}[e_r|e_s] = 0,$$

it would be transformed into

$$\sum \alpha_{r,r}[e_1|e_1] = 0,$$

whence, since $[e_1|e_1]$ equals 1, into

$$\sum \alpha_{r,r} = 0.$$

But if this is the case, then the equation

$$\sum \alpha_{r,s}[e_r|e_s] = 0$$

already results from the above groups, whence those two groups include the complete system of defining equations.

REMARK. For the inner multiplication of two arbitrary magnitudes of orders p and q $(q \geq p)$ the system of defining equations is included in the two groups of equations

(a) $$[E|F] = 0$$

if E is not incident on F, and

(b) $$[E|EG] = [E'|E'G],$$

where E, F, G, E' are combinatorial products of the original units, E, E' are of pth order, F, $[EG]$, and $[E'G]$ of qth order and the last two are not zero.[67]

189. $[a|b] = [b|a]$ (special case of No. 144).

190. *One has*

$$[(\alpha_1 a_1 + \alpha_2 a_2 + \cdots)|(\beta_1 a_1 + \beta_2 a_2 + \cdots)] = \alpha_1\beta_1 a_1^{\underline{2}} + \alpha_2\beta_2 a_2^{\underline{2}} + \cdots,$$

if $a_1, a_2, \ldots$ *are normal to one another.*

PROOF. For, if $\alpha_1 a_1 + \alpha_2 a_2 + \cdots$ is denoted by $\sum \alpha_r a_r$ and $\beta_1 a_1 + \beta_2 a_2 + \cdots$ by $\sum \beta_s a_s$, then

$$\begin{aligned}[(\alpha_1 a_1 + \alpha_2 a_2 + \cdots)|(\beta_1 a_1 + \beta_2 a_2 + \cdots)] &= \left[\sum \alpha_r a_r \middle| \sum \beta_s a_s\right] \\ &= \sum \alpha_r \alpha_s [a_r|a_s] \qquad [42] \\ &= \sum \alpha_r \beta_r [a_r|a_r],\end{aligned}$$

since a_r and a_s are by assumption normal to one another when r is different from s, and consequently (according to 152) their inner product is zero. But the last expression is

$$= \sum \alpha_r \beta_s a_r^{\underline{2}} = \alpha_1\beta_1 a_1^{\underline{2}} + \alpha_2\beta_2 a_2^{\underline{2}} + \cdots.$$

191. *One has*

$$[(\alpha_1 a_1 + \alpha_2 a_2 + \cdots)|(\beta_1 a_1 + \beta_2 a_2 + \cdots)] = \alpha_1\beta_1 + \alpha_2\beta_2 + \cdots,$$

if $a_1, a_2, \ldots$ *form an elementary normal system.*

PROOF. For, according to 190,

$$[(\alpha_1 a_1 + \alpha_2 a_2 + \cdots)|(\beta_1 a_1 + \beta_2 a_2 + \cdots)] = \alpha_1\beta_1 a_1^{\underline{2}} + \alpha_2\beta_2 a_2^{\underline{2}} + \cdots,$$

but if $a_1, a_2, \ldots$ form an elementary normal system, then $a_1^{\underline{2}} = a_2^{\underline{2}} = \cdots = 1$, whence the expression found

$$= \alpha_1\beta_1 + \alpha_2\beta_2 + \cdots.$$

192. *If* $a_1, a_2, \ldots$ *are normal to one another, then*

$$(a_1 + a_2 + \cdots)^{\underline{2}} = a_1^{\underline{2}} + a_2^{\underline{2}} + \cdots.$$

PROOF. From 190.

193. *One has*

$$(a + b)^{\underline{2}} = a^{\underline{2}} + 2[a|b] + b^{\underline{2}}.$$

PROOF. One has

$$(a+b)^2 = [(a+b)|(a+b)]$$
$$= [a|a] + [a|b] + [b|a] + [b|b], \qquad [42]$$

whence (according to 189)

$$= a^2 + 2[a|b] + b^2.$$

194. *One has*

$$(a+b+c)^2 = a^2 + b^2 + c^2 + 2[b|c] + 2[c|a] + 2[a|b].$$

PROOF. As in 193.

REMARK. Geometrically interpreted, theorems 192–194 represent the Pythagorean theorem, including its extension for the plane and for space.

5. Introduction of the angle

195. DEFINITION. By $\angle AB$ (the angle AB), where A and B are of the same but not zeroth order, and α and β are their numerical values, I mean that angle between 0 and π (including these limits)[68] whose cosine is equal to the inner product of those magnitudes divided by their numerical values, that is I set

$$\cos \angle AB = \frac{[A|B]}{\alpha\beta}, \quad \angle AB = 0, \ldots, \pi.$$

Furthermore, if a, b, c, ... are magnitudes of first order, and α, β, γ, ... their numerical values, I mean by $\sin(abc\cdots)$ the numerical magnitude equal to $\dfrac{[abc\cdots]}{\alpha\beta\gamma\cdots}$ and which is not negative,[69] that is

$$\sin(abc\cdots) \geq 0, \text{ and numerically } = \frac{[abc\cdots]}{[\alpha\beta\gamma]},$$

that is

$$\sin^2(abc\cdots) = \frac{[abc\cdots]^2}{[\alpha^2\beta^2\gamma^2]}.$$

196. *If a, b, are magnitudes of first order, then*

$$\sin(ab) = \sin \angle ab.$$

PROOF. For according to 195,

$$\sin^2(ab) = \frac{[ab]^2}{\alpha^2\beta^2} = \frac{a^2b^2 - [a|b]^2}{\alpha^2\beta^2} \qquad [177]$$
$$= \frac{\alpha^2\beta^2 - [a|b]^2}{\alpha^2\beta^2} \qquad [151]$$
$$= 1 - \left(\frac{[a|b]}{\alpha\beta}\right)^2$$
$$= 1 - \cos^2 \angle ab \qquad [195]$$
$$= \sin^2 \angle ab.$$

And since by definition $\sin(ab)$ is never negative and $\angle ab$ is an angle between 0 and π, so that $\sin \angle ab$ is also not negative, it follows that $\sin(ab) = \sin \angle ab$.

197. *$[A|B] = \alpha\beta \cos \angle AB$, if A and B are of the same order and α and β are their numerical values.*

PROOF. From 195.

198. *$[ab]^{\underline{2}} = (\alpha\beta \sin \angle ab)^2$, where α and β are the numerical values of a and b.*[70]

PROOF. According to 177 {and 151}

$$\begin{aligned} [ab]^{\underline{2}} &= \alpha^2\beta^2 - [a|b]^2 \\ &= \alpha^2\beta^2 - (\alpha\beta\cos\angle ab)^2 && [197] \\ &= \alpha^2\beta^2(1 - \cos^2\angle ab) \\ &= \alpha^2\beta^2\sin^2\angle ab. \end{aligned}$$

REMARK. In these formulas the contrast between the outer and inner products appears most simply. While the inner product of two magnitudes of first order equals the product of the numerical values with the *cosine* of the angle between them, the numerical value of their outer product is equal to the product of their numerical values with the *sine* of the angle between them.

199. *One has*

$$[ab|cd] = \alpha\beta\gamma\delta \sin\angle ab \sin\angle cd \cos\angle(ab \cdot cd),$$

where α, β, γ, δ are the numerical values of a, b, c, d.

PROOF. The numerical value of $[ab]$ is $([ab]^{\underline{2}})^{1/2}$ and that of $[cd]$ is $([cd]^{\underline{2}})^{1/2}$, whence (according to 197)[71]

$$\begin{aligned} [ab|cd] &= ([ab]^{\underline{2}}[cd]^{\underline{2}})^{1/2}\cos\angle(ab \cdot cd) \\ &= [(\alpha\beta\sin\angle ab \cdot \gamma\delta\sin\angle cd)^2]^{1/2}\cos\angle(ab \cdot cd). && [198] \end{aligned}$$

But, since the product $\alpha\beta\sin\angle ab \cdot \gamma\delta\sin\angle cd$ is positive, the progressive powers 2 and 1/2 of these magnitudes cancel, and one has

$$[ab|cd] = \alpha\beta\gamma\delta\sin\angle ab\sin\angle cd\cos(ab \cdot cd).$$

200. *The normal shadow of A on a magnitude of the same order B is numerically equal to $A\cos\angle AB$.*

PROOF. If A' is the normal shadow of A on B, then (according to 166)

$$\begin{aligned} A' &= \frac{[A|B]B}{\beta^2} = \frac{\alpha\beta\cos\angle AB \cdot B}{\beta^2} && [197] \\ &= \alpha\cos\angle AB \cdot \frac{B}{\beta}, \end{aligned}$$

and thus is numerically equal to

$$A\cos\angle AB. \qquad \{151\}$$

201. *If a, b, c, ... are normal to one another, then for every magnitude k numerically derivable from them*

$$\frac{k}{\kappa} = \frac{a}{\alpha}\cos\angle ak + \frac{b}{\beta}\cos\angle bk + \cdots,$$

where κ, α, β, ... are the numerical values of k, a, b, ...

PROOF. Let $k = xa + yb + \cdots$; then since $[b|a]$, ... are zero, upon inner multiplication by a we obtain

$$[a|k] = x[a|a] = x\alpha^2, \qquad [151]$$

and thus

$$x = \frac{[a|k]}{\alpha^2} = \frac{\alpha\kappa\cos\angle ak}{\alpha^2} \qquad [197]$$
$$= \frac{\kappa}{\alpha}\cos\angle ak.$$

On the same grounds $y = \frac{\kappa}{\beta}\cos\angle bk$, ... Substitution of these values into the above equation yields

$$k = \frac{\kappa}{\alpha}\cos\angle ak\cdot a + \frac{\kappa}{\beta}\cos\angle bk\cdot b + \cdots,$$

that is

$$\frac{k}{\kappa} = \frac{a}{\alpha}\cos\angle ak + \frac{b}{\beta}\cos\angle bk + \cdots.$$

202. *If a, b, ... are normal to one another and k and l are numerically derivable from them, then*

$$\cos\angle kl = \cos\angle ak\cdot\cos\angle al + \cos\angle bk\cos\angle bl + \cdots.$$

PROOF. According to 195, if κ, λ, α, β, ... are the numerical values of k, l, a, b, ..., then

$$\cos\angle kl = \frac{[k|l]}{\kappa\lambda} = \left[\frac{k}{\kappa}\middle|\frac{l}{\lambda}\right]$$
$$= \left[\left(\frac{a}{\alpha}\cos\angle ak + \frac{b}{\beta}\cos\angle bk + \cdots\right)\middle|\left(\frac{a}{\alpha}\cos\angle al + \frac{b}{\beta}\cos\angle bl + \cdots\right)\right] \qquad [201]$$
$$= \frac{a^2}{\alpha^2}\cos\angle ak\cos\angle al + \frac{b^2}{\beta^2}\cos\angle bk\cos\angle bl + \cdots,$$

since $[a|b]$, ... are zero. Now since $a^2 = \alpha^2$, $b^2 = \beta^2$, ..., one obtains

$$\cos\angle kl = \cos\angle ak\cos\angle al + \cos\angle bk\cos\angle bl + \cdots.$$

COROLLARY. One can also express this theorem thus: *Instead of shadowing a magnitude of first order (k) on another l, one can first shadow it on the magnitudes of a normal system and then shadow the shadows so obtained on l, and add these shadows, since all the shadows are normal here.*

203. *If a, b, ... are normal to one another, then for every k numerically derivable from them*

$$1 = \cos^2\angle ak + \cos^2\angle bk + \cdots.$$

PROOF. The formula follows from 202 when one sets $l = k$.

204. *If a, b, ... are normal to one another and k and l are numerically derivable from them and likewise normal to one another, then*

$$0 = \cos\angle ak \cos\angle al + \cos\angle bk \cos\angle bl + \cdots .$$

PROOF. The formula follows from 202 when one assumes k and l are normal to one another; for then (according to 152) $[k|l] = 0$, whence (according to 195) $\cos\angle kl = 0$.

205. *If $a + b + \cdots = 0$, and α, β, ... are the numerical values of a, b, ..., then*

(a) $$\alpha : \beta : \cdots = \sin a' : \sin b' : \cdots ,$$

where a', b', ... are the supplementary combinations to a, b, ..., {and further}

(b) $$\alpha \cos\angle ax + \beta \cos\angle bx + \cdots = 0,$$

where x is an arbitrary magnitude, {and finally}

(c) $$\sin a' \cos\angle ax + \sin b' \cos\angle bx + \cdots .$$

PROOF. 1. If one multiplies the equation

$$a + b + \cdots = 0$$

combinatorially by $[cd\ldots]$, one obtains

$$[acd\cdots] + [bcd\cdots] = 0,$$

whence

$$[acd\cdots]^2 = [bcd\cdots]^2,$$

where $[acd\cdots]$ is the product of all the magnitudes a, b, c, ... except b, and $[bcd\cdots]$ is the product of all the magnitudes except a. Consequently (according to 195)

$$(\alpha\gamma\delta\cdots)^2 \sin^2(acd\cdots) = (\beta\gamma\delta\cdots)^2 \sin^2(bcd\cdots),$$

or

$$\alpha^2 \sin^2(acd\cdots) = \beta^2 \sin^2(bcd\cdots).$$

Now $[cad\cdots]$ is the supplementary combination to b, and thus $= b'$, and $[bcd\cdots]$ the supplementary to a, thus $= a'$, whence $\sin b' = \sin(cad\cdots)$ and $\sin a' = \sin(bcd\cdots)$, whence, since α, β, $\sin a'$, $\sin b'$ are positive,

$$\alpha \sin b' = \beta \sin a', \text{ that is } \alpha : \beta = \sin a' : \sin b',$$

and thus generally

$$\alpha : \beta : \cdots = \sin a' : \sin b' : \cdots .$$

2. If one inner multiplies the equation $a + b + \cdots = 0$ by an arbitrary nonzero magnitude of first order x, one obtains

$$[a|x] + [b|x] + \cdots = 0,$$

whence, if ξ is the numerical value of x, then (according to 197)

$$\alpha\xi \cos\angle ax + \beta\xi \cos\angle bx + \cdots = 0,$$

that is

$$\alpha \cos\angle ax + \beta \cos\angle bx + \cdots = 0.$$

3. If one substitutes into the equation just derived the value previously found for $\alpha : \beta : \gamma : \cdots$, one obtains

$$\sin a' \cos \angle ax + \sin b' \cos \angle bx + \cdots = 0.$$

REMARK. The formulas developed only have a meaning if between the magnitudes a, b, ... there is no other relation except that represented by $a + b + \cdots = 0$, that is, if the n magnitudes a, b, ... are united in no domain of less than $(n-1)$th order. For three magnitudes the first part includes the well known theorem that in a triangle the lengths of the sides are as the sines of the opposite angles.

206–213. From the formulas of 172, 175–178, 183, 185, and with the help of the angle formulas just derived the following formulas result:

206. $$\sin(AB)\sin(\mathrm{AB})\cos\angle(AB\cdot\mathrm{AB}) = \sum \sin A_r \sin B_r \sin \mathrm{A} \sin \mathrm{B} \cos\angle A_r\mathrm{A}\cos\angle B_r\mathrm{B},$$

where {A is of the same order as A *and B is of the same order as* B*, and further} A_r are the combinations of the elementary factors of $[AB]$ of such a class as amounts to the order of A, and B_r are the supplementary combinations.*

207. $$\sin(abc\cdots)\sin(a'b'c'\cdots)\cos\angle(abc\cdots)(a'b'c'\cdots) = \textit{Determ.}\begin{cases}\cos\angle aa' & \cos\angle ab' & \dots\\ \cos\angle ba' & \cos\angle bb' & \dots\\ \dots & \dots & \dots\end{cases}$$

208. $\sin\angle ab\sin\angle cd\cos\angle(ab\cdot cd) = \cos\angle ac\cos\angle bd - \cos\angle bc\cos\angle ad$, and, if a and c are substituted for c and d here,

209. $\sin\angle ab\sin\angle ac\cos\angle(ab\cdot ac) = \cos\angle bc - \cos\angle ab\cos\angle ac$, a well-known formula of spherical trigonometry, and further

210. $\sin^2\angle ab = 1 - \cos^2\angle ab$, which enters here as the transformation of 177.

211.[72] $\sin^2(abc) = 1 - \cos^2\angle bc - \cos^2\angle ca - \cos^2\angle ab + 2\cos\angle bc\cos\angle ca\cos\angle ab$.

212. $\sin A\sin B\cos\angle AB + \sin A_1\sin B_1\cos\angle A_1B_1 + \cdots = 0$, *if A, A_1, ... are the combinations of $2n$ magnitudes of first order, of nth class, and B, B_1, ... are the supplementary combinations.*

213. $$\sin\angle ab\sin\angle cd\cos\angle(ab\cdot cd) + \sin\angle ac\sin\angle db\cos\angle(ac\cdot db) + \sin\angle ad\sin\angle bc\cos\angle(ad\cdot bc) = 0.$$

REMARK. Likewise the remaining formulas of §3 would have to be restructured if one had set

$$\frac{[abc|d]}{\alpha\beta\gamma\delta} = \cos\angle abcd, \text{ etc.}$$[73]

214–215. Further, from 193, 194, there results

214. $(a+b)^{\underline{2}} = \alpha^2 + 2\alpha\beta\cos\angle ab + \beta^2$.

215. $(a+b+c)^{\underline{2}} = \alpha^2 + \beta^2 + \gamma^2 + 2\beta\gamma\cos\angle bc + 2\gamma\alpha\cos\angle ca + 2\alpha\beta\cos\angle ab$, *where α, β, γ are the numerical values of a, b, c.*[74]

{REMARK. In this connection cf. Nos. 337–340.}

CHAPTER 5

Applications to Geometry

1. Addition, subtraction, multiples and fractions of points and displacements

216. DEFINITION. If a point E and three mutually orthogonal and equally long lines are chosen as original units, and α, α_1, α_2, α_3 are arbitrary numbers, then

a) I mean by

$$E + \alpha_1 e_1 + \alpha_2 e_2 + \alpha_3 e_3$$

the *point* A at which one arrives if {see Fig. 1} one proceeds from E first by a displacement EB equal to $\alpha_1 e_1$, that is which is directed the same as or opposite to e_1 according as α_1 is positive or negative, and whose length is to that of e_1 as α_1 to 1; then from B by a *displacement* BC, which is in the same sense equal to $\alpha_2 e_2$, and finally from C by a displacement CA, which in the same sense is equal to $\alpha_3 e_3$;

b) second, I mean by

$$\alpha_1 e_1 + \alpha_2 e_2 + \alpha_3 e_3$$

a displacement, that is a straight line of fixed length and direction, and specifically that displacement that has the same length and direction as the line drawn from E to the point $E + \alpha_1 e_1 + \alpha_2 e_2 + \alpha_3 e_3$;

c) third, I mean by

$$\alpha(E + \alpha_1 e_1 + \alpha_2 e_2 + \alpha_3 e_3) = \alpha E + \alpha\alpha_1 e_1 + \alpha\alpha_2 e_2 + \alpha\alpha_3 e_3$$

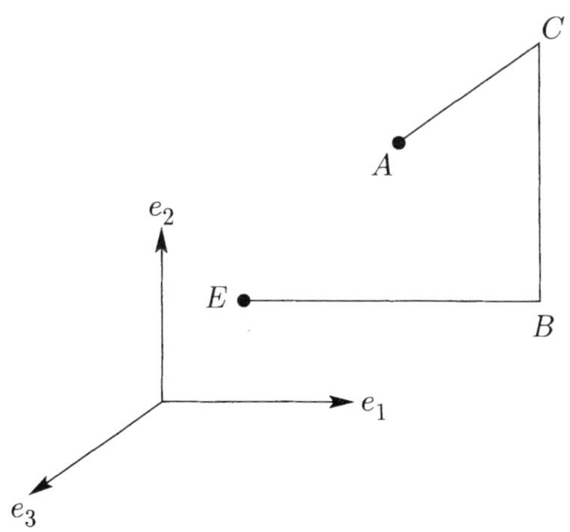

FIGURE 1

the *α-fold of the point* $E + \alpha_1 e_1 + \alpha_2 e_2 + \alpha_3 e_3$ and I require that, for all these magnitudes and their conjunctions, the definitions given in Chapter 1, as well as the theorems derived therefrom, be valid.

REMARK. Thus magnitudes of first order are the simple and multiple points and the displacements of fixed length and direction here. With the definitions in §1 {of Chapter 1} addition, subtraction, multiples and fractions of these magnitudes are then defined, and by the theorems of this paragraph the validity of the algebraic laws of conjunction are verified for them, and in the following paragraphs the special properties belonging to them, as extensive magnitudes. Next, from these formal definitions we derive the constructions from which the resultants of the various conjunctions follow.

217. *If*

$$A = E + \alpha_1 e_1 + \alpha_2 e_2 + \alpha_3 e_3,$$

then $\alpha_1 e_1$, $\alpha_2 e_2$, $\alpha_3 e_3$ *are the orthogonal projections of* EA *on the three axes extending from* E *and directed along* e_1, e_2, e_3.

PROOF. Follows at once from the Definition.

218. LOAN THEOREM FROM GEOMETRY. Similarly directed displacements orthogonally projected on the same straight line yield similarly directed projections whose lengths are as the projected displacements; and conversely, if the orthogonal projections of two straight lines on three mutually orthogonal axes are equally long and similarly directed, then the projected lines are themselves equally long and similarly directed.

219. LOAN THEOREM FROM ANALYTIC GEOMETRY. If A, B, C are three arbitrary points of a straight line, and AB, BC, AC, measured by a segment DE of this line, yield the quotients α, β, γ respectively, where each quotient is taken as positive or negative according as the measured line is directed the same as or opposite to the measuring line (DE), then it is always true that

$$\alpha + \beta = \gamma,$$

which one can also write for short as

$$AB + BC = AC.$$

220. *To add several displacements (of the same direction and length) one lays them, without changing their lengths or directions, end to end, that is lays them so that where one stops the next one begins; then the straight line from the initial point of the first to the final point of the last is the same length and in the same direction as the sum sought.*

PROOF. *First*, for *two* displacements a and b {see Fig. 2}. Let

$$a = \alpha_1 e_1 + \alpha_2 e_2 + \alpha_3 e_3,$$
$$b = \beta_1 e_1 + \beta_2 e_2 + \beta_3 e_3,$$

whence (according to 6)

$$a + b = (\alpha_1 + \beta_1)e_1 + (\alpha_2 + \beta_2)e_2 + (\alpha_3 + \beta_3)e_3.$$

Further, let $E + a = A$, $E + b = B$, $E + (a + b) = C$; then (according to 216) the straight line EA is the same length and direction as a, EB as b, EC as $a + b$. Finally, let FG be the same length and direction as EA, and GH the same as EB; then it is to be shown that FH is the same length and direction as EC.

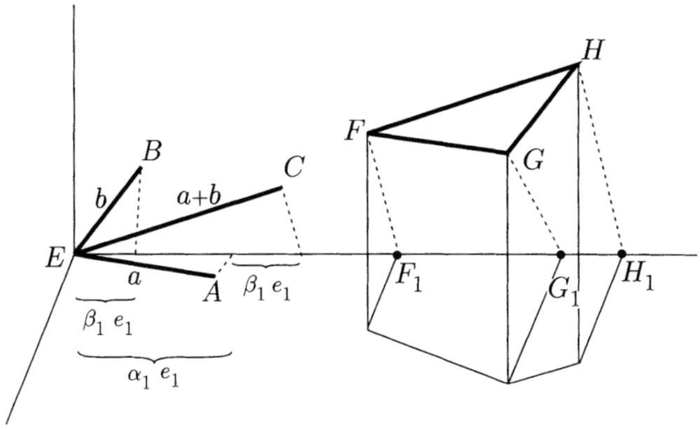

FIGURE 2

Since FG is the same length and direction as EA, this also holds (according to 218) for their projections; but according to 217, the projections of EA are the same lengths and directions as $\alpha_1 e_1$, $\alpha_2 e_2$, $\alpha_3 e_3$, consequently this also holds for the projections of FG; on the same grounds, the projections of GH are the same lengths and directions as $\beta_1 e_1$, $\beta_2 e_2$, $\beta_3 e_3$. Now let F_1, G_1, H_1 be the projections of F, G, H on the axis extending from E along e_1; then F_1G_1 is the same length and direction as $\alpha_1 e_1$, G_1H_1 the same as $\beta_1 e_1$, that is, F_1G_1 and G_1H_1, measured by e_1, yield the quotients α_1 and β_1, and consequently (according to 219), since F_1, G_1, H_1 lie on the same straight line, F_1H_1, measured by e_1, yields the quotient $\alpha_1 + \beta_1$, that is F_1H_1 is the same length and direction as $(\alpha_1 + \beta_1)e_1$. But F_1H_1 is the projection of FH on the axis laid through E in the direction of e_1. If one applies the same procedure to the other axes, then it follows that the projections of FH are the same lengths and directions as $(\alpha_1+\beta_1)e_1$, $(\alpha_2+\beta_2)e_2$, $(\alpha_3+\beta_3)e_3$, that is as the projections of EC, whence (according to 218) FH is the same length and direction as EC, that is as $a + b$, which was to be proved.

Second. Now if one has *several* displacements a, b, c, ..., and if a is the same length and direction as FG, b as GH, c as HI, ..., then according to the first part of the Proof $a + b$ is the same length and direction as FH, and thus, since $a + b$ is the same length and direction as FH, c as HI, $a + b + c$ is the same length and direction as FI, ...

221. *The product of a displacement a with a number α is again a displacement (b), which is directed along or opposite to the first (a) according as the number α is positive or negative, and whose length is to that of a as α is to* 1.

PROOF. Let E, e_1, e_2, e_3 be taken as units, and let

$$a = \alpha_1 e_1 + \alpha_2 e_2 + \alpha_3 e_3;$$

then

$$\begin{aligned} b = \alpha a &= \alpha(\alpha_1 e_1 + \alpha_2 e_2 + \alpha_3 e_3) \\ &= \alpha\alpha_1 e_1 + \alpha\alpha_2 e_2 + \alpha\alpha_3 e_3. \end{aligned}$$

But (according to 216) the last expression denotes a displacement that has the same length and direction as straight line drawn from E to the point $B = E + \alpha\alpha_1 e_1 + \alpha\alpha_2 e_2 + \alpha\alpha_3 e_3$. The projections of this line on the three axes extending from E parallel to e_1, e_2, e_3 are (according to 217) $\alpha\alpha_1 e_1$, $\alpha\alpha_2 e_2$, $\alpha\alpha_3 e_3$. Likewise $\alpha_1 e_1 + \alpha_2 e_2 + \alpha_3 e_3$ is a displacement with the same direction and length as the line drawn from E to the point $A = E + \alpha_1 e_1 + \alpha_2 e_2 + \alpha_3 e_3$, and $\alpha_1 e_1$, $\alpha_2 e_2$, $\alpha_3 e_3$ are the projections of EA on those three axes. First, if α is positive, then $\alpha\alpha_1 e_1$, $\alpha\alpha_2 e_2$, $\alpha\alpha_3 e_3$ are directed similarly to $\alpha_1 e_1$, $\alpha_2 e_2$, $\alpha_3 e_3$ respectively, and are to them as $\alpha : 1$; thus (according to 218) the same is true for the projected lines, that is, EB is directed similarly to EA, and its length is to that of EA as $\alpha : 1$. Now since b is the same length and direction as EB and a the same as EA, a and b are also directed similarly to one another, and their lengths are as $1 : \alpha$.

If however α should be negative, $= -\beta$, then $\alpha\alpha_1 e_1$, $\alpha\alpha_2 e_2$, $\alpha\alpha_3 e_3$, that is $-\beta\alpha_1 e_1$, $-\beta\alpha_2 e_2$, $-\beta\alpha_3 e_3$, are the projections of EB, and (according to 216) they are oppositely directed to those of EA, that is $\alpha_1 e_1$, $\alpha_2 e_2$, $\alpha_3 e_3$, whence those of BE, that is $\beta\alpha_1 e_1$, $\beta\alpha_2 e_2$, $\beta\alpha_3 e_3$, are directed similarly to those of EA, and their lengths are as $\beta : 1$, whence EB and EA and consequently also b and a are oppositely directed, while their lengths remain as $\beta : 1$.

222. *The sum $\alpha A + \beta B + \cdots$, in which A, B, ... are points, α, β, ... numbers, is a displacement or a multiple point according as $\alpha + \beta + \cdots$ is zero or not, and specifically* ***in the first case***

$$\alpha A + \beta B + \cdots = \alpha(A - R) + \beta(B - R) + \cdots,$$

and ***in the second***

$$\alpha A + \beta B + \cdots = (\alpha + \beta + \cdots)S,$$

where

$$(*) \qquad S - R = \frac{\alpha(A - R) + \beta(B - R) + \cdots}{\alpha + \beta + \cdots},$$

and R is an arbitrary point.[75]

PROOF. Let

$$A = R + A - R, \quad B = R + B - R, \quad \ldots .$$

If one inserts these values into the expression $\alpha A + \beta B + \cdots$, one obtains

$$\alpha A + \beta B + \cdots = (\alpha + \beta + \cdots)R + \alpha(A - R) + \beta(B - R) + \cdots .$$

Thus first, if $\alpha + \beta + \cdots = 0$,

$$= \alpha(A - R) + \beta(B - R) + \cdots .$$

If on the other hand $\alpha + \beta + \cdots$ is different from zero, say equal to σ, then

$$\begin{aligned} \alpha A + \beta B + \cdots &= \sigma R + \alpha(A - R) + \beta(B - R) + \cdots \\ &= \sigma\left(R + \frac{\alpha(A - R) + \beta(B - R) + \cdots}{\sigma}\right) \\ &= \sigma S, \end{aligned}$$

if one sets

$$S = R + \frac{\alpha(A - R) + \beta(B - R) + \cdots}{\sigma},$$

that is

$$S - R = \frac{\alpha(A-R) + \beta(B-R) + \cdots}{\sigma}.$$

COROLLARY. *If A and B are points, then $A - B$ is the displacement of the same length and direction as the line BA.*

PROOF. According to 216 $A - E$ is a displacement with the same length and direction as the straight line EA, and $B - E$ is one with the same length and direction as EB; now

$$A - B = (A - E) - (B - E),$$

whence

$$= (A - E) + (E - B) = (E - B) + (A - E).$$

Now since $E - B$ and $A - E$ are displacements with the same lengths and directions as BE and EA respectively, their sum is (according to 220) the same length and direction as BA, that is, $A - B$ is the same as BA.

REMARK. Thus displacements are reduced to the differences of points, and their sum, formed by laying them one after the other, is represented as the sum of such differences, in which the end point of each displacement is cancelled by the initial point of the next.

223. *If one draws lines from a* ***moving*** *point (R) to a series of fixed points (A, B, ...), and these evolve by fixed ratios ($1 : \alpha$, $1 : \beta$, ... , so that thereby the lines RA', RB', ... result, which are directed either similarly or opposite to RA, RB, ... , respectively, according as α, β, ... is positive or negative, and their lengths are to those of RA, RB, ... as the numbers α, β, ... to unity), and then the lines (RA', RB', ...) so obtained are placed end to end without altering their magnitudes or directions, then the line (RP) from the initial point (R) of the first to the final point (P) of the last has the following properties:*

1) *If the sum of the ratios (α, β, ...) is zero, the line (RP) is of fixed length and direction,*

2) *If the sum of the ratios is different from zero, this line (RP) goes through a fixed point (S) which cuts from this line (RP) as many parts as that sum ($\alpha+\beta+\cdots$) amounts to.*

PROOF. The theorem is just another way of expressing 222.

REMARK. The point S is well known as the center of gravity of the points A, B, ... if their weights are as $\alpha : \beta : \ldots$. Here the natural name summation point is introduced.

224. *The summation point S of the sum $\alpha A + \beta B + \cdots$, in which $\alpha + \beta + \cdots$ is not zero, has the property that*

$$\alpha(A - S) + \beta(B - S) + \cdots = 0,$$

and no other point has this property.

PROOF. For if, in (*) of 222, one sets the point $R = S$, then

$$S - S = \frac{\alpha(A-S) + \beta(B-S) + \cdots}{\alpha + \beta + \cdots},$$

that is

$$0 = \alpha(A - S) + \beta(B - S) + \cdots .$$

Should this equation hold for some other point R, thus

$$0 = \alpha(A - R) + \beta(B - R) + \cdots ,$$

then by subtraction one obtains

$$\begin{aligned} 0 &= \alpha(R - S) + \beta(R - S) + \cdots \\ &= (\alpha + \beta + \cdots)(R - S), \end{aligned}$$

whence, since $\alpha + \beta + \cdots$ is not zero (by hypothesis),

$$0 = R - S,$$

or $R = S$, that is, there is no second point different from S possessing that property.

225. *The sum of two simple points is twice their mean, and the sum of two multiple points whose coefficients have the same sign is a multiple point whose coefficient is the sum of the coefficients of the summands and whose position lies between those of the summands in such a way that it is distant from them in inverse ratio to their coefficients. If on the other hand their coefficients have opposite signs and are not numerically equal, the sum is a multiple point whose coefficient is the algebraic sum of the coefficients of the summands, and whose position lies on the continuation of the line connecting the summands in such a way that it is distant from these positions in inverse ratio to their coefficients.*

PROOF. Follows directly from 222.

226. *The sum of a simple point and a displacement is the final point of the straight line equal in length and direction to the displacement, whose initial point is the given point.*

PROOF. Let the straight line AB be the same length and direction as the displacement p; then (according to the Corollary to 222)

$$B - A = p.$$

Thus

$$A + p = A + B - A = B.$$

227. *The sum of an α-fold point (αA) and a displacement (p) is the α-fold final point of a straight line (AB) whose α-fold is the same length and direction as the displacement (p) and whose initial point (A) is the given point.*

PROOF.

$$\alpha A + p = \alpha\left(A + \frac{p}{\alpha}\right),$$

whence, if the α-fold of AB is the same length and direction as p, thus AB with $\frac{p}{\alpha}$,

$$B - A = \frac{p}{\alpha},$$

and therefore

$$\alpha A + p = \alpha(A + B - A) = \alpha B.$$

REMARK. The addition of points was first given (1827) by MÖBIUS in his *Barycentrische Kalkül.* The addition of displacements seems to have been published first by BELLAVITIS in several articles (1835, 1837) in the *Annali delle scienze del regno Lombardo-Veneto.* Completely independent of these is the development in my *Ausdehnungslehre* of 1844 (§§24, 101–102), where in addition the connection between the two additions was first brought to light. Missing in both their works and mine, however, is the demonstration that there is no other addition of points and lines than those given here, and yet this demonstration seems necessary if that addition is to be interpreted as a genuine addition of those magnitudes, and not just an abbreviated notation, as MÖBIUS would have it. It is therefore to be shown that, if the general concept of addition is applied specifically to points (or to displacements of fixed length and direction), it can yield nothing but the addition presented above.

To this end it is first necessary to establish the general condition that no conjunction of geometric objects as such may be bound to a specific position in space; or, to express this condition purely mathematically: "All conjunctions of spatial magnitudes must be such that every equation obtaining between an assembly of points must also remain valid if one replaces these points with the corresponding points of a congruent assembly." Addition and subtraction are thus defined, since first of all the four fundamental formulas

$$\begin{aligned}
&1)\quad a + b = b + a,\\
&2)\quad a + (b + c) = (a + b) + c,\\
&3)\quad a + b - b = a,\\
&4)\quad a - b + b = a
\end{aligned}$$

hold; and in addition the magnitudes resulting from the conjunction must be, to the greatest extent possible, of the same species as those conjoined. This last condition is yet to be individualized.

Since, according to the third fundamental formula, if both A and B are {simple} points, then

$$A + B - B = A$$

is likewise a {simple} point, and according to the first and third,

$$A + B - A = B$$

is also a {simple} point, the natural assumption is that $A + B - C$ is to be a {simple} point. Yet this assumption is only satisfied for the case that C is made the mean between A and B. Thus, in order to satisfy the assumption made, we make the assumption that "if C is the mean between the {simple} points A and B, {and C is to be interpreted as a simple point}, then $A + B - C$ is always again a {simple} point." Thus the necessary assumptions are exhausted.

Next, from the validity of the four fundamental formulas follows the validity of all the general laws of addition and subtraction. Next I will prove that, if the point C is the mean between the points A and B, then $A + B - C = C$.

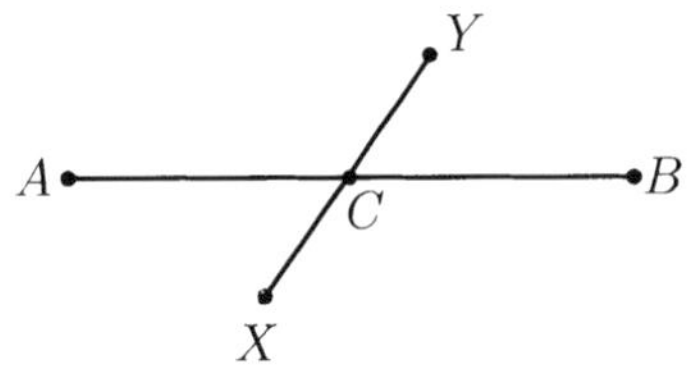

FIGURE 3

Let $A + B - C = X$; then X cannot be different from C. For assume that X were different from C {see Fig. 3}, and prolong XC by itself to Y, so that $XC = CY$. If one now rotates the figure that includes the points A, B, C, X, within the plane in which the figure lies, about the point C until it has described an angle of $180°$, then A will coincide with where B lay, B with where A lay, and X with Y, that is, the assembly A, B, C, X is congruent with the assembly B, A, C, Y. Now since by assumption

$$A + B - C = X,$$

so also, according to the above condition, to which all geometric conjunctions are subject, this equation must continue to hold if one replaces A, B, C, X with B, A, C, Y respectively, whence

$$B + A - C = Y.$$

Thus one has

$$\begin{aligned} Y &= B + A - C = A + B - C && \text{(according to fundamental formula 1)} \\ &= X, && \text{(by assumption)} \end{aligned}$$

whence $Y = X$. But Y was produced from X when XC was prolonged by itself to Y; thus should Y coincide with X, then X must coincide with C, that is $X = C$, whence $A + B - C = C$.

If in this equation one brings C to the right side, then one obtains

$$A + B = 2C,$$

that is, "the sum of two {simple} points is twice the {simple} point lying midway between the two points."

Now let AB and CD be two arbitrary straight lines of the same length and direction; then the quadrilateral $ABCD$ is a parallelogram. Let the diagonals intersect in E {see Fig. 4}. Since the diagonals of a parallelogram bisect one another, E is the mean between both A and D and between B and C, that is

$$A + D = 2E = B + C, \quad \text{whence} \quad A + D = B + C.$$

If in this last equation one brings D and B to the opposite sides, one obtains

$$A - B = C - D.$$

Conversely, if this last equation holds, then so also does the previous one, $A + D = B + C$, that is, the mean between A and D must simultaneously be the mean between B and C, that is the quadrilateral $ABCD$ must be a parallelogram, whence AB must be of the same length and direction as CD.

From this follows the theorem: "A difference $A - B$ of two {simple} points is equal to a difference $C - D$ of two other {simple} points if and only if AB and CD are of the same length and direction." If for short one calls the difference $A - B$ or $-B + A$ a displacement, B its initial point, A its final point, then there follows at once the theorem: "Displacements (of given direction and length) are added if (without changing their directions or lengths) one lays them continuously together, that is, lays them end to end so that the final point of each of them coincides with the initial point of the next; then the displacement that

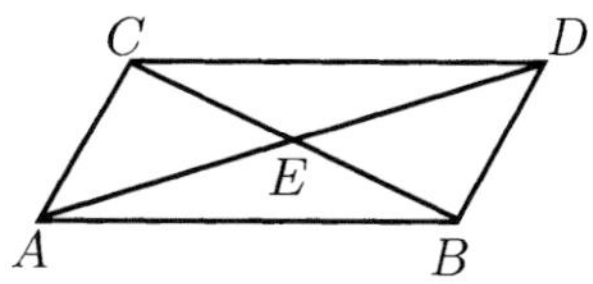

FIGURE 4

has the initial point of the first displacement as its initial point, and the final point of the last as its final point, is the sum of those displacements." Thus if, for example, the first displacement is equal to $-A+B$, the second equal to $-B+C$, the third to $-C+D$, then the sum is

$$= -A + B - B + C - C + D = -A + D,$$

which was to be proved.

For the division of a displacement by a whole positive number the condition has to be imposed that the quotient is again a displacement (where by a displacement is always meant the difference between two points, and thus a displacement of given length and direction). It then follows according to the well-known procedure that the product of a displacement with an arbitrary whole or fractional rational or irrational number a is again a displacement, directed similarly or opposite to the one given according as a is positive or negative, and whose length is to the length of the given line as a to 1.

Thus is solved the general problem of finding the sum $aA + bB + \cdots$, where $a, b, \ldots$ are numerical magnitudes, $A, B, \ldots$ {simple} points. Thus for any arbitrary {simple} point R one has

$$aA + bB + \cdots = (a + b + \cdots)R + a(A - R) + b(B - R) + \cdots .$$

We distinguish two cases, according as $a + b + \cdots$ is equal to zero or not. In the first case

$$aA + bB + \cdots = a(A - R) + b(B - R) + \cdots ,$$

that is, it equals a displacement, which is constructible according to the above. In the second, when $a + b + \cdots = s \neq 0$,

$$aA + bB + \cdots = sR + a(A - R) + b(B - R) + \cdots .$$

Here $a(A - R) + b(B - R) + \cdots$ is a displacement; let the sth part of this displacement be laid down so that R is its initial point; if its final point is denoted S, then

$$a(A - R) + b(B - R) + \cdots = s(S - R).$$

Substituting this value in the above equation yields

$$\begin{aligned} aA + bB + \cdots &= sR + s(S - R) \\ &= s(R + S - R) \\ &= sS, \end{aligned}$$

whereby the problem is completely solved, and the concept of the addition of simple and multiple points and displacements is completely defined, and indeed is in harmony with the definitions given in the main text.[76]

2. Spatial domains

228. DEFINITION. By an *infinitely distant point* let the direction of a straight line be meant, by an *infinitely distant straight line* all the directions of a plane, and by an *infinitely distant plane* all the directions of space, that is, let it be said of two parallel straight lines that they have an infinitely distant point in common, of two parallel planes that they have an infinitely distant line in common, and of

all infinitely distant points and straight lines that they lie in an infinitely distant plane.

In order to be able to handle all the spatial magnitudes of first order, that is the simple or multiple points and displacements (of given length and direction), in the same way, I will say that the *position of a displacement* is the infinitely distant point that the lines parallel to this displacement have in common, or alternatively that that displacement is a magnitude of first order that lies on these lines at an infinite distance. In addition I will for the sake of simplicity replace the expression "spatial magnitude of first order" by a simpler one, and call both simple and multiple points, as well as displacements, "points" for short, and in particular the latter I will call "infinitely distant points".

REMARK. The essence of spatial magnitudes, as of magnitudes in general, is a definite metric value, by virtue of which they can be increased or decreased (in definite ratios). This is represented by numerical coefficients for simple and multiple points, by their lengths for displacements.

It would still be possible in principle to accept infinitely distant points whose metric value is not represented by the length of a displacement but by a numerical coefficient, that is, that result from the multiple point αA when one, without changing α, removes the point A to infinity. But, as one easily sees, what results thereby bears with it all the characteristics of the infinite, in that it would not change at all by the addition of a finite magnitude (a finitely distant point, or even a displacement). With such infinities nothing whatever can be calculated in general, since no algebraic laws hold for infinity, and the analysis of the infinite can only lead to correct results if the error that one introduced with the assumption of the infinite is thrown out again before the derivation of any result. On the other hand the displacement is still a finite magnitude, even though for convenience one permitted the expression that its position is infinitely distant, since it is changed by the addition of any nonzero magnitude.

229. *All the displacements in space can be numerically derived from three arbitrary displacements that are not parallel to the same plane.*

PROOF. Let a, b, c be three displacements not parallel to the same plane, and e be an arbitrary displacement, which will be shown to be numerically derivable from a, b, c.

If one draws from an arbitrary point D the lines DA, DB, DC, DE with the same lengths and directions as a, b, c, e {see Fig. 5}, then (according to the Corollary to 222) $A - D = a$, $B - D = b$, $C - D = c$, $E - D = e$. Further, if one draws through E the parallel to DC, which strikes the plane ABD in F, and through F the parallel to DB, which strikes DA in G, then $DG // DA$, $GF // DB$, $FE // DC$.

If one lets

$$DG : DA = \alpha : 1, \quad GF : DB = \beta : 1, \quad FE : DC = \gamma : 1$$

algebraically (that is, including the algebraic sign), then (according to 221)

$$\begin{aligned} G - D &= \alpha(A - D) = \alpha a, \\ F - G &= \beta(B - D) = \beta b, \\ E - F &= \gamma(C - D) = \gamma c. \end{aligned}$$

Thus, adding,

$$E - D = \alpha a + \beta b + \gamma c,$$

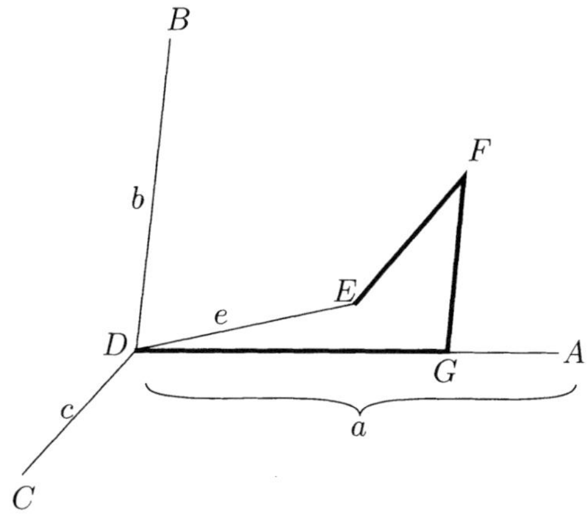

FIGURE 5

that is

$$e = \alpha a + \beta b + \gamma c.$$

230. *All the displacements of a plane can be numerically derived from two arbitrary nonparallel displacements of the plane.*

PROOF. Let a, b be two arbitrary nonparallel displacements of a plane, and d be an arbitrary displacement of the plane, which will be shown to be numerically derivable from a and b.

If one draws from an arbitrary point C of the plane the lines CA, CB, CD with the same lengths and directions as a, b, d {see Fig. 6}, draws through D a parallel to CB, which strikes CA in E, then $CE//CA$, $ED//CB$. If they are related algebraically as

$$CE : CA = \alpha : 1, \quad ED : CB = \beta : 1,$$

then (according to 221)

$$E - C = \alpha(A - C) = \alpha a,$$
$$D - E = \beta(B - C) = \beta b,$$

whence

$$D - C = \alpha a + \beta b, \quad \text{that is} \quad d = \alpha a + \beta b.$$

{230A. *All the displacements of a straight line can be numerically derived from an arbitrary displacement of the straight line.*

PROOF. Let a and b be two displacements of the same straight line; it will be shown that the displacement b is numerically derived from a.

If one draws from an arbitrary point C of the straight line the straight lines CA and BC with lengths equal to a and b, then (according to the Corollary to 222)

$$a = A - C, \quad b = B - C.$$

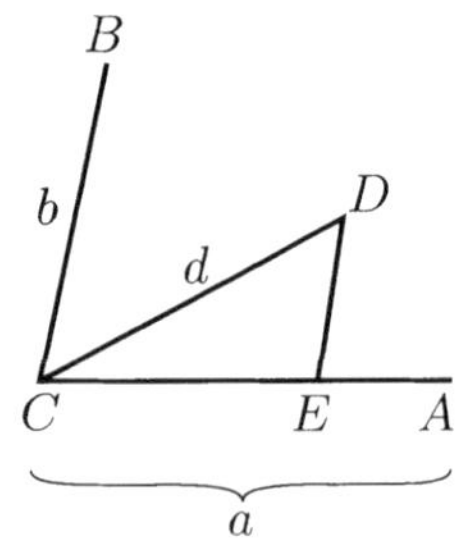

FIGURE 6

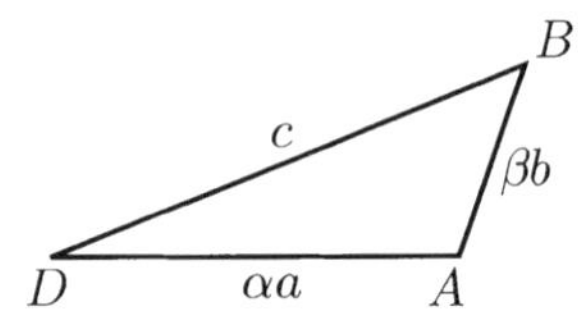

FIGURE 7

Now if algebraically $CB : CA = 1$, then (according to 221)

$$B - C = \alpha(A - C) = \alpha a,$$

that is

$$b = \alpha a;$$

thus b is really numerically derivable from a.}

231. *If there is a numerical relation between three displacements, they are coplanar.*

PROOF. Let a, b, c be the three displacements and

$$c = \alpha a + \beta b$$

be their numerical relation. Should a and b be parallel, then c would also be parallel to them, and so there would be infinitely many planes with which a, b, c are simultaneously parallel. If a is not parallel to b, then one draws {see Fig. 7} from an arbitrary point D a line DA parallel to a and related to a as $\alpha : 1$, and from A a line AB parallel to b, and related to b as $\beta : 1$; then

$$A - D = \alpha a,$$
$$B - A = \beta b.$$

Thus, adding,

$$B - D = \alpha a + \beta b = c.$$

Consequently c is, like a and b, parallel to the plane ABD.

232. *All the points of space can be numerically derived from any four points that do not lie in a single plane; in particular*

1) *from a finitely distant point and three displacements not parallel to the same plane;*

2) *from two finitely distant, noncoincident points and two displacements, neither of which is parallel to any plane laid on those two points;*

3) *from three finitely distant points that do not lie on a single line, and one displacement that is not parallel to the plane laid on the three points;*

4) *from four finitely distant points that do not lie in the same plane.*

PROOF. 1. Let a, b, c be three displacements not parallel to the same plane and $d' = \delta D$ be a finitely distant point, D its position and $e' = \varepsilon E$ be a finitely distant point with E its position; it is then to be shown that e' is numerically derivable from a, b, c, d'. According to 229 the displacement $E - D$ is numerically derivable from a, b, c; let

$$E - D = \alpha a + \beta b + \gamma c;$$

then

$$E = D + \alpha a + \beta b + \gamma c,$$

that is

$$\frac{c'}{\varepsilon} = \frac{d'}{\delta} + \alpha a + \beta b + \gamma c,$$

whence

$$e' = \frac{\varepsilon}{\delta} d' + \varepsilon\alpha a + \varepsilon\beta b + \varepsilon\gamma c,$$

that is, e' is numerically derivable from a, b, c, d'. If the point to be derived is infinitely distant, that is a displacement, then (according to 229) it is already numerically derivable from a, b, c, and thus also from a, b, c, d' $(= \alpha a + \beta b + \gamma c + 0d')$.

2. Let a, b be two displacements, $c' = \gamma C$, $d' = \delta D$ two finitely distant points, C and D their positions, and let it be assumed that through C and D no plane can be laid parallel to a and b. If one sets $C - D = c$, then a, b, c are three displacements not parallel to the same plane, whence (according to Part 1 of the Proof) any point e' is numerically derivable from a, b, c, d'. If in the expression for this derivation one replaces c by its value $C - D$, that is $\frac{c'}{\gamma} - \frac{d'}{\delta}$, one obtains an expression by which e' is numerically derived from a, b, c', d'.

3. Let a be a displacement, $b' = \beta B$, $c' = \gamma C$, $d' = \delta D$ three finitely distant points, B, C, D their positions, and let it be assumed that a is not parallel to the plane BCD. If one sets $B - D = b$, then (according to Part 2 of the Proof) any point e' is numerically derivable from a, b, c', d'. If in the expression for this derivation one replaces b by its value $B - D$, that is $\frac{b'}{\beta} - \frac{d'}{\delta}$, one obtains an expression by which e' is numerically derived from a, b', c', d'.

4. Let $a' = \alpha A$, $b' = \beta B$, $c' = \gamma C$, $d' = \delta D$ be four finitely distant points, A, B, C, D their positions, and let it be assumed that these points do not lie in the same plane. If one sets $A - D = a$, then (according to Part 3 of the Proof) any point e' is numerically derivable from a, b', c', d'. If in the expression for this derivation one replaces a by its value $\frac{a'}{\alpha} - \frac{d'}{\delta}$, one obtains an expression by which e' is numerically derived from a', b', c', d'.

REMARK. If the three displacements are equally long, the first of the four derivation systems specified in the theorem is the usual parallel coordinate system; the last, if the points are simple, is the barycentric system of MÖBIUS, and, if they are arbitrary, the most general linear coordinate system, as it is treated by PLÜCKER and others.

233. *All the points of a plane can be numerically derived from any three points not lying in a straight line.*

PROOF. As in 232.

234. *All points of a straight line can be numerically derived from any two spatially separated points of it.*

PROOF. As in 232.

235. *If three points stand in a numerical relation to one another, they lie in a single straight line.*

PROOF. Let a, b, c be the three points, and

$$a = \beta b + \gamma c$$

be their numerical relation.

If b and c are *infinitely distant*, that is displacements (according to 220), then (according to 221 and 220) a is also a displacement. According to 231 the three displacements are parallel to the same plane, that is (according to 228) a, b, c are infinitely distant points that lie in an infinitely distant straight line.

If on the other hand b and c *are not both infinitely distant at the same time*, then one connects them by the straight line DE, and takes D and E as two simple finitely distant points of this line. Then, since b and c lie in the straight line laid through D and E, they are (according to 234) numerically derivable from D and E, and thus also $a = \beta b + \gamma c$. Let $a = \delta D + \varepsilon E$.

Now if {*first*} $\delta + \varepsilon = 0$, and thus $\delta = -\varepsilon$, then $a = \varepsilon(E - D)$. But $\varepsilon(E - D)$ is a displacement parallel to DE, that is, an infinitely distant point of the line DE, whence a, b, c lie in DE.

But if {*second*} $\delta + \varepsilon = \sigma$, not equal to zero, then

$$a = \delta D + \varepsilon E = \sigma D + \varepsilon(E - D),$$

that is, $a = \sigma A$, where $A = D + \frac{\varepsilon}{\sigma}(E - D)$, that is $A - D = \frac{\varepsilon}{\sigma}(E - D)$. Thus $A - D$ is parallel to $E - D$, that is A is a point of the line DE, whence in this case as well a, b, c lie in a straight line.

REMARK. The last part of the Proof only verifies that the center of gravity of two points with arbitrary weights lies on the straight line connecting these points.

236. *If four points stand in a numerical relation to one another, they lie in the same plane.*

PROOF. Let a, b, c, d be four points and

$$a = \beta b + \gamma c + \delta d$$

be the numerical relation. First, if b, c, d *are all three infinitely distant at the same time*, then it is to be shown that a lies in the infinitely distant plane, that is that it is also infinitely distant, which is to say a displacement. This follows from 228, since then b, c, d and thus also {according to 221} their multiples are displacements, and consequently (according to 220) their sum as well.

If *not all three of the* b, c, d *are infinitely distant*, then let DEF be the plane laid upon them and D, E, F be three simple, finitely distant points of this plane.

Then (according to 233) b, c, d can be numerically derived from D, E, F, whence also $\beta b+\gamma c+\delta d$, that is a. Let

$$a=\delta D+\varepsilon E+\zeta F.$$

First, if $\delta+\varepsilon+\zeta=0$, then

$$a=\delta D+\varepsilon E+\zeta F-(\delta+\varepsilon+\zeta)D=\varepsilon(E-D)+\zeta(F-D),$$

whence a is numerically derivable from $E-D$ and $F-D$, that is (according to 231) the displacements a, $D-E$, and $F-D$ are parallel to the same plane, whence a is parallel to the plane DEF, that is, an infinitely distant point of this plane.

{*Second,*} if $\delta+\varepsilon+\zeta=\sigma$ is not zero, then

$$\begin{aligned}a&=\delta D+\varepsilon E+\zeta F=\sigma D+\varepsilon(E-D)+\zeta(F-D)\\&=\sigma A,\end{aligned}$$

if

$$A=D+\frac{\varepsilon}{\sigma}(E-D)+\frac{\sigma}{\zeta}(F-D),$$

whence (according to 231) $A-D$ is parallel to the plane DEF, that is, A is a point of the plane DEF, and so a is also a point of this plane.

237. *The spatial domain of first order is a point (regarded as a position), that of second order an unbounded straight line, that of third order an unbounded plane, and that of fourth order unbounded space.*

PROOF. A domain of nth order is (according to 14) the collection of magnitudes that are numerically derivable from n magnitudes, assuming that those magnitudes cannot all be numerically derived from less than n magnitudes. Now (according to 232) all points of space are numerically derivable from four magnitudes of first order; according to 236 the points derivable from three such magnitudes form a plane, whence the points of space are not derived from less than four magnitudes of first order. Thus space is a domain of fourth order. Likewise it follows from 233 and 235 that the domain of third order is a plane, and from 234, and that from a point only positionally identical points are derivable, it follows that the domain of second order is a straight line, as well as that the domain of first order is a point.

238. PROBLEM. To express the derivation numbers (coordinates), by which a point (p) results from four points (a, b, c, d) not lying in the same plane, in terms of the derivation numbers by which the same point (p) is derivable from four new points (a', b', c', d'); assuming that these four new points are expressed by the four old ones.

SOLUTION. Let

1) $$a'=\alpha a+\beta b+\gamma c+\delta d,$$
2) $$b'=\alpha' a+\beta' b+\gamma' c+\delta' d,$$
3) $$c'=\alpha'' a+\beta'' b+\gamma'' c+\delta'' d,$$
4) $$d'=\alpha''' a+\beta''' b+\gamma''' c+\delta''' d,$$
5) $$p=x'a+y'b+z'c+u'd,$$
6) $$p=xa'+yb'+zc'+ud'.$$

If in 6) one sets for a', b', c', d', p the values in 1) through 5), then upon ordering according to a, b, c, d, one obtains

$$7)\ x'a + y'b + z'c + u'd = (x\alpha + y\alpha' + z\alpha'' + u\alpha''')a + (x\beta + y\beta' + z\beta'' + u\beta''')b + \cdots .$$

Since here a, b, c, d do not lie in the same plane, they stand in no numerical relation to one another (according to 236). Consequently (according to 29) the corresponding coefficients in the equation found are equal, thus

$$\begin{aligned} x' &= x\alpha + y\alpha' + z\alpha'' + u\alpha''', \\ y' &= x\beta + y\beta' + z\beta'' + u\beta''', \\ z' &= x\gamma + y\gamma' + z\gamma'' + u\gamma''', \\ u' &= x\delta + y\delta' + z\delta'' + u\delta'''. \end{aligned}$$

REMARK. This is the solution of the general problem of coordinate transformations.

3. Combinatorial multiplication of points

239. DEFINITION. The parallelogram in which AB and BC are two sides I will call the *parallelogram ABC* for short, and if it is identified thus, I will call AB its *first side*, BC its *second side.* Further, all parallelograms whose first sides are the same length and direction as the displacement a and whose second sides are the same length and direction as b I will call the *parallelograms ab*.

I will regard two parallelograms ABC and DEF that lie in parallel planes as having the *same sign* if and only if one can, by parallel progression of their planes and movement of the parallelograms within their planes, bring them into such a position that, while AB and DE lie on the same straight line in the same direction, C and F are found on the same side of this straight line.

240. DEFINITION. The parallelepiped {Spat} in which AB, BC, CD are three edges not lying in the same plane I will call the *parallelepiped ABCD*, AB its *first edge*, BC its *second*, CD its *third.* And all parallelepipeds whose first edge is the same length and direction as a, whose second is the same as b, and whose third is the same as c I will call the *parallelepipeds abc*.

I will regard two parallelepipeds $ABCD$ and $EFGH$ as having the *same sign* if and only if one can bring them to such positions that, while ABC and EFG are parallelograms of the same sign in the same plane, D and H lie on the same side of this plane.

COROLLARY. The parallelepipeds *abc*, *bca*, *cab* are equal to one another (including their signs).

241. LOAN THEOREM. Two parallelograms, whose first and whose second sides are the same lengths and similarly directed, are equal to one another (including their signs) and lie in parallel* planes. Two parallelepipeds, whose corresponding edges (first, second, third) are the same lengths and similarly directed, are equal to one another (including their signs); that is, all parallelograms denoted by the same symbol *ab* are equal to one another, and likewise for all parallelepipeds denoted by the same symbol *abc*.

*The identity is to be reckoned among the parallels.

242. DEFINITION. I say of two parallelograms that lie in parallel planes, and likewise of two arbitrary parallelepipeds, that they are related as two numbers α and β if they have the same or opposite signs according as α and β do, and if, apart from sign, they are related as α to β (cf. 221).

243. LOAN THEOREM. Two parallelograms ABC and ABD (with the same base side AB) are equal (including their signs) if and only if CD is parallel to AB.

244. LOAN THEOREM. Two parallelepipeds $ABCD$ and $ABCE$ (with the same base surface ABC) are equal (including their signs) if and only if DE is parallel to the plane ABC.

REMARK. From these preliminary theorems, which are borrowed from geometry, we can now derive the concept of the combinatorial product of points directly from the general concept of the combinatorial product.

245. *The combinatorial product of two points is zero if and only if the two points coincide, the combinatorial product of three points if they lie on a straight line, the combinatorial product of four points if they lie in the same plane; and the combinatorial product of five points is always zero.*

PROOF. According to 61 and 66 the combinatorial product of two or more magnitudes is zero if and only if they stand in a numerical relation to one another; according to 216 two points stand in a numerical relation if and only if they coincide, three points (according to 234 and 235) when they lie on the same straight line, four points (according to 233, 236) when they lie in the same plane, and according to 232 five points always stand in a numerical relation. Thus the proof.

246. *If A is a finitely distant point, and b, c, d are infinitely distant points, that is displacements, then*

from	$[Ab] = 0$	*follows the equation*	$b = 0,$
from	$[Abc] = 0$	*follows the equation*	$[bc] = 0,$
from	$[Abcd] = 0$	*follows the equation*	$[bcd] = 0.$

PROOF. Let $[Abcd] = 0$. Assume now that $[bcd]$ is not zero; then (according to 61) b, c, d cannot stand in a numerical relation to one another. But since $[Abcd] = 0$, there must {according to 66} exist a numerical relation between A, b, c, d, and since b, c, d stand in no numerical relation to one another, A must (according to 2) be numerically derivable from b, c, d. But (according to {221}, 220) under numerical derivation from the infinitely distant points or displacements b, c, d there can only result displacements, that is, points of the infinitely distant plane, and therefore not the finitely distant point A. Consequently the assumption that $[bcd]$ is different from zero is contrary to hypothesis, that is $[bcd]$ must be zero. The other parts of the theorem follow in exactly the same way.

247. *A combinatorial product $[AB]$ of two simple points A and B is equal to a combinatorial product $[CD]$ of two simple points C and D if and only if the infinite straight lines AB and CD coincide, and AB is the same length and direction as CD.*

PROOF. 1. If the infinite straight lines AB and CD coincide, and AB is the same length and direction as CD, then it is to be proved that $[AB] = [CD]$.

Since AB and CD have the same length and direction, it follows (according to the Corollary to 222) that

$$(*) \qquad B - A = D - C.$$

Further, since A, B, C lie on the same straight line (by hypothesis), $B - A$ and $C - A$ are displacements on the same line, and therefore (according to 230A) they stand in a numerical relation to one another. Let

$$(**) \qquad C - A = \alpha(B - A).$$

Now

$$\begin{aligned} [CD] &= [C(D - C)] && [67] \\ &= [C(B - A)] && [*] \\ &= [(A + C - A)(B - A)] \\ &= [(A - \alpha(B - A))(B - A)] && [**] \\ &= [A(B - A)] && [67] \\ &= [AB]. && [67] \end{aligned}$$

2. Assuming

$$[AB] = [CD],$$

it is then to be proved that A, B, C, D lie on the same straight line, and that AB is the same length and direction as CD.

If $[AB] = [CD]$, then (according to 76) C and D must be derivable from A and B by a linear evolution. The elementary linear evolution of two magnitudes consists (according to 71) in the addition to one of them a multiple of the other, thus for example A and B are transformed into A and $B + \alpha A$. The resulting new magnitude is therefore numerically derived from the two original magnitudes and therefore (according to 235) it lies on the straight line connecting those magnitudes, whence from A and B continued linear evolutions can only result in points of the straight line AB; consequently C and D lie on the straight line AB. Now let E be a point of the straight line AB such that CE is the same length as AB; then (according to Part 1 of the Proof)

$$[CE] = [AB],$$

and by hypothesis

$$[AB] = [CD],$$

and thus in addition

$$[CE] = [CD];$$

consequently

$$0 = [CD] - [CE] = [C(D - E)].$$

As a result (according to 246),

$$D - E = 0,$$

that is

$$D = E.$$

Now since by assumption CE is the same length and direction as AB, so also is CD, which is identical to CE, the same length and direction as AB.

248. COROLLARY. *If A, B, C, D are simple points, from the equation*

$$[AB] = [CD]$$

it follows that

$$A - B = C - D,$$

but the converse does not hold.

249. DEFINITION. We call the product $[AB]$ a *line element*, and say that it is an element of the unbounded straight line AB, and is the same length and direction as the bounded straight line AB.

250. COROLLARY. *Two line elements are equivalent if and only if they have the same length and direction, and are elements of the same unbounded straight line.*

251. *The combinatorial product of a simple point with a displacement is a line element that lies on the straight line drawn parallel to the displacement through the point, and is the same length and direction as the displacement.*

PROOF. Let A be a point and p be a displacement. If one draws through A a straight line AB the same length and direction as p, then (according to the Corollary to 222), $p = B - A$, whence

$$[Ap] = [A(B - A)] = [AB], \qquad [67]$$

and $[AB]$ is a line element lying on the straight line AB, and thus on the line parallel to p drawn through A, and having the same length and direction as AB, and therefore p.

252. *The product of a line element $[AB]$ with a number α is a line element lying on the same unbounded straight line, which is algebraically related to $[AB]$ as $\alpha : 1$.*

PROOF.

$$\alpha[AB] = \alpha[A(B - A)] \qquad [67]$$

$$= [A \cdot \alpha(B - A)]. \qquad [40]$$

This last product is (according to 251) a line element lying on the straight line drawn through A parallel to $\alpha(B - A)$, that is it lies on the straight line AB, and has the same length and direction as $\alpha(B - A)$, that is (according to 221), it is to AB as $\alpha : 1$.

253. *If A and B are simple points and α and β are numbers, $[\alpha A \cdot \beta B]$ is a line element lying on the unbounded straight line AB and is to the bounded straight line AB as $\alpha\beta$ to 1.*

PROOF. $[\alpha A \cdot \beta B] = \alpha\beta[AB]$ (according to 46), whence (according to 252) it is a line element of the unbounded straight line AB, which is to the bounded line AB as $\alpha\beta$ to 1.

254. *Two different nonzero combinatorial products $[ab]$ and $[cd]$ of any two displacements a and b, c and d, are equal to one another if and only if the parallelograms ab and cd are equal in content and sign, and lie in parallel planes.*

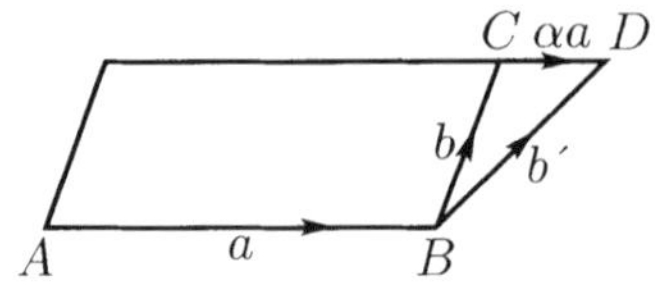

FIGURE 8

PROOF. In 72 and 76 it was proved that two combinatorial products $[ab]$ and $[cd]$ are equal to one another if and only if c and d are derivable from a and b by linear evolutions; and the elementary linear evolution of two magnitudes consists (according to 71) in the addition to one of them a multiple of the other, while this other one remains unchanged, that is, therefore, that a and b are transformed either into a and $b + \alpha a$, or $a + \beta b$ and b, where α and β are arbitrary numbers.

Now let AB be the same length and direction as a, BC the same as b {see Fig. 8}, and let b be evolved into $b' = b + \alpha a$, and further let CD be drawn parallel to AB and be algebraically related to AB as $\alpha : 1$; then (according to {the Corollary to} 222)

$$B - A = a, \quad C - B = b, \quad D - C = \alpha a.$$

Then

$$D - B = D - C + C - B = \alpha a + b = b',$$

that is, BD is the same length and direction as b'. Further, since AB and CD are parallel, (according to 243) the parallelograms ABC and ABD are equal to one another (including their signs) and lie in *the same* plane. Thus the parallelograms ab and ab' are equal and have the same sign and lie in parallel planes. The same is true when a and b are evolved into $a + \beta b$ and b. It therefore follows that, if c and d result from a and b by elementary linear evolutions, the parallelograms ab and cd are equal (including their signs) and lie in parallel planes. The same is also true when c and d result from a and b by the application of several elementary linear evolutions, that is by an arbitrary linear evolution. Consequently it follows that:

First. If $[ab] = [cd]$, then c and d must be derivable from a and b by a linear evolution (76); and if c and d are derivable from a and b by linear evolutions, the parallelograms ab and cd must be equal (including their signs) and lie in parallel planes.

Second. Conversely, supposing ab and cd are equal (including their signs) and lie in parallel planes, then, since a, b, c, d are parallel to the same plane, c and d are (according to 230) numerically derivable from a and b, it follows that (according to 63) the combinatorial products $[ab]$ and $[cd]$ stand in a numerical relation to one another.

Let $[cd] = \alpha[ab]$ be the expression for this numerical relation. If we set $\alpha b = b'$, then $[cd] = [a \cdot \alpha b] = [ab']$. Thus (according to Part 1 of the Proof) the parallelograms cd and ab' are equal and have the same sign, and therefore, since by hypothesis cd and ab are equal and have the same sign, ab and ab' are also equal and have the same sign.

Now let AB have the same length and direction as a, BD the same as b'; then the parallelogram ABC is a parallelogram with the same sign as ab and ABD a parallelogram with the same sign as ab'. Thus ABC and ABD are equal and have the same sign, and consequently, since ABC and ABD also lie in the same plane, (according to 243) C and D lie on a line parallel to AB. Now BC and BD are both parallel to b, and thus to each other, an thus, since they have a point (B) in common, they lie on the same straight line, whence, since D also lies on the straight line drawn parallel to AB through C, C and D coincide. Consequently BC and BD are identical, and thus b and b', the first of which has the same length and direction as BC, the second the same as BD, are also equal to one another; it therefore follows, since one set $\alpha b = b'$, that $\alpha = 1$. Now one had also set

$$[cd] = \alpha[ab],$$

whence, since $\alpha = 1$,

$$[cd] = [ab].$$[77]

255. *Two different nonzero combinatorial products $[ABC]$ and $[DEF]$, each of three simple points A, B, C and D, E, F, are equal to one another if and only if the parallelograms ABC and DEF are equal and have the same sign and lie in the same plane.*

PROOF. 1. Let ABC and DEF be equal and have the same sign and lie in the same plane; then it is to be shown that $[ABC] = [DEF]$.

Let AB be the same length and direction as a, BC be the same as b, DE the same as c, EF the same as d, that is

$$(*) \qquad B - A = a, \quad C - B = b, \quad E - D = c, \quad F - E = d;$$

then (according to 254)

$$(**) \qquad [ab] = [cd].$$

Further, since D lies in the plane ABC, and in addition $B - A = a$ and $C - B = b$ are displacements in this plane, (according to 230) $D - A$ must be numerically derivable from a and b. Let

$$(***) \qquad D - A = \alpha a + \beta b;$$

then

$$\begin{aligned}
[DEF] &= [DE(F-E)] = [D(E-D)(F-E)] && [67]\\
&= [Dcd] && [*]\\
&= [D(cd)] && [80]\\
&= [D(ab)] && [**]\\
&= [(\alpha a + \beta b + A)ab] && [***, 80]\\
&= [Aab] && [67]\\
&= [A(a+A)(b+A)] && [67]\\
&= [ABC]. && [*]
\end{aligned}$$

2. Conversely, let it be assumed that

$$[ABC] = [DEF].$$

Then (according to 76) D, E, F must be derivable from A, B, C by linear evolutions, and thus must be numerically derivable from them. But then (according to 236) D, E, F must lie in the plane ABC. Now on the straight line BC let a point G be chosen such that ABG and DEF are equal and have the same sign; then (according to Part 1 of the Proof), since ABG and DEF lie in the same plane (ABC),

$$[ABG] = [DEF].$$

But by hypothesis

$$[ABC] = [DEF].$$

Thus $[ABG] = [ABC]$. Now since G is a point on BC, $G - B$ is numerically derivable from $C - B$; let $G - B = \alpha(C - B)$. Then $G = B + \alpha(C - B)$, and therefore

$$[ABC] = [ABG] = [AB(B + \alpha(C - B))] = [AB\alpha(C - B)]$$
$$= \alpha[ABC]. \qquad [40, 67]$$

Thus $\alpha = 1$. Consequently, since $G - B = \alpha(C - B)$, $G - B = C - B$, that is $G = C$, or, the points G and C coincide; therefore the parallelograms ABG and ABC coincide. It therefore follows, since ABG and DEF are equal parallelograms with the same sign, lying in the same plane, that the same is also true of ABC and DEF.

256. COROLLARY. *If A, B, C, D, E, F are simple points, then from the equation*

$$[ABC] = [DEF]$$

there follows the equation

$$[(B - A)(C - B)] = [(E - D)(F - E)];$$

the converse holds, however, only if the condition is added that the planes ABC and DEF are not just parallel, but ***identical.***

257. DEFINITION. We call the product $[ABC]$ a *surface element* and the area of the parallelogram ABC its *content*, and say that the surface element $[ABC]$ lies in the plane ABC.

REMARK. The more precise name for the product $[ABC]$ would be a plane element {Ebenenteil} instead of a surface element {Flächenteil}. But the former expression must be rejected on account of the similarity of its plural {die Ebenenteile} to the expression "the plane areas" {die ebenen Teile}.

258. COROLLARY. *Two surface elements are equal to one another if and only if they lie in the same plane and their contents are equal and have the same sign.*

REMARK. One could also have defined the content of the surface element $[ABC]$ as the area of the triangle ABC. But it will be shown below {cf. No. 331} that then the content of the inner square of a displacement would only be half of the content of the square {of the length} of this displacement, whereas the two are in agreement with our nomenclature.[78]

259. *The combinatorial product of two simple points A, B and a displacement c is a surface element lying in the plane containing AB parallel to c, whose content is equal to that of a parallelogram ABC in which BC has the same length and direction as c, that is*

$$[ABc] = [ABC],$$

if $c = C - B$.

PROOF.

$$[ABc] = [AB(C - B)] = [ABC]. \qquad [67]$$

260. *The combinatorial product of a simple point A with two displacements b and c is a surface element lying in the plane through A parallel to b and c, having for content the area of a parallelogram ABC whose first side (AB) is the same length and direction as b, and whose second (BC) the same as c, that is*

$$[Abc] = [ABC],$$

where $b = B - A$, $c = C - B$.

PROOF.

$$\begin{aligned}[Abc] &= [A(B - A)(C - B)] = [AB(C - B)] && [67]\\ &= [ABC]. && [67]\end{aligned}$$

261A. *The product $\alpha[ABC]$ of a surface element $[ABC]$ with a number α is a surface element of the same plane whose content is to that of $[ABC]$ as $\alpha : 1$.*

PROOF.

$$\begin{aligned}\alpha[ABC] &= \alpha[AB(C - B)] && [67]\\ &= [AB \cdot \alpha(C - B)] && [40]\\ &= [AB(D - B)],\end{aligned}$$

if BD is parallel to BC, and is to it as $\alpha : 1$. But (according to 67) this is

$$= [ABD],$$

that is, it is equal to surface element of the same plane (ABC), whose content is equal to the area of the parallelogram ABD. But since BD and BC are parallel and are related algebraically as $\alpha : 1$, so also the parallelograms ABD and ABC are related as $\alpha : 1$, that is the contents of $\alpha[ABC]$ and $[ABC]$ are as $\alpha : 1$.

261B. *If A, B, C are elementary points, and α, β, γ are numbers, then*

$$[\alpha A \cdot \beta B \cdot \gamma C]$$

is a surface element of the plane ABC, whose content is algebraically related to that of the parallelogram ABC as $\alpha\beta\gamma : 1$.

PROOF. $[\alpha A \cdot \beta B \cdot \gamma C] = \alpha\beta\gamma[ABC]$ (No. 46), and is thus proved according to 261A.

262. *Two different nonzero combinatorial products $[abc]$ and $[def]$, each of any three displacements a, b, c and d, e, f, are equal to one another if and only if the parallelepipeds abc and def are equal and have the same sign.*

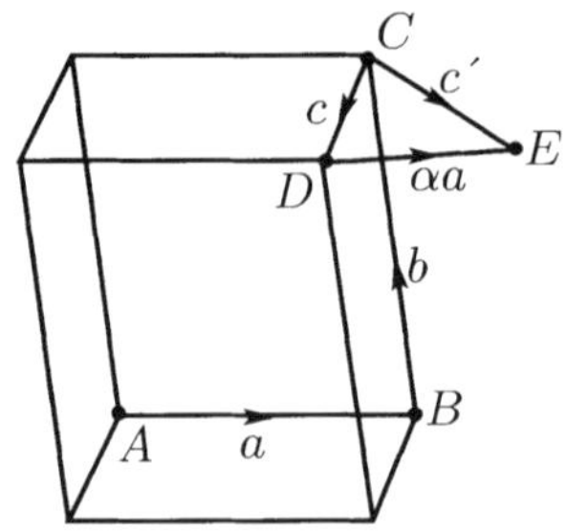

FIGURE 9

PROOF. 1. If it is assumed that

$$[abc] = [def],$$

then it is to be shown that the parallelepipeds abc and def are equal and have the same sign.

Since $[abc] = [def]$, d, e, f must (according to 76) result from a, b, c by linear evolutions. Now we can show that an elementary linear evolution[79] of the three sides a, b, c of a parallelepiped abc always produces an equal parallelepiped with the same sign. The elementary linear evolution of the three magnitudes consists (according to 71) in adding to one of them a multiple of one of the other two, while these other two remain unchanged. Thus if to the third, c, there is added a multiple of one of the first two, for example a, then a, b, c are transformed into a, b, c', where $c' = c + \alpha a$. Thus {see Fig. 9} let AB, BC, CD, DE be the same length and direction as a, b, c, αa, respectively; that is

$$B - A = a, \quad C - B = b, \quad D - C = c, \quad E - D = \alpha a;$$

then

$$E - C = E - D + D - C = \alpha a + c = c',$$

whence CE is the same length and direction as c'. Further, since DE is parallel to a, and thus also to AB, and consequently to the plane ABC as well, (according to 244) the parallelepipeds $ABCD$ and $ABCE$, or, what is the same thing, the parallelepipeds abc and abc', are equal and have the same sign, that is, the parallelepiped remains unchanged when to the third side a multiple of one of the two others is added. Furthermore (according to the Corollary to 240), $abc = bca = cab$, and likewise $abc' = bc'a = c'ab$. Thus in addition, since $abc = abc'$, one has $bca = bc'a$ and $cab = c'ab$, that is, a parallelepiped remains unchanged if the second edge or the first is changed by adding to it a multiple of one of the other two edges.

Consequently a parallelepiped remains equal to itself, including sign, under repeated progressive elementary linear evolutions of its edges, that is, under arbitrary linear evolutions. Since, however, according to the above d, e, f are derivable from a, b, c by linear evolutions, the parallelepiped def must be equal to and have the same sign as abc.

2. Now conversely let it be assumed that the parallelepipeds def and abc are equal and have the same signs; then it is to be shown that $[def] = [abc]$.

Since it is assumed that the combinatorial products are not zero, a, b, c are not parallel to the same plane, and thus (according to 229) d, e, f are numerically derivable from them, and also (according to 63) $[def]$ is numerically derivable from

$[abc]$. Let $[def] = \alpha[abc]$; then if one sets $\alpha c = c'$, $[def] = [abc']$, whence (according to Part 1 of the Proof) the parallelepipeds def and abc' are equal. Now by hypothesis the parallelepipeds def and abc are equal; consequently the parallelepipeds abc and abc' are equal. Let AB, BC, CD, CD' be the same length and direction as a, b, c, c', respectively. Then the parallelepipeds

$$ABCD = abc, \quad ABCD' = abc',$$

and therefore

$$ABCD = ABCD'.$$

Consequently (according to 244) D and D' lie in a plane parallel to the plane ABC, but D and D' also lie on the straight line CD, since CD' is parallel to c', that is to αc, and thus also to c, that is to CD. Therefore D and D' lie on the intersection of that plane and that line, that is they coincide. Thus CD and CD' are identical, whence $c = c'$, and thus, by virtue of the equation $c' = \alpha c$, $\alpha = 1$; consequently the equation $[def] = \alpha[abc]$ is transformed into

$$[def] = [abc].$$

263. *Two different nonzero combinatorial products $[ABCD]$ and $[EFGH]$, each of any four simple points A, B, C, D and E, F, G, H, are equal to one another if and only if the parallelepipeds $ABCD$ and $EFGH$ are equal and have the same sign.*

Proof. 1. Let $ABCD$ and $EFGH$ be equal parallelepipeds with the same sign, and let AB, BC, CD, EF, FG, GH be the same length and direction as b, c, d, f, g, h respectively, that is $B - A = b, \dots$; then (according to 262)

$$(*) \qquad [bcd] = [fgh].$$

Further, since (according to 229) all the displacements of space are numerically derivable from b, c, d, this must also be true of the displacement $E - A$; let

$$E - A = \beta b + \gamma c + \delta d,$$

that is

$$E = A + \beta b + \gamma c + \delta d.$$

Then one obtains

$$[EFGH] = [EFG(H-G)] = [EF(G-F)(H-G)]$$
$$= [E(F-E)(G-F)(H-G)]. \qquad [67]$$

Thus, since $F - E = f$, $G - F = g$, $H - G = h$, one finds that the last expression found

$$= [Efgh] = [E(FGH)] \qquad [80]$$
$$= [E(bcd)]. \qquad [*]$$

Further, the expression found

$$
\begin{aligned}
&= [Ebcd] && [79]\\
&= [(A + \beta b + \gamma c + \delta d)bcd] = [Abcd] && [67]\\
&= [A(B - A)(C - B)(D - C)],
\end{aligned}
$$

if one replaces b, c, d by their values, and thereby one obtains, upon application of 67,

$$= [AB(C - B)(D - C)] = [ABC(D - C)] = [ABCD].$$

Thus

$$[EFGH] = [ABCD].$$

2. Conversely, let it be assumed that

$$[ABCD] = [EFGH],$$

and on the straight line CD choose a point D' such that the parallelepiped $ABCD'$ is equal to (and has the same sign as) $EFGH$; then (according to Part 1 of the Proof)

$$[ABCD'] = [EFGH].$$

Thus also, since it is hypothesized that $[EFGH] = [ABCD]$,

$$[ABCD'] = [ABCD].$$

Now, since $D - C$ and $D' - C$ are parallel, $D' - C$ is numerically derivable from $D - C$. Let

$$D' - C = \alpha(D - C);$$

then

$$
\begin{aligned}
[ABCD] = [ABCD'] = [ABC(D' - C)] &= [ABC \cdot \alpha(D - C)]\\
&= \alpha[ABC(D - C)] && [40]\\
&= \alpha[ABCD]. && [67]
\end{aligned}
$$

Thus, since $[ABCD]$ is not zero, $\alpha = 1$, and consequently, from the equation $(D' - C) = \alpha(D - C)$, it follows that

$$D' - C = D - C,$$

whence $D' = D$, that is D and D' coincide, and since the parallelepiped $ABCD'$ is equal to and has the same sign as $EFGH$, the same is true of the parallelepipeds $ABCD$ and $EFGH$.

264. Corollary. *The equations*

$$[ABCD] = [EFGH]$$

and

$$[(B - A)(C - B)(D - C)] = [(F - E)(G - F)(H - G)],$$

or

$$[(B - A)(C - A)(D - A)] = [(F - E)(G - E)(H - E)]$$

are mutually replaceable, that is, from each of them the two others follow.

PROOF. The equation

$$[ABCD] = [EFGH]$$

holds (according to 263) if and only if the parallelepipeds $ABCD$ and $EFGH$ are equal and have the same sign. Likewise, the equation

$$[(B - A)(C - B)(D - C)] = [(F - E)(G - F)(H - G)]$$

holds (according to 262) if and only if the parallelepiped whose three edges are the same length and direction as AB, BC, CD is equal to the parallelepiped whose edges are the same length and direction as EF, FG, GH, that is, if and only if the parallelepiped $ABCD$ is equal in content and sign to $EFGH$. Consequently the two equations always hold in the same cases.

Finally, the third equation is just a transformation of the second, for

$$\begin{aligned}
&[(B - A)(C - B)(D - C)] \\
&\quad = [(B - A)(C - B)(D - C + C - B + B - A)] && [67] \\
&\quad = [(B - A)(C - B)(D - A)] = [(B - A)(C - B + B - A)(D - A)] && [67] \\
&\quad = [(B - A)(C - A)(D - A)],
\end{aligned}$$

and on the same grounds

$$[(F - E)(G - F)(H - G)] = [(F - E)(G - E)(H - E)].$$

Thus the two equations are equivalent.

265. DEFINITION. We call the product $[ABCD]$ of four simple points a *volume element*, and the volume of the parallelepiped $ABCD$ (including its sign, $\pm$), its *content*.

266. *The combinatorial product of three simple points A, B, C and a displacement d is a volume element whose content is equal to that of a parallelepiped $ABCD$ in which CD is the same length as d, that is*

$$[ABCd] = [ABCD], \quad \textit{where} \quad d = D - C.$$

PROOF.

$$[ABCd] = [ABC(D - C)] = [ABCD]. \qquad [67]$$

267. *The combinatorial product of two simple points A, B and two displacements c and d is equal in content to the parallelepiped $ABCD$ in which BC is the same length and direction as c, CD the same as d, that is*

$$[ABcd] = [ABCD],$$

if $c = C - B$, $d = D - C$.

PROOF.

$$\begin{aligned}
[ABcd] &= [AB(C - B)(D - C)] \\
&= [ABC(D - C)] = [ABCD]. \qquad [67]
\end{aligned}$$

268. *The combinatorial product of a simple point A with three displacements b, c, d is equal in content to the parallelepiped bcd, or*

$$[Abcd] = [ABCD],$$

where $b = B - A$, $c = C - B$, $d = D - C$.

PROOF.

$$\begin{aligned} [Abcd] &= [A(B-A)(C-B)(D-C)] \\ &= [AB(C-B)(D-C)] && [67] \\ &= [ABC(D-C)] = [ABCD]. && [67] \end{aligned}$$

269. *The product $\alpha[ABCD]$ of a volume element $[ABCD]$ and a number $\{\alpha\}$ is a volume element whose content is to that of $[ABCD]$ as $\alpha : 1$.*

PROOF.

$$\begin{aligned} \alpha[ABCD] &= \alpha[ABC(D-C)] && [67] \\ &= [ABC \cdot \alpha(D-C)] && [40] \\ &= [ABC(E-C)], \end{aligned}$$

where CE is parallel to CD and is to it as $\alpha : 1$. This again is (according to 67) equal to

$$= [ABCE].$$

Since however CE and CD are parallel and related as $\alpha : 1$, the parallelepipeds $ABCE$ and $ABCD$ are also algebraically related as $\alpha : 1$, that is, the contents of $\alpha[ABCD]$ and $[ABCD]$ are as $\alpha : 1$.

270. *If A, B, C, D are simple points and α, β, γ, δ are numbers, then*

$$[\alpha A \cdot \beta B \cdot \gamma C \cdot \delta D]$$

is a volume element which is to $[ABCD]$ as $\alpha\beta\gamma\delta$ to 1.

PROOF. $[\alpha A \cdot \beta B \cdot \gamma C \cdot \delta D] = \alpha\beta\gamma\delta[ABCD]$ (according to 46), whence (according to 269) the proof.

REMARK. If we look back over the various combinatorial products whose concepts we have more precisely defined, then for two, three and four simple points there results the one-fold, two-fold, and six-fold of the line, surface and volume elements lying between them, and the domains belonging to them are the unbounded straight line, plane, and unbounded space. Further, just as the infinitely distant point appears as a displacement of fixed length and direction, so the infinitely distant line element[80] appears as a bounded plane of fixed area and direction, and the infinitely distant surface element as a volume element of fixed content.

If onto a displacement or onto a product of two or three displacements a point is introduced as the first factor, this product yields the same content and direction as if the point had not been introduced. By the introduction of the point there is added to the previous definitions (content and direction), in the first case the line laid through the point parallel to the displacement, in the second the plane laid through the point parallel to the two displacements, which plane forms the domain of those magnitudes; thus the displacement is transformed into a line element, the surface of fixed area and direction into what we have called a surface element. The product of three displacements is only formally changed by the introduction of the point.

271. *If A, B, C, D, E, F are points and a, b, c, d displacements, then*

1) $$A \equiv B$$

means that A coincides with B,

2) $$[AB] \equiv [CD]$$

that the unbounded straight lines AB and CD coincide,

3) $$[ABC] \equiv [DEF]$$

that the unbounded planes ABC and DEF coincide,

4) $$a \equiv b$$

that a is parallel to b, and

5) $$[ab] \equiv [cd]$$

that the plane incorporating the directions a and b is parallel to the plane incorporating the directions c and d.

PROOF. According to No. 2 the congruence $p \equiv q$ of two extensive magnitudes means that p and q stand in a numerical relation to one another, and that neither of them is zero. Now if 1) holds for A and B, then (according to 216) their positions must coincide; if 2) holds for $[AB]$ and $[CD]$, then (according to {252 and} 247) the unbounded straight lines AB and CD must coincide; if 3) holds for $[ABC]$ and $[DEF]$, then (according to {261A and} 255) the planes ABC and DEF must coincide. Finally, 4) and 5) follow from 1) and 2) if one displaces the points to an infinite distance.

4. Addition of lines and surfaces

272. *Upon summation, two line elements in the same plane yield again a line element in the same plane, and two surface elements {in space} yield again, upon summation, a surface element.*

PROOF. Since (according to 249) the line element is a combinatorial product of two points, and (according to 257) the surface element a combinatorial product of three points, and (according to 228) points are magnitudes of first order, (according to 77B) the line element and the surface element are respectively elementary magnitudes of second and third order. Furthermore (according to 237) the plane is a domain of third and space a domain of fourth order. But according to 88 the magnitudes of $(n-1)$th order in a principal domain of nth order yield upon summation an elementary (that is, representable as a combinatorial product) magnitude of $(n-1)$th order in the same principal domain, whence the line elements

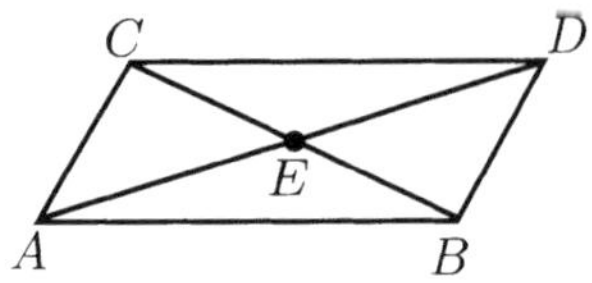

FIGURE 10

of a single plane a line element of the same plane, the surface elements in space a surface element.

COROLLARY. *The same is also true for more than two line elements in the same plane, and for more than two surface elements {in space}.*

273. *Two finitely distant line elements,* ***whose lines intersect,*** *yield upon summation a finitely distant line element whose line goes through the same point of intersection and which is the same length and direction as the diagonal of a parallelogram whose sides produced from the same vertex, are the same length and direction as the summed line elements.*

PROOF. Let A be the point of intersection of the two lines, and $[AB]$ and $[AC]$ the two line elements, where A, B, C are simple points {see Fig. 10}; then

$$[AB] + [AC] = [A(B + C)] = 2[AE],$$

if E is the mean between B and C. But AE is half the diagonal of the parallelogram CAB, whence $2AE$ is the whole of it.

274. *Two finitely distant,* ***similarly directed*** *line elements yield upon summation another similarly directed line element {in their plane}, whose length is the sum of the lengths of the summands and whose straight line lies between the straight lines of the summands and is distant from these lines in reciprocal proportion to the lengths of the summands.*

PROOF. Let $[Ap]$ and $[Bq]$ be these line elements, where A and B are simple points and p and q are similarly directed displacements, and let $1 : \alpha$ be the ratio of their lengths, that is (according to 251) the ratio of p to q, whence $q = \alpha p$ {where α is positive}; then

$$\begin{aligned} [Ap] + [Bq] &= [Ap] + [B \cdot \alpha p] = [Ap] + [\alpha B \cdot p] && [40] \\ &= [(A + \alpha B)p] = [(1 + \alpha)S \cdot p] && [\{39\}, 225] \end{aligned}$$

if S is the summation point of A and αB,

$$\begin{aligned} &= [S \cdot (1 + \alpha)p] && [40] \\ &= [S(p + \alpha p)] = [S(p + q)], \end{aligned}$$

that is, the sum is a line element, directed similarly to the summands, whose length is equal to that of the displacement $(p + q)$, and thus is equal to the sum of the lengths of the summands, and whose line goes through S. But (according to 225) S lies on the straight line AB, and is distant from A and B in the proportion $\alpha : 1$, that is, in inverse proportion to the summands (p and q), and therefore the straight line $[S(p + q)]$ is distant from the straight lines $[Ap]$ and $[Bq]$ in this ratio.

FIGURE 11

275. *Two finitely distant,* ***oppositely directed but not equally long*** *line elements yield upon summation a finitely distant line element {in their plane}, which is directed similarly to the greater of the summands, whose length is the difference between the lengths of the summands, and whose line lies outside the lines of the summands (on the side of the greater summand), and is distant from these lines in inverse ratio to the lengths of the summands.*

PROOF. As in 274, except that one replaces α with $-\alpha$.

276. *The sum of two* ***oppositely directed and equally long*** *line elements* $[AB]$ *and* $[CD]$ *is a product of displacements whose content is the same and has the same sign as that of the parallelogram* ABC *{or* CDA*}, which has one of the line elements as its base side, the other as its covering side.*

PROOF. If AB is the same length as and is oppositely directed to CD {see Fig. 11}, then (according to the Corollary to 222)

$$B - A = C - D.$$

Thus

$$\begin{aligned} [AB] + [CD] &= [A(B-A)] + [(C-D)D] && [67] \\ &= [A(B-A)] + [(B-A)D] && [\text{Hyp.}] \\ &= -[(B-A)A] + [(B-A)D] && [55] \\ &= [(B-A)(D-A)], \end{aligned}$$

that is, it is equal to a product of displacements whose content is equal to that of a parallelogram whose first side is the same length and direction as AB, and whose second side is the same as AD. But this is the parallelogram ABC {which is equal to and has the same sign as the parallelogram CDA}, whence the proof.

277. *The sum of a finitely distant line element* $[AB]$ *and a combinatory product* $[ab]$ *of two displacements* a *and* b*, one of which is parallel to the plane laid through the line element* $[AB]$*, is a finitely distant line element* $[DC]$ *in the same plane which is the same length and direction as the former {see Fig. 11} and is laid so that the parallelogram* ABC *having the first line element as its base side, the second as its covering side, is opposite to the combinatorial product* $[ab]$ *(that is, it is equal in content but opposite in sign).*

PROOF. If we denote by P the product of displacements equal to {the parallelogram} ABC, then according to the previous theorem

$$[AB] + [CD] = P,$$

whence

$$\begin{aligned}[DC] &= [AB] - P \\ &= [AB] + [ab],\end{aligned}$$

since by hypothesis

$$-P = [ab].$$

278. *The sum of two combinatorial products $[ab]$ and $[cd]$ of any two displacements is again a combinatorial product of two displacements, and is such that, if those are represented in the form of two parallelograms upon the same (or an equally long and similarly directed) base line, the sum can be represented as a parallelogram upon the same (or an equally long and similarly directed) base line, in which the second side is the displacement sum of the second sides of those parallelograms.*

PROOF. One lays a plane parallel to a and b, and another parallel to c and d, and lets e be a displacement parallel to the line of intersection of the two planes (and, if they do not intersect, parallel to any one of their lines). Then one can (according to 254) cast $[ab]$ into the form $[ef]$ and $[cd]$ into the form $[eg]$, so that then

$$[ab] + [cd] = [ef] + [eg] = [e(f+g)],$$

and this is the required form of the sum.

279. *Two finitely distant surface elements* ***whose planes intersect*** *yield upon summation a surface element whose plane goes through the line of intersection of those planes so that, if the summands are represented as parallelograms with a common base side, the sum can be represented as a parallelogram having the same base side, and in which the second side is the displacement sum of the second sides of the summands, or expressed otherwise: The summands are equal to the projections of the sum on the two planes of the summands when each plane is parallel projected on the other.*

PROOF. Let A and B be two simple points on the line of intersection of the planes, and c and d be two displacements such that the two surface elements to be added are equal to $[ABc]$ and $[ABd]$; then

$$[ABc] + [ABd] = [AB(c+d)],$$

that is, the sum is represented by a parallelogram in which AB is the base side and $c+d$ is the second side.

280. *The sum of two* ***parallel and similarly signed*** *(finitely distant) surface elements (E_1 and E_2) is a parallel and similarly signed surface element whose content is the sum of the contents of the summands, and whose plane lies between those of the summands so that it is distant from them in inverse proportion to the contents of the summands.*

PROOF. Let $E_1 = [Abc]$, where A is a {simple} point, b and c are displacements, and drop from A a perpendicular AD on the plane E_2; then, since the planes are parallel, $[Dbc]$ is a surface element of the plane of E_2, and thus stands in a numerical relation to E_2. Let $E_2 = \alpha[Dbc]$; then {α being positive, since E_1 and E_2 are similarly signed,}

$$E_1 + E_2 = [Abc] + \alpha[Dbc] = [(A + \alpha D)bc] = [(1 + \alpha)Sbc],$$

where (according to 225) S lies on AD and is distant from A and D in the ratio $\alpha : 1$. Consequently the plane of the sum is a plane laid through S parallel to b and c, and thus is also parallel to the planes of E_1 and E_2, which lies between those two planes and which is distant from them in the ratio $\alpha : 1$, that is, in inverse proportion to their contents (bc and αbc). The content of the sum is (according to 260) equal to the content of $(1 + \alpha)bc$, that is $= bc + \alpha bc$, and thus is equal to the sum of the contents of the summands.

281. *The sum of two* **parallel and oppositely signed but not equal** *in content (finitely distant) surface elements E_1 and E_2 is a surface element parallel to them, with the same sign as the greater, whose content is the difference of the contents of the summands (E_1 and E_2), and whose plane lies outside the space between the two planes of the summands {on the side of the greater summand} in such a way that it is distant from these planes in inverse proportion to the summands.*

PROOF. As in 280, but with $-\alpha$ replacing α.

282. *The sum of two* **parallel, oppositely signed and equal in content** *(finitely distant) surface elements E_1 and E_2 is equal to the combinatorial product of three displacements, and is equal to the content of this product, but oppositely signed to that of a parallelepiped that has E_1 as its base surface and whose covering surface lies in the plane of E_2.*

PROOF. Let $E_1 = [Abc]$, $E_2 = -[Dbc]$; then

$$\begin{aligned} E_1 + E_2 &= [Abc] - [Dbc] = [(A - D)bc] \\ &= -[(D - A)bc]. \end{aligned}$$

But (according to 58) $[(D - A)bc] = [bc(D - A)]$, and this last is (according to 262) equal to the content of a parallelepiped whose first side is the same length and direction as b, whose second side is the same as c, and whose third side is the same as $(D - A)$, and thus whose base surface is $[Abc]$ and whose covering surface goes through D; thus the proof.

283. *The sum of a finitely distant surface element E_1 and a combinatorial product P of three displacements is a surface element E_2 equal in content and sign, and parallel to, the first, which is laid so that the parallelepiped having the first surface element as base side {and the plane of E_2 as covering surface} is equal in content and sign to the given combinatorial product P.*

PROOF. According to 282,

$$E_1 - E_2 = -P,$$

whence

$$E_2 = E_1 + P.$$

284. *The sum of three surface elements* (E_1, E_2, E_3) *whose planes intersect in a vertex* (D) *is a surface element* (E_4) *whose plane goes through the same vertex* (D) *and is so constructed that, if one projects this surface element* (E_4) *on each of the three planes parallel to the line of intersection of the other two, one by one, these projections are equal to the summands* (E_1, E_2, E_3).

PROOF. Let the edges in which the planes E_2 and E_3, E_3 and E_1, and E_1 and E_2 intersect be parallel to the directions a, b, c respectively; then it is first to be proved that E_1, E_2, E_3 are the projections of E_4 on the planes E_1, E_2, E_3 along the directions a, b, c.

In order to prove this first for E_1, let $E_2 + E_3 = E'$; then a is (by hypothesis) parallel to the line of intersection of the two planes E_2 and E_3, and thus also (according to 279) to E'. If one now projects E_4 on the plane E_1 along the direction a, then since a is parallel to E' and $E_4 = E' + E_1$, this projection $= E_1$ (according to 279). In the same way it follows that the projection of E_4 on the plane E_2 along the direction b equals E_2, and that on the plane E_3 along the direction c equals E_3.

Finally, E_4 must also go through D; for (according to 279) by virtue of the equation $E' = E_2 + E_3$, E', E_2, E_3 have the same edge in common, and thus also the point D, which (by hypothesis) lies in E_2 and E_3; further, according the same theorem, E_4, E', E_1, by virtue of the equation $E_4 = E' + E_1$, have the same edge in common, and thus also the point D, which, as we have shown, lies in E' and by hypothesis also in E_1; that is, E_4 also goes through D.

285. *A sum S of line elements can always be reduced to a sum of two line elements, and in fact for one of these line elements one can always arbitrarily choose a point* (A) *through which its line is to go, and for the other a plane BCD in which its line is to lie, except that the point A may not lie in the plane BCD.*

PROOF. Since A, B, C, D do not lie in the same plane, one can (according to 232) numerically derive all the points of space from them, and thus also the points whose multiplication leads to any two line elements whose sum is S. If, after introducing these derivation expressions, one removes all parentheses, and (according to 55) sets

$$[BA] = -[AB],\quad [CA] = -[AC],\quad [DA] = -[AD],$$
$$[CB] = -[BC],\quad [DB] = -[BD],\quad [DC] = -[CD],$$

then one obtains an expression of the form

$$S = \alpha[AB] + \beta[AC] + \gamma[AD] + \delta[BC] + \varepsilon[BD] + \zeta[CD],$$

where α, β, γ, δ, ε, ζ are numbers. But this is

$$= [A(\alpha B + \beta C + \gamma D)] + \delta[BC] + \varepsilon[BD] + \zeta[CD].$$

Here (according to 222, {251} and 253) the first term yields a line element, and (according to the Corollary to 272) the sum $\delta[BC]+\varepsilon[BD]+\zeta[CD]$ yields a (finitely or infinitely distant) line element of the plane BCD, whence the proof.

286. *A sum S of line elements is again a line element if and only if*

$$[SS] = 0.$$

PROOF. 1. If S is a line element $= [AB]$, then

$$\begin{aligned}[SS] &= [AB \cdot AB] \\ &= [ABAB] && [80] \\ &= 0. && [60]\end{aligned}$$

2. If $[SS] = 0$, then (according to 285) S can be reduced to two line elements; let $S = [AB] + [CD]$, whence

$$0 = [SS] = [(AB + CD)(AB + CD)] = [AB \cdot CD] + [CD \cdot AB],$$

since $[AB \cdot AB]$ and $[CD \cdot CD]$ are zero (according to {80 and} 60). But (according to 58) $[CD \cdot AB] = [AB \cdot CD]$, whence

$$0 = 2[AB \cdot CD] \quad \text{or} \quad 0 = [ABCD], \qquad \{80\}$$

that is, A, B, C, D lie in the same plane (according to {66 and} 236), and thus (according to 272) $[AB] + [CD]$ is a line element.

REMARK. One perceives that in statics the center of gravity appears as a sum of points, the static force as a line element, the resultant of static forces as the sum of line elements, and the static moment as a surface element, and from this one can already infer what fruitful applications the analysis developed here permits in statics and mechanics, as I hope to show in a later work. {Cf. also Nos. 346 and 347.}

5. Planimetric and stereometric multiplication

{287. This Number now appears in its proper place as No. 119C. In the original edition, *Grassmann* himself said, regarding No. 287: "This theorem should have followed 119B, and was only omitted there due to an oversight."}

288. DEFINITION. By *planimetric* multiplication I mean relative multiplication in the plane {as a domain of third order}, by *stereometric*, relative multiplication in space {as a domain of fourth order}.

289. *The* ***planimetric*** *product of two line elements $[AB]$ and $[AC]$, whose lines intersect at a finite distance, is a point whose location is the point of intersection (A) of those lines, and whose coefficient, if A, B, C are simple points, is equal to the content of the parallelogram ABC, that is*

$$[AB \cdot AC] = [ABC \cdot A].$$

PROOF. According to 103.

REMARK. Since in planimetric multiplication a surface element must be taken as the unit (by virtue of 94), the coefficient $[ABC]$ is a number, whence $[ABC]A$ is in fact a (simple or multiple) point.

290. *The* ***planimetric*** *product of two parallel line elements $[AB]$ and $[CD]$ is a displacement parallel to the two lines, which is algebraically related to the displacement AB as the parallelogram BCD is to unity.*

PROOF. Since AB is parallel to CD, the displacements $A-B$ and $C-D$ stand in a numerical relation (according to 230A). Let $C-D=\alpha(A-B)$; then

$$\begin{aligned}[AB\cdot CD] &= [(A-B)B\cdot(C-D)D] && [67]\\ &= \alpha[(A-B)B\cdot(A-B)D] && [\text{by hypothesis}]\\ &= \alpha[(A-B)BD](A-B) && [103]\\ &= [(C-D)BD](A-B) && [\text{by hypothesis}]\\ &= [CBD](A-B) && [67]\\ &= [BCD](B-A), && [55]\end{aligned}$$

that is, $[AB\cdot CD]$ is equal to a displacement parallel to AB, and is algebraically related to AB as is $[BCD]$ to 1.

291. *The* ***planimetric*** *product of a line element* $[AB]$ *and a point* C *is, if* A, B, C *are simple points, equal to the content of the parallelogram* ABC, *and therefore is zero* {*if and*} *only if* A, B, C *lie on a straight line.*

PROOF. According to 257 {and 245}.

292. *The* ***planimetric*** *product of three line elements* $[AB]$, $[AC]$, $[BC]$, *that form the sides of a triangle is, if* A, B, C *are simple points, four times as great as the square of this triangle, or, it is equal to the square of the parallelogram* ABC, *that is*

$$[AB\cdot AC\cdot BC]=[ABC]^2.$$

PROOF. Let $[ABC]=\alpha$. If one then sets $A_1=A:\alpha$, it follows that $[A_1BC]=1$, whence (according to 112)

$$[A_1B\cdot A_1C\cdot BC]=1,$$

and thus

$$[AB\cdot AC\cdot BC]=\alpha^2=[ABC]^2.$$

REMARK. We could also have written the formula as

$$[AB\cdot BC\cdot CA]=[ABC]^2.$$

293. *The* ***planimetric*** *product of two magnitudes of first or second order is zero if and only if the magnitudes are incident, that is, for two points when their positions coincide, for two line elements when their lines coincide, and for a line element and a point when the position of the point falls on the line (of that line element).*

PROOF. According to 119C.

294. *The* ***planimetric*** *product of two non-incident line elements is congruent to the point of intersection of their lines, that is*

$$[AB\cdot AC]\equiv A.$$

PROOF. $[AB\cdot AC]=[ABC]A$ (according to 289), whence, since $[ABC]$ is a number, $\equiv A$ (according to 2).

295. *The* ***planimetric*** *product of three line elements is zero if and only if their lines intersect in a (finitely or infinitely distant) point.*

PROOF. According to 119C.

296. *The* ***stereometric*** *product of two surface elements* $[ABC]$ *and* $[ABD]$, *whose planes intersect at a finite distance, is an element of this line of intersection, which, if* A, B, C, D *are simple points, is algebraically related to* $[AB]$ *as is the parallelepiped* $ABCD$ *to unity.*

PROOF. $[ABC \cdot ABD] = [ABCD][AB]$.

REMARK. Since in stereometric multiplication a volume element is taken as unit (according to 94, 288), $[ABCD]$ is a number, and thus $[ABCD][AB]$ is in fact a line element.

297. *The* ***stereometric*** *product of two surface elements* $[ABC]$ *and* $[DEF]$, *whose planes are parallel, is a product of two displacements parallel to this plane, and, if* A, B, C, D, E, F *are simple points, the content of this product is algebraically related to that of the parallelogram* ABC *as is the parallelepiped* $ADEF$ *to unity.*

PROOF.

$$[ABC \cdot DEF] = [A(B-A)(C-A) \cdot D(E-D)(F-D)]. \qquad [67]$$

Now since by hypothesis the planes ABC and DEF are parallel, (according to 230) $E-D$ and $F-D$ are derivable from $B-A$ and $C-A$, whence also the product of the former from that of the latter. If $B-A$ is denoted by p and $C-A$ by q, then $[(E-D)(F-D)]$ is derivable from $[pq]$; let it be $= \alpha[pq]$; then

$$\begin{aligned}[ABC \cdot DEF] &= [Apq \cdot \alpha(Dpq)] = \alpha[Apq \cdot Dpq] && [40]\\ &= \alpha[ADpq][pq] && [107]\\ &= [AD(E-D)(F-D)][pq] && [\{80\}, \text{hypothesis}]\\ &= [ADEF][pq]. && [67]\end{aligned}$$

298. *The* ***stereometric*** *product of two line elements* $[AB]$ *and* $[CD]$, *and likewise that of a surface element* $[ABC]$ *and a point* $\{D\}$, *is, if* A, B, C, D *are simple points, equal to the parallelepiped* $ABCD$.

PROOF.

$$[AB \cdot CD] = [ABC \cdot D] = [ABCD]. \qquad [80]$$

299. *The* ***stereometric*** *product of three surface elements* $[ABC]$, $[ABD]$, $[ACD]$ *that intersect in a finitely distant point* A *is a multiple point whose position is the point of intersection* A, *and, if* A, B, C, D *are the simple vertices of a tetrahedron, then the coefficient belonging to that point is equal to the square of the content of the parallelepiped* $ABCD$.

PROOF. Let $[ABCD] = \alpha$, and let $B_1 = B : \alpha$; then $[AB_1CD] = 1$, whence (according to 112)

$$[AB_1C \cdot AB_1D \cdot ACD] = A,$$

and thus

$$[ABC \cdot ABD \cdot ACD] = \alpha^2 A = [ABCD]^2 A.$$

300. *The* ***stereometric*** *product of four surface elements* $[ABC]$, $[ABD]$, $[ACD]$, $[BCD]$ *is, if* A, B, C, D *are the simple vertices of a tetrahedron, equal to the cube of the parallelepiped* $ABCD$.[81]

PROOF. Let $[ABCD] = \alpha$, and let $A_1 = A : \alpha$; then $[A_1BCD] = 1$, whence (according to 112)

$$[A_1BC \cdot A_1BD \cdot A_1CD \cdot BCD] = 1,$$

and therefore

$$[ABC \cdot ABD \cdot ACD \cdot BCD] = \alpha^3 = [ABCD]^3.$$

301. *The* ***stereometric*** *product of two line elements is zero if and only if their lines lie in the same plane; the stereometric product of two magnitudes of first, second, or third order, but not both simultaneously of second order, is zero if and only if the magnitudes are incident, whence for two points if their positions coincide; for two surface elements if their planes coincide; for a point and a line or surface element if the point lies on the line or the plane of the latter; and for a line element and a surface element if the line of the former lies in the plane of the latter.*

PROOF. According to 119C.

302. *The* ***stereometric*** *product of two non-incident surface elements is congruent to the line of intersection of their planes.*

PROOF. Let a, b, c, d be multiple points; then

$$\begin{aligned} [abc \cdot abd] &= [abcd][ab] && [103] \\ &\equiv [ab], \end{aligned}$$

since $[abcd]$ is a number.

303. *The* ***stereometric*** *product of a surface element and a line element that does not lie in the plane of the former is congruent to the point of intersection of the plane and the line.*

PROOF.

$$\begin{aligned} [abc \cdot ad] &= [abcd]a && [103] \\ &\equiv a, \end{aligned}$$

where again a, b, c, d are multiple points.

304. DEFINITION. In planimetric multiplication I denote by U the product $[ab]$ of two displacements a and b if the parallelogram ab is equal to the surface content chosen as the unit, and likewise in stereometric multiplication I denote by U the product $[abc]$ of three displacements a, b, and c if the parallelepiped abc is equal to the volume chosen as the unit. If these two need to be distinguished I will denote the first by U_2, the second by U_3.

305. *If a is a multiple point, $[AB]$ a line element, $[ABC]$ a surface element, and A, B, C are simple points, then*

$$[aU]$$

is the coefficient of a,

$$[ABC]$$

is equal to the displacement the same length and direction as AB, $= B - A$, and

$$[ABCU]$$

is equal to the displacement product equal and laid parallel to the parallelogram ABC,

$$= [(B - A)(C - A)].$$

PROOF. Let $a = \alpha A$ and $U = [bcd]$, where (according to 304) b, c, d are displacements and the parallelepiped bcd is equal to unity; then

$$[aU] = [\alpha Abcd] = \alpha[Abcd] = \alpha,$$

since $[Abcd]$ is (according to 268) equal in content and sign to $[bcd]$, and thus is equal to one. Further,

$$[ABU] = [AB \cdot bcd] = [A(B - A) \cdot bcd]. \qquad [67]$$

Since here $B - A$, as a displacement, is numerically derivable from b, c, d (according to 229), whence {$B - A$ is subordinate to $[bcd]$, so that} (according to 108)

$$[A(B - A) \cdot bcd] = [Abcd](B - A) = B - A,$$

since $[Abcd] = 1$.

Further,

$$[ABCU] = [ABC \cdot bcd] = [A(B - A)(C - A) \cdot bcd]. \qquad [67]$$

Since $B - A$ and $C - A$ are displacements and thus (according to 229) numerically derivable from b, c, d, $[(B - A)(C - A)]$ is subordinate to $[bcd]$, whence

$$\begin{aligned}[A(B - A)(C - A) \cdot bcd] &= [Abcd][(B - A)(C - A)] \qquad [108]\\ &= [(B - A)(C - A)],\end{aligned}$$

since $[Abcd] = 1$.

REMARK. It is these magnitudes, $[aU]$, $[ABU]$, $[ABCU]$, that in the first edition of the *Ausdehnungslehre* of 1844 I called the *divergences* {Ausweichungen} of the magnitudes a, $[AB]$, $[ABC]$, and therefore a separate name for them is henceforth made superfluous by the use of the *infinitely distant unit* (U).

6. Special laws for a planimetric {and stereometric} product set to zero. Plane {algebraic} curves. {Algebraic surfaces}

306. *The equation of a point x lying in a straight line with the points a and b is*

$$[xab] = 0.$$

PROOF. For (according to 245) $[xab]$ is zero if and only if x lies on a straight line with a and b.

REMARK. Since the metric value of the factors never appears in products set to zero, there is no longer any need to distinguish simple and multiple and infinitely distant points, and I will therefore generally choose to symbolize all of them by the same notation of lower case Latin letters, whereas to symbolize line elements, or, since their magnitudes are never of importance here, straight lines, I will choose upper case Latin letters.

307. *If A and B are straight lines in a plane, the equation of a straight line X, going through the same point as A and B and lying in the same plane, reads*

$$[XAB] = 0,$$

{*where $[XAB]$ is a planimetric product*}.

PROOF. According to 295.

308. *The order of a* ***planimetric*** *product of arbitrarily many factors, whether they be magnitudes first or second order, is congruent to the sum of the orders of all the factors modulo* 3.

PROOF. According to 96.

309. *If $\mathfrak{P}_{n,x}$ is a* ***planimetric*** *product of zeroth order, which includes the point x n times, and otherwise only constant points and lines as factors, then* {*the equation*}

$$\mathfrak{P}_{n,x} = 0,$$

if it is not satisfied by every point x, is the point-equation of an algebraic curve of nth order, that is, the equation says that the point x lies on an algebraic curve of nth order.

PROOF. Let a, b, c be three arbitrary points not lying on the same line, for example a a simple point, b and c two mutually orthogonal and equally long displacements (infinitely distant points); then all points in the plane are numerically derivable from a, b, c, and thus in particular the point x; let

$$x = x_1 a + x_2 b + x_3 c.$$

If one replaces x with this expression in the equation

$$\mathfrak{P}_{n,x} = 0,$$

and removes all the parentheses that enclose the expression $(x_1 a + x_2 b + x_3 c)$ in the product $\mathfrak{P}_{n,x}$, one obtains an equation homogeneous of nth degree with respect to x_1, x_2, x_3, all the terms of which have the form $\mathfrak{A} x_1^{\mathfrak{a}} x_2^{\mathfrak{b}} x_3^{\mathfrak{c}}$, where $\mathfrak{a} + \mathfrak{b} + \mathfrak{c} = n$, and $\mathfrak{A}$ is a product of constant lines and points, and is in fact a product of zeroth order, since the orders of the factors is not changed. Thus $\mathfrak{A}$ is, as a magnitude of zeroth order, a number, and the equation is therefore an ordinary numerical equation, which with respect to x_1, x_2, x_3 is homogeneous of nth degree. But if a is a simple point and b and c are mutually orthogonal and equally long lines, then $\frac{x_2}{x_1}$ and $\frac{x_3}{x_1}$ are the ordinary coordinates of the point x, and therefore, if the equation is not satisfied identically, the curve it represents is an algebraic curve of nth order.[82]

310. *If $\mathfrak{P}(n, X)$ is a* ***planimetric*** *product of zeroth order that includes the straight line X n times and otherwise only constant points and lines as factors, then*

$$\mathfrak{P}(n, X) = 0$$

is the line equation of an algebraic curve of nth class, or, expressed more simply, the geometric locus for the line X that satisfies this equation is a locus of nth degree.

PROOF. Precisely as in 309.

311. *If $\mathfrak{P}_{n,x}$ is a* ***stereometric*** *product of zeroth order that includes the point x n times and otherwise only constant points, lines, and planes as factors, then*

$$\mathfrak{P}_{n,x} = 0$$

is the point equation of an algebraic surface of nth order, or, expressed more simply, the geometric locus of the point x that satisfies the above equation is a locus of nth degree; assuming however that not every point x satisfies the above equation.

PROOF. Let a, b, c, d be four arbitrary points that do not lie in the same plane, for example a a simple point, b, c, d three mutually orthogonal and equally long displacements (infinitely distant points); then (according to 232) x can be numerically derived from a, b, c, d.

Let

$$x = x_1 a + x_2 b + x_3 c + x_4 d,$$

where x_1, x_2, x_3, x_4 are numbers. If one replaces x with this expression everywhere in the product $\mathfrak{P}_{n,x}$, and removes the parentheses, one obtains pure terms of the form $\mathfrak{A} x_1^{\mathfrak{a}} x_2^{\mathfrak{b}} x_3^{\mathfrak{c}} x_4^{\mathfrak{d}}$, where $\mathfrak{a} + \mathfrak{b} + \mathfrak{c} + \mathfrak{d} = n$, and $\mathfrak{A}$ is a product of zeroth order, and thus a number. Consequently the resulting equation is a numerical equation which, with respect to x_1, x_2, x_3, x_4, is homogeneous of nth degree, unless it should happen that all the coefficients $\mathfrak{A}, \ldots$ are zero, that is, unless the equation is satisfied by every point x, which was excluded above. Now $\frac{x_2}{x_1}$, $\frac{x_3}{x_1}$, $\frac{x_4}{x_1}$, are the ordinary coordinates of the point x, whence since

$$x = x_1 \left(a + \frac{x_2}{x_1} b + \frac{x_3}{x_1} c + \frac{x_4}{x_1} d \right),$$

it follows that

$$x \equiv a + \frac{x_2}{x_1} b + \frac{x_3}{x_1} c + \frac{x_4}{x_1} d.$$

Consequently the geometric position of x is a surface of nth order.

312. *If $\mathfrak{P}(n, \xi)$ is a* ***stereometric*** *product of zeroth order that includes the plane ξ n times as a factor, and otherwise only constant points, lines, and planes, then*

$$\mathfrak{P}(n, \xi) = 0$$

is the plane equation of an algebraic surface of nth class; assuming that not every plane ξ satisfies the equation.

PROOF. As in 311, {for every surface element is in fact a product of three points and therefore (according to 232, 65) can be numerically derived from surface elements that do not go through the same point}.

313. *A planimetric, and likewise a stereometric product remains congruent to itself if one replaces any one of its factors by one congruent to it, or, {what is the same thing}, if one multiplies or divides it by an arbitrary nonzero number (a magnitude of zeroth order).*

PROOF. Two magnitudes A and B are called congruent (according to 2) if there is between them an equation of the form $\mathsf{A} = n\mathsf{B}$, in which n denotes an arbitrary nonzero number, (positive {or negative}, integer or fraction, rational or irrational). Now if in a product $P(\mathsf{A})$ that includes the factor A one replaces A by its congruent factor $n\mathsf{A}$, then (according to 46) $P(n\mathsf{A}) = nP(\mathsf{A})$, whence it is congruent to $P(\mathsf{A})$.

REMARK. According to the concepts introduced a product of zeroth order is congruent to another if and only if they are either both simultaneously zero, or both simultaneously different from zero. Thus the theorem includes this, since if in a product of zeroth order one replaces an arbitrary factor with another congruent to it, the product remains zero if it was zero, and remains different from zero if it was different from zero. Since in all of the following considerations only congruence is important, I will generally replace line elements and surface elements by straight lines and planes.

314. *A planimetric and likewise a stereometric product remains congruent to itself if one interchanges the two factors of which it consists, that is* $[\mathsf{AB}] \equiv [\mathsf{BA}]$, *whatever the magnitudes* A *and* B *may be.*

PROOF. According to {114 and} 120.

315. *A* ***planimetric*** *product of three points or three lines, and likewise a* ***stereometric*** *product of four points or planes remains congruent to itself if one arbitrarily orders and combines its factors.*

PROOF. For since the product is then (according to 114) pure, theorems 120 and 119 hold for it.

316. *A planimetric and likewise a stereometric product remains congruent to itself if one interchanges two of its immediately sequential and mutually incident factors (thus a straight line or a plane and a point lying in it).*

PROOF. According to 123.

317. *A planimetric and likewise a stereometric product of zeroth order remains congruent to itself if one reverses the order of the factors, or encloses an arbitrary series of the later factors in parentheses and reverses them.*

PROOF. According to 126.

318. *A* ***stereometric*** *product of three or four points, or of three or four planes, or of two points and a straight line, or of two planes an a straight line, remains congruent to itself if one arbitrarily orders and combines its factors.*

PROOF. For since the product is then (according to 114) pure in each case, theorems 119 and 120 are applicable.

319. *A **stereometric** product $[aBC]$ of a point a and two straight lines B and C **that intersect** remains congruent to itself if one interchanges these straight lines, that is*

$$[aBC] \equiv [aCB],$$

if B and C intersect. {This also holds if one replaces the point a with a plane α.}

PROOF. According to 124e.

REMARK. All the cases of interchangeability for planimetric and stereometric products are thus exhausted. (Cf. 124.)

320. *If in a **planimetric** product of the form*

$$[xaBcD\cdots],$$

that is, in which, after the point x, alternating points and straight lines follow, or in the planimetric product

$$[XBcD\cdots],$$

in which, after the line X, alternating lines and points follow, no factor is incident on the next in sequence, the product is different from zero.

PROOF. Let it be assumed that among the magnitudes x, a, B, c, D, ... no two sequential factors are incident; then x and a are two non-incident points, whence their product is a nonzero straight line. This straight line is not incident on B, and thus their product $[xaB]$ is a nonzero point on the straight line B. This cannot coincide with c, since c does not lie on B, whence their product $[xaBc]$ is a nonzero straight line going through c. This cannot coincide with D since c does not lie on D, whence their product $[xaBcD]$ is a nonzero point on the straight line D, and so forth. If one sets $xa = X$, the second part of the Proof follows.

321. *If in a stereometric product of the form*

$$[xa\beta c\delta\cdots],$$

that is, one in which, after the point x, alternating points and planes (denoted here by Greek letters) follow, or in the stereometric product

$$[\xi\beta c\delta\cdots],$$

in which, after the plane ξ, alternating planes and points follow, no factor is incident on the next in sequence, the product is different from zero.

PROOF. As in 320.

322. *If in a stereometric product of the form*

$$[xABC\cdots] \quad \text{or} \quad [\xi BC\cdots],$$

that is, in which, after the point x or the plane ξ, pure straight lines follow as factors, the first two factors are not incident on one another, and none of the lines intersect the next in sequence, the product is different from zero.

PROOF. Since x does not lie on the straight line A, $[xA]$ is different from zero, and is in fact congruent to the plane laid through x and A. The straight line B cannot lie in this plane, since otherwise it must intersect the straight line A of this plane (although {perhaps} at an infinite distance), contrary to hypothesis; therefore the product $[xAB]$ of the plane $[xA]$ and the straight line B is different from zero, and in fact (according to 303) it is congruent to their point of intersection. Since this lies on B, and therefore not on C (since B and C do not intersect), the product $[xABC]$ is a nonzero plane going through C, and so forth. The second part of the theorem follows if one sets $[xA] = \xi$.

REMARK. The theorems introduced here suffice to discuss the formulas for curves and surfaces previously presented with the greatest of ease, as I will illustrate with the following two examples.

323. *The equation of a conic section that goes through the five points a, b, c, d, e, no three of which lie on the same straight line, is*

$$[xa(cd)(ab \cdot de)(bc)ex] = 0,$$

or

$$[xaBc_1Dex] = 0,$$

where $B = [cd]$, $c_1 = [ab \cdot de]$, $D = [bc]$.

PROOF. The product on the left side is of zeroth order, since the sum of the orders is twelve, and thus divisible by three. One easily perceives that not every point x satisfies the equation.

If for example one draws a line ap that does not go through e {see Fig. 12}, and assumes that x is to lie on this straight line, but not on a, then $[xa] \equiv [pa]$. Thus (according to 313) we can replace xa in the above equation with pa, and obtain

$$[paBc_1Dex] = 0,$$

that is {according to 306}, the point x must lie on the straight line that goes through the point $[paBc_1D]$, call it q, and the point e; at the same time, by hypothesis, it is to lie on the straight line ap, and thus simultaneously on qe and ap.[83] These two straight lines are {however} necessarily different, since e does not lie on ap, whence they intersect in only one point, that is, the straight line ap includes only one point other than the point a that satisfies the above equation. Thus not every point satisfies it.

Consequently the geometric locus for x is (according to 309) a curve of second order, that is a conic section. Thus it is only left to show that it goes through the five points $a, \ldots, e$, that is, that if x coincides with any of the five points $a, \ldots, e$, the equation is satisfied.

If x coincides with a, then $[xa] = 0$, and thus the whole product is also zero; the same holds for $x \equiv e$, if one writes the equation in the form

$$[xeDc_1Bax] = 0.$$

If $x \equiv c$, then $[ca(cd)] = [cad]c$ (according to 103), and this is again $\equiv c$, since $[cad]$ is a nonzero number; thus

$$\begin{aligned}[ca(cd)(ab \cdot de)(bc)ec] &\equiv [c(ab \cdot de)(bc)ec] && [313]\\ &\equiv [(ab \cdot de)c(bc)ec] && [314]\\ &\equiv [(ab \cdot de)(bc)cec] && [316]\\ &\equiv [(ab \cdot de)(bc)][cec], && [40]\end{aligned}$$

since $[(ab \cdot de)(bc)]$ is of zeroth order, that is a number. Here (according to 60) $[cec] = 0$, and thus also the whole product.

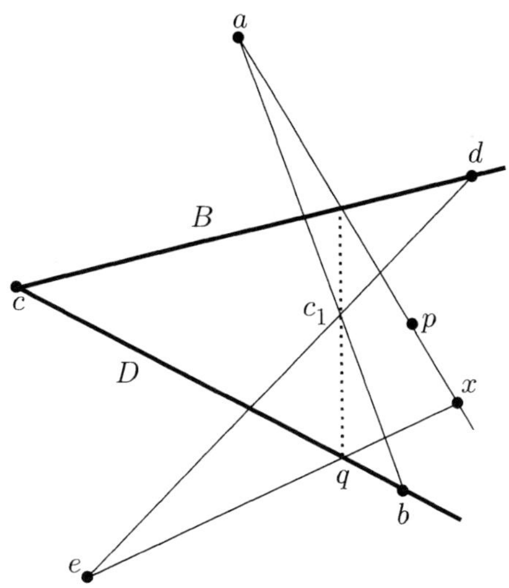

FIGURE 12

If $x \equiv d$, then

$$[da \cdot cd] \equiv [da \cdot dc] \text{ (No.314)} = [dac]d \text{ (No. 103)} \equiv d.$$

Consequently

$$\begin{aligned}[da(cd)(ab \cdot de)(bc)ed] &\equiv [d(ab \cdot de)(bc)ed] && [313]\\ &\equiv [ab(de)d(bc)ed] && [314]\\ &\equiv [abd(de)(bc)ed] && [316]\\ &\equiv [abd][(de)(bc)(de)], && [40, \{317\}]\end{aligned}$$

since $[abd]$ is a number. But $[de \cdot bc \cdot de]$ is zero (according to 295), whence the whole product $= 0$. If $x \equiv b$, then from the reversed form of the equation it follows in the same way that the equation is satisfied. Therefore all five points $a, \ldots, e$ are points of the conic section.

324. *If A, B, C are three lines in space, no two of which intersect, then*

$$[xABCx] = 0$$

is the equation of that surface of second order on which the three lines A, B, C lie.

PROOF. The sum of the orders is eight, and is therefore divisible by four, whence the product, as stereometric, is of zeroth order. Not every point x satisfies the equation. For if one lays through the straight line A a plane α {see Fig. 13}, and assumes that the point x lies on this plane, but outside of A, then $[xA] \equiv \alpha$, whence $[xABC] \equiv [\alpha BC]$, and is different from zero (according to 322). But $[\alpha B]$ is a point and $[\alpha BC]$ the plane laid through this point and the straight line C. The equation

$$[\alpha BCx] = 0$$

{therefore} says that the point x must lie in this plane; but by hypothesis it also lies in the plane α, and thus in both simultaneously. The two planes do not however coincide, since otherwise A and C lie in these planes and must then intersect,

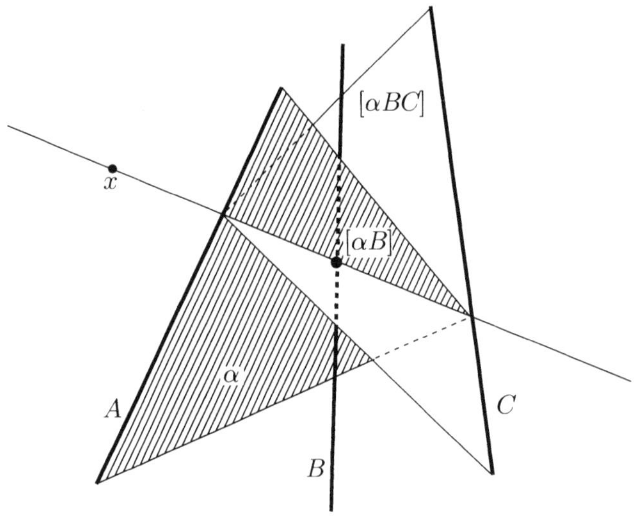

FIGURE 13

contrary to hypothesis. Thus x must lie on the line of intersection of the two planes in order to satisfy the equation. Thus not every point satisfies it.

Now since the above product is of zeroth order, includes x twice as factor, and is not satisfied by every point, the locus of x is (according to 311) a surface of second order. In it lie A and C, for if x lies on A, then $[xA] = 0$, and thus the product is zero, and likewise if x lies on C {according to 317}. Finally, if x lies on B, one has

$$[xABCx] \equiv [AxBCx] \qquad [314]$$
$$\equiv [ABxCx] \qquad [316]$$
$$\equiv [AB][xCx], \qquad [40]$$

since $[AB]$ is of zeroth order; finally $[xCx]$ is (according to 60) zero, and therefore the whole product is zero, that is, every point that lies on B satisfies the equation.

REMARK. For the converse theorem, that every algebraic curve in the plane can be represented in the form of a planimetric product equal to zero, it comes down to representing every algebraic function of the coordinates of a point in the form of a planimetric product. This problem would be solved if, by any given method of representing the numbers spatially, both the product and the sum of two spatially represented numbers can be represented by a planimetric product. The corresponding applies for algebraic surfaces. For the first case we will pursue the development far enough so that from any given algebraic equation its planimetric correspondent can be read off immediately.

325. *Let c be an arbitrary simple point, a and b be two nonparallel displacements, and x be an arbitrary simple point of the plane cab {see Fig. 14}, and in particular let $x = x_1a + x_2b + c$. Let $d = a + b + c$ (that is, on a parallelogram, one vertex of which is c, and whose sides proceeding from this vertex are the same length and direction as a and b, d is the vertex diagonally opposite c). Then if (x_1) and (x_2) are those {simple} points on the diagonal cd for which the equations*

$$[c(x_1)] : [cd] = x_1, \quad [c(x_2)] : [cd] = x_2$$

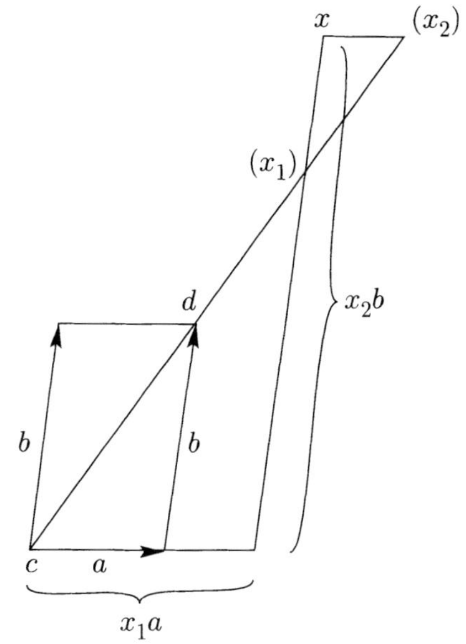

FIGURE 14

hold, then

$$(x_1) \equiv [xbC], \quad (x_2) \equiv [xaC]$$

and

$$[(x_1)b] = [xb], \quad [(x_2)a] = [xa],$$

where for short $[cd]$ *is denoted by* C.

PROOF. According to 225, since the points c, (x_1), d lie on a straight line, and $[c(x_1)] : [cd] = x_1 : 1$,

$$(x_1) - c = x_1(d - c) = x_1(a + b),$$

since (by hypothesis) $d = a + b + c$; thus

$$(x_1) = x_1a + x_1b + c,$$

whence

$$[(x_1)b] = [(x_1a + c)b] \qquad [67]$$

and

$$[xb] = [(x_1a + x_2b + c)b] \qquad \text{[by hypothesis]}$$
$$= [(x_1a + c)b]. \qquad [67]$$

Consequently

$$[(x_1)b] = [xb].$$

Here $[xb]$ is a straight line drawn through x parallel to b. According to the last equation the point (x_1) lies on this line; but by hypothesis it also lies on the straight line cd or C, and thus on the intersection of these two, whence

$$(x_1) \equiv [xbC].$$

On the same grounds $[(x_2)a] = [xa]$ and $(x_2) \equiv [xaC]$.

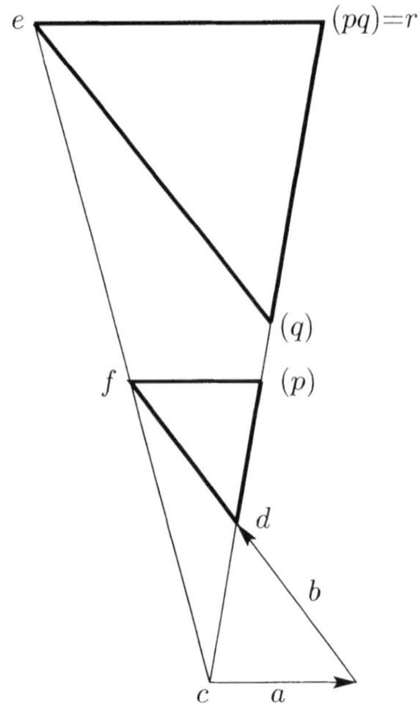

FIGURE 15

326. *If a, b, c, d have the same meaning as in the previous theorem, and p, q are two arbitrary numbers, and, just as in the previous theorem, one means by* (p), (q), (pq) *those {simple} points of the straight line cd for which the equations*

$$(a) \qquad [c(p)] : [cd] = p, \quad [c(q)] : [cd] = q, \quad [c(pq)] : [cd] = pq$$

hold {see Fig. 15}, then, if one sets

$$(b) \qquad [da] = A, \quad [db] = B, \quad [dc] = C$$

for short, one has

$$(pq) \equiv [(p)aBc((q)b)aC] \equiv [(p)bAc((q)a)bC],$$
$$[(pq)a] \equiv [(p)aBc((q)b)a],$$
$$[(pq)b] \equiv [(p)bAc((q)a)b].$$

PROOF. If in the third of equations (a) one replaces p and q by their values from the first two, one obtains the proportion

$$[c(pq)] : [c(p)] = [c(q)] : [cd].$$

From this proportion and the fact that the five points c, d, (p), (q), (pq) lie on a straight line it follows immediately that, if we denote the point (pq) by r for short, c is the point of similarity of the two assemblies of points r, (q) and (p), d. Thus if one draws upon the base sides $r(q)$ and $(p)d$ the parallel triangles $r(q)e$ and $(p)df$, then c is also the point of similarity of these triangles, and it follows that the corresponding points e and f lie with c on a straight line, whereby r can be found.

In particular, if one takes re and $(p)f$ parallel to a, and $(q)e$ and df parallel to b, then

$$f \equiv [(p)a \cdot db] \equiv [(p)aB],$$

since $[db] = B$; furthermore

$$e \equiv [fc \cdot (q)b], \quad r \equiv [ea \cdot cd] \equiv [eaC],$$

since $[dc] = C$, and finally

$$[ra] \equiv [ea],$$

all of which follows at once if one draws the appropriate figure {see Fig. 15}.

If one inserts the values from the first two of these equations into the last one, one obtains

$$r \equiv [(p)aBc((q)b)aC],$$
$$[ra] \equiv [(p)aBc((q)b)a],$$

and if in the above demonstration one everywhere replaces b with a and conversely, and A with B, one obtains

$$r \equiv [(p)bAc((q)a)bC],$$
$$[rb] \equiv [(p)bAc((q)a)b],$$

and, since $r = (pq)$, these are the equations to be proved.

327. *If a, b, c, d, A, B, C have the same meaning as in the previous theorem and p, q, $\mathfrak{a}$, $\mathfrak{b}$ are arbitrary numbers, but with the condition that $\mathfrak{a} + \mathfrak{b}$ is* ***different from zero****, and if further*

$$\text{(a)} \qquad r = \frac{\mathfrak{a}p + \mathfrak{b}q}{\mathfrak{a} + \mathfrak{b}},$$

and as before (p), (q), (r) are those {simple} points of the straight line $[cd] \equiv C$ {see Fig. 16} that satisfy the equations

$$\text{(b)} \qquad [c(p)] : [cd] = p, \quad [c(q)] : [cd] = q, \quad [c(r)] : [cd] = r,$$

then

$$r \equiv [(p)a((q)b)(\mathfrak{a}a - \mathfrak{b}b)C].$$

PROOF. If one substitutes the values of p, q, r from (b) into the equation $(\mathfrak{a} + \mathfrak{b})r = \mathfrak{a}p + \mathfrak{b}q$, one obtains

$$(\mathfrak{a} + \mathfrak{b})[c(r)] = \mathfrak{a}[c(p)] + \mathfrak{b}[c(q)],$$

or, since they all lie on the same straight line,

$$(\mathfrak{a} + \mathfrak{b})((r) - c) = \mathfrak{a}((p) - c) + \mathfrak{b}((q) - c) \qquad \text{[Corollary to 222]}$$

whence

$$(*) \qquad (\mathfrak{a} + \mathfrak{b})(r) = \mathfrak{a}(p) + \mathfrak{b}(q),$$

that is {according to the Remark following 223}, (r) is the center of gravity between the multiple points $\mathfrak{a}(p)$ and $\mathfrak{b}(q)$. Now if one draws from (q) the parallel to b, and from (p) the parallel to a, which intersect in e, and likewise from d and c the parallels to b and a, which intersect in f, then the triangles dfc and $(q)c(p)$ are parallel and thus similar, and in addition $d - f = b$ and $f - c = a$. Thus if $(q) - e = m(d - f)$, where m denotes a number, then {also} $e - (p) = m(f - c)$, that is, $(q) - e = mb$,

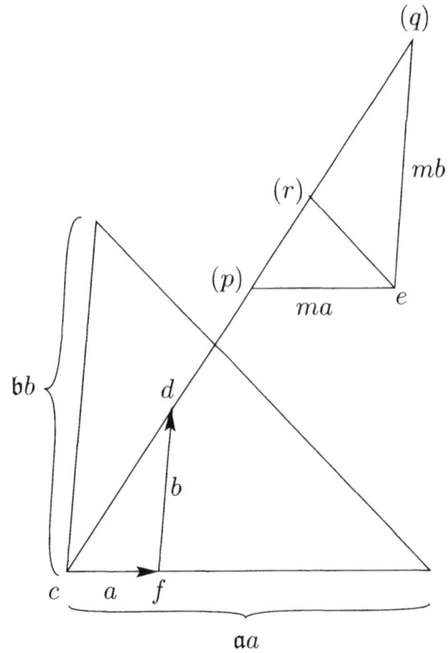

FIGURE 16

$e-(p)=ma$. Furthermore, if one adds $-(\mathfrak{a}+\mathfrak{b})e$ to both sides of the above equation $(*)$, one obtains

$$\begin{aligned}(\mathfrak{a}+\mathfrak{b})((r)-e)&=\mathfrak{a}((p)-e)+\mathfrak{b}((q)-e)\\&=-m\mathfrak{a}a+m\mathfrak{b}b=-m(\mathfrak{a}a-\mathfrak{b}b).\end{aligned}$$

If one multiplies both sides planimetrically with e one obtains (according to 67)

$$(\mathfrak{a}+\mathfrak{b})[e(r)]=-m[e(\mathfrak{a}a-\mathfrak{b}b)],$$

whence

$$[e(r)]\equiv[e(\mathfrak{a}a-\mathfrak{b}b)], \tag{2}$$

that is, (r) lies on the straight line $[e(\mathfrak{a}a-\mathfrak{b}b)]$, but (by hypothesis) also on the straight line C, and thus on the intersection of the two lines, that is

$$(r)\equiv[e(\mathfrak{a}a-\mathfrak{b}b)C].$$

But now, according to the construction given, e is the intersection of the straight lines $[(p)a]$ and $[(q)b]$, whence $e\equiv[(p)a\cdot(q)b]$, and consequently

$$(r)\equiv[(p)a(q)b(\mathfrak{a}a-\mathfrak{b}b)C].$$

REMARK. If the numbers are represented by points of the straight line cd in the way given, then according to these last two theorems both the sum and the product of two numbers, and therefore in addition any complete function of numbers, can be represented by a planimetric product composed of the points a, b, c, d and the points representing the given numbers. Herewith the theorem is already proven that every algebraic curve in the plane can be represented by a planimetric product set equal to zero. Even so, it will be shown in the following how one can immediately derive that planimetric product from the given algebraic equation of the curve.

328. *If a and b are {two nonparallel} displacements, c a simple point and*

$$d = a + b + c, \quad A \equiv [da], \quad B \equiv [db], \quad C \equiv [dc], \quad x = x_1 a + x_2 b + c,$$

then the equation

$$f(x_1, x_2) = \mathfrak{a} x_1^m x_2^n + \mathfrak{b} x_1^r x_2^s + \cdots + \mathfrak{k} x_1^v x_2^w = 0,$$

in which ***the sum of the coefficients is different from zero****, is,* ***for all finitely distant points*** *x, equivalent to the equation*

$$[Px] = 0,$$

where, if the terms of $f(x_1, x_2)$ are so ordered that the sums $\mathfrak{a} + \mathfrak{b}$, $\mathfrak{a} + \mathfrak{b} + \mathfrak{c}$, ... are all different from zero,

(*a*) $$P \equiv [L L_1 a_1 C a L_2 a_2 C a L_3 a_3 C a \cdots L_k a_k]$$

(*b*) $$\begin{cases} a_1 = \mathfrak{a} a - \mathfrak{b} b, \quad a_2 = (\mathfrak{a} + \mathfrak{b}) a - \mathfrak{c} b, \ldots, \\ \qquad a_k = (a + b + \cdots + i) a - k b, \end{cases}$$

(*c*) $$L \equiv [x b \mathfrak{R}^n C a \mathfrak{R}_1^{m-1}],$$

(*d*) $$L_1 \equiv [x a \mathfrak{R}_1^p C b \mathfrak{R}^{q-1}], \; L_2 \equiv [x a \mathfrak{R}_1^r C b \mathfrak{R}^{s-1}], \; \ldots, \; L_k \equiv [x a \mathfrak{R}_1^v C b \mathfrak{R}^{w-1}],$$

and $\mathfrak{R}$ denotes the progressive series of factors A, c, xa, b, and $\mathfrak{R}_1$ is the progressive series of factors B, c, xb, a, so that for every line X

(*e*) $$[X\mathfrak{R}] \equiv [XAc(xa)b], \quad [X\mathfrak{R}] \equiv [XBc(xb)a].$$

PROOF. If, for any arbitrary number p, we denote by (p) that {simple} point of the line cd for which

(∗) $$[c(p)] : [cd] = p,$$

then (according to 325) we obtain

(∗∗) $$[(x_1)b] = [xb], \quad [(x_2)a] = [xa],$$

and (according to 326), if p is an arbitrary number,

$$[(px_2)b] \equiv [(p)bAc((x_2)a)b] \equiv [(p)bAc(xa)b] \qquad [**]$$
$$\equiv [(p)b\mathfrak{R}]. \qquad [e]$$

If another factor x_2 is appended to px_2, then the series of factors $\mathfrak{R}$ is once more appended to $(p)b\mathfrak{R}$, and so forth, whence

(∗∗∗) $$[(px_2^n)b] \equiv [(p)b\mathfrak{R}^n];$$

and likewise, if one interchanges x_2 and b with x_1 and a, and thus $\mathfrak{R}$ with $\mathfrak{R}_1$, one obtains

(∗∗∗∗) $$[(px_1^n)a] \equiv [(p)a\mathfrak{R}_1^n].$$

When we multiply this expression by C we obtain the point $(x_1 x_2^n)$, that is

$$(x_1 x_2^n) \equiv [xb\mathfrak{R}^n C].$$

Thus if, after replacing n by $m - 1$ in formula (∗∗∗∗), one substitutes $x_1 x_2^n$ for p in this equation, one therefore obtains

$$[(x_1^m x_2^n)a] \equiv [xb\mathfrak{R}^n Ca\mathfrak{R}_1^{m-1}] \equiv L. \qquad [e]$$

Likewise one obtains

$$[(x_1^p x_2^q)a] \equiv [xb\mathfrak{R}^q Ca\mathfrak{R}_1^{p-1}],$$

or, when one interchanges a with b, and thus also x_1 with x_2, p with q, $\mathfrak{R}$ with $\mathfrak{R}_1$,

$$[(x_1^p x_2^q)b] \equiv [xa\mathfrak{R}_1^p Cb\mathfrak{R}^{q-1}] \equiv L_1,$$

and likewise

$$[(x_1^r x_2^s)b] \equiv L_2, \ \ldots, \ [(x_1^v x_2^w)b] \equiv L_k.$$

Now in order to find the expression for $((\mathfrak{a}x_1^m x_2^n + \mathfrak{b}x_1^p x_2^q) : (\mathfrak{a}+\mathfrak{b}))$, one has only to substitute $x_1^m x_2^n$ and $x_1^p x_2^q$ for p and q and thus L and L_1 for $[(p)a]$ and $[(q)b]$ in 327, and obtain

$$\begin{aligned}((\mathfrak{a}x_1^m x_2^n + \mathfrak{b}x_1^p x_2^q) : (\mathfrak{a}+\mathfrak{b})) &\equiv [LL_1(\mathfrak{a}a - \mathfrak{b}b)C] \\ &\equiv [LL_1a_1C]. \qquad \text{[b]}\end{aligned}$$

In order to find the expression for $((\mathfrak{a}x_1^m x_2^n + \mathfrak{b}x_1^p x_2^q + \mathfrak{c}x_1^r x_2^s) : (\mathfrak{a}+\mathfrak{b}+\mathfrak{c}))$, one has only to substitute the expression $(\mathfrak{a}x_1^m x_2^n + \mathfrak{b}x_1^p x_2^q) : (\mathfrak{a}+\mathfrak{b})$ for p, and $x_1^r x_2^s$ for q, and thus L_2 for $(q)b$ and at the same time $\mathfrak{a}+\mathfrak{b}$ for $\mathfrak{a}$, and $\mathfrak{c}$ for $\mathfrak{b}$ in 327, and obtain

$$((\mathfrak{a}x_1^m x_2^n + \mathfrak{b}x_1^p x_2^q + \mathfrak{c}x_1^r x_2^s) : (\mathfrak{a}+\mathfrak{b}+\mathfrak{c})) \equiv [LL_1a_1CaL_2a_2C],$$

since $(\mathfrak{a}+\mathfrak{b})a - \mathfrak{c}b = a_2$, and so forth; and finally

$$\begin{aligned}((\mathfrak{a}x_1^m x_2^n + \mathfrak{b}x_1^p x_2^q + \mathfrak{c}x_1^r x_2^s + \cdots + \mathfrak{k}x_1^v x_2^w) &: (\mathfrak{a}+\mathfrak{b}+\mathfrak{c}\cdots+\mathfrak{k})) \\ &\equiv [LL_1a_1CaL_2a_2Ca\ldots L_ka_kC] \\ &\equiv [PC]. \qquad \text{[a]}\end{aligned}$$

Thus

$$(f(x_1, x_2) : (\mathfrak{a}+\mathfrak{b}+\mathfrak{c}+\cdots+\mathfrak{k})) \equiv [PC].$$

Now if $f(x_1, x_2) = 0$, then one has $(0) \equiv [PC]$. But according to $(*)$ one has $[c(0)] : [cd] = 0$, whence the dividend $[c(0)]$ equals zero, that is the point (0) coincides with c, whence we obtain $c \equiv [PC]$, that is, c is the point of intersection of the straight lines P and C; therefore it also lies on P, that is (according to 293)

$$[Pc] = 0.$$

Conversely, if this last equation is satisfied, then c lies on P, but (by hypothesis) also on C, whence $c \equiv [PC]$, that is, the point belonging to the number $f(x_1 x_2) : (\mathfrak{a}+\mathfrak{b}+\cdots)$ lies on c; consequently that number is zero, and thus its numerator $f(x_1, x_2) = 0$.

329. *If all the other assumptions of the previous theorem remain in place, but it is now assumed that* ***the sum of the coefficients*** $(\mathfrak{a}+\mathfrak{b}+\cdots+\mathfrak{k})$ ***is zero****, then the equation*

$$f(x_1, x_2) = 0$$

is {for all ***finitely distant*** *points} equivalent to the equation*

$$[LL_1a_1CaL_2a_2Ca\cdots L_{k-1}a_{k-1}L_kC] = 0,$$

where the letters have the same meaning as in the previous theorem, and the terms in $f(x_1, x_2)$ *are also so ordered here that the sums* $\mathfrak{a}+\mathfrak{b}$, $\mathfrak{a}+\mathfrak{b}+\mathfrak{c}$, $\ldots$ *all, with the exception of the last,* $(\mathfrak{a}+\mathfrak{b}+\cdots+\mathfrak{k})$*, are different from zero.*

PROOF. With division by the coefficient of the last term one can cast the equation $f(x_1, x_2) = 0$ into a form in which the coefficient ($\mathfrak{k}$) of the last term is -1; then the sum of the other coefficients is $+1$. Let the {resulting} last term {equal} $-h$, and the sum of the rest be g; then the equation $f(x_1, x_2) = 0$ is equivalent to $g - h = 0$ or

$$g = h.$$

Then, from the development in the previous theorem,

$$(g) \equiv [LL_1a_1CaL_2a_2Ca \cdots L_{k-1}a_{k-1}C],$$
$$(h) \equiv [L_kC].$$

Now since $g = h$, the two points (g) and (h) are congruent, and thus

$$[LL_1a_1CaL_2a_2Ca \cdots L_{k-1}a_{k-1}C] \equiv [L_kC].$$

This congruence expresses that the point represented by the left side lies on the intersection of the lines L_k and C, that is (according to 293)

$$[LL_1a_1CaL_2a_2Ca \cdots L_{k-1}a_{k-1}CL_k] = 0.$$

Here one can (according to 315) also interchange the last two factors, whereby the equation sought results; conversely, from this last equation it follows again that $g = h$, and thus $f(x_1, x_2) = 0$.[84]

REMARK. It is easy to see that, instead of taking a and b as displacements and c as a simple point in the foregoing theorems, one could more generally have taken a, b and c as three arbitrary points not lying on the same straight line, whence the condition that x be a finitely distant point would be replaced by another, that x not lie on the straight line ab.

The basis for the admissibility of this generalization is that (according to 110) the laws of relative multiplication in a principal domain remain unchanged if one replaces the original units a, b, c by three others numerically derivable from them, whereby the condition that the combinatorial product of the former as well as the latter be 1 drops out here, where it is only a question of congruence.

If one considers the forms of the equations in 328, it is easy to show that with respect to x the planimetric equation $[Pc] = 0$ is of the same degree as the sum of all the exponents in the algebraic equation $f = 0$; for the series of factors $\mathfrak{R}$ each include the point x only once, and the magnitudes L are therefore of the same degree as the terms of the function f in the series. The product $[Pc]$, which includes each of the magnitudes L once as factor, and otherwise only constant factors, is therefore with respect to x of the same degree as the sum of the degrees of all terms of f, whence, as soon as f includes more than one variable term, it is of higher degree than f. Let n be the degree of the function f, and $n + p$ be the degree of the product $[Pc]$. Now since the equations $[Pc] = 0$ and $f = 0$ are (according to 328) completely equivalent for all points x that do not lie on the (infinitely distant) straight line ab, the difference in degree can only be because the equation $[Pc] = 0$ represents yet other lines that coincide with ab.

In order to examine this situation more precisely I substitute the point y for x, and set $y = ua + vb + wc$, where u, v, w are numbers, and set the original coordinates of x respectively equal to $u : w$ and $v : w$. Then $y = wx$, whence $y \equiv x$, provided w is not zero. This case, where $w = 0$, is however that for which $y = ua + vb$, that is, y is a point on the line ab, and thus the same as was excluded in 328. Thus, in all the cases admitted there, the product that results from $[Pc]$ when one substitutes y for x, and which I will denote by Q, is congruent to $[Pc]$ (according to 113). If one multiplies the function of nth degree f by w^n, then since the variables included therein were $u : w$ and $v : w$, there results from f a homogeneous function of nth degree, which I will denote by F, that is zero in the same cases for which f is zero. Thus for the case that y does not lie on ab, that is that w is not zero, the equation $Q = 0$ is equivalent to the equation $F = 0$. Since however the former is of $(n+p)$th degree, the latter of nth degree, the equation $Q = 0$ must be equivalent to the equation $w^p F = 0$ in all cases, whence

$$Q \equiv w^p F \equiv [aby]^p F;$$

the latter since from $y = ua + vb + wc$ follows $w \equiv [aby]$. Therefore Q must be divisible by $[aby]^p$.

It is therefore a question of transforming the planimetric product Q into a congruent product that is freed of this factor $[aby]$. But since this reduction, although generally executable, is as a rule beset with great difficulties, it is expedient first to restructure the algebraic equation by transformation of the coordinate system so that it includes the least possible variable terms before one proceeds to the derivation of the planimetric formula. Thus for example the equation of third degree can be reduced to the form $pqr = ms^3$, where p, q, r, s are linear functions of the coordinates, and m denotes a constant number. If one then shifts the straight line whose equation is $s = 0$ to infinity by projection, the equation

$$pqr = m,$$

transformed by the above rules, yields a geometric equation of third degree that has the form

$$[xaBc(xb)aCdEfx] = 0$$

and which continues to hold under any projection, and therefore also if one gives the line shifted to infinity its original position by projection.

There is no difficulty in applying the principles developed here to surfaces in space; but in order not to be led too far afield here, I must refer the reader to my articles in *Crelles Journal*, namely in vol. 49, p. 1f. {Articles on the linear generation of surfaces.}

7. Inner multiplication in geometry

330. DEFINITION. For inner multiplication I always take as the original units in space three mutually orthogonal and equally long displacements (e_1, e_2, e_3), and in the plane two of them (e_1 and e_2), and in particular I take the length of these displacements as the unit of length, and $[e_1e_2e_3]$, and in the plane $[e_1e_2]$ as units of volume and surface area.

REMARK. Hereby all definitions and theorems depending on the concept of inner multiplication (Nos. 137–215) are applicable to geometry.

331. *For the plane the concept of length coincides with that of numerical value, the concept of orthogonality with that of the normal, and the concept of rotation by an angle α with that of circular evolution by an angle α, the angle being taken as positive if its second leg lies on the same side of the first as the second unit (e_2) lies with respect to the first (e_1).*

PROOF. 1. All the analytical concepts named in the theorem (normal, numerical value, circular evolution) are (in 151–154) associated with the concepts of the inner product, and these again (according to 137) with that of the supplement (89 and 90). The supplement of a was symbolized by $|a$. I therefore show first that, if $a = x_1e_1 + x_2e_2$ is an arbitrary displacement in the plane, $|a$ is orthogonal to and the same length as a, and lies on the same side of a as e_2 lies with respect to e_1.

Let AB be the same length and direction as x_1e_1, BC the same as x_2e_2, AD the same as $x_1|e_1$, and DE the same as $x_2|e_2$ {see Fig. 17}. According to 89, $|e_1 = e_2$ and $|e_2 = -e_1$, and therefore $AD = x_1e_2$, $DE = -x_2e_1$. Now since (according to 330) e_1 and e_2 are equally long, x_1e_2 and x_1e_1 are equally long, that

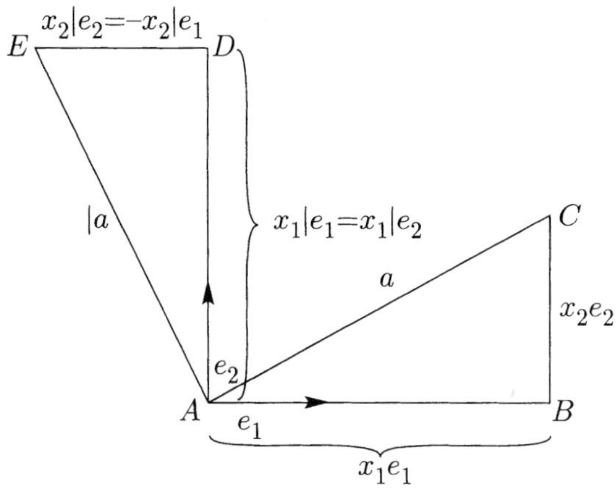

FIGURE 17

is, AD and AB, and likewise $-x_2e_1$ and x_1e_2 are equally long, that is, DE and BC. Since in addition (according to 330) e_2 is orthogonal to e_1, so also is x_2e_2 to x_1e_1 and $-x_2e_1$ to x_1e_2, that is, BC to AB and DE to AD. As a consequence the triangles ABC and ADE are congruent (by two sides and the included angle), whence AC is equal in length to AE. Further, e_2 lies on the same side of e_1 as $-e_1$ with respect to e_2, whence also BC is on the same side of AB as DE with respect to AD, that is, the angles BAC and DAE are equal including sign. Now $\angle CAE = \angle CAB + \angle DAE = \angle CAD + \angle BAC$ (as shown above) $= \angle BAD$, that is, AE is orthogonal to AC, and on the same side as AD with respect to AB, and thus also as e_2 with respect to e_1. But (according to 220) AC is the same length and direction as $x_1e_1 + x_2e_2$, that is, as a, and AE the same as $x_1|e_1 + x_2|e_2$, that is, as $|a$ (according to 90). Thus $|a$ is equally as long as a, and is orthogonal to a on the same side as e_2 with respect to e_1.

2. The numerical value of a is (according to 151) the positive square root of $[a|a]$, whereby (according to 89) the product $[e_1e_2]$ is set to unity. Now (according to 254) $[a|a]$ is equal (including sign) to a parallelogram whose first side is the same length and direction as a, and whose second is the same as $|a$. This parallelogram is (according to Part 1 of the Proof) a square which (according to 330) has the same sign as the chosen unit square $[e_1e_2]$. Now if the length of a (with e_1 taken as the unit of length) equal to α, then the content of the square over a is α^2, and the positive square root is α, that is $\sqrt{[a|a]} = \alpha$, and thus is actually the numerical value of the length.

3. According to 152, two displacements a and b are called normal to one another if $[a|b] = 0$, that is (according to 245) if a is parallel to $|b$; now (according to Part 1 of the Proof) $|b$ is orthogonal to b, and thus also the parallel a to $|b$ is orthogonal to b. Likewise it follows conversely that if a is orthogonal to b, a is parallel to $|b$; but then (according to 245) $[a|b]$ is zero, whence a is normal to b. The concept of orthogonality thus coincides (in the plane) with that of the normal.

4. If a and b are numerically equal and normal to one another {see Fig. 18}, and they have been transformed from

$$a \text{ into } a' = a\cos\alpha + b\sin\alpha,$$

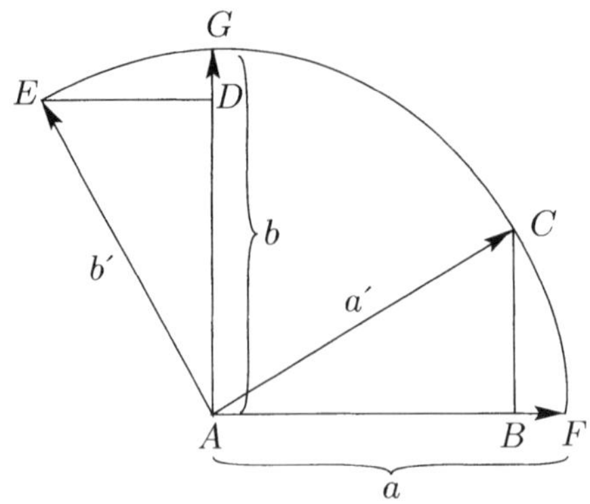

FIGURE 18

and from

$$b \text{ into } b' = b\cos\alpha - a\sin\alpha,$$

then this says (according to 154) that the assembly of displacements a and b have been circularly evolved by the angle α. Let AF be the same length and direction as a and AG be the same as b; then since a and b are numerically equal and normal to one another, AG is (according to Parts 2 and 3 of the Proof) the same length as AF and orthogonal to AF. If one takes the angle α on the same side on AF and on AG, and makes the second legs AC and AE the same length as AF, drops from C the perpendicular CB on AF and from E the perpendicular ED on AG, then AB is the same length and direction as $a\cos\alpha$, BC the same as $b\sin\alpha$, AD the same as $b\cos\alpha$, and DE the same as $-a\sin\alpha$, whence (according to 220) AC is the same length and direction as $a\cos\alpha + b\sin\alpha$, that is as a', and AE the same as $b\cos\alpha - a\sin\alpha$, that is as b'. But a' and b' result from a and b through a rotation by the angle α; consequently the concept of rotation by an angle α coincides with that of circular evolution by the angle α.

REMARK. This same series of conclusions is applicable to any plane in which two displacements a and b are included for which the same assumptions are made as for e_1 and e_2.

332. *The normal system* ***in space*** *is identical to the assembly of three mutually orthogonal and equally long displacements, and the length of these displacements is equal to the numerical value of the normal system.*

PROOF. 1. The system of original units e_1, e_2, e_3 forms (according to 162) an elementary normal system, and e_1, e_2, e_3 are (according to 330) mutually orthogonal and of unit length. Every other elementary normal system of three displacements can (according to 161, {229}) be derived from that normal system by a circular evolution. Now if one has an elementary normal system a, b, c in which a, b, c are mutually orthogonal and of unit length, then the circular evolution consists in evolving any two of them, for example a and b, by an angle α lying in the plane ab. Now in 331 (cf. the Remark) we have shown that the displacements a' and b' resulting thereby are again mutually orthogonal and have unit length; but in addition c is orthogonal to them, for since by hypothesis c is orthogonal to a and b, c is also orthogonal to all lines in the plane ab, and thus is also to a' and b'. Thus, from a normal system whose three displacements are mutually orthogonal and of unit length, there results through a circular evolution another normal system of the same type, and thus also for repeated circular evolutions. Thus an arbitrary circular evolution of the normal system e_1, e_2, e_3 always results in an assembly of three displacements that are mutually orthogonal and of unit length, that is, every elementary normal system consists of three such displacements.

2. Conversely it is to be shown that, if a, b, c are any three mutually orthogonal displacements of length 1, they form a normal system.

According to 160 one can always find an elementary normal system of three displacements, one of which is in the direction of c. Then (according to Part 1 of the Proof) its length must be equal to unity, and therefore the displacement in the direction of c is also the same length as c, and thus be the same generally. Supposing the other two are a' and b', then (according to Part 1 of the Proof) c is orthogonal to a' and b', and thus also (by hypothesis) to a and b, whence a', b', a, b lie in the same plane. Thus (according to 160) one can again find an elementary normal system of two displacements in this plane, one displacement of which is b; let the other be a''. Then a'' is orthogonal to c, since it lies in the plane ab, but (according to Part 1 of the Proof) it is also orthogonal to b and is of length 1, whence a'' is orthogonal to the plane bc. But a is also orthogonal to the same plane and of length 1, whence either $a'' = a$ or $= -a$; now since a'', b, c form an elementary normal system, in both cases so also do a, b, c.

3. Now if one has an *arbitrary* normal system a, b, c of numerical value n, then $a : n$, $b : n$, $c : n$ form an *elementary* normal system, and are therefore (according to Part 1 of the Proof) mutually orthogonal and of length 1, whence a, b, c are mutually orthogonal and of length n. And likewise it follows conversely that if a, b, c are mutually orthogonal and of length n, then $a : n$, $b : n$, $c : n$ form an elementary normal system, and thus a, b, c a normal system of numerical value n.

333. *For space as well, the concept of length coincides with that of numerical value, that of orthogonality with that of the normal {for displacements}.*

PROOF. In 332.

334. *The numerical value of a product P of two displacements p and q is equal to the area of the parallelogram in which two adjacent sides are parallel to those displacements, assuming that this content is positive, and that the square of the unit of length is taken as the unit of area.*

PROOF. Let a, b, c be an elementary normal system of three displacements such that a and b are parallel to the same plane to which p and q are parallel, and such that $[abc] = +1$; then (according to 230) p and q are numerically derivable from a and b, whence also (according to 70) $P = [pq]$ from $[ab]$. Let $P = \alpha[ab]$. Now (according to 151) the numerical value of P is $= \sqrt{[P|P]} = \alpha\sqrt{[ab|ab]}$; but (according to 167) $|[ab] = c$, whence

$$\sqrt{[P|P]} = \alpha\sqrt{[abc]} = \alpha.$$

Since in addition $P = \alpha[ab]$, and $[ab]$ is the square of the unit of length, and thus to be chosen as unit area, the content of P, that is the content of the parallelogram whose first side is p and whose second side is q, is equal to α, that is, it is equal to the numerical value of P.

335. *The supplement of a displacement {in space} is the surface whose plane is orthogonal to that displacement, whose numerical value is equal to that of the displacement, and whose positive sense is defined so that the outer product of the displacement and the surface is positive.*

PROOF. Let αa be the displacement, and let the numerical value of a equal one, whence that of αa is α, and let a, b, c be a normal system such that $[abc] = +1$; then (according to 167) $|\alpha a = \alpha[bc]$, but (according to 332, 334) $\alpha[bc]$ is a surface with the properties stated in the theorem.

336. *If A is the supplement of a displacement a {in space}, then a is also the supplement of A, or*

$$||a = a.$$

PROOF. According to 92, $||a = (-1)^{pq}a$, where p is the order of a and q is that if its supplement. In our case these two numbers are 1 and 2, whence

$$||a = (-1)^2 a = a.$$

REMARK. The theorem does not hold in the corresponding way for planimetry, where no more than two mutually orthogonal displacements can be found. Rather, in planimetry $||a = -a$.

{336A. *For a displacement and a surface area (that is, a product of two displacements) as well, the concept of orthogonality coincides with that of the normal.*

PROOF. *First*, let a be a displacement and B be a surface area *orthogonal* to it; then (according to 336 and 335) the supplement of B is a displacement orthogonal to B, and thus parallel to a, and consequently (according to 230A) $|B = \alpha a$. Thus

$$[a|B] = [a \cdot \alpha a] = 0, \qquad [61]$$

that is (according to 152), the displacement a and the surface area B are normal to one another.

Conversely, if *a is normal to B*, that is (according to 152)

$$[a|B] = 0,$$

one sets $|B = b$; then $[ab] = 0$, and thus (according to 66) $a = \beta b$. This equation signifies however (according to 221) that the displacement a is parallel to b. But the displacement b is (according to 336, 335) orthogonal to B, and consequently the same holds for the displacement a parallel to b.}

337. *If a and b are displacements, then $\angle ab$ is equal to the angle whose legs are directed similarly to a and b, and if A and B are surfaces (products of displacements), $\angle AB$ is equal to the angle of inclination formed by two planes parallel and similarly signed to those surfaces, assuming that the angle is always taken as positive (between 0 and π).*

REMARK. If in the second case one represents A and B as rectangles whose first sides are the same, then the angle included by the second sides of these rectangles is the angle of inclination formed by the planes parallel and similarly signed to A and B.

PROOF. 1. The angle is independent of the numerical values of the angle forming magnitudes, so we can set the numerical values of a, b, A, B equal to 1; in this case (according to 195)

$$\cos\angle ab = [a|b], \quad \cos\angle AB = [A|B].$$

Further, if the angle from a to b is equal to α, b results from a through a circular evolution by the angle α (according to 154, 331), that is, if a' lies orthogonal to a in the plane ab and lies on the side of b,

$$b = a\cos\alpha + a'\sin\alpha,$$

whence

$$[b|a] = [a|(a\cos\alpha + a'\sin\alpha)] = [a|a]\cos\alpha + [a|a']\sin\alpha,$$

but, since a' is normal to a, $[a|a'] = 0$, and since the numerical value of a is equal to one, the last expression

$$= \cos\alpha.$$

Thus $\cos\angle ab = \cos\alpha$. Now (according to 195) $\angle ab$ lies between 0 and π, but by hypothesis so also does α, whence $\angle ab = \alpha$.

2. Let A and B be the supplements of a and b respectively, whence $A = |a$ and $B = |b$, then (according to 335) a and b are orthogonal to A and B respectively, and lie on the same side, whence the angle of inclination α between A and B is equal to the angle between a and b, that is (according to Part 1 of the Proof) $\cos\alpha = \cos\angle ab = [a|b] = |[a|b]$ (according to 89), since $[a|b]$ (according to 141) is a number. But $|[a|b] = [|a||b]$ (according to 97), and this again (by hypothesis) $= [A|B] = \cos\angle AB$; thus $\cos\alpha = \cos\angle AB$.

{COROLLARY. *For two surface areas (that is, products of any two displacements) as well the concepts of orthogonality and the normal are equivalent.*}

REMARK. The concept of the normal can in fact be applied to magnitudes of any type, and thus also to points. In particular, in order to be able to derive all the magnitudes in space, one must, in addition to the three mutually orthogonal displacements of unit length that we have taken as the original units, take a finitely distant {simple} point as the fourth unit. Thus one would be led to four original units a, b, c, d of which, say, the last three are mutually orthogonal displacements of unit length, and the first is a simple point. One can apply the concept of the normal to this assembly.

If instead of these one takes another assembly a', b, c, d, where a' denotes a simple point different from a, as the system of original units, then it is clear that a and a' have the same supplement $[bcd]$, since (according to 268) $[abcd] = [a'bcd]$, and likewise that $[a'b]$ and $[ab]$ have the same supplement $[cd]$, and $[a'bc]$ and $[abc]$ have the same supplement d, and generally *that the supplements of points, line elements, surface elements are independent of the positions of the points chosen as original units.* On the other hand *this is no longer the case for the supplements of displacements or products of displacements.*

Thus for example the supplement of $[bcd]$ is with the first assumption equal to $-a$, and with the second equal to $-a'$. And thus the concept of the supplements of displacements and products of displacements would no longer have a sense independent of the position of the original units.

Since for the normal shadow (164) on points, lines, and planes {according to 165} only the first type of supplement appears, we can immediately apply this shadow to geometry. Here it yields {insofar as it is concerned only with the shadow on *lines* and *planes*} the orthogonal projection, {whereas the normal shadow on a *point* amounts to a parallel displacement maintaining the numerical value}, which however I will not develop further here.[85]

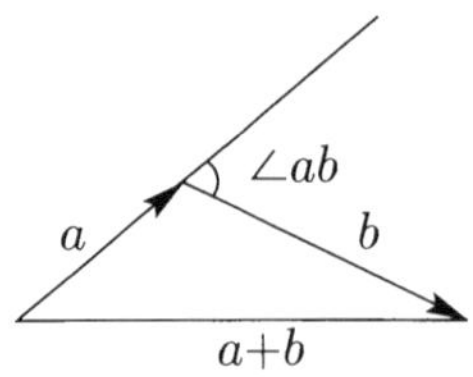

FIGURE 19

In the following I will apply the concept of the supplement only in the sense given in the text.

338. *The formula*

$$(a+b)^2 = a^2 + 2[a|b] + b^2 = \alpha^2 + 2\alpha\beta\cos\angle ab + \beta^2, \qquad \{193, 214\}$$

where α and β are the lengths of a and b, represents the generalization of the Pythagorean theorem, namely: The square of the base side of a triangle is equal to the sum of the squares of the legs and twice the product of the legs with the cosine of the exterior angle at the apex {see Fig. 19}.

339. *The formula* {Cf. 194, 215}

$$\begin{aligned}(a+b+c)^2 &= a^2 + b^2 + c^2 + 2[b|c] + 2[c|a] + 2[a|b] \\ &= \alpha^2 + \beta^2 + \gamma^2 + 2\beta\gamma\cos\angle bc + 2\gamma\alpha\cos\angle ca + 2\alpha\beta\cos\angle ab,\end{aligned}$$

where α, β, γ are the lengths of the displacements a, b, c, represents the generalization of that theorem for space.

340. *The formula*

$$\begin{aligned}(A+B+C)^2 &= A^2 + B^2 + C^2 + 2[B|C] + 2[C|A] + 2[A|B] \\ &= \alpha^2 + \beta^2 + \gamma^2 + 2\beta\gamma\cos\angle BC + 2\gamma\alpha\cos\angle CA + 2\alpha\beta\cos\angle AB,\end{aligned}$$

where α, β, γ are the areas of the surface areas A, B, C, and $\angle BC$, ... are the angles of inclination of their planes, represents the theorem:

The square of the base surface of a tetrahedron is equal to the sum of the squares of the sides, diminished by twice the products of each pair of these sides with the cosine of the angle of inclination enclosed by them {*i.e.* *by the supplements of* $\angle BC$, $\angle CA$, $\angle AB$}.

PROOF. If a, b, c are in length and direction the edges leading from the apex to the vertices of the base side, then $a-b$, $b-c$, $c-a$ are the edges of the base surface. The base surface is therefore (according to 254) equal to

$$\frac{[(a-b)(b-c)]}{2},$$

whereas the sides are equal to

$$\frac{bc}{2}, \frac{ca}{2}, \frac{ab}{2}.$$

If we denote the latter respectively by A, B, C, and the base surface

$$\frac{[(a-b)(b-c)]}{2}$$

by D, and consider that

$$[(a-b)(b-c)] = [ab] - [ac] - [bb] + [bc] = [ab] + [ca] + [bc],$$

then we have

$$D = A + B + C,$$

and thus

$$\begin{aligned} D^{\underline{2}} &= (A + B + C)^{\underline{2}} \\ &= A^{\underline{2}} + B^{\underline{2}} + C^{\underline{2}} + 2[B|C] + 2[C|A] + 2[A|B] \\ &= \alpha^2 + \beta^2 + \gamma^2 + 2\beta\gamma \cos \angle BC + 2\gamma\alpha \cos \angle CA + 2\alpha\beta \cos \angle AB. \end{aligned}$$

But $\angle BC$ is the angle between the planes $[ca]$ and $[ab]$, that is, between $-[ac]$ and $[ab]$. The angle between $[ac]$ and $[ab]$ is however that enclosed between the corresponding sides of the tetrahedron, and thus the angle between $-[ac]$ and $[ab]$, that is $\angle BC$, is the adjacent angle, and the same holds for the angles $\angle CA$ and $\angle AB$.

REMARK. From this presentation one perceives how our analysis must produce the solution of the tetrahedron in a simple way by virtue of the relation that one of its sides can be represented as the geometric sum of the others. And since the spherical triangle or three-edged vertex can again be reduced to a tetrahedron in which the three edges are radii of a sphere, it is shown how closely spherical trigonometry is connected to this.

Yet I observe here that all the formulas of spherical trigonometry would be symmetric if, as already shown above, the angle of inclination of the surfaces were replaced by their exterior angle. This is particularly apparent here, since then the angles of the polar vertex would be *equal* to the sides of the original vertex, and consequently all the formulas of trigonometry would retain their validity directly if one interchanged angles and sides. It then follows immediately that, if a, b, c are displacements directed along the edges of a vertex, the supplements of these displacements, that is the surface areas $|a$, $|b$, $|c$, are parallel to the planes of the polar vertex. In order not to be led too far afield, however, I must leave the further development of these ideas to the reader.[86]

341. PROBLEM. *To express the multiple sum of the squares of the distances of a variable point x from several fixed points a, b, ... in the simplest form, if the coefficients α, β, ... of that multiple sum are given.*

SOLUTION. Thus the expression

$$S = \alpha(x - a)^{\underline{2}} + \beta(x - b)^{\underline{2}} + \cdots$$

is to be represented in the simplest form possible.

Since x, a, b, ... are points, and for them the inner product no longer has a simple meaning, we take an arbitrary simple point s as an auxiliary, and set

$$x - a = x - s + s - a, \quad x - b = x - s + s - b, \ldots,$$

where $x - s$, $s - a$, $s - b$, ... are displacements; then

$$S = \alpha(x - s + s - a)^{\underline{2}} + \beta(x - s + s - b)^{\underline{2}} + \cdots .$$

Now since

$$(x - s + s - a)^{\underline{2}} = (x - s)^{\underline{2}} + 2[(x - s)|(s - a)] + (s - a)^{\underline{2}}$$

(according to 193), we obtain

$$\begin{aligned} S = {} & (\alpha + \beta + \cdots)(x - s)^{\underline{2}} \\ & + 2[(x - s)|(\alpha(s - a) + \beta(s - b) + \cdots)] \\ & + \alpha(s - a)^{\underline{2}} + \beta(s - b)^{\underline{2}} + \cdots , \end{aligned}$$

or, if we set

$$\alpha + \beta + \cdots = \sigma,$$

we find

$$(*)\ S = \sigma(x-s)^{\underline{2}} + 2[(x-s)|(\sigma s - \alpha a - \beta b - \cdots)] + \alpha(s-a)^{\underline{2}} + \beta(s-b)^{\underline{2}} + \cdots.$$

Now we can distinguish two cases, according as σ is zero or not.

First, assuming the latter, we can choose σ so that the second term is zero, which is achieved if we set

$$\sigma s = \alpha a + \beta b + \cdots,$$

that is,

$$s \equiv \alpha a + \beta b + \cdots,$$

which is to say (according to 222, {Remark following 223}) that s lies on the center of gravity of the system of points αa, βb, ... Then

$$S = \sigma(x-s)^{\underline{2}} + \mu,$$

if for short we set the constant magnitude

$$\alpha(s-a)^{\underline{2}} + \beta(s-b)^{\underline{2}} + \cdots = \mu,$$

that is:

The multiple sum S of the squares of the distances of a variable point x from several fixed points a, b, ..., where α, β, ... are the coefficients of this sum and the sum σ of these coefficients is positive, achieves its smallest value when x lies at the center of gravity s of this point-assembly αa, βb, ... If however x is removed from this center of gravity s by the distance ρ, then the multiple sum increases by the σ-fold of the square of this distance, and is therefore constant on the surface of a sphere having s as its center. If σ is negative, everything remains the same, except that instead of the minimum a maximum enters.

342. CONTINUATION. *Second*, if $\alpha + \beta + \cdots = 0$, the formula $(*)$ of the preceding number is transformed into

$$S = -2[(x-s)|(\alpha a + \beta b + \cdots)] + \alpha(s-a)^{\underline{2}} + \beta(s-b)^{\underline{2}} + \cdots.$$

In this case (according to 222) the sum $\alpha a + \beta b + \cdots$ is a displacement of fixed length and direction, which we will denote by r; then

$$S = -2[(x-s)|r] + \alpha(s-a)^{\underline{2}} + \beta(s-b)^{\underline{2}} + \cdots.$$

Now let it be assumed that r is *not zero*, and let its length be equal to ρ. Then in order to reduce the expression even further, we take a point s' on the straight line drawn parallel to r from s {see Fig. 20}, and set $s' - s = zr$, where z is a number, since by construction $s' - s$ is parallel to r. Then

$$[(x-s)|r] = [(x-s'+s'-s)|r] = [(x-s')|r] + [(s'-s)|r].$$

But $[(s'-s)|r]$ equals $[zr|r] = z[r|r]$, since z is a number, and thus $= zr^{\underline{2}} = z\rho^2$, since ρ is the numerical value of r. Consequently

$$[(x-s)|r] = [9x - s')|r] + z\rho^2,$$

and

$$S = -2[(x-s')|r] - 2z\rho^2 + \alpha(s-a)^{\underline{2}} + \beta(s-b)^{\underline{2}} + \cdots.$$

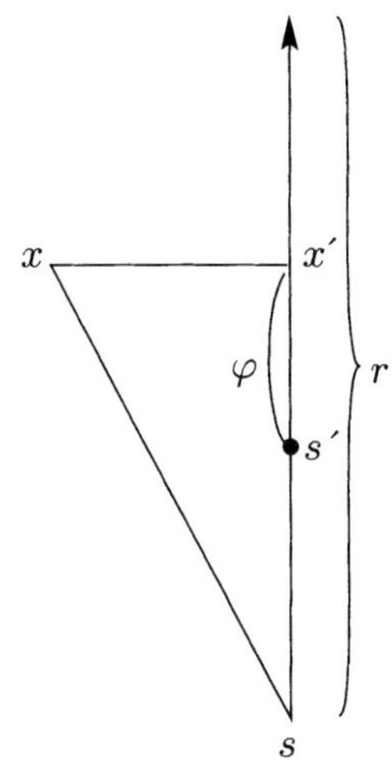

FIGURE 20

Since s' is an arbitrary point on the straight line $[sr]$, z is not yet specified; if z is defined such that

$$z = \frac{\alpha(s-a)^{\underline{2}} + \beta(s-b)^{\underline{2}} + \cdots}{2\rho^2},$$

then

$$S = -2[(x - s')|r].$$

Now if one drops the perpendicular xx' onto the straight line $[sr]$ or $[s'r]$, then $x - x'$ is normal to r, that is $[(x - x')|r] = 0$, whence

$$[(x - s')|r] = [(x - x' + x' - s')|r] = [(x' - s')|r],$$

and thus

$$S = -2[(x' - s')|r] = \pm 2\varphi\rho,$$

where φ is the length of $x' - s'$, and the lower or upper sign is chosen according as $x' - s'$ is directed the same as or opposite to r, that is:

The multiple sum S of the squares of the distances of a variable point x from several fixed points a, b, ... , where α, β, ... are the coefficients of the sum, and the sum of these coefficients is zero, is equal to zero for all the points of a certain plane, which is orthogonal to the displacement $\alpha a + \beta b + \cdots$. If however the point x is removed from this plane by the distance φ, the sum $= -2\varphi\rho$ or $+2\varphi\rho$, according as it is removed from that plane in the direction of the displacement r or in the direction opposite to it, where ρ represents the length of r.

343. CONCLUSION. *Finally, if even r is zero*, then

$$S = \alpha(s-a)^{\underline{2}} + \beta(s-b)^{\underline{2}} + \cdots,$$

{and is therefore independent of x,} that is a constant, which is to say:

The multiple sum S of the squares of the distances of a variable point x from several fixed points a, b, ... is constant when the corresponding multiple sum of points a, b, ... is zero, that is

$$\alpha(s-a)^{\underline{2}} + \beta(s-b)^{\underline{2}} + \cdots = \text{const},$$

when

$$\alpha a + \beta b + \cdots = 0.$$

In particular (according to 222) the equation $\alpha a + \beta b + \cdots = 0$ already includes the equation $\alpha + \beta + \cdots = 0$.[87]

344. PROBLEM. *To express the multiple sum S of the distances of a variable point x from several fixed planes A, B, ... in the simplest form possible, if the coefficients α, β, ... of the multiple sum are given, and for each plane the side on which the positive distance lies is specified.*

SOLUTION. In each of the planes one takes a surface element whose content is equal to the unit of area, and which (by the way it is generated) is so constituted that the outer product of this surface element forms a positive product with a displacement directed toward the side of the plane taken as positive. If these surface elements, regarded as magnitudes of third order (255), are denoted by A', B', ... , then (according to 263) the product $[A'x]$ is equal to the content of a parallelepiped whose base surface is A', and whose covering surface (the surface above the base surface) goes through the point x, and thus is equal to the content of A' times the height, or, since the content of A' is equal to unity, it is equal to the height, that is, it is equal to the distance of the point x from the plane, including the sign, and likewise for the other planes. Therefore

$$\begin{aligned} S &= \alpha[A'x] + \beta[B'x] + \cdots \\ &= [(\alpha A' + \beta B' + \cdots)x] \\ &= \rho[Rx], \end{aligned}$$

where ρR is the corresponding multiple sum of the surface elements A', B', ... , and the content of R is unity, but ρ is positive.

The multiple sum S of the distances of a variable point x from several fixed planes A, B, ... with the coefficients α, β, ... stands in a constant proportion $\rho : 1$ to the distance of the same point from a fixed plane R. And one finds R and ρ if on the planes A, B, ... one takes surface elements A', B', ... of unit content that give positive products when outer multiplied by points lying on the sides of the respective planes that are taken as positive; for then R is the plane of the surface element $\alpha A' + \beta B' + \cdots$, and ρ is its content.

If however this sum gives an infinitely distant plane, that is a volume (262), *whence $[Rx]$ is a volume element, then* (*according to* 268) *$\rho[rx]$ is a constant, and thus also the multiple sum S is a constant.*

Finally, should $\alpha A' + \beta B' + \cdots$ itself be zero, then S would also be zero for every point x.

REMARK. The theorems found in the above problems can be combined into a *single* theorem if one replaces the points and planes by spherical surfaces which, if the radii are zero, are transformed into points, and if they are infinite, into planes, and indeed, if one replaces the squares of the distances from a point and the simple distances from a plane by the product of the least and greatest distance from the spherical surface, then, when the spherical surface is diminished to a point, this product goes over into the square of the distance, and when the spherical surface is opened into a plane, then one of the distances becomes infinite, and therefore can be regarded as equal for all the planes and thus can be divided out, whereby the simple distances result. This generalization will be carried out in the following theorem.

345. *If by the* **double distance** *of a point from a spherical surface one means the product of the least and greatest distances of the point from the spherical surface (the product being taken as positive if the distances are similarly directed, that is, if the point lies outside the spherical surface, negative otherwise), then the multiple sum S of the double distances α', β', ... of several fixed spherical surfaces whose centers are a, b, ... , and whose radii are a', b', ... , that is the multiple sum*

$$\alpha\alpha' + \beta\beta' + \cdots,$$

if α, β, ... represent their coefficients, is a maximum or a minimum if x lies on the center of gravity s of the assembly of points αa, βb, ... And if the point x is removed from this center of gravity s by a distance φ, then S increases by the product {of the square} of this distance with the sum σ of the coefficients α, β, ... , that is by $\varphi^2\sigma$ or $(\alpha + \beta + \cdots)\varphi^2$.

But if the assembly of points αa, βb, ... has no center of gravity, that is if $\alpha+\beta+\cdots$ is zero, there is a plane E orthogonal to the displacement $r = \alpha a+\beta b+\cdots$ for whose points S is zero; then if x is removed from this plane by a distance φ, S becomes $\pm 2\varphi\rho$, where ρ is the numerical value of r and the lower or upper sign is chosen according as x is moved toward the side lying in the direction of r from E, or in the opposite direction.

If however $\alpha a + \beta b + \cdots = 0$, then S is constant.

PROOF. If one draws from x the line through the center a of the first sphere, which intersects its surface in x_1 and x_2, then the double distance is

$$\alpha' = (x - x_1)(x - x_2) = (x - a - a')(x - a + a')$$

if $x - x_1$ denotes the displacement from x_1 to x, and so forth, and a' is the radius. Thus $a' = (x - a)^2 - a'^2$, or if we now mean by $x - a$ a displacement of definite length and direction,

$$a' = (x - a)^{\underline{2}} - a'^2.$$

Thus

$$\begin{aligned} S &= \alpha a' + \beta b' + \cdots \\ &= \alpha[(x - a)^{\underline{2}} - a'^2] + \beta[(x - b)^{\underline{2}} - b'^2] + \cdots . \end{aligned}$$

Now if we take a constant auxiliary point s, which will serve for the simplification of the expression, then

$$\begin{aligned} (x - a)^{\underline{2}} &= (x - s + s - a)^{\underline{2}} \\ &= (x - s)^{\underline{2}} + 2[(x - s)|(s - a)] + (s - a)^{\underline{2}}, \end{aligned}$$

and similarly for the other squares. If in addition we set $\alpha + \beta + \cdots = \sigma$, then

$$S = \sigma(x - s)^{\underline{2}} + 2[(x - s)|(\sigma s - \alpha a - \beta b - \cdots)] + \mu,$$

where for the magnitude depending only on the choice of s we set

$$\alpha(s - a)^{\underline{2}} + \beta(s - b)^{\underline{2}} + \cdots - \alpha a'^2 - \beta b'^2 - \cdots = \mu.$$

Now if σ is different from zero, then the factor

$$\sigma s - \alpha a - \beta b - \cdots$$

is zero if s is the center of gravity of the assembly of points αa, βb, ... Thus if we assume that s lies on this center of gravity, then

$$S = \sigma(x-s)^2 + \mu,$$

whereby the first part is proved.

But if $\sigma = 0$, then $\alpha a + \beta b + \cdots$ is a displacement; let it be r, and its numerical value ρ {see Fig. 20}. Then

$$S = 2[(s-x)|r] + \mu.$$

Since S is of first degree with respect to x, one can easily find points x for which S is zero. Let s' be such a point,* that is

$$2[(s-s')|r] + \mu = 0;$$

then

$$S = 2[(s-s'+s'-x)|r] + \mu = 2[(s'-x)|r]$$

by virtue of the previous equation, and this

$$= \pm 2\phi\rho,$$

if ϕ is the numerical value of the orthogonal projection of $s'-x$ on r, and the lower or upper sign is chosen according as $x-s'$ and r are directed to the same or the opposite side of the plane erected on s' perpendicular to r, whereby the second part of the theorem is proved.

Finally, if in addition $\alpha a + \beta b + \cdots = 0$, then

$$S = \mu,$$

whence {it is independent of x, that is} it is constant.

{REMARK. In this connection, cf. Nos. 392–409.}

346. PROBLEM. *To reduce the sum S of several line elements in space to the sum of a line element and a surface area orthogonal to it.*[89]

SOLUTION. If one takes a system of four units in space: a, a_1, a_2, a_3, of which a is a simple point and a_1, a_2, a_3 are displacements, then (according to 222) every point in space can be numerically derived from them, whence every line element, as a product of two points (according to 65), is numerically derivable from the multiplicative combinations aa_1, aa_2, aa_3, a_1a_2, a_1a_3, a_2a_3; thus also S, as the sum of such line elements. Let

$$S = \alpha_1[aa_1] + \alpha_2[aa_2] + \alpha_3[aa_3] + \beta_1[a_1a_2] + \beta_2[a_1a_3] + \beta_3[a_2a_3]$$

or

$$= [a(\alpha_1a_1 + \alpha_2a_2 + \alpha_3a_3)] + \beta_1[a_1a_2] + \beta_2[a_1a_3] + \beta_3[a_2a_3].$$

Let

$$\alpha_1a_1 + \alpha_2a_2 + \alpha_3a_3 = b;$$

then b is a displacement, and $\beta_1[a_1a_2] + \beta_2[a_1a_3] + \beta_3[a_2a_3]$, as a sum of products of every pair of displacements, is again such an object, and can therefore be interpreted as the supplement of a displacement, say the displacement c; that is, let

$$\beta_1[a_1a_2] + \beta_2[a_1a_3] + \beta_3[a_2a_3] = |c,$$

whence

$$S = [ab] + |c.$$

*If one sets the point $s + (\mu r : 2\rho^2) = p$ then s' is an arbitrary point that lies in the plane through p orthogonal to r.[88]

Let a' be some other, yet to be specified {simple} point, and let $a - a' = d$, where d is a displacement; then $a = a' + d$, whence

$$S = [(a' + d)b] + |c = [a'b] + [db] + |c.$$

Here $[db] + |c$ is a surface area; and d is to be defined so that, as required by the problem, this surface area is orthogonal to the line element $[a'b]$. Since however a' is a point and b a displacement, this line element has the direction of b, whence that surface area must also be orthogonal to b, that is (according to 152, 336A) it must be that $[(db + |c)|b] = 0$, or

$$[db|b] + |[cb] = 0, \qquad [99]$$

that is

$$|[cb] = [bd|b]. \qquad [55]$$

If we assume that d is orthogonal to b, that is, that it lies in the plane $|b$, then (according to 108) the last expression found

$$= [b|b]d = \beta^2 d,$$

if β is the numerical value of b, whence

$$d = |[cb] : \beta^2.$$

Conversely, if d has this value, then by the reversed transformations it follows that the surface area $[db] + |c$ is orthogonal to b, whence the problem is solved.

347. *If two sums of line elements are each recast as a sum, one term of which is a line element and the other is a surface area orthogonal to the line of the first, then those sums are equal only if both the line elements and the surface areas are equal to one another; that is, if*

$$[ab] + \gamma|b = [a_1 b_1] + \gamma_1|b_1,$$

where a and a_1 are simple points, b and b_1 {two different nonzero} displacements, γ and γ_1 are numbers, then

$$[ab] = [a_1 b_1], \quad \gamma|b = \gamma_1|b_1.$$

PROOF. If one first multiplies the given equation by a product U of three displacements whose parallelepiped is taken as the unit volume, then (according to 301), since $|b$, $|b_1$ are products of every pair of displacements, $[|b \cdot U] = 0 = [|b_1 \cdot U]$, whereas (according to 305) $[abU] = b$ and $[ab_1U] = b_1$; thus one obtains $b = b_1$. Consequently the given equation is transformed into

$$[ab] + \gamma|b = [a_1 b] + \gamma_1|b.$$

If one multiplies this equation by b, then since $[abb] = [a_1bb] = 0$ (according to 60), $\gamma[b|b] = \gamma_1[b|b]$, whence, since $[b|b] = \underline{b^2}$ is a nonzero number, $\gamma = \gamma_1$. Consequently the original equation is transformed into $[ab] = [a_1b]$, that is, the line elements $[ab]$ and $[a_1b_1]$ and the surface areas $\gamma|b$ and $\gamma_1|b_1$ are equal to one another.

REMARK. Thus every sum of line elements, that is, every geometric magnitude of second order, can be represented in a specific, completely unambiguous way in the form of a line element and a surface area orthogonal to its line. It is interesting that this general magnitude of second order is completely represented by the motion of a body in space.

Thus if a body is transferred from one position into any other, it is well known that it is always possible to effect this transfer if one lets a certain line of the body progress in its own direction by a certain amount, while the body completes a certain rotation about this line as about its axis, and these two partial motions are completely specified by the initial and final positions of the body. The former motion can be completely represented by a line element, the latter by a surface area orthogonal to it, and thus both together by the general spatial magnitude of second order.

Naturally in statics one could have represented the resultant of several forces in space in the same way, where just the line elements represent the static forces and their sum the static resultant of the corresponding forces (cf. *Ausdehnungslehre* {of 1844}, §121ff).

Part 2

The Theory of Functions

CHAPTER 1

Functions in General

1. Concept of a function, and reduction of several functions of several variables to a single function of a single variable

348. DEFINITION. If a magnitude u depends on one or more magnitudes x, $y, \ldots$ in such a way that, as soon as $x, y, \ldots$ assume definite values, u assumes a definite (unique) value, we call u a *function* of $x, y, \ldots$.

REMARK. It is to be noted here that the above definition holds even if $u, x, y, \ldots$ are extensive magnitudes. It is also to be noted that many-valued functions, that is such that for definite values of the independent variables $x, y, \ldots$, the magnitude u can assume several different values, without these different values being specified by a new variable — are here completely excluded; thus, then, in general all functions many-valued in this sense are to be banned from mathematics, since *no* mathematical formula can be applied to them with certainty.

As soon as the relation between the independent variables and the dependent variable u is given by means of an equation, by which, for definite values of the former, the latter u can assume several different values, one can recognize u as a function of those variables and a new {variable}, r, that runs through a definite series of values, say that of the integers, so that if, in addition to the original variables, the value of r as well is specified, u is also uniquely specified. Or if a single such new variable r does not suffice, one can adopt several such as auxiliaries.

If for example one has the equation $u^n = x$, then for each value of x the magnitude u can still assume n different values. Let one of them be denoted $x^{1/n}$; then, as is well known,

$$u = (-1)^{2r/n} x^{1/n} = \left(\cos\frac{2r\pi}{n} + \sqrt{-1}\sin\frac{2r\pi}{n}\right) x^{1/n},$$

where r can assume any integer value, one by one. Thus u is in this way represented as a function of x and a new variable r, whereby the many-valuedness of the function vanishes.

349. DEFINITION. I call a function a *numerical function* if, for arbitrary values of the variables on which it depends, it always assumes a numerical value (real or imaginary, integer or fraction, rational or irrational). I call a function an *extensive function* if, for all (or certain) values of its variables, it is equal to an extensive magnitude. I will always denote numerical functions by lower-case letters $f, \phi, \psi, \ldots$, extensive functions by upper-case letters $F, \Phi, \Psi, \ldots$.

REMARK. The imaginary numerical magnitude $p + q\sqrt{-1}$ stands at the boundary of extensive magnitudes. It is interpreted as an extensive magnitude as soon as one permits only *real* numerical coefficients for the expression of the numerical relation between several magnitudes, since then 1 and $\sqrt{-1}$ appear as units standing in no numerical relation. On the other hand it is interpreted as a numerical magnitude as soon as imaginary coefficients as well are permitted for the expression of the numerical relation. Thus if the function $y = f(x)$ assumes imaginary values, say $u + v\sqrt{-1}$, for real x, then it can be interpreted as an extensive magnitude in the former sense; but in the theory of functions we will always adopt the latter sense, thus even if y is imaginary y is interpreted as a numerical function, as is done in the Definition.

350. *Every numerical function of arbitrarily many numerical magnitudes can be represented as a numerical function of a single extensive magnitude, and in fact, if*

$$y_1 = f(x_1, x_2, \ldots, x_n),$$

then this expression is equivalent to

$$y_1 = f([x|e_1], [x|e_2], \ldots, [x|e_n]) = \phi(x),$$

where $e_1, \ldots, e_n$ *form an elementary normal system* (*see* 153), *and*

$$x = x_1e_1 + x_2e_2 + \cdots + x_ne_n.$$

PROOF. If $x = x_1e_1 + x_2e_2 + \cdots + x_ne_n$, and $e_1, \ldots, e_n$ form an elementary normal system, then

$$[x|e_1] = [(x_1e_1 + x_2e_2 + \cdots + x_ne_n)|e_1] = x_1,$$

since $[e_1|e_2], \ldots, [e_1|e_n]$ are zero (according to 188), and $[e_1|e_1] = 1$. On the same grounds

$$[x|e_2] = x_2, \ \ldots, \ [x|e_n] = x_n.$$

Thus

$$y_1 = f(x_1, x_2, \ldots, x_n) = f([x|e_1], [x|e_2], \ldots, [x|e_n]),$$

whence y_1 is a function of the extensive magnitude x. If this function is denoted by $\phi(x)$, then one has

$$y_1 = \phi(x).$$

351. *Every system of numerical functions of arbitrarily many numerical magnitudes can be represented as a single function of a single extensive magnitude, and in fact, if*

$$\begin{aligned} y_1 &= f_1(x_1, x_2, \ldots, x_n), \\ y_2 &= f_2(x_1, x_2, \ldots, x_n), \\ &\cdots\cdots\cdots\cdots \\ y_m &= f_m(x_1, x_2, \ldots, x_n), \end{aligned}$$

then this system of equations is equivalent to the equation

$$y = F(x),$$

where

$$\begin{gathered} x = x_1e_1 + \cdots + x_ne_n, \ y = y_1e_1 + \cdots + y_me_m, \\ F(x) = e_1\phi_1(x) + e_1\phi_2(x) + \cdots + e_m\phi_m(x), \\ \phi_r(x) = f_r([x|e_1], [x|e_2], \ldots, [x|e_n]), \end{gathered}$$

and $e_1, \ldots, e_n$ *and* $e_1, \ldots, e_m$ *form elementary normal systems.*

PROOF. According to 350, if one sets $x = x_1e_1 + \cdots + x_ne_n$, then

$$y_r = f_r(x_1, x_2, \ldots, x_n)$$

is equivalent to

$$y_r = f_r([x|e_1], [x|e_2], \ldots, [x|e_n]).$$

If this function of x is denoted by $\phi_r(x)$, then

$$y_r = \phi_r(x) = f_r([x|e_1], [x|e_2], \dots, [x|e_n]).$$

Furthermore (according to 29) the system of equations

$$y_1 = \phi_1(x),\ y_2 = \phi_2(x),\ \dots,\ y_m = \phi_m(x)$$

is equivalent to the equation

$$y_1 e_1 + \dots + y_m e_m = e_1\phi_1(x) + e_2\phi_2(x) + \dots + e_m\phi_m(x),$$

that is, to the equation

$$y = e_1\phi_1(x) + e_2\phi_2(x) + \dots + e_m\phi_m(x).$$

Denoting this {extensive} function by $F(x)$, that is

$$F(x) = e_1\phi_1(x) + e_2\phi_2(x) + \dots + e_m\phi_m(x),$$

the m given equations

$$y_1 = f_1(x_1, x_2, \dots, x_n),\ \dots,\ y_m = f_m(x_1, x_2, \dots, x_n)$$

are equivalent to the equation

$$y = F(x).$$

352. *Every system of functions of arbitrarily many variables can be replaced by a single function of a single variable, assuming that all the independent variables can be numerically derived from a single system of units, and similarly for the dependent variables.*

PROOF. 1. The theorem is proved in 351 for an arbitrary system of numerical functions of arbitrarily many numerical magnitudes.

2. According to 157 all magnitudes numerically derivable from a system of n units can be numerically derived from an elementary normal system of n magnitudes. Let the normal system from which the independent variables x, y, ... can be derived consist of the magnitudes e_1, e_2, ... , and that from which the dependent variables u, v, ... can be derived consist of the magnitudes $e^{(1)}$, $e^{(2)}$, ... , $e^{(m)}$. Further, let

$$\begin{aligned} x &= x_1e_1 + \dots + x_ne_n, & u &= u_1e^{(1)} + \dots + u_me^{(m)},\\ y &= y_1e_1 + \dots + y_ne_n, & v &= v_1e^{(1)} + \dots + v_me^{(m)},\\ &\dots\dots\dots \end{aligned}$$

where all the coefficients are numerical magnitudes, and let

$$u = F(x, y, \dots),\ v = \Phi(x, y, \dots),\ \dots$$

be the given functions; then, when one inserts the above values into the first of these equations, one obtains

$$u_1e^{(1)} + \dots + u_me^{(m)} = F(x_1e_1 + \dots + x_ne_n,\ y_1e_1 + \dots + y_ne_n, \dots).$$

Here the right side is an extensive function of the numerical magnitudes x_1, ... , x_n, y_1, ... , y_n, ... This function is to be equal to a magnitude that is numerically derivable from the units $e^{(1)}$, ... , $e^{(m)}$, whence it must itself be numerically derivable from them, that is, it must appear in the form $e^{(1)}\phi_1 + \dots + e^{(m)}\phi_m$, where $\phi_1, \ldots, \phi_m$ are numerical functions of $x_1, \ldots, x_n, y_1, \ldots, y_n, \ldots$. Thus one obtains

$$u_1 e^{(1)} + \cdots + u_m e^{(m)} = e^{(1)}\phi_1 + \cdots + e^{(m)}\phi_m.$$

This equation, which is equivalent to $u = F(x, y, \ldots)$, is (according to 29) replaced by the system of numerical equations

$$u_1 = \phi_1, \ u_2 = \phi_2, \ \ldots, \ u_m = \phi_m.$$

In the same way the equation

$$v = \Phi(x, y, \ldots)$$

is replaced by a system of numerical equations of the form

$$v_1 = \psi_1, \ v_2 = \psi_2, \ \ldots, \ v_m = \psi_m,$$

where $\psi_1, \psi_2, \ldots$ are likewise numerical functions of the numerical magnitudes $x_1, \ldots, x_n, y_1, \ldots, y_n, \ldots$. Consequently the given equations

$$u = F(x, y, \ldots), \ v = \Phi(x, y, \ldots), \ \ldots$$

are replaced by the system of numerical equations

$$\begin{array}{llll} u_1 = \phi_1, & u_2 = \phi_2, & \ldots, & u_m = \phi_m, \\ v_1 = \psi_1, & v_2 = \psi_2, & \ldots, & v_m = \psi_m, \\ \ldots & \ldots & \ldots & \ldots \end{array}$$

where $\phi_1, \ldots, \phi_m, \psi_1, \ldots, \psi_m, \ldots$ are numerical functions of the numerical magnitudes $x_1, \ldots, x_n, y_1, \ldots, y_n, \ldots$. But according to 151 such a system can be replaced by a single function of a single magnitude, whence also the given system can be replaced by a single such function of a single magnitude.

REMARK. The reduction to a single function of a single variable is also of importance for the treatment of numerical functions, since the theorems on numerical functions, differential calculus, series, and even, with certain restrictions, those of integral calculus, can be applied to such extensive functions of extensive magnitudes, as will be shown below. In particular there follows most easily and directly JACOBI's theorem on functional determinants, as well as the agreement (suggested or hinted at by JACOBI in his celebrated article (*Crelles Journal* **22**, 319f)) of this theorem with the theorem on the differentiation of a single elementary function of a single variable.

2. Complete functions and their representation by open products

353. DEFINITION. If $P_{a_1, a_2, \ldots, a_n}$ is an arbitrary product in which the magnitudes of first order $a_1, a_2, \ldots, a_n$ occur as factors, then by

$$P_{l,l,\ldots,l}(x_1 x_2 \cdots x_n),$$

where $x_1, x_2, \ldots, x_n$ are arbitrary magnitudes of first order, I mean the arithmetic mean between all the expressions that result from

$$P_{x_1, x_2, \ldots, x_n}$$

when one gives the magnitudes $x_1, x_2, \ldots, x_n$ every different sequence possible. Here I call the expression $P_{l,l,\ldots,l}$ a *product with* n {*interchangeable*} openings.[90]

Thus for example

$$P_{l,l,l}(xyz) = \frac{1}{6}(P_{x,y,z} + P_{x,z,y} + P_{y,x,z} + P_{y,z,x} + P_{z,x,y} + P_{z,y,x}).$$

REMARK. By the arithmetic mean of several magnitudes is meant here, as usual, the sum of the magnitudes divided by the number of them. It is self-evident that if such an open expression is to be bound to another expression by multiplication, that expression is then to be enclosed between parentheses.

It is yet to be noted that in the above the magnitudes entered into the openings have been set as magnitudes of first order. One easily sees that one could have extended this concept, and could in addition have adopted openings of higher order, that is openings in which magnitudes of higher order are entered. But one can almost always eliminate such cases in which the openings of higher order would appear. Thus in the most frequent case, when the principal domain is of nth order, and magnitudes of $(n-1)$th order are to be entered in the openings, one can represent these magnitudes as supplements of magnitudes of first order, and thus in the form $|a, \ldots$, and instead of openings of $(n-1)$th order write $|l$, whereby the opening is then reduced to first order, and is thus subject to the Definition in 353.

{Since further, according to the Definition, the factors $x_1, x_2, \ldots, x_n$ are inserted in the openings in all possible orderings, and the arithmetic mean of all the resulting expressions is then taken, it is of no importance what order one gives the filling magnitudes $x_1, x_2, \ldots, x_n$ standing outside the product $P_{l,l,\ldots,l}$ (cf. No. 362). In addition it is not necessary to distinguish the individual openings of the product $P_{l,l,\ldots,l}$ by different symbols $l_1, l_2, \ldots, l_n$, since the interchange of two opening symbols would not change the meaning of the open product; on these grounds the openings are called interchangeable in the Definition. Only later on (cf. No. 485) will open products also enter, in which each opening is assigned a definite filling magnitude, and whose openings are therefore not interchangeable.}

354. *One has*

$$P_{l,l,\ldots,l}(x_1x_2\cdots x_n) = \sum P_{x_r,x_s,\ldots} : (1\cdot 2\cdots n),$$

where the expression $P_{l,l,\ldots,l}$ is a product with n {interchangeable} openings, and x_r, $x_s, \ldots$ denote the same magnitudes as $x_1, x_2, \ldots, x_n$, but in an arbitrarily changed sequence, and where the sum relates to all the different sequences. In particular,

$$P_{l,l,\ldots,l}x^n = P_{x,x,\ldots}.$$

PROOF. According to 353, $P_{l,l,\ldots,l}(x_1x_2\cdots x_n)$ is the arithmetic mean of the expressions $P_{x_r,x_s,\ldots}$, which result from $P_{x_1,x_2,\ldots,x_n}$ by arbitrary orderings of the magnitudes $x_1, \ldots, x_n$, that is, it is equal to the sum of those expressions divided by the number of them, and thus by $1\cdot 2\cdots n$. In particular, if $x_1 = x_2 = \cdots = x_n = x$, then all those expressions are equal to $P_{x,x,\ldots}$, whence their arithmetic mean is equal to one of them, and thus the special formula is proven.

355. DEFINITION. If P_m is a product with m {interchangeable} openings, and $m > n$, then by $P_m(x_1x_2\cdots x_n)$ I mean the expression one obtains when one develops the expression $P_m(x_1x_2\cdots x_m)$ (according to 353), and then replaces each of the magnitudes $x_{n+1}, x_{n+2}, \ldots, x_n$ with an opening; for example,

$$P_{l,l,l}x = \frac{1}{3}(P_{x,l,l} + P_{l,x,l} + P_{l,l,x});$$

$$P_{l,l,l}(xy) = \frac{1}{6}(P_{x,y,l} + P_{y,x,l} + P_{x,l,y} + P_{y,l,x} + P_{l,x,y} + P_{l,y,x}).$$

356. *If $P_{l,l,\ldots,l}$ is a product with m {interchangeable} openings, and m is greater than n, then*

$$P_{l,l,\ldots,l}(x_1x_2\cdots x_n) = \frac{S}{m(m-1)\cdots(m-n+1)},$$

where S is the sum of all terms that result when one distributes $x_1, x_2, \ldots, x_n$ in all possible different ways among the m openings of $P_{l,l,\ldots,l}$; in particular,

$$P_{l,l,\ldots,l}x = \frac{1}{m}(P_{x,l,\ldots,l} + P_{l,x,\ldots,l} + \cdots).$$

PROOF. By $P_{l,l,\ldots,l}(x_1x_2\cdots x_n)$ is meant (according to 355) the expression one obtains when one replaces each of the magnitudes x_{n+1}, x_{n+2}, ..., x_n in $P_{l,l,\ldots,l}(x_1x_2\cdots x_m)$ by an opening. Now (according to 354)

$$P_{l,l,\ldots,l}(x_1x_2\cdots x_m) = \sum P_{x_r,x_s,\ldots} : (1\cdot 2\cdots m).$$

If in this one replaces x_{n+1}, x_{n+2}, ..., x_m with openings, then all those terms are equal that are distinguished only by the sequence of the magnitudes x_{n+1}, x_{n+2}, ..., x_m. One can therefore, instead of having these equal terms enter as often as their number amounts to, multiply one of them by that number, that is $1\cdot 2\cdots(m-n)$; consequently one obtains each of the different terms multiplied by $1\cdot 2\cdots(m-n) : (1\cdot 2\cdots m)$. But the sum of these different terms is S, whence

$$P_{l,l,\ldots,l}(x_1x_2\cdots x_n) = \frac{S\cdot 1\cdot 2\cdots(m-n)}{1\cdot 2\cdots m} = \frac{S}{m(m-1)\cdots(m-n+1)}.$$

Since the number of terms included in S is equal to $m(m-1)\cdots(m-n+1)$, the right-hand expression is at once the arithmetic mean of these terms.

357. DEFINITION. If A, B, ..., A', B', ... are products, each with n openings, and α, β, ..., α', β', ... are numbers, I set

$$\alpha\mathsf{A} + \beta\mathsf{B} + \cdots = \alpha'\mathsf{A}' + \beta'\mathsf{B}' + \cdots$$

if and only if, *for every series of n magnitudes of first order $x_1, x_2, \ldots, x_n$,*

$$\alpha\mathsf{A}x_1x_2\cdots x_n + \beta\mathsf{B}x_1x_2\cdots x_n + \cdots = \alpha'\mathsf{A}'x_1x_2\cdots x_n + \beta'\mathsf{B}'x_1x_2\cdots x_n + \cdots.$$

We call such a multiple sum of open products (with n openings) an *open expression* (with n openings).

358. *Every complete numerical function of nth degree of arbitrarily many (variable) numerical magnitudes can be represented in the form*

$$\mathsf{A}x^n,$$

where A is an expression with n {interchangeable} openings, and in fact

$$\sum \alpha_{\mathfrak{a},\mathfrak{b},\ldots}x_1^{\mathfrak{a}}x_2^{\mathfrak{b}}\cdots = \mathsf{A}x^n, \quad [\mathfrak{a}+\mathfrak{b}+\cdots \le n],$$

where

$$x = e_0 + x_1e_1 + x_2e_2 + \cdots$$

and

$$\mathsf{A} = \sum \alpha_{\mathfrak{a},\mathfrak{b},\ldots}[l|e_o]^{\mathfrak{r}}[l|e_1]^{\mathfrak{a}}[l|e_2]^{\mathfrak{b}}\cdots$$

$$[\mathfrak{r}+\mathfrak{a}+\mathfrak{b}+\cdots = n],$$

and where further e_0, e_1, e_2, ... *form an elementary normal system, and the sum relates to all possible values of* $\mathfrak{r}$, $\mathfrak{a}$, $\mathfrak{b}$, ... *that satisfy the condition in brackets without being negative.*

PROOF. By assumption $x = x_0e_0 + x_1e_1 + x_2e_2 + \cdots$, where $x_0 = 1$. Further, since $x_0 = 1$ and $\mathfrak{a} + \mathfrak{b} + \cdots \leq n$,

$$\sum \alpha_{\mathfrak{a},\mathfrak{b},\ldots} x_1^{\mathfrak{a}} x_2^{\mathfrak{b}} \cdots = \sum \alpha_{\mathfrak{a},\mathfrak{b},\ldots} x_0^{\mathfrak{r}} x_1^{\mathfrak{a}} x_2^{\mathfrak{b}} \cdots,$$

with the condition that $\mathfrak{r} + \mathfrak{a} + \mathfrak{b} + \cdots = n$. But the equation thus found is

$$= \sum \alpha_{\mathfrak{a},\mathfrak{b},\ldots} [x|e_0]^{\mathfrak{r}} [x|e_1]^{\mathfrak{a}} [x|e_2]^{\mathfrak{b}} \cdots$$
$$= \mathsf{A}x_n, \qquad [354]$$

where A represents the open expression given in the theorem.

359. *Every system of complete functions of* n*th degree of arbitrarily many variables can be represented in the form*

$$\mathsf{A}x^n,$$

where A *is a constant open expression with* n *{interchangeable} openings.*

PROOF. Follows from 358 by virtue of 352.

360. *Instead of multiplying an open expression {whose openings are interchangeable} by a series of factors (according to the Definition in 353), one can multiply it by the factors progressively, that is:*

$$\mathsf{A}(x_1x_2\cdots x_n) = \mathsf{A}x_1x_2\cdots x_n.$$

PROOF. 1. Let A consist of only *one* product with m {interchangeable} openings; then (according to 356)

$$(*) \qquad \mathsf{A}(x_1x_2\cdots x_n) = \frac{S}{m(m-1)\cdots(m-n+1)},$$

where S is the sum of all the terms that result when one distributes x_1, x_2, ... , x_n in the m openings in A in all possible different ways. Likewise let S_1 be the sum of all terms that result when one inserts x_1 into each individual opening in the product A one by one; further, let S_2 result from S_1 when, in each term of S_1, one inserts the magnitude x_2 into each of the $m-1$ openings one by one, and adds all the terms thus produced. Then S_2 is at the same time the sum of all terms that result when one introduces x_1, x_2 in two of the openings in A in all possible different ways. In a corresponding way S_3 may be derived from S_2, and so forth. Then (according to 356)

$$\mathsf{A}x_1 = \frac{S_1}{m}, \quad \mathsf{A}x_1x_2 = \frac{S_2}{m(m-1)}, \quad \ldots,$$

and finally

$$\mathsf{A}x_1x_2\cdots x_n = \frac{S}{m(m-1)\cdots(m-n+1)} = \mathsf{A}(x_1x_2\cdots x_n). \qquad \text{[according to (*)]}$$

2. Let A be an arbitrary open expression $= \sum P_\mathfrak{a}$, where every $P_\mathfrak{a}$ is a product with m {interchangeable} openings; then

$$\begin{aligned}\mathsf{A}(x_1x_2\cdots x_n) &= \sum P_\mathfrak{a}\cdot(x_1x_2\cdots x_n)\\ &= \sum P_\mathfrak{a}(x_1x_2\cdots x_n) && [357]\\ &= \sum P_\mathfrak{a}x_1x_2\cdots x_n && \text{[Part 1 of the Proof]}\\ &= \sum P_\mathfrak{a}\cdot x_1x_2\cdots x_n && [357]\\ &= \mathsf{A}x_1x_2\cdots x_n.\end{aligned}$$

361. *In the expression $\mathsf{A}x_1x_2\cdots x_n$, in which A is an arbitrary open expression with n or more {interchangeable} openings, one can, without changing its value, enclose an arbitrary group of the magnitudes $x_1, x_2, \ldots, x_n$ in parentheses.*

PROOF. From 360.

362. *The order of the factors to be entered in an open expression {with only interchangeable openings} is of no importance for the resultant, that is*

$$\mathsf{A}x_1x_2\cdots = \mathsf{A}x_rx_s\cdots,$$

where $x_1, x_2, \ldots$ and $x_r, x_s, \ldots$ contain the same factors but in a different order, and A denotes an open expression {with interchangeable openings}.

PROOF. Let $\mathsf{A} = \sum P_\mathfrak{a}$, where every $P_\mathfrak{a}$ is an open product; then

$$\begin{aligned}\mathsf{A}x_1x_2\cdots &= \mathsf{A}(x_1x_2\cdots) && [360]\\ &= \sum P_\mathfrak{a}(x_1x_2\cdots). && [357]\end{aligned}$$

But (according to 356) $P_\mathfrak{a}(x_1x_2\cdots)$ is the arithmetic mean of all the expressions that result when one introduces $x_1, x_2, \ldots$ into the openings of $P_\mathfrak{a}$ in all possible orderings, and thus it is not important in what order the magnitudes $x_1, x_2, \ldots$ appear in the expression $P_\mathfrak{a}(x_1x_2\cdots)$, that is $P_\mathfrak{a}(x_1x_2\cdots) = P_\mathfrak{a}(x_rx_s\cdots)$, if $x_1x_2\cdots$ and $x_rx_s\cdots$ include the same factors in different sequences. Consequently

$$\begin{aligned}\mathsf{A}x_1x_2\cdots &= \sum P_\mathfrak{a}(x_1x_2\cdots) = \sum P_\mathfrak{a}\cdot(x_rx_s\cdots) && [357]\\ &= \mathsf{A}(x_rx_s\cdots) = \mathsf{A}x_rx_s\cdots. && [360]\end{aligned}$$

363. *If any of the factors that are multiplied with an open expression is a sum, then one can replace the sum by the individual summands and add the resulting open expressions, that is*

$$\mathsf{A}p(x+y+\cdots)q = \mathsf{A}pxq + \mathsf{A}pyq + \cdots,$$

where p and q denote factors, and A an open expression {with interchangeable openings}.

PROOF. According to 362,

$$\mathsf{A}p(x+y+\cdots)q = \mathsf{A}pq(x+y+\cdots).$$

Here $\mathsf{A}pq$ is again an open expression, and therefore $\mathsf{A}pq(x+y+\cdots)$ has the form of a sum of products, each of which includes $(x+y+\cdots)$ as a factor, and thus has the form

$$P_{x+y+\cdots} + Q_{x+y+\cdots} + \cdots.$$

But (according to 39) this is equal to

$$\begin{aligned} P_x + P_y + \cdots + Q_x + Q_y + \cdots &= P_x + Q_x + \cdots + P_y + Q_y + \cdots \\ &= \mathsf{A}pqx + \mathsf{A}pqy + \cdots = \mathsf{A}pxq + \mathsf{A}pyq + \cdots. \end{aligned}$$

REMARK. Hereby the product of the factors that enter into an open expression {with interchangeable openings} is completely subject to the laws of algebraic multiplication, and thus the general problem is presented to us of considering those products of extensive magnitudes that are subject to the laws of algebraic multiplication, which will be the subject of the following section.

3. Algebraic multiplication

364. DEFINITION. By *algebraic multiplication* I mean that multiplication whose defining equations are

$$e_r e_s = e_s e_r \quad \text{and} \quad E(Fe_r) = EFe_r,$$

where e_r, e_s are original units and E, F are algebraic products of original units. I denote them as ordinary products in algebra without enclosing parentheses.

REMARK. The name and notation are based on the fact that for this multiplication, as will be shown below, all laws of the multiplication utilized in algebra hold, and no others. This multiplication was not treated in Part One because it did not yield any elementary magnitude, and because it is most closely connected with the theory of functions. It can be regarded as the characteristic multiplication of Part Two, while the combinatorial multiplication characteristic of Part One will from now on recede more and more into the background. {Only in expressions with *noninterchangeable openings* does combinatorial multiplication once again achieve greater significance. Cf. No. 504ff.}

365. *In an* ***algebraic*** *product of two elementary factors one can interchange the factors, that is*

$$ab = ba.$$

PROOF. Let $a = \sum \alpha_{\mathfrak{a}} e_{\mathfrak{a}}, b = \sum \beta_{\mathfrak{b}} e_{\mathfrak{b}}$; then

$$\begin{aligned} ab &= \sum \alpha_{\mathfrak{a}} e_{\mathfrak{a}} \cdot \sum \beta_{\mathfrak{b}} e_{\mathfrak{b}} = \sum \alpha_{\mathfrak{a}} \beta_{\mathfrak{b}} (e_{\mathfrak{a}} e_{\mathfrak{b}}) && [42] \\ &= \sum \beta_{\mathfrak{b}} \alpha_{\mathfrak{a}} (e_{\mathfrak{b}} e_{\mathfrak{a}}) && [364] \\ &= \sum \beta_{\mathfrak{b}} e_{\mathfrak{b}} \cdot \sum \alpha_{\mathfrak{a}} e_{\mathfrak{a}} && [42] \\ &= ba. \end{aligned}$$

366. *Instead of appending an elementary factor c to the second factor of an algebraic product, one can append it to the whole product, that is*

$$A(Bc) = ABc.$$

PROOF. Here A and B are algebraic products of magnitudes numerically derived from the units $e_1, e_2, \dots$, whence (according to 45) they are representable in the forms

$$A = \sum \alpha_{\mathfrak{a}} E_{\mathfrak{a}}, \quad B = \sum \beta_{\mathfrak{b}} F_{\mathfrak{b}},$$

where $E_{\mathfrak{a}}$ and $F_{\mathfrak{a}}$ are algebraic products of the units. Finally, let $c = \sum \gamma_{\mathfrak{c}} e_{\mathfrak{c}}$; then one has

$$\begin{aligned} A(Bc) &= \sum \alpha_{\mathfrak{a}} E_{\mathfrak{a}} \cdot \left(\sum \beta_{\mathfrak{b}} F_{\mathfrak{b}} \cdot \sum \gamma_{\mathfrak{c}} e_{\mathfrak{c}}\right) = \sum \alpha_{\mathfrak{a}} \beta_{\mathfrak{b}} \gamma_{\mathfrak{c}} E_{\mathfrak{a}} (F_{\mathfrak{b}} e_{\mathfrak{c}}) && [45] \\ &= \sum \alpha_{\mathfrak{a}} \beta_{\mathfrak{b}} \gamma_{\mathfrak{c}} E_{\mathfrak{a}} F_{\mathfrak{b}} e_{\mathfrak{c}} && [364] \\ &= \sum \alpha_{\mathfrak{a}} E_{\mathfrak{a}} \cdot \sum \beta_{\mathfrak{b}} F_{\mathfrak{b}} \cdot \sum \gamma_{\mathfrak{c}} e_{\mathfrak{c}} && [45] \\ &= ABc. \end{aligned}$$

REMARK. Thus algebraic multiplication is shown to be a *linear multiplication*, that is, one whose defining equations continue to hold when one replaces the units with arbitrary magnitudes numerically derivable from them.

367. *The order in which one {in an* ***algebraic*** *product} progressively multiplies by elementary factors is of no importance for the resultant, that is*

$$Abcd \cdots = Acbd \cdots = \cdots.$$

PROOF. One has

$$\begin{aligned} Abc &= A(bc) && [366] \\ &= A(cb) && [365] \\ &= Acb. && [366] \end{aligned}$$

Thus in a product without parentheses one can interchange two sequential {elementary} factors. By repetition of this process one can bring each elementary factor to any position in the product, and thus one can give the progressive elementary factors an arbitrary order.

368. *Instead of {****algebraically****} multiplying with several elementary factors, one can multiply by their product, that is*

$$Abc \cdots = A(bc \cdots).$$

PROOF. According to 366,

$$\begin{aligned} A(bc \cdots pq) &= A(bc \cdots p)q = A(bc \cdots)pq = \cdots \\ &= Abc \cdots pq. \end{aligned}$$

369. *In an* ***algebraic*** *product of two arbitrary factors one can interchange the factors, that is*

$$AB = BA.$$

PROOF. Let $A = a_1 \cdots a_m$, $B = b_1 \cdots b_n$; then

$$\begin{aligned} AB &= A(b_1 \cdots b_n) = Ab_1 \cdots b_n && [368] \\ &= b_1 \cdots b_n a_1 \cdots a_m && [367] \\ &= b_1 \cdots b_n (a_1 \cdots a_m). && [368] \end{aligned}$$

370. *In an* ***algebraic*** *product of three arbitrary progressive factors one can enclose the second and third factors in parentheses, that is*

$$ABC = A(BC).$$

PROOF. Let $B = b_1 \cdots b_m$, $C = c_1 \cdots c_n$; then (according to 368)

$$\begin{aligned} A(BC) &= A(B(c_1 \cdots c_n)) = A(Bc_1 \cdots c_n) \\ &= Ab_1 \cdots b_m c_1 \cdots c_n \\ &= A(b_1 \cdots b_m) c_1 \cdots c_n \\ &= A(b_1 \cdots b_m)(c_1 \cdots c_n) \\ &= ABC. \end{aligned}$$

371. *If, from the original units $e_1, \ldots, e_n$, one forms the* ***combinations with repetition*** *of mth class, and sets each of these combinations as an* ***algebraic*** *product of the elements it includes, then these products stand in no numerical relation to one another, and every algebraic product of m magnitudes that can be numerically derived from the units $e_1, \ldots, e_n$ can be numerically derived from those products.*

PROOF. According to the Definition in 364 the equations

$$e_r e_s = e_s e_r$$

and

$$E(Fe_r) = EFe_r$$

are to be the complete defining equations of algebraic multiplication, that is, no relation between the products is to obtain, as such, that can be derived from those defining equations. It must therefore be possible to make every equation existing between them identically equal to zero by the application of those equations. Since however those fundamental equations always include the same units on both sides, but only in a different sequence or grouping, no new units can be introduced into the product by the application of those equations. On the other hand one can cast all the terms of any equation that might obtain into a form in which they are well-ordered combinations (with repetitions permitted) of the units $e_1, \ldots, e_n$. Once this is done the equation must be identically zero, that is, all the coefficients {of those combinations} must be zero, which is to say that no numerical relation between them exists.

The second part of the theorem follows at once from 49.

REMARK. Consequently the combinations *with repetition* of the original units of any given (mth) class, the combinations being regarded as *algebraic* products, form a system of units of higher order, from which all algebraic products of m factors numerically derived from the original units can again be numerically derived. It can very easily be shown that this also holds when one replaces the n original units with n arbitrary magnitudes numerically derived from them, that stand in no numerical relation to one another. I will however leave the proof of this to the reader and proceed at once to the following theorem, indispensable for further developments.

372. *If an* ***algebraic*** *product is zero, then necessarily one of its factors must be zero, that is, if*

$$AB = 0, \qquad A \neq 0,$$

then necessarily

$$B = 0.$$

PROOF. In the most general case A and B are multiple sums of algebraic products whose factors are numerically derived from the original units a, b, c, $\ldots$. Then (according to 371) the combinations with repetition of a, b, c, $\ldots$ form the units from which A and B are numerically derivable, if each of those combinations is treated as an algebraic product of the units included in it. One supposes that the elements of each combination are alphabetically ordered, and the combinations themselves are then ordered by class so that each lower class precedes the higher, and further that within each class the combinations are lexicographically ordered. We will call this arrangement the *well-ordered arrangement of the combinations.*

Since by hypothesis A is different from zero, at least one among the coefficients by which A is numerically derived from those combinations must necessarily be different from zero. Of all the combinations that have a nonzero coefficient in the derivation expression for A, let E_1 be the one that comes first in the well-ordered arrangement; then A appears in the form

$$A = \alpha_1 E_1 + \alpha_2 E_2 + \cdots = \sum \alpha_{\mathfrak{a}} E_{\mathfrak{a}},$$

where α_1 is different from zero. Further, let F_1 , F_2, $\ldots$ be the combinations according to the series in the above arrangement {from which B is derived}, and

$$B = \sum \beta_{\mathfrak{b}} F_{\mathfrak{b}} = \beta_1 F_1 + \beta_2 F_2 + \cdots ;$$

then

$$AB = \sum \alpha_{\mathfrak{a}} \beta_{\mathfrak{b}} E_{\mathfrak{a}} F_{\mathfrak{b}}. \qquad [42]$$

Each product $E_{\mathfrak{a}} F_{\mathfrak{b}}$ again yields an algebraic product of the original units, and if we denote these products as G_1, G_2, $\ldots$, then AB is a multiple sum of these products of units; thus, since by hypothesis AB is zero, all the coefficients of this multiple sum must be zero, that is

$$\sum \alpha_{\mathfrak{a}} \beta_{\mathfrak{b}} [E_{\mathfrak{a}} F_{\mathfrak{b}} = G_m] = 0,$$

where the condition inside the brackets says that one has taken the sum of all the products $\alpha_{\mathfrak{a}} \beta_{\mathfrak{b}}$ whose associated products of units $E_{\mathfrak{a}} F_{\mathfrak{b}}$ have the constant value G_m.

Here we have only to consider those equations of this type that include the factor α_1 in any given term; we can write these equations in the form

$$0 = \alpha_1 \beta_1 + \sum \alpha_{\mathfrak{a}} \beta_{\mathfrak{b}} [E_{\mathfrak{a}} F_{\mathfrak{b}} = E_1 F_k],$$

where the condition is to be added that none of the products in the summation is identical to $\alpha_1 \beta_k$. I now show that one can express this condition as: $\mathfrak{b}$ must be less than k.

In fact it follows first that $\mathfrak{a}$ cannot be 1; for then the product of the two combinations E_1 and $F_\mathfrak{b}$ would have to produce the same combination as the product of the two combinations E_1 and F_k, that is, $F_\mathfrak{b}$ would have to be identical to F_k, whence $\mathfrak{b} = k$, whence $\alpha_\mathfrak{a}\beta_\mathfrak{b}$ would be identical to $\alpha_1\beta_k$, in conflict with the stated condition. Thus necessarily $\mathfrak{a} > 1$, that is, $E_\mathfrak{a}$ must come later than E_1 in the well-ordered arrangement of the combinations.

This is possible in two ways. *First*, if E_1 includes fewer elements than $E_\mathfrak{a}$; then, since $E_1 F_k$ is to yield the same combination as $E_\mathfrak{a} F_\mathfrak{b}$, F_k must include more elements than $F_\mathfrak{b}$, that is, F_k must come later in that well-ordered arrangement than $F_\mathfrak{b}$, which is to say, $\mathfrak{b} < k$. Or *second*, if E_1 and $E_\mathfrak{a}$ include the same number of elements, then $E_\mathfrak{a}$ must come later in the lexicographic arrangement than E_1.

But implicit in the lexicographic arrangement of combinations is that the first of the elements $a, b, \ldots$, which has unequal exponents in two combinations, has the *smaller* exponent in the combinations following later, so that, if this element has the exponent α in E_1, in $E_\mathfrak{a}$ the exponent γ, then necessarily $\gamma < \alpha$. Further, from the constraint equation $E_\mathfrak{a} F_\mathfrak{b} = E_1 F_k$, it follows that, if α, β, γ, δ are the exponents that any given element has in the combinations E_1, F_k, $E_\mathfrak{a}$, $F_\mathfrak{b}$, respectively, then necessarily $\alpha + \beta = \gamma + \delta$. From this it follows immediately that, if an element in E_1 has the same exponent as in $E_\mathfrak{a}$, it must also have the same exponent in F_k as in $F_\mathfrak{b}$, and therefore the first of the elements $a, b, \ldots$, which in $E_\mathfrak{a}$ has another exponent than it has in E_1, is also the first which in $F_\mathfrak{b}$ has another exponent than it has in F_k. In particular, if α, β, γ, δ are the exponents of this element in E_1, F_k, $E_\mathfrak{a}$, $F_\mathfrak{b}$, then we say that necessarily $\gamma < \alpha$; but from the equation $\alpha + \beta = \gamma + \delta$, that necessarily $\delta > \beta$, and therefore that according to the principle of lexicographic arrangement the combination $F_\mathfrak{b}$ must come before F_k, that is, necessarily $\mathfrak{b} < k$, since the combinations $F_1, F_2, \ldots$ are to be lexicographically ordered within each class.

We have therefore found that in both possible cases it is necessary that $\mathfrak{b} < k$. Thus we have the equation

$$(*) \qquad 0 = \alpha_1\beta_1 + \sum \alpha_\mathfrak{a}\beta_\mathfrak{b}[E_\mathfrak{a} F_\mathfrak{b} = E_1 F_k,\ \mathfrak{b} < k].$$

First let $k = 1$; then the sum $\sum \alpha_\mathfrak{a}\beta_\mathfrak{b}$ completely disappears, since its condition, $\mathfrak{b} < 1$, cannot be satisfied, whence one has $\alpha_1\beta_1 = 0$, and since by hypothesis $\alpha_1 \neq 0$, and α_1 and β_1 are numbers, it is necessary that

$$\beta_1 = 0.$$

If one sets $k = 2$, then since $\mathfrak{b} < 2$ and $\beta_1 = 0$, the sum $\sum \alpha_\mathfrak{a}\beta_\mathfrak{b}$ likewise drops out, and one obtains $\alpha_1\beta_2 = 0$, whence

$$\beta_2 = 0.$$

And on the same grounds it follows that $\beta_3 = 0$, and so forth. And therefore also B, which $= \beta_1 F_1 + \beta_2 F_2 + \cdots, = 0$.

373. *If two* ***algebraic*** *products, each of two factors, are equal, and have an equal nonzero factor, then their other factors must be equal, that is, if*

$$AB = CB,$$

and $B \neq 0$, then necessarily

$$A = C.$$

PROOF. From $AB = CB$ it follows that

$$0 = AB - CB = (A - C)B.$$

Thus, since $B \neq 0$, necessarily (according to 372) $A - C$ must be zero, that is $A = C$.

374. DEFINITION. If B is different from zero, then I mean by the *algebraic quotient* $A : B$ that expression which, when multiplied with B, gives A, and thus that satisfies the equation

$$(A : B)B = A.$$

375. *{If B is different from zero} one has*

$$AB : B = A.$$

PROOF. According to 374, if one replaces A by AB, then

$$(AB : B)B = AB.$$

Thus {since $B \neq 0$,} (according to 373)

$$AB : B = A.$$

376. *All the algebraic laws of multiplication and division hold for the* ***algebraic multiplication*** *and* ***division*** *of extensive magnitudes.*

PROOF. All these laws are based on the formulas

$$AB = BA,$$
$$ABC = A(BC),$$
$$(A : B)B = A,$$
$$AB : B = A.$$

and on the relation of all types of multiplication (according to 42, 45) to addition and subtraction.

REMARK. From the identity of the laws of calculation for the multiplication just considered with those of the ordinary multiplication of algebra the identity of notation and nomenclature is justified. Their only distinction lies in the conjoined magnitudes, which there are numbers, here arbitrary extensive magnitudes, for example points, lines, and so forth.

It would be perverse if one were to transfer this difference in the conjoined magnitudes to the notation or nomenclature, whereby the terminology would be uselessly increased and the connections would be obscured. In addition the properties of this product that are based on the special characteristics of extensive magnitudes are found again in the algebra, and indeed in the theory of complete functions. Thus for example the theorem that every complete function of a single variable of nth degree can be decomposed into n linear factors, applied here to displacements in the plane, yields the theorem:

Every sum of algebraic products, each of n displacements in a single plane, can be reduced to a single such product.

Naturally it is assumed here that the adoption of imaginary displacements is allowed.

4. Complete functions of first degree. Quotient

377. DEFINITION. If a_1, a_2, ... , a_n are magnitudes of first (or $(n-1)$th) order in a principal domain of nth order, and *stand in no numerical relation to one another*, I mean by the *fraction* (*quotient*)

$$Q = \frac{(b_1, b_2, \ldots, b_n)}{(a_1, a_2, \ldots, a_n)}$$

the expression which, multiplied by a_1, a_2, ... , a_n, yields the values b_1, b_2, ... , b_n respectively, so that

$$\frac{(b_1, b_2, \ldots, b_n)}{(a_1, a_2, \ldots, a_n)} a_r = b_r.$$

I call $(a_1, a_2, \ldots, a_n)$ the *denominators* of the fraction, $(b_1, b_2, \ldots, b_n)$ their corresponding *numerators*, and set two fractions, or two expressions numerically derived from fractions, equal to one another if and only if they yield equals when they are multiplied by every magnitude of first {or $(n-1)$th} order. If in addition the numerators are magnitudes of first (or $(n-1)$th) order, *and stand in no numerical relation to one another*, I call the fraction *invertible*, and in this case, if

$$Q = \frac{(b_1, b_2, \ldots, b_n)}{(a_1, a_2, \ldots, a_n)},$$

I denote the inverted fraction by $\frac{1}{Q}$, that is, I set

$$\frac{1}{Q} = \frac{(a_1, a_2, \ldots, a_n)}{(b_1, b_2, \ldots, b_n)}.$$

REMARK. The idea behind this definition is easy to see. Thus, in *algebra*, by the quotient $a : b$ is meant the expression that gives a when multiplied by b, and this suffices (if b is not zero) for the definition of the numerical quotient. For *extensive magnitudes* it does not {however} suffice to know what resultant multiplication by any given (nonzero) magnitude yields, since thereby multiplication with it is *only* defined for such magnitudes as are numerically derivable from that magnitude. One therefore perceives that in a domain of nth order the resultants of multiplication of a quotient by n magnitudes standing in no numerical relation must be specified for the quotient to be completely defined. The quotient, understood in this sense, is indispensable both for the differential and integral calculus of extensive magnitudes, as well as for the treatment of geometric relationships.

Yet I note that one could also have allowed as denominators arbitrary magnitudes of higher order; but one would then exchange the indubitable advantage of greater simplicity for the doubtful advantage of barren generality. On the same grounds I will always treat the numerators, where not otherwise expressly stated, as magnitudes of first {or $(n-1)$th} order.

378. *Two fractions or multiple sums of fractions that yield equals when multiplied by n magnitudes of first order standing in no numerical relation to one another are equal to one another only if n is the order of the principal domain.*

PROOF. Let Q and Q_1 be the two fractions or multiple sums of fractions which, when multiplied by n magnitudes $a_1, \ldots, a_n$ of first order standing in no numerical relation to one another, yield equals; thus let

$$(*) \qquad Q a_r = Q_1 a_r$$

for every value r between 1 and n. Then it is to be shown that $Q = Q_1$, that is (according to 377), Q and Q_1 multiplied by *any* magnitude x of first order yield equals.

Now (according to 24) each such magnitude in a principal domain of nth order can be numerically derived from n magnitudes standing in no numerical relation to one another, and thus from $a_1, \ldots, a_n$. Let

$$x = x_1 a_1 + \cdots + x_n a_n;$$

then

$$\begin{aligned} Qx = Q(x_1 a_1 + \cdots + x_n a_n) &= x_1 Q a_1 + \cdots + x_n Q a_n && [44] \\ &= x_1 Q_1 a_1 + \cdots + x_n Q_1 a_n && [(*)] \\ &= Q_1(x_1 a_1 + \cdots + x_n a_n) && [44] \\ &= Q_1 x. \end{aligned}$$

379. *One multiplies a fraction by a number when one multiplies each numerator by this number, and one adds fractions with the same denominators when one adds the corresponding numerators, whence the denominators remain unchanged in both cases, that is, combining the two cases,*

$$\beta \frac{(b_1, b_2, \ldots)}{(a_1, a_2, \ldots)} + \gamma \frac{(c_1, c_2, \ldots)}{(a_1, a_2, \ldots)} + \cdots = \frac{((\beta b_1 + \gamma c_1 + \cdots), (\beta b_2 + \gamma c_2 + \cdots), \ldots)}{(a_1, a_2, \ldots)}.$$

PROOF. It is to be shown that both sides of the above equation, multiplied by each of the magnitudes $a_1, \ldots, a_n$, yield the same resultant. Now (according to 44)

$$\begin{aligned} &\left(\beta \frac{(b_1, b_2, \ldots)}{(a_1, a_2, \ldots)} + \gamma \frac{(c_1, c_2, \ldots)}{(a_1, a_2, \ldots)} + \cdots \right) a_r \\ &\qquad = \beta \frac{(b_1, b_2, \ldots)}{(a_1, a_2, \ldots)} a_r + \gamma \frac{(c_1, c_2, \ldots)}{(a_1, a_2, \ldots)} a_r + \cdots \\ &\qquad = \beta b_r + \gamma c_r + \cdots . && [377] \end{aligned}$$

Further,

$$\frac{((\beta b_1 + \gamma c_1 + \cdots), (\beta b_2 + \gamma c_2 + \cdots), \ldots)}{(a_1, a_2, \ldots)} a_r = \beta b_r + \gamma c_r + \cdots . \qquad [377]$$

If for short we denote the left side of the equation to be proved by L, its right side by R, then, for every index r,

$$(*) \qquad L a_r = R a_r.$$

Consequently $L = R$ (according to 378).

380. *One can cast every fraction into the form in which its denominators are n arbitrary magnitudes of first {or $(n-1)$th} order standing in no numerical relation to one another (where n is the order of the principal domain), and in fact*

$$\frac{(b_1, b_2, \ldots)}{(a_1, a_2, \ldots)} = \frac{(\sum \alpha_{1,\mathfrak{a}} b_{\mathfrak{a}}, \sum \alpha_{2,\mathfrak{b}} b_{\mathfrak{b}}, \ldots)}{(\sum \alpha_{1,\mathfrak{a}} a_{\mathfrak{a}}, \sum \alpha_{2,\mathfrak{a}} a_{\mathfrak{a}}, \ldots)},$$

if

$$\sum \alpha_{1,\mathfrak{a}} a_{\mathfrak{a}}, \sum \alpha_{2,\mathfrak{a}} a_{\mathfrak{a}}, \ \ldots$$

are n magnitudes standing in no numerical relation to one another, and $\alpha_{r,s}$ *are numbers.*

PROOF. One has

$$\frac{(b_1, b_2, \dots)}{(a_1, a_2, \dots)} \sum \alpha_{r,\mathfrak{a}} a_{\mathfrak{a}} = \sum \alpha_{r,\mathfrak{a}} \frac{(b_1, b_2, \dots)}{(a_1, a_2, \dots)} a_{\mathfrak{a}} \quad [44]$$

$$= \sum \alpha_{r,\mathfrak{a}} b_{\mathfrak{a}}. \quad [377]$$

But also

$$\frac{(\sum \alpha_{1,\mathfrak{a}} b_{\mathfrak{a}}, \sum \alpha_{2,\mathfrak{a}} b_{\mathfrak{a}}, \dots)}{(\sum \alpha_{1,\mathfrak{a}} a_{\mathfrak{a}}, \sum \alpha_{2,\mathfrak{a}} a_{\mathfrak{a}}, \dots)} \sum \alpha_{r,\mathfrak{a}} a_{\mathfrak{a}} = \sum \alpha_{r,\mathfrak{a}} b_{\mathfrak{a}}. \quad [377]$$

Thus both expressions, multiplied by $\sum \alpha_{r,\mathfrak{a}} a_{\mathfrak{a}}$, yield the same resultant for every value r from 1 to n, and therefore, since by hypothesis $\sum \alpha_{1,\mathfrak{a}} a_{\mathfrak{a}}, \sum \alpha_{2,\mathfrak{a}} a_{\mathfrak{a}}, \dots$ stand in no numerical relation to one another, they are equal.

381. *If* $e_1, e_2, \dots, e_n$ *are the original units, and if* $E_{r,s}$ *denotes for short the fraction whose denominators are the original units, and of whose numerators the one that corresponds to the denominator* e_r *is equal to* e_s*, while all its other numerators are zero, that is, when*

$$(*) \qquad E_{r,s} e_r = e_s \ \text{ and } \ E_{r,s} e_t = 0 \ \ [t \neq r],$$

the n^2 *expressions that result from* $E_{r,s}$ *when one replaces r and s one by one by any of the numbers* $1, \dots, n$*, can be set as fractional units, that is, all fractions {whose numerators and denominators are magnitudes of first order of the principal domain} can be numerically derived from them, while they themselves stand in no numerical relation to one another, and in fact*

$$\frac{(\sum \alpha_{1,\mathfrak{b}} e_{\mathfrak{b}}, \sum \alpha_{2,\mathfrak{b}} e_{\mathfrak{b}}, \dots)}{(e_1, e_2, \dots)} = \sum \alpha_{\mathfrak{a},\mathfrak{b}} E_{\mathfrak{a},\mathfrak{b}}.$$

PROOF. 1. One has

$$\frac{(\sum \alpha_{1,\mathfrak{b}} e_{\mathfrak{b}}, \sum \alpha_{2,\mathfrak{b}} e_{\mathfrak{b}}, \dots)}{(e_1, e_2, \dots)} e_r = \sum \alpha_{r,\mathfrak{b}} e_{\mathfrak{b}}. \quad [377]$$

Further,

$$\sum \alpha_{\mathfrak{a},\mathfrak{b}} E_{\mathfrak{a},\mathfrak{b}} \cdot e_r = \sum \alpha_{\mathfrak{a},\mathfrak{b}} E_{\mathfrak{a},\mathfrak{b}} e_r \quad [44]$$

$$= \sum \alpha_{\mathfrak{a},\mathfrak{b}} E_{r,\mathfrak{b}} e_r = \sum \alpha_{r,\mathfrak{b}} e_{\mathfrak{b}}, \quad [*]$$

that is, since $E_{\mathfrak{a},\mathfrak{b}} e_r = 0$ for $\mathfrak{a} \neq r$, and $E_{r,\mathfrak{b}} e_r = e_{\mathfrak{b}}$. Thus the two expressions are (according to 378) equal, since they yield equals when multiplied by each of the magnitudes $e_1, \dots, e_n$.

2. Assuming, second, that there exists between the magnitudes $E_{r,s}$ an equation of the form

$$\sum \alpha_{\mathfrak{a},\mathfrak{b}} E_{\mathfrak{a},\mathfrak{b}} = 0,$$

one then has for each index r

$$0 = \sum \alpha_{\mathfrak{a},\mathfrak{b}} E_{\mathfrak{a},\mathfrak{b}} \cdot e_r = \sum \alpha_{\mathfrak{a},\mathfrak{b}} E_{\mathfrak{a},\mathfrak{b}} e_r. \quad [44]$$

Thus, since $E_{\mathfrak{a},\mathfrak{b}} e_r = 0$ if r is different from $\mathfrak{a}$,

$$0 = \sum \alpha_{r,\mathfrak{b}} E_{r,\mathfrak{b}} e_r = \sum \alpha_{r,\mathfrak{b}} e_{\mathfrak{b}}. \quad [*]$$

But if

$$0 = \sum \alpha_{r,\mathfrak{b}} e_{\mathfrak{b}} = \alpha_{r,1} e_1 r_1 + \alpha_{r,2} e_2 + \cdots ,$$

then, since e_1, e_2, ... stand in no numerical relation to one another, (according to 28) each of the coefficients of e_1, e_2, ... must be zero, that is

$$\alpha_{r,s} = 0$$

for every r and s, and therefore the equation posited is identically zero, that is (according to 2) the magnitudes $E_{r,s}$ stand in no numerical relation to one another.

382. *Every fraction can be represented as an open expression with one opening, and in fact, if the denominators $a_1, \ldots, a_n$ form an elementary normal system,*

$$\frac{(b_1, b_2, \ldots)}{(a_1, a_2, \ldots)} = [l|a_1]b_1 + [l|a_2]b_2 + \cdots .$$

PROOF. It is to be shown that the two expressions, multiplied by any magnitude x of first order, yield equals.

Now if $x = x_1 a_1 + x_2 a_2 + \cdots$,

$$([l|a_1]b_1 + [l|a_2]b_2 + \cdots)x = [x|a_1]b_1 + [x|a_2]b_2 + \cdots \qquad [\{357\}, 353]$$
$$= x_1 b_1 + x_2 b_2 + \cdots . \qquad [350]$$

But also

$$\frac{(b_1, b_2, \ldots)}{(a_1, a_2, \ldots)} x = \frac{(b_1, b_2, \ldots)}{(a_1, a_2, \ldots)} (x_1 a_1 + x_2 a_2 + \cdots)$$
$$= x_1 b_1 + x_2 b_2 + \cdots . \qquad [377]$$

Consequently the two sides of the equation to be proved yield equals when multiplied by any magnitude of first order, and so are themselves equal (according to 377).

REMARK. Thus the fraction is just a simpler form of the open expression with *one* opening, and thus any system of arbitrarily many *linear* numerical functions of arbitrarily many numerical magnitudes can be represented as the product of a quotient with an extensive variable.

383. DEFINITION. The relative product of the numerators of a fraction whose denominators form the system of original units I call the *power* of the fraction, and I denote the power of the fraction Q, if the number of its numerators is n, by $[Q^n]$, that is, if

$$Q = \frac{(a_1, a_2, \ldots, a_n)}{(e_1, e_2, \ldots, e_n)},$$

and e_1, e_2, ..., e_n form the system of original units, I set

$$[Q^n] = [a_1 a_2 \cdots a_n].$$

REMARK. If each numerator is the p-fold of the corresponding denominator, then by virtue of the Definition the fraction is equal to the number p; the relative product of the numerators is then

$$[pe_1 \cdot pe_2 \cdots pe_n] = p^n [e_1 e_2 \cdots e_n],$$

whence, since the relative product of the original units is one, it is $= p^n$, that is, in a principal domain of nth order the power of a number is equal to p^n. The deeper reason for the nomenclature and notation is implicit in a characteristic conjunction of the extensive fractions that corresponds perfectly to relative multiplication, and whose essence I will attempt to develop in a more intuitive way here. I base the concept of this conjunction on that of the open product.

Thus if A, B, ... , A_1, B_1, ... are fractions whose denominators may perhaps be the original units, and a, b, ... are arbitrary magnitudes of first order, l being an opening into which fractions of the said type (and thus magnitudes of zeroth order) are to be entered, then I set $[AB\cdots] = [A_1B_1\cdots]$ if and only if, with respect to an arbitrary series of magnitudes of first order a, b, ... , whose number is equal to the number of factors in that product,

$$[la \cdot lb \cdots]AB \cdots = [la \cdot lb \cdots]A_1B_1 \cdots .$$

I call the product $[AB\cdots]$ a relative product (on the principal domain) of the fractions A, B, ...

From this concept it follows immediately (according to 362) that *the order of the factors in this product is of no importance for its value*, a law that is in agreement with that expressed in 58, since the fractions of the said type are to be regarded as magnitudes of zeroth order.

Furthermore, it follows easily that if, in the product $[la \cdot lb \cdots]$, two of the factors (for example the first two) are equal, then always

$$[la \cdot lb \cdots]AB \cdots = 0.$$

For (according to 353) $[la \cdot la \cdot lc \cdots]ABC \cdots$ is equal to a fraction whose numerator is the sum of all the expressions that one obtains when one lets A, B, C, ... enter into the openings in all possible orderings, and whose denominator is the number of these expressions. Now it is shown that the terms of the numerator cancel pairwise. For if $PQR\cdots$ represents another ordering of the factors $ABC\cdots$, such that P is to enter into the first opening, Q into the second, and so forth, the resulting term is equal to $[Pa \cdot Qa \cdot Rc \cdots]$. If one interchanges P and Q, then from this ordering the term $[Qa \cdot Pa \cdot Rc \cdots]$ results; but the sum of these two terms is zero, since Pa and Qa are magnitudes of first order whose interchange (according to 55) produces the opposite value. Consequently the terms in the numerator cancel pairwise, that is the numerator is zero, whence the fraction is zero. The same holds when any two factors in the product $[la \cdot lb \cdots]$ are zero, so that in this case it is always true that $[la \cdot lb \cdots]AB \cdots = 0$.

But then it again follows immediately that if the series of magnitudes a, b, ... is linearly evolved, for example if b is evolved into $b' = b + \alpha a$, the product $[la \cdot lb \cdots]AB \cdots$ maintains the same value, and therefore (according to 76), *if a, b, ... are each so evolved that their product $[ab\cdots]$ remains constant, $[la \cdot lb \cdots]AB$ ... remains constant.* We can therefore regard this expression as a conjunction, and indeed as a multiplicative conjunction of $[ab\cdots]$ and $[AB\cdots]$, and thus write*

$$[la \cdot lb \cdots]AB \cdots = [AB\cdots][ab\cdots].$$

Here, since $[AB\cdots]$, conjoined with the product $[ab\cdots]$, again yields a sum of such products with the same number of factors, and thus a magnitude of the same order, it has precisely the meaning of an extensive fraction, but such that, if its number of factors is m, it only appears together with magnitudes of mth order.

Let E_1, E_2, ... be the units of mth order, and let

$$[AB\cdots]E_1 = A_1, \ [AB\cdots]E_2 = A_2, \ \ldots ;$$

then it is clear that $[AB\cdots]$ is equal to a fraction Q whose denominators are E_1, E_2, ... , and whose corresponding numerators are A_1, A_2, ... For when multiplied by the units

---

*{In No. 504 the notation

$$[AB\cdots ab\cdots]$$

will be introduced for the right side of the product.}

of mth order, and thus also by every magnitude derivable from them, that is by any magnitude of mth order, that product $[AB\cdots]$ gives the same resultant as does this fraction Q conjoined in the same way, that is, by virtue of the definitions of that product and this quotient $[AB\cdots] = Q$, that is, the relative product of m fractions whose denominators and numerators are of first order, gives a fraction whose denominators and numerators are of mth order.

From the reciprocity between magnitudes of first and $(n-1)$th order (in a principal domain of nth order) it also follows that the relative product of m fractions, whose denominators and numerators are of $(n-1)$th order, yields a fraction whose denominators and numerators are of $(n-m)$th order. This latter product would therefore be treated as a *regressive*, the former as a *progressive product of fractions.*

Here we stay with the former, that is the progressive product of fractions, in order to consider as well the progressive powers of fractions.

Since the product of equal factors permits just one ordering of these factors, it follows immediately from the concept that

$$[A^m][ab\cdots] = [Aa \cdot Ab\cdots],$$

that is, it is equal to the relative product of the numerators, and thus (according to 383) it is equal to the power of A. But, since (according to 94) the relative product of the n original units is equal to one, $[A^m]$ itself is equal to this power, wherein is implicit the complete basis for the above-chosen notation. {Cf. the further development of the product structure considered here in Nos. 504–510.}

384. *The power of a fraction is equal to the relative product of the numerators divided by that of the denominators, that is, if*

$$Q = \frac{(a_1, a_2, \ldots, a_n)}{(b_1, b_2, \ldots, b_n)},$$

then

$$[Q^n] = \frac{[a_1, a_2, \ldots, a_n]}{[b_1, b_2, \ldots, b_n]}.$$

PROOF. Let $Qe_r = c_r$ for every index r from 1 to n; then

$$Q = \frac{(c_1, c_2, \ldots)}{(e_1, e_2, \ldots)}$$

(according to 378). Further, let b_1, b_2, ... be expressed as multiple sums of the original units e_1, e_2, ... , and for every index r from 1 to n let $b_r = \sum \beta_{r,\mathfrak{a}} e_{\mathfrak{a}}$; then

$$a_r = Qb_r = \sum \beta_{r,\mathfrak{a}} Q e_{\mathfrak{a}} = \sum \beta_{r,\mathfrak{a}} c_{\mathfrak{a}}.$$

Consequently

$$\frac{[a_1, a_2, \ldots]}{[b_1, b_2, \ldots]} = \frac{([\sum \beta_{1,\mathfrak{a}} c_{\mathfrak{a}} \cdot \sum \beta_{2,\mathfrak{a}} c_{\mathfrak{a}} \cdots])}{([\sum \beta_{1,\mathfrak{a}} e_{\mathfrak{a}} \sum \beta_{2,\mathfrak{a}} e_{\mathfrak{a}}])}$$

$$= \frac{(\sum \beta_{1,\mathfrak{a}} \beta_{2,\mathfrak{b}} \cdots [c_{\mathfrak{a}} c_{\mathfrak{b}} \cdots])}{(\sum \beta_{1,\mathfrak{a}} \beta_{2,\mathfrak{b}} \cdots [e_{\mathfrak{a}} e_{\mathfrak{b}} \cdots])} \qquad [45]$$

$$= \frac{(\sum (-1)^r \beta_{1,\mathfrak{a}} \beta_{2,\mathfrak{b}} \cdots)[c_1 c_2 \cdots]}{(\sum (-1)^r \beta_{1,\mathfrak{a}} \beta_{2,\mathfrak{b}} \cdots)[e_1 e_2 \cdots]}, \qquad [57]$$

where $\mathfrak{a}$, $\mathfrak{b}$, ... represent all possible different orderings of the numbers 1, 2, ... , and r is the number of pairs of numbers that, in the two series of numbers

$$\begin{array}{ccc} \mathfrak{a}, & \mathfrak{b}, & \ldots, \\ 1, & 2, & \ldots, \end{array}$$

are oppositely ordered. If the two sums are canceled, then since $[e_1 e_2 \cdots] = 1$, one obtains

$$\frac{[a_1, a_2, \ldots]}{[b_1, b_2, \ldots]} = [c_1 c_2 \cdots] = [Q^n]. \qquad \text{[according to 383]}$$

385. *If Q and Q_1 are fractions with n denominators, and stand in the numerical relation*

$$Q = \alpha Q_1$$

to one another, then their powers stand in the numerical relation

$$[Q^n] = \alpha^n [Q_1^n].$$

PROOF. Let

$$Q_1 = \frac{(a_1, \ldots, a_n,)}{(a_1, \ldots, a_n,)};$$

then

$$Q = \frac{(\alpha a_1, \ldots, \alpha a_n)}{(e_1, \ldots, e_n)},$$

whence (according to 383)

$$\begin{aligned} [Q^n] &= [\alpha a_1 \cdot \alpha a_2 \cdots \alpha a_n] = \alpha^n [\alpha_1 a_2 \cdots a_n] && [46] \\ &= \alpha^n [Q_1^n]. && [383] \end{aligned}$$

386. *If there is a numerical relation between the numerators of a fraction, then the fraction can always be cast into a form in which one or more of its numerators is zero, and no numerical relation obtains between the rest of its numerators; and in fact, if $e_1, \ldots, e_n$ are the denominators, $a_1, \ldots, a_n$ the numerators of the fraction Q, and no numerical relation obtains between $a_1, \ldots, a_m$, but the remaining $n-m$ numerators are numerically derivable from them, so that*

$$(*) \qquad a_{m+r} = \alpha_{r,1} a_1 + \alpha_{r,2} a_2 + \cdots + \alpha_{r,m} a_m,$$

then

$$Q = \frac{(a_1, \ldots, a_m, 0, 0, \ldots, 0)}{(e_1, \ldots, e_m, c_{m+1}, c_{m+2}, \ldots, c_n)},$$

where

$$c_{m+r} = \alpha_{r,1} e_1 + \alpha_{r,2} e_2 + \cdots + \alpha_{r,m} e_m - e_{m+r},$$

that is,

$$= \frac{(e_1, \ldots, e_m)}{(a_1, \ldots, a_m)} a_{m+r} - e_{m+r}.$$

And all magnitudes numerically derivable from $c_{m+1}, \ldots, c_n$, but no others, give zero when multiplied with Q.

PROOF. {One has}

$$\begin{aligned} Qc_{m+r} &= \alpha_{r,1} Q e_1 + \cdots + \alpha_{r,m} Q e_m - Q e_{m+r} \\ &= \alpha_{r,1} a_1 + \cdots + \alpha_{r,m} a_m - a_{m+r}, \end{aligned}$$

since by hypothesis $a_1, \ldots, a_n$ are the numerators of the fraction Q belonging to the denominators $e_1, \ldots, e_n$, {that is}

$$= 0. \qquad [(*)]$$

Thus the numerators belonging to the denominators $c_{m+1}, \ldots, c_n$ are zero.

Next it is to be shown that there is no numerical relation between the n magnitudes $e_1, \ldots, e_m, c_{m+1}, \ldots, c_n$.

For short I set

$$\alpha_{r,1}e_1 + \cdots + \alpha_{r,m}e_m = q_r,$$

so that $c_{m+r} = q_r - e_{m+r}$, whence

$$[e_1e_2 \cdots e_m c_{m+1} \cdots c_n] = [e_1e_2 \cdots e_m(q_1 - e_{m+1}) \cdots (q_{n-m} - e_n)].$$

Since $c_1, q_2, \ldots$ are numerically derivable from $e_1, e_2, \ldots, e_m$, we can (according to 67) drop them and thus obtain the product

$$= \pm[e_1e_2 \cdots e_n],$$

whence it is different from zero; consequently $e_1, e_2, \ldots, e_m, c_{m+1}, \ldots, c_n$ stand in no numerical relation; thus (according to 378) Q can be represented as a fraction in the way given in the theorem.

Now therefore, since $c_{m+1}, \ldots, c_n$ multiplied by Q gives zero, it follows at once that this also holds for any multiple sum of these magnitudes. But conversely every magnitude p which, when multiplied by Q, gives zero, must be a multiple sum of $c_{m+1}, \ldots, c_n$. For, however p may be constituted, it must (according to 24) be derived from $e_1, e_2, \ldots, e_m, c_{m+1}, \ldots, c_n$, and thus it can be represented in the form

$$p = \alpha_1e_1 + \cdots + \alpha_me_m + q,$$

where q is a multiple sum of $c_{m+1}, \ldots, c_n$. Thus, should $Qp = 0$, then, since $Qe_1 = a_1, \ldots, Qc_{m+1} = 0, \ldots,$ one has

$$0 = Qp = \alpha_1e_1 + \cdots + \alpha_me_m,$$

whence (according to 28) $\alpha_1, \alpha_2, \ldots, \alpha_m = 0$, and therefore $p = q$, that is, it is a multiple sum of $c_{m+1}, \ldots, c_n$.

387. Definition. If a fraction Q multiplied by a nonzero magnitude of first order yields a multiple of this magnitude, say the ρ-fold of it, so that

$$Qx = \rho x,$$

I call the coefficient ρ (whether it is real or imaginary) a *principal number* of the fraction Q, and the domain to which all those magnitudes x belong that satisfy that equation the *principal domain* of that principal number.

388. Problem. *To find the principal numbers of a fraction and the principal domains belonging to them.*

Solution. Let ρ be a principal number of a fraction Q with n denominators $e_1, e_2, \ldots, e_n$, and let $x = \sum x_{\mathfrak{a}}a_{\mathfrak{a}}$ nonzero magnitude which, multiplied by Q, yields its ρ-fold; then one has

$$Qx = \rho x,$$

that is

$$0 = (\rho - Q)x.$$

If one replaces x by its value, and sets

$$(a) \qquad (\rho - Q)e_{\mathfrak{a}} = c_{\mathfrak{a}},$$

one obtains

$$(b) \qquad 0 = \sum x_{\mathfrak{a}} c_{\mathfrak{a}}.$$

Now since x is different from zero, at least one of the numbers $x_1, \ldots, x_n$ must also be different from zero, whence (according to 16) there must be a numerical relation between $c_1, \ldots, c_n$, and therefore

$$(c) \qquad 0 = [c_1 c_2 \cdots c_n],$$

that is (according to 384), the power of the fraction $\rho - Q$ must be zero.

Conversely, if this equation (c) is satisfied, then (according to 66) an equation of the form (b) also holds, whence there is a magnitude $x \neq 0$ that satisfies the equation $Qx = \rho x$, that is, ρ is then a principal number.

If in the last equation one replaces $c_1, c_2, \ldots$ by their values from (a), one obtains

$$0 = [(\rho e_1 - Qe_1)(\rho e_2 - Qe_2) \cdots (\rho e_n - Qe_n)],$$

or, when one removes the parentheses,

$$(d) \qquad \alpha_0 \rho^n - \alpha_1 \rho^{n-1} + \alpha_2 \rho^{n-2} - \cdots + (-1)^n \alpha_n = 0,$$

where α_r results from the product $[e_1 e_2 \cdots e_n]$ when one interchanges r of the magnitudes $e_1, e_2, \ldots, e_n$ in all possible different ways in the corresponding magnitudes $Qe_1, Qe_2, \ldots, Qe_n$, while in each case leaving the rest unchanged {and finally adding the products thus produced}. The n roots $\rho_1, \ldots, \rho_n$ of this equation (d) are therefore the principal numbers sought; their product is, according to *Newton's* theorem, equal to $\alpha_n : \alpha_0$, that is it is equal to the power of Q (according to 384).

The magnitudes x are then defined by equation (b). According to this equation $c_1, c_2, \ldots, c_n$ stand in a numerical relation to one another. Consequently (according to 17) an assembly of less than n magnitudes, say $c_1, c_2, \ldots, c_m$ can be separated from the magnitudes $c_1, \ldots, c_n$, which are subject to no numerical relation, and from which the remaining magnitudes $(c_{m+1}, \ldots, c_n)$ are numerically derivable. But then the fraction $\rho - Q$ can, with the numerators $c_1, \ldots, c_n$ belonging to the denominators $e_1, \ldots, e_n$ (according to (a)), (according to 386) be cast into the form in which $n - m$ of the numerators are zero. Let the denominators belonging to them be $a_{m+1}, \ldots, a_n$; then (according to 386) all magnitudes x derivable from $a_{m+1}, \ldots, a_n$, but no others, have the property that $(\rho - Q)x = 0$, that is, that $Qx = \rho x$, which is to say, the domain $a_{m+1}, \ldots, a_n$ is the principal domain belonging to the principal number ρ. Thus:

Every fraction Q with n denominators has n principal numbers, and in fact these are the roots ρ of the equivalent equations (c) or (d). The product of these n roots is equal to the power of Q, and one obtains the principal domain belonging to the principal number ρ when (according to 386) one represents $\rho - Q$ as a fraction, one or more of whose numerators are zero, while the remaining numerators stand in no numerical relation to one another; then the domain of that numerator of this fraction $\rho - Q$, whose corresponding numerator is zero, is the principal domain sought.

389. *If the n principal numbers $\rho_1, \dots, \rho_n$ of a fraction Q are* ***all different from one another****, then the n principal domains belonging to them are all of first order and stand in no numerical relation to one another.*

PROOF. According to 388, for each of the magnitudes $\rho_1, \dots, \rho_n$, for example ρ_r, it is possible to find a nonzero magnitude of first order which, multiplied with Q, yields its ρ_r-fold. Let $a_1, \dots, a_n$ be such magnitudes, so that $Qa_r = \rho_r a_r$.

Assuming now that $a_1, \dots, a_n$ stand in a numerical relation, it must (according to 17) be possible to separate from them m magnitudes, say $a_1, \dots, a_m$, that stand in no numerical relation to one another, and from which the rest, for example a_r, are numerically derivable. Let $a_r = \alpha_1 a_1 + \cdots + \alpha_m a_m$; then, since a_r is different from zero, at least one of the coefficients $\alpha_1, \dots, \alpha_m$ must also be different from zero. Let this for example be α_1. Now one has

$$Qa_r = Q \sum \alpha_{\mathfrak{a}} a_{\mathfrak{a}} = \sum \alpha_{\mathfrak{a}} Q a_{\mathfrak{a}} = \sum \alpha_{\mathfrak{a}} \rho_{\mathfrak{a}} a_{\mathfrak{a}},$$

since by hypothesis $Qa_r = \rho_r a_r$, whence

$$\begin{aligned} \alpha_1 \rho_1 a_1 + \cdots + \alpha_m \rho_m a_m &= Qa_r = \rho_r a_r \\ &= \rho_r \alpha_1 a_1 + \cdots + \rho_r \alpha_m a_m, \end{aligned}$$

and consequently (according to 29) the corresponding coefficients are equal, namely $\alpha_1 \rho_1 = \alpha_1 \rho_r$, that is, since $\alpha_1 \neq 0$, $\rho_r = \rho_1$, which is contrary to the assumption.

Thus $a_1, \dots, a_n$ can stand in no numerical relation to one another. In addition therefore none of the principal domains can be of higher than first order. For if for example the principal domain belonging to ρ_1 is of higher order, then this domain must (according to 26) have at least one domain of first order in common with the domain of $(n-1)$th order of the magnitudes $a_2, \dots, a_n$. Let c be a (nonzero) magnitude of this common domain; then c would be numerically derivable from $a_2, \dots, a_n$, and yet would still, since it lies in the domain belonging to ρ_1, be transformed into its ρ_1-fold, which was shown to be impossible.

390. PROBLEM. *To investigate the case of equal principal numbers.*

SOLUTION. If one retains the notation of the previous theorem, then (according to 388) one has

$$(a) \qquad [c_1 c_2 \cdots c_n] = 0,$$

where

$$(b) \qquad c_r = (\rho - Q) e_r.$$

Thus $c_1, c_2, \dots, c_n$ are the numerators belonging to the denominators $e_1, e_2, \dots, e_n$ of the fraction $\rho - Q$, and equation (a) says that the combinatorial product of the n numerators is zero, that is, that the power of $\rho - Q$ is zero.

According to the above, equations (a) and (b) maintain their meaning if one replaces the denominators $e_1, \dots, e_n$ with any n magnitudes standing in no numerical relation to one another. The n values of ρ that satisfy equation (a) are (according to 388) the n principal numbers of Q. Now if several, say $\mathfrak{a}$, of these are equal to α, then this means that equation (a) presents, in all, $\mathfrak{a}$ values for ρ that are equal to α. But if a value for ρ is equal to α, then (according to 388) a nonzero magnitude of first order a_1 can be found which, multiplied by Q, yields its α-fold. Let this magnitude be $a_1 - x_1 e_1 + \cdots + x_n e_n$; then at least *one* of the coefficients $x_1, \dots, x_n$ must be different from zero, since otherwise, contrary to assumption,

a_1 itself would be zero. Let x_1 be different from zero; then (according to 19) a_1, $e_2, \ldots, e_n$ stand in no numerical relation to one another, and thus according to the above they can replace $e_1, e_2, \ldots, e_n$ in equations (a) and (b). Then the new numerator, belonging to a_1, of the fraction $\rho - Q$ obtains the value

$$c_1' = (\rho - Q)a_1 = \rho a_1 - Qa_1 = \rho a_1 - \alpha a_1 = (\rho - \alpha)a_1, \qquad \text{[by assumption]}$$

and equation (a) is replaced by the equation

$$(\rho - \alpha)[a_1 c_2 \cdots c_n] = 0.$$

We will assume that, if equation (a) has in all $\mathfrak{a}$ roots $\rho = \alpha$, one has little by little transformed equation (a) into the forms

$$(\rho - \alpha)^2 [a_1 a_2 c_3 \cdots c_n] = 0$$

and so forth, and finally into the form

$$(c) \qquad (\rho - \alpha)^r [a_1 a_2 \cdots a_r c_{r+1} \cdots c_n] = 0,$$

where $r < \mathfrak{a}$, and in fact so that

$$Qa_1 = \alpha a_1, \ [a_1 \cdot Qa_2] = \alpha[a_1 a_2], \ \ldots,$$

and finally

$$(d) \qquad [a_1 a_2 \cdots a_{r-1} Q a_r] = \alpha [a_1 a_2 \cdots a_r],$$

and the n magnitudes $a_1, \ldots, a_r, e_{r+1}, \ldots, e_n$ stand in no numerical relation to one another. It can then be shown that one can continue to apply these transformations until at last $r = \mathfrak{a}$.

In fact, so long as r is smaller than $\mathfrak{a}$, that is that there are still more than r roots $\rho = \alpha$ that satisfy equation (c), then, as is well known from the theory of equations, if one divides that equation by $(\rho - \alpha)^r$ the quotient still has a root $\rho = \alpha$, that is, it must still be that

$$(e) \qquad [a_1 \cdots a_r c_{r+1} \cdots c_n] = 0$$

for $\rho = \alpha$. Let $d_{r+1}, \ldots, d_n$ be the values into which $c_{r+1}, \ldots, c_n$ are transformed when one replaces ρ by α; then one obtains

$$(f) \qquad [a_1 \cdots a_r d_{r+1} \cdots d_n] = 0,$$

where

$$(g) \qquad d_{r+1} = (\alpha - Q)e_{r+1}, \ldots, d_n = (\alpha - Q)e_n.$$

Here equation (f) says that between the magnitudes $a_1, \ldots, a_r, d_{r+1}, \ldots, d_n$ there is a numerical relation (according to 66). Let

$$(h) \qquad \alpha_1 a_1 + \cdots + \alpha_r a_r + \alpha_{r+1} d_{r+1} + \cdots + \alpha_n d_n = 0$$

be the expression for this numerical relation; then one of the coefficients $\alpha_{r+1}, \ldots$, α_n must be different from zero, for, were they all zero, there would be a numerical relation between $a_1, \ldots, a_r$, contrary to hypothesis. Let α_{r+1}, say, be different from zero, and let

$$\alpha_{r+1} e_{r+1} + \cdots + \alpha_n e_n = a_{r+1};$$

then (according to 19) $a_1, \ldots, a_{r+1}, e_2, \ldots, e_n$ stand in no numerical relation to one another, and thus they can replace $e_1, \ldots, e_n$ in equations (a) and (b), or $a_1, \ldots, a_r, e_{r+1}, \ldots, e_n$ in equation (c), if only one simultaneously replaces the magnitudes $c_1, \ldots, c_{r+1}$ by the numerators $c_1', \ldots, c_{r+1}'$ corresponding to the denominators $a_1, \ldots, a_{r+1}$ in the fraction $\rho - Q$. Hereby equation (c) goes over into

$$(c') \qquad (\rho - \alpha)[a_1 a_2 \cdots a_r c'_{r+1} c_{r+2} \cdots c_n] = 0,$$

where

$$c_{r+1} = (\rho - Q) a_{r+1}.$$

Furthermore, then,

$$\begin{aligned}(\alpha - Q)a_{r+1} &= \alpha_{r+1}(\alpha - Q)e_{r+1} + \cdots + \alpha_n(\alpha - Q)e_n \\ &= \alpha_{r+1}d_{r+1} + \cdots + \alpha_n d_n \qquad &&\text{[according to } (g)] \\ &= -\alpha_1 a_1 - \cdots - \alpha_r a_r. &&\text{[according to } (h)]\end{aligned}$$

Thus

$$Q a_{r+1} = \alpha_1 a_1 + \cdots + \alpha_r a_r + \alpha a_{r+1}.$$

Consequently

$$\begin{aligned}(i) \qquad [a_1 \cdots a_r \cdot Q a_{r+1}] &= [a_1 \cdots a_r(\alpha_1 a_1 + \cdots + \alpha_r a_r + \alpha a_{r+1})] \\ &= \alpha[a_1 \cdots a_r a_{r+1}].\end{aligned}$$ [67]

Now, as shown above,

$$c'_{r+1} = (\rho - Q)a_{r+1} = \rho a_{r+1} - Q a_{r+1},$$

whence

$$\begin{aligned}[a_1 \cdots a_r c'_{r+1}] &= \rho[a_1 \cdots a_r a_{r+1}] - [a_1 \cdots a_r Q a_{r+1}] \\ &= \rho[a_1 \cdots a_{r+1}] - \alpha[a_1 \cdots a_{r+1}] \qquad &&\text{[according to } (i)] \\ &= (\rho - \alpha)[a_1 \cdots a_{r+1}].\end{aligned}$$

If one inserts this into equation (c'), one obtains

$$(k) \qquad (\rho - \alpha)^{r+1}[a_1 a_2 \cdots a_{r+1} c_{r+2} \cdots c_n] = 0,$$

that is, equations (c) and (d) continue to hold, even if, so long as $r < \mathfrak{a}$, one replaces r by $r+1$ in them, whence they continue to hold if $r = \mathfrak{a}$, that is:

If among the principal numbers of Q several, say $\mathfrak{a}$, are equal to one another and in fact $= \alpha$, then $\mathfrak{a}$ magnitudes of first order $a_1, \ldots, a_{\mathfrak{a}}$ can be found such that these stand in no numerical relation to $n - \mathfrak{a}$ of the magnitudes $e_1, \ldots, e_n$, say with $e_{\mathfrak{a}+1}, \ldots, e_n$, and if

$$(*) \qquad \begin{cases} Q a_1 = \alpha a_1, \\ [a_1 \cdot Q a_2] = \alpha[e_1 e_2], \\ \dots\dots\dots\dots\dots\dots\dots\dots \\ [a_1 a_2 \cdots a_{a-1} \cdot Q a_a] = \alpha[a_1 a_2 \cdots a_a], \end{cases}$$

then the equation for the principal numbers ρ is the following:

$$(**) \qquad (\rho - \alpha)^{\mathfrak{a}}[a_1 a_2 \cdots a_{\mathfrak{a}} c_{\mathfrak{a}+1} \cdots c_n] = 0,$$

where

$$(***) \qquad c_r = (\rho - Q)e_r$$

for each index r from $\mathfrak{a} + 1$ to n.

It is thus important to find the numerators of the fraction Q belonging to the denominators $a_1, \ldots, a_n$.

First, according to the above the numerator belonging to the denominator a_1 is equal to αa_1. Let the numerator belonging to the denominator a_r be x, that is let $Qa_r = x$; then (from $(*)$) one has

$$[a_1 a_2 \cdots a_{r-1} x] = \alpha [a_1 a_2 \cdots a_r],$$

that is,

$$[a_1 a_2 \cdots a_{r-1}(x - \alpha a_r)] = 0,$$

whence (according to 66) between the magnitudes $a_1, a_2, \ldots, a_{r-1}, x - \alpha a_r$ there is a numerical relation, that is

$$x - \alpha a_r = a_r',$$

where a_r' is any magnitude numerically derivable from $a_1, a_2, \ldots, a_{r-1}$, and one obtains $x = \alpha a_r + a_r'$, that is

$$(l) \qquad Qa_r = \alpha a_r + a_r',$$

where a_r' is a multiple sum of $a_1, a_2, \ldots, a_{r-1}$.

Now let the product of Q with any magnitude p numerically derivable from $a_1, \ldots, a_{\mathfrak{a}}$ be analyzed, and indeed let

$$p = \alpha_1 a_1 + \cdots + \alpha_r a_r,$$

where $r \leq \mathfrak{a}$ and $\alpha_r \neq 0$; then one has

$$\begin{aligned} Qp &= Q(\alpha_1 a_1 + \cdots + \alpha_r a_r) \\ &= \alpha_1 Q a_1 + \cdots + \alpha_r Q a_r \\ &= \alpha_1 \alpha a_1 + \cdots + \alpha_r \alpha a_r + p', \qquad \text{[according to } (l)] \end{aligned}$$

where p' represents a multiple sum of $a_1, a_2, \ldots, a_{r-1}$,

$$= \alpha p + p',$$

that is:

The magnitudes $a_1, a_2, \ldots, a_{\mathfrak{a}}$ specified by the above equation $(*)$ have the property that every multiple sum of them is subject to the equation

$$(****) \qquad Qp = \alpha p + p',$$

where, if p is derivable from the first r of those magnitudes, p' is derivable from the first $(r-1)$ of them.

It is immediately clear that if several of the magnitudes $a_2', a_3', \ldots, a_{\mathfrak{a}}'$ of equation (l) are zero, say the magnitudes $a_2', \ldots, a_r'$, then every multiple sum of $a_1, \ldots, a_r$ is transformed by multiplication with Q into its α-fold, and thus the principal domain of Q belonging to α (see 387 {and 388}) is of rth order.

If among the principal numbers of Q not only $\mathfrak{a}$ of them are found that equal α, but also $\mathfrak{b}$ that equal β, $\mathfrak{c}$ that equal γ, $\ldots$, where $\alpha, \beta, \gamma, \ldots$ are all different from one another, then, according to the above, $\mathfrak{b}$ magnitudes $b_1, \ldots, b_{\mathfrak{b}}$ standing in no numerical relation to one another can be specified such that every multiple sum q of $b_1, \ldots, b_{\mathfrak{b}}$ satisfy the equation $Qq = \beta q + q'$, where, if q is derivable from the first m of those magnitudes, q' is derivable from the first $(m-1)$, and likewise $\mathfrak{c}$ magnitudes $c_1, \ldots, c_{\mathfrak{c}}$ standing in no numerical relation to one another can be specified such that every multiple sum r of $c_1, \ldots, c_{\mathfrak{c}}$ satisfy the equation $Qr = \gamma r + r'$, where, if r is derivable from the first m of the magnitudes $c_1, \ldots, c_{\mathfrak{c}}$, r' is derivable from the first $m-1$, and so forth.

It can be shown that the magnitudes $a_1, \dots, a_{\mathfrak{a}}, b_1, \dots, b_{\mathfrak{b}}, c_1, \dots, c_{\mathfrak{c}}, \dots$ stand in no numerical relation to one another, and thus can be set as denominators of the fraction Q.

Thus if for example we assume that there is not yet any numerical relation between the magnitudes $a_1, \dots, a_{\mathfrak{a}}, b_1, \dots, b_{\mathfrak{b}}, c_1, \dots, c_{m-1}$, but that between these magnitudes and the magnitude c_m a numerical relation appears, then this will have the form

$$(m) \qquad p + q + r = 0,$$

where p is a multiple sum of $a_1, \dots, a_{\mathfrak{a}}$, q a multiple sum of $b_1, \dots, b_{\mathfrak{b}}$, r a multiple sum of $c_1, \dots, c_m$, whence also

$$0 = Q(p + q + r).$$

But, as shown above, this is

$$= \alpha p + p' + \beta q + q' + \gamma r + r',$$

or, when we replace r by its value $= -p - q$ from equation (m),

$$0 = (\alpha - \gamma)p + (\beta - \gamma)q + p' + q' + r'.$$

Since according to the above r' is numerically derivable from $c_1, \dots, c_{m-1}$, all the magnitudes appearing in this last equation are numerically derivable from the magnitudes $a_1, \dots, a_{\mathfrak{a}}, b_1, \dots, b_{\mathfrak{b}}, c_1, \dots, c_{m-1}$, which by assumption stand in no numerical relation to one another. The right side of the last equation can therefore be represented as a multiple sum of these latter magnitudes, and since the left side is zero, all the coefficients of this multiple sum must (according to 28) be zero.

Now if p were different from zero, say $= x_1 a_1 + \dots + x_s a_s$, where $x_s \neq 0$, then according to the above p' would be numerically derivable from $a_1, \dots, a_{s-1}$, whence, since a_s is also not included in q, q', r', it could only appear once in that multiple sum set to zero, namely bound with the coefficients $(\alpha - \gamma)x_s$; this must therefore be zero, which is impossible, since x_s is not equal to zero by assumption, and α is not equal to γ. Consequently the assumption that p is different from zero is impossible, that is p is equal to zero. On the same grounds $q = 0$. But then it follows from equation (m) that $r = 0$. But now since the magnitudes $a_1, \dots, a_{\mathfrak{a}}$ stand in no numerical relation to one another, it follows from $p = 0$ that all the coefficients of the expression by which p is derived from $a_1, \dots, a_{\mathfrak{a}}$ are zero, and the same follows for q and r. Thus equation (m) includes no expression for a numerical relation at all; therefore no such expression between the magnitudes $a_1, \dots, a_{\mathfrak{a}}, b_1, \dots, b_{\mathfrak{b}}, c_1, \dots, c_{\mathfrak{c}}, \dots$ exists, as was to be shown. Thus

If the equation

$$[(\rho e_1 - Qe_1)(\rho e_2 - Qe_2) \cdots (\rho e_n - Qe_n)] = 0,$$

which is of nth degree with respect to ρ, has $\mathfrak{a}$ roots $= \alpha$, $\mathfrak{b}$ roots $= \beta, \dots$, where $\alpha, \beta, \dots$ are different from one another, one can specify n magnitudes $a_1, \dots, a_{\mathfrak{a}}, b_1, \dots, b_{\mathfrak{b}}, \dots$ standing in no numerical relation to one another such that, if p is a multiple sum of the first m of the magnitudes $a_1, \dots, a_{\mathfrak{a}}$, or of the magnitudes $b_1, \dots, b_{\mathfrak{b}}$, or of the magnitudes of any following group, then Qp in the first case $= \alpha p + p'$, in the second $= \beta p + p', \dots$, where p' is numerically derivable from the first $m - 1$ of the same group.

REMARK 1. If among the roots ρ a pair or several pairs of imaginary roots occur, this changes nothing in the results obtained, since the reasoning holds just as well for imaginary as for real roots. There is in addition not the least difficulty in converting the definitions flowing from imaginary pairs of roots into real form, which I will therefore leave to the reader.

REMARK 2. The laws developed above are important for the theory of geometric relationships. Thus, geometrically interpreted, every quotient, if it includes no more than four denominators, represents a specific collinear relationship such that every point system multiplied by the quotient yields a point system collinear with the first one, and conversely every point system collinearly related to an original point system can be derived from it by multiplication by a quotient.

The quotient provides the advantage over every other analytic expression of that relationship that the essential characteristics of the relationship are symbolically represented by it in the simplest way. If, for example, the four principal numbers of a quotient with four denominators are all real and different from one another, then every two collinearly related point systems in space that can be represented by that quotient present four noncollinear points that coincide with those corresponding to them, and aside from these there is no fifth point that coincides with the point corresponding to it in the other system. Likewise the foregoing theorems include the special relations of the collinear relationship for the case where several of the principal numbers of the quotients to which they belong are equal.

The special geometric relationships subordinate to collineation result from special assumptions. Thus for example affinity, from the assumption that to the infinitely distant points of each system there also correspond infinitely distant points of the other. Further the special type of affinity[91] in which corresponding volumes have equal and similarly signed spatial content, by the assumption that in addition {simple points are again assigned to simple points, and that} the product of the principal numbers, that is the power of the quotient, is equal to one. Congruence, from the {further} assumption that the corresponding displacements are to be equally long (for which it is necessary that two of the principal numbers of the quotient be $= 1$, and the other two either $= 1$, or $= -1$, or $= \cos\alpha \pm i \sin\alpha$). Congruence is transformed into symmetry if the product of the principal numbers is $= -1$ instead of $+1$. Finally, similarity results from affinity through the assumption that the corresponding displacements stand in the same numerical ratio.

In the following we treat yet another special form of the quotient, which stands in the closest relation to the relationship of reciprocity.[92]

391. *If a fraction Q has the property that, for any nonzero magnitudes of first order a and b,*

$$(a) \qquad [Qa|b] = [Qb|a] \quad \textit{and} \quad [Qa|a] \neq 0,$$

then it is always possible to find n magnitudes of first order $c_1, \ldots, c_n$ standing in no numerical relation to one another, such that

$$(b) \qquad [Qc_r|c_s] = 0$$

whenever r is different from s. Further, the n principal numbers of the fraction Q are then all real, and as many of them are positive as there are positive values among the products

$$(c) \qquad [Qc_1|c_1], \ldots, [Qc_n|c_n].$$

Finally it is always possible to find n magnitudes $e_1, \ldots, e_n$ normal to one another such that each of them multiplied by Q yields a multiple sum of it, thus

$$(d) \qquad Qe_r = \rho_r e_r,$$

where

(e) $$[e_r|e_s] = 0$$

for every r different from s.

PROOF. First, I show that n magnitudes $c_1, \ldots, c_n$ of the type sought can be found. It suffices to show that they satisfy equation (b) for the case that $r < s$, for since by assumption (a) $[Qc_r|c_s] = [Qc_s|c_r]$, it follows that the condition continues to hold when conversely the first index is greater than the second.

For short I set $Qc_r = k_r$, and show next that one can find n nonzero magnitudes of first order $c_1, \ldots, c_n$ that satisfy the equations $[k_r|c_s] = 0$ for every $r < s$.

According to 189 we can also write this equation as $[c_s|k_r] = 0$. Next we choose for c_1 an arbitrary (nonzero) magnitude (of first order). Then c_2 must satisfy the equation $[c_2|k_1] = 0$, that is, c_2 must be normal to k_1 (according to 152), or in other words c_2 must belong to the domain $|k_1$, which is of $(n-1)$th order. Otherwise c_2 is arbitrary. Further, c_3 must satisfy the equations $[c_3|k_1] = [c_3|k_2] = 0$, that is, c_3 must belong to the domains $|k_1$ and $|k_2$, and thus to their common domain, this being (according to 26) at least of $(n-2)$th order, within which c_3 is arbitrary. On the same grounds c_4 must satisfy the equations $[c_4|k_1] = [c_4|k_2] = [c_4|k_3] = 0$, and thus belongs to that domain common to the three domains of $(n-1)$th order $|k_1$, $|k_2$, $|k_3$; this domain is at least of $(n-3)$th order, within which c_4 is arbitrary, and thus one continues. Finally c_n must satisfy the equations

$$[c_n|k_1] = [c_n|k_2] = \cdots = [c_n|k_{n-1}] = 0,$$

that is, c_n must belong to the $(n-1)$ domains of $(n-1)$th order $|k_1, |k_2, \ldots, |k_{n-1}$, which have at least a domain of first order in common, within which c_n may be chosen arbitrarily (but different from zero).

Thus we now have n nonzero different magnitudes satisfying the equations $[c_s|k_r] = 0$, that is, satisfying $0 = [k_r|c_s] = [Qc_r|c_s]$, at first for every $r < s$, and thus also according to equation (a) for *every* s different from r. It is still to be shown that $c_1, \ldots, c_n$ stand in no numerical relation to one another.

Assuming there exists between them the numerical relation

$$x_1c_1 + \cdots + x_nc_n = 0,$$

where at least one of the coefficients, for example x_r, is nonzero, one has

$$0 = [Qc_r|(x_1c_1 + \cdots + x_nc_n)] = x_r[Qc_r|c_r],$$

since all the other products are zero, whence, since $x_r \neq 0$, one has $[Qc_r|c_r] = 0$, which contradicts the assumption (in (a)) and thus it is impossible that $c_1, \ldots, c_n$ stand in a numerical relation to one another.

Now let $[Qc_r|c_r] = \alpha_r$, and set $a_r = c_r : \sqrt{\alpha_r}$. Then the magnitudes $a_1, \ldots, a_n$ form an assembly subject to the conditions

(f) $$[Qa_r|a_s] = 0$$

if $r \neq s$, and

(g) $$[Qa_r|a_r] = 1,$$

and in fact a_r is real if $[Qc_r|c_r]$ is positive; on the other hand a_r is a simple imaginary, that is, is representable as the product of a real magnitude and the square root of -1, if $[Qc_r|c_r]$ is negative. Therefore among the magnitudes $a_1, \ldots, a_n$ as many are real as there are positive values among the products $[Qc_1|c_1], \ldots, [Qc_n|c_n]$.

I will call each such assembly that satisfies equations (f) and (g), and in which each of the magnitudes $a_1, \dots, a_n$ is real or simple imaginary, a *conjugate assembly* for short.

{It will now be shown that such an assembly continues to be such an assembly under a circular evolution of the magnitudes appearing in it, and indeed such that the number of reals among the n magnitudes of the one assembly is the same as that in the other. Here I will mean by the circular evolution of two magnitudes a_1 and a_2, when both are real or both simple imaginaries, their transformation into two other magnitudes b_1 and b_2 for which

$$(h) \qquad b_1 = xa_1 + ya_2, \qquad b_2 = xa_2 - ya_1,$$

while x and y are both real, and the sum of their squares is one, whence $x^2 + y^2 = 1$. On the other hand if one of the two magnitudes a_1 and a_2 is real, the other imaginary, then

$$b_1 = xa_1 + yia_2, \quad b_2 = xa_2 - yia_1,$$

where $i = \sqrt{-1}$, x and y are both real, and $x^2 - y^2$, that is $x^2 + (yi)^2 = 1$, or in other words, equations (h) represent every circular evolution of a_1 and a_2 if x is always real, and y is only imaginary, and indeed simple imaginary, if one of the magnitudes a_1 and a_2 is real and the other is simple imaginary.}[93]

It is clear at once that b_1 and b_2 are both real, or both simple imaginary, or one real and the other simple imaginary, according as this was the case for a_1 and a_2, and thus that the number of real magnitudes of the assembly remains the same under the circular evolutions {of the type described}. Further, by virtue of equations (h), for every $r > 2$,

$$[Qb_1|a_r] = x[Qa_1|a_r] + y[Qa_2|a_r] = 0,$$

since $[Qa_1|a_r]$ and $[Qa_2|a_r]$ are zero (according to (f)); on the same grounds, then,

$$[Qb_2|a_2] = 0.$$

Further,

$$[Qb_1|b_2] = x^2[Qa_1|a_2] - y^2[Qa_2|a_1] + xy([Qa_2|a_2] - [Qa_1|a_1]),$$

or, since $[Qa_1|a_2] = [Qa_2|a_1] = 0$ and $[Qa_1|a_1] = [Qa_2|a_2] = 1$,

$$[Qb_1|b_2] = 0.$$

In addition,

$$[Qb_1|b_1] = x^2[Qa_1|a_1] + y^2[Qa_2|a_2] + 2xy[Qa_1|a_2],$$

or, since $[Qa_1a_2]$ is zero, $[Qa_1|a_1] = [Qa_2|a_2] = 1$ and $x^2 + y^2 = 1$,

$$[Qb_1|b_1] = 1,$$

and on the same grounds $[Qb_2|b_2] = 1$. Thus the assembly that results from a_1, a_2, $\dots$, a_n by a circular evolution of two magnitudes still always satisfies equations (f) and (g), and thus so does every assembly that results from $a_1, a_2, \dots, a_n$ by repeated circular evolutions.

I now show that, if any two of the magnitudes $a_1, \dots, a_n$, say a_1 and a_2, are not yet normal to one another, one can by a circular evolution {of the type described} make them normal to one another, and that thereby the product of the numerical values of these magnitudes is always reduced.

Thus, if b_1 and b_2 are to be normal to one another, that is (according to 152) $[b_1|b_2] = 0$, then (according to (h)) one has

$$0 = [(xa_1 + ya_2)|(xa_2 - ya_1)] = x^2[a_1|a_2] - y^2[a_1|a_2] + xy(a_{\overline{2}}^2 - a_{\overline{1}}^2)$$

or

$$0 = x^2 - y^2 - 2\gamma xy,$$

where $\gamma = (a_{\overline{1}}^2 - a_{\overline{2}}^2) : 2[a_1|a_2]$, whence it follows that

$$\frac{x}{y} = \gamma \pm \sqrt{1 + \gamma^2}.$$

Now if a_1 and a_2 are both real or both simple imaginary, then γ is real, whence $x : y$ is also real, and so we can assume that x and y are both real, and the evolution is circular {of the type described}. If on the other hand one of the magnitudes a_1 and a_2 is real {no matter which}$= r$, the other simple imaginary $= r'i$ (where $i = \sqrt{-1}$), then

$$\gamma = \pm(r^{\underline{2}} + r'^{\underline{2}}) : 2i[r|r'],$$

whence

$$\begin{aligned} 1 + \gamma^2 &= 1 - \left(\frac{r^{\underline{2}} + r^{\underline{2}}}{2[r|r']}\right)^2 = \left(1 + \frac{r^{\underline{2}} + r^{\underline{2}}}{2[r|r']}\right)\left(1 - \frac{r^{\underline{2}} + r^{\underline{2}}}{2[r|r']}\right) \\ &= \frac{r^{\underline{2}} + 2[r|r'] + r'^{\underline{2}}}{2[r|r']} \cdot \frac{2[r|r'] - r^{\underline{2}} - r'^{\underline{2}}}{2[r|r']} \\ &= -\frac{(r + r')^{\underline{2}}(r - r')^{\underline{2}}}{4[r|r']^2}, \end{aligned}$$

whence $\sqrt{1 + \gamma^2}$ is simple imaginary, but so is γ, and thus their sum, that is $x : y$. Thus if we assume x is real, then y is simple imaginary, but this was precisely the condition for the {special} circular evolution for this case. Thus in every case each pair of magnitudes of the assembly that are not yet normal to one another are, through {such} a circular evolution, made normal to one another.

It is still to be shown that under this evolution the product of the numerical values of the magnitudes is smaller. Here by the numerical value of a magnitude ri, where r is real and $i = \sqrt{-1}$, the numerical value of r will be meant.*

In this sense let α_1 and α_2 be the numerical values of a_1 and a_2, and β_1 and β_2 be those of b_1 and b_2; then (according to 156) $[a_1a_2]^{\underline{2}} = [b_1b_2]^{\underline{2}}$, but (according to 198) $[a_1a_2]^{\underline{2}} = (\alpha_1\alpha_2 \sin\angle a_1a_2)^2$, if a_1 and a_2 are real; now the same will also be the case if a_1 and a_2 are simple imaginary, and by $\angle a_1a_2$ is always meant the angle between the corresponding real magnitudes; if on the other hand one of the magnitudes a_1 and a_2 is real, the other simple imaginary, then

$$[a_1a_2]^{\underline{2}} = -(\alpha_1\alpha_2 \sin\angle a_1a_2)^2,$$

but then also

$$[b_1b_2]^{\underline{2}} = -(\beta_1\beta_2 \sin\angle b_1b_2)^2,$$

whence, since $[a_1a_2]^{\underline{2}} = [b_1b_2]^{\underline{2}}$, in every case

$$(\alpha_1\alpha_2 \sin\angle a_1a_2)^2 = (\beta_1\beta_2 \sin\angle b_1b_2)^2.$$

*{This stipulation agrees with the definition given later (cf. No. 413 and 414) of the numerical value of an imaginary extensive magnitude.}

Now if a_1 and a_2 are not normal to one another, but b_1 and b_2 are normal to one another, then $(\sin\angle a_1a_2)^2 < 1$, $(\sin\angle b_1b_2)^2 = 1$, whence $(\beta_1\beta_2)^2 < (\alpha_1\alpha_2)^2$, that is, since α_1, α_2, β_1, β_2 are positive, $\beta_1\beta_2 < \alpha_1\alpha_2$.

Thus if any two of the magnitudes of a conjugate assembly are not yet normal to one another, the assembly can be circularly evolved so that the product of the numerical values of all the magnitudes of the assembly is smaller. Now since there must be a minimum for this product, and this minimum can only arise if all the magnitudes of the assembly are normal to one another, it must be possible by circular evolutions {of the type described} of the assembly $a_1, a_2, \dots, a_n$ to derive an assembly for which every pair of its n magnitudes are normal to one another.

Let $r_1, r_2, \dots, r_n$ be this assembly; then since it is derived from the assembly $a_1, a_2, \dots, a_n$, as shown above it still satisfies equations (f) and (g), and includes as many real magnitudes as the latter assembly, and thus as many real magnitudes as there are positive products among the products $[Qc_1|c_1], \dots, [Qc_n|c_n]$. Now if one lets $Qr_1 = x_1r_1 + \cdots + x_nr_n$, then the equation $0 = [Qr_1|r_2]$ included in group (f) is transformed into

$$0 = [(x_1r_1 + x_2r_2 + \cdots + x_nr_n)|r_2] = x_2[r_2|r_2],$$

since all the other products $[r_1|r_2]$, $[r_3|r_2]$, ... are zero by virtue of the normal condition (according to 152). Now since further r_2, and thus also $[r_2|r_2]$, is nonzero, it follows from the equation $0 = x_2[r_2|r_2]$ that x_2 is zero; in the same way it also follows that $x_3 = 0, \dots, x_n = 0$, and thus that $Qr_1 = x_1r_1$. Then the equation $1 = [Qr_1|r_1]$ included in group (g) is transformed into $1 = x_1[r_1|r_1] = x_1r_1^2$, that is, $x_1 = 1 : r_1^2$, whence

$$Qr_1 = \frac{1}{r_1^2}r_1,$$

and on the same grounds

$$Qr_2 = \frac{1}{r_2^2}r_2, \quad \dots, \quad Qr_n = \frac{1}{r_n^2}r_n.$$

If therefore one sets

$$(i) \qquad \frac{1}{r_1^2} = \rho_1, \quad \frac{1}{r_2^2} = \rho_2, \quad \dots, \quad \frac{1}{r_n^2} = \rho_n,$$

and sets $r_1, r_2, \dots, r_n$ as the denominators of the fraction Q, the numerators belonging to them will be $\rho_1r_1, \rho_2r_2, \dots, \rho_nr_n$. The numerators {belonging to the same denominators} of the fraction $\rho - Q$ are then $(\rho-\rho_1)r_1, (\rho-\rho_2)r_2, \dots, (\rho-\rho_n)r_n$; the power of the fraction $\rho - Q$, which (according to 384) is equal to the combinatorial product of its numerators divided by that of its denominators, thus equals $(\rho-\rho_1)(\rho-\rho_2)\cdots(\rho-\rho_n)$.

But the equation by which the principal numbers of a quotient {Q} are specified says that the power of $\rho - Q$ is zero {cf. No. 388}, whence one has

$$(\rho-\rho_1)(\rho-\rho_2)\cdots(\rho-\rho_n) = 0,$$

that is, $\rho_1, \dots, \rho_n$ are the principal numbers of Q, which (according to (i)) are equal to

$$\frac{1}{r_1^2}, \frac{1}{r_2^2}, \dots, \frac{1}{r_n^2}.$$

Now according as r is real or simple imaginary, $1 : r^2$ is positive or negative, whence among the principal numbers there are as many positive as there are reals among the magnitudes $r_1, r_2, \ldots, r_n$, that is, as shown above, as there are positives among the products $[Qc_1|c_1], \ldots, [Qc_n|c_n]$. Thus the theorem is completely proved.[94]

REMARK. The principal numbers of the quotient Q and the magnitudes $r_1, r_2, \ldots, r_n$ belonging to them can be found immediately by the procedure in 388; it is only a question of deriving the particular elementary relations that appear between those magnitudes, among the special assumptions we have made for Q. Incidentally, in the above Proof there appears the so-called *law of inertia* of quadratic forms; in addition STURM's theorem on the roots of algebraic equations can easily be derived. Applied to geometry, our theorem includes the theorem that every algebraic surface of second order includes three real principal axes, and the theorem 388 shows how to find them directly.

5. Functions as extensive magnitudes

392. DEFINITION. I say a function f is numerically derivable from one or more functions $f_1, f_2, \ldots$ if f can be represented in the form

$$f = \alpha_1 f_1 + \alpha_2 f_2 + \cdots,$$

where $\alpha_1, \alpha_2, \ldots$ express numerical magnitudes that are either constant or are independent of the variables of the functions, and where the equal sign expresses that the equality holds for arbitrary values of these latter variables.

REMARK. Once this definition is established, all definitions and theorems previously formulated also relate to functions, which hereafter appear as extensive magnitudes. This method of treatment is of great significance for the theory of functions and therefore for the theory of curves and surfaces as well, as will be shown below. It can also serve to establish the concept of addition of points, lines, surfaces and so forth, independent of the earlier geometric developments.

For sharpening the conception it is yet to be noted that it must always be established what are the variables upon which the functions are thought to be dependent, and consequently to which the above defining equation refer.

In order to provide an example for the special formulation of the general concept when functions are set as units, I consider the six functions x^2, y^2, xy, x, y, 1, in which x and y are the variables. These stand in no numerical relation to one another, since, if

$$ax^2 + by^2 + cxy + dx + ey + f = 0$$

for every value of x and y, all the coefficients a, b, c, d, e, f must be zero. We can therefore set those six functions as a system of units. The functions numerically derivable from them are the functions of second degree with two variables. These therefore form a functional domain of sixth order. Therefore, from six functions of second degree with two variables standing in no numerical relation to one another, all functions of second degree with two variables can be numerically derived.

393. DEFINITION. Let x and y be the coordinates of a point in the plane (or x, y, z be the coordinates of a point in space), and further, let $f_1, f_2, \ldots, f_n$ be functions of these coordinates standing in no numerical relation to one another, and let

$$f = x_1 f_1 + x_2 f_2 + \cdots + x_n f_n,$$

that is, $x_1, \ldots, x_n$ are the derivation numbers by which f is derivable from $f_1, f_2, \ldots, f_n$. Finally, suppose there is between these derivation numbers the {homogeneous} equation

$$(a) \qquad \phi(x_1, x_2, \ldots, x_n) = 0;$$

then I call the totality of curves (surfaces) $f = 0$ for which the derivation numbers of f satisfy equation (a) a *curve structure* (*surface structure*) *belonging to this equation*, and in fact a *curve structure* (*surface structure*) *of mth degree*, if equation (a) is of mth degree.

In particular, if f_1, f_2, f_3 are functions of first degree of x and y, and

$$f = x_1 f_1 + x_2 f_2 + x_3 f_3,$$

then the homogeneous equation {of mth degree}

$$f(x_1, x_2, x_3) = 0$$

defines a *linear structure* {of first degree} *in the plane.*

If f_1, f_2, f_3, f_4 *are functions of first degree* of x, y, z, and

$$f = x_1 f_1 + x_2 f_2 + x_3 f_3 + x_4 f_4,$$

then the homogeneous equation {of mth degree}

$$\phi(x_1, x_2, x_3, x_4) = 0$$

defines a *planar structure* {of mth degree} *in space.*

REMARK. The whole idea introduced here is nothing but the natural broadening of the concept of coordinates, which arises of itself from the nature of this concept.

The coordinates are, reduced to their original idea, the derivation numbers of a magnitude, through which this magnitude results from several units standing in no numerical relation to one another. If we now substitute for these units functions of the original coordinates, the above more general concept arises. These functions set as units are then each represented in the plane by a curve and a coefficient, namely by the curve that includes all points whose original coordinates make that function zero, and by the coefficient of any nonzero term of the function, whereby on the one hand both the curve and the coefficient are specified by the function, and on the other the latter is specified by the former. These curves themselves represent the spatial position of the units and these coefficients their metric values.

Besides the case suggested in the theorem, where f_1, f_2, ... are functions of first degree, I will consider the case where f_1, f_2, ... are circular functions.

394. DEFINITION. I call the function

$$x^2 + y^2 + \beta x + \gamma y + \delta$$

a *simple circular function*, the function

$$\alpha(x^2 + y^2) + \beta x + \gamma y + \delta$$

an *α-fold circular function.* And if $f(x, y)$ is a circular function, I call the circle whose equation is

$$f(x, y) = 0,$$

in rectangular coordinates, the circle belonging to this function. {It will be said of a circle that it is numerically derivable from a number of circles, or that it stands in a numerical relation to another circle, if the corresponding holds for the circular functions belonging to them.}

395. *All circular functions are numerically derivable from four circular functions standing in no numerical relation to one another.*

PROOF. Every circular function is numerically derivable from the four functions $x^2 + y^2$, x, y, 1, which stand in no numerical relation to one another. Consequently (according to 24) also from any four functions, standing in no numerical relation to one another, that are derivable from x^2+y^2, x, y, 1, that is, from four such circular functions.

396. *The double distance (see 345) of a point (x', y') from a circle whose equation is*

$$f(x,y)=0,$$

where $f(x,y)$ denotes a ***simple*** *circular function, is equal to*

$$f(x',y').$$

REMARK. The proof by coordinates is well known. The proof depending on the concept of extensive magnitudes is however much simpler. This is based on the fact that, if p is a variable point, α the center of the circle and a' its radius,

$$(p-a)^2 - a'^2 = 0$$

is the equation of the circle, and at the same time its left side represents the double distance of the point p from the circle, both of which are directly implicit in the concept.

397. *Three circles stand in a numerical relation to one another if and only if all three go through the same two (real or imaginary) points.*

PROOF. Let f_1, f_2, f_3 be three circular functions of x and y, k_1, k_2, k_3 the three circles whose equations are

$$f_1 = 0, \quad f_2 = 0, \quad f_3 = 0,$$

respectively. First suppose there is a numerical relation between them, say

$$(a) \qquad f_3 = f_1 + f_2.$$

The points of intersection of the circles k_1 and k_2 are now those points for which f_1 and f_2 are simultaneously zero; but then, by virtue of equation (a), f_3 is also zero, that is, these points of intersection also lie on k_3.

Now assume conversely that the points of intersection of k_1 and k_2 also lie on k_3. Given any third point (x', y') on k_3, if one substitutes its coordinates for x and y in the functions f_1 and f_2, these may respectively take on the values α_1 and α_2, whence the circle whose equation is

$$\alpha_2 f_1 - \alpha_1 f_2 = 0$$

has, besides the above points of intersection, the point (x', y') in common with k_3; they are therefore identical, that is, the circle k_3 is numerically derivable from k_1 and k_2.

REMARK. If the points of intersection of the circles k_1 and k_2 are imaginary, then every circle that includes the same two imaginary points {still always} has the property that its center lies on the straight line with the centers of those circles, and the three circles have the same line of equal double distances (equal "power" according to STEINER).

398. *Two circles always have a straight line of equal double distances, and three circles always have a point of equal double distances, and in fact, if f_1, f_2, f_3 are three simple circular functions, then*

$$f_1 - f_2 = 0$$

is the equation for the straight line of equal double distances from the circles be-

longing to f_1 and f_2, and the point defined by the equations

$$f_1 - f_2 = 0, \quad f_1 - f_3 = 0$$

is the point of equal double distances from the three circles belonging to f_1,f_2,f_3.

PROOF. For the points of equal double distances from the circles belonging to the simple functions f_1, f_2 one has (according to 396)

$$f_1 = f_2, \quad \text{that is,} \quad f_1 - f_2 = 0.$$

Since however f_1 and f_2 are simple circular functions, in the difference the quadratic terms cancel out and $f_1 - f_2$ is a linear function, whence $f_1 - f_2 = 0$ is the equation of a straight line. For the point P of equal double distances from the three circles belonging to f_1,f_2,f_3 one has, on the same grounds, $f_1 = f_2 = f_3$, that is $f_1 - f_2 = 0$ and $f_1 - f_3 = 0$; thus P is the point of intersection of the straight lines represented by the last two equations.

REMARK. {The line of equal double distances from two circles is orthogonal to the central of these circles.} For two concentric circles that line is infinitely distant, and undefined if they are identical. For three circles whose centers lie on a straight line the point of equal double distances is either infinitely distant or undefined along a straight line or completely undefined, according as between the three circles there exist zero, one, or two numerical relations, in which latter case the three circles are identical.

399. *Four circles stand in a numerical relation to one another if all four have a point a of equal double distances, and in fact, if a is finitely distant, they stand in the same numerical relation to one another as do their centers.*

PROOF. 1. Let f_1, f_2, f_3, f_4 be four simple circular functions, k_1, k_2, k_3, k_4 the circles belonging to them. First let it be assumed that those four functions stand in a numerical relation, thus, say, that

$$f_4 = \alpha_1 f_1 + \alpha_2 f_2 + \alpha_3 f_3;$$

then, since all four functions are simple circular functions, necessarily $\alpha_1 + \alpha_2 + \alpha_3 = 1$. For the point a of equal double distances from the three circles k_1, k_2, k_3 one has (according to 398) $f_1 = f_2 = f_3$, whence for this point

$$f_4 = (\alpha_1 + \alpha_2 + \alpha_3) f_1 = f_1,$$

since $\alpha_1 + \alpha_2 + \alpha_3 = 1$, that is, the point a is a point of equal double distances from the four circles f_1, f_2, f_3, f_3.

2. *Conversely*, let it be assumed that a is a finitely distant point that has equal double distances from the four circles k_1, k_2, k_3, k_4; let a_1, a_2, a_3, a_4 be the displacements drawn from the point a to the centers of those circles, and r_1, r_2, r_3, r_4 be the four radii, and p the variable displacement, drawn from a to an arbitrary point of the plane {see Fig. 21}; then f_1, f_2, f_3, f_4 assume the form

$$f_s = (p - a_s)^2 - r_s^2 = p^2 - 2[a_s|p] + \delta, \tag{144}$$

where $\delta = a_s^2 - r_s^2$. According to 396, at the same time δ represents the double distance of the point for which $p = 0$, that is the point a from the circle k_s. By hypothesis this double distance is the same for the four circles $k_1, \ldots, k_4$. Further (according to 233) there is between the simple centers of the circles $k_1, \ldots, k_4$ a numerical relation; the same numerical relation also obtains (according to 222)

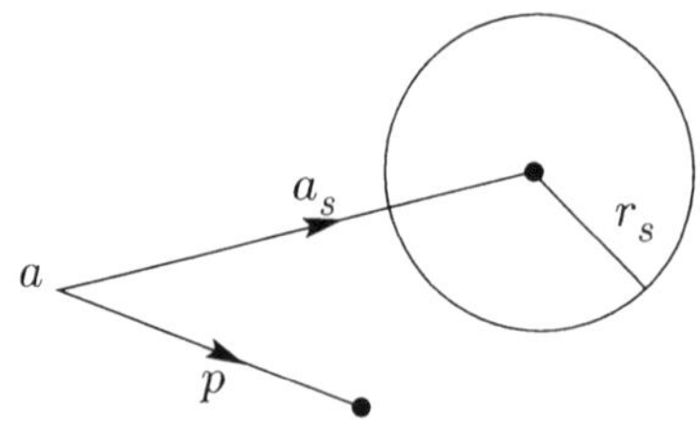

FIGURE 21

between the displacements drawn from an arbitrary point to those centers, and thus also between $a_1, \ldots, a_4$. Let

$$a_4 = \alpha_1 a_1 + \alpha_2 a_2 + \alpha_3 a_3$$

be this numerical relation; then, since the centers are simple points, it is necessary that $\alpha_1 + \alpha_2 + \alpha_3 = 1$; then one has

$$\begin{aligned} f_4 &= p^2 - 2[a_4|p] + \delta \\ &= p^2 - 2[(\alpha_1 a_1 + \alpha_2 a_2 + \alpha_3 a_3)|p] + \delta \\ &= (\alpha_1 + \alpha_2 + \alpha_3)(p^2 + \delta) - 2[(\alpha_1 a_1 + \alpha_2 a_2 + \alpha_3 a_3)|p] \\ &= \alpha_1 f_1 + \alpha_2 f_2 + \alpha_3 f_3. \end{aligned}$$

3. If the point a of equal double distances is infinitely distant, then (according to {the Remark following} 398) the centers of the four circles lie on a straight line. But four such circles always stand in a numerical relation to one another; for if one makes this line the abscissa axis (of x), the four functions f_1, f_2, f_3, f_4 are of the form

$$f_s = x^2 + y^2 - 2\beta_s x + \delta_s,$$

since the term with y drops out. Thus the four functions $f_1, \ldots, f_u$ are numerically derivable from the three functions x^2+y^2, x, and 1 {according to 392}, and therefore (according to 22) stand in a numerical relation to one another.

REMARK. If the point a of equal double distances from four circles lies outside one of the circles, then, as a consequence of the definition, the double distance from this circle is positive, whence also the double distances from the other circles are also positive, and so a simultaneously lies outside of the other circles. If one draws from a the tangents to the four circles, then these must be equal, since for every circle the square of the tangent drawn from an exterior point is equal to the double distance of the point. Thus if one scribes a circle around a whose radius is that tangent, all four circles will intersect this last circle orthogonally. On the other hand, if a lies within one circle, it must also lie within the others. If in any one of the circles one then draws from a the chord that is bisected by a, the {taken as negative} square of the half chord is equal to the double distance of the point a from this circle, whence if in all four circles one draws the chords bisected by a, they must all be equal to one another. Finally, if one scribes a circle around a with the half chord s, this circle is intersected by each of the four circles in a diameter, that is they intersect so that the two points of intersection are the end points of the same diameter drawn through this circle.

One can treat this circle intersected in its diameters as a circle intersected orthogonally, whose radius is imaginary, $= s\sqrt{-1}$, while a remains the center, and one thereby acquires the advantage of a common expression. Our theorem would then read thus: Four circles stand in a numerical relation to one another if and only if they are orthogonally intersected by a single circle, and the center of this circle is the point of equal double distances from any three of the four circles, and the radius is equal to the square root of this distance. We can also express this as follows: The domain derivable from three circles is the totality of circles that are orthogonally intersected by a single circle. In this sense the last-named circle represents that domain, that is, the domain of third order derivable from circles; while the domain of second order is represented by an assembly of two points {according to 97}.[95]

400. PROBLEM. *To find the circle derivable from n given simple circles by n given numbers.*

SOLUTION. Let $\alpha_1, \alpha_2, \ldots, \alpha_n$ be the n given numbers, their sum being α; further let an arbitrary point O be taken as the initial point of all displacements, and let the displacements drawn from it to the n centers be $a_1, a_2, \ldots, a_n$, and the n radii be $\beta_1, \beta_2, \ldots, \beta_n$; further, let the displacement drawn from O to a variable point be r; then the n simple functions belonging to the circles are

$$(r - a_{\mathfrak{a}})^{\underline{2}} - \beta_{\mathfrak{a}}^2$$

for every value of the index $\mathfrak{a}$ from 1 to n. Thus the circular function f sought is

$$\begin{aligned} f &= \sum \alpha_{\mathfrak{a}}[(r - a_{\mathfrak{a}})^{\underline{2}} - \beta_{\mathfrak{a}}^2] = \sum \alpha_{\mathfrak{a}}(r^{\underline{2}} - 2[r|a_{\mathfrak{a}}] + a_{\mathfrak{a}}^{\underline{2}} - \beta_{\mathfrak{a}}^2) \\ &= \alpha r^{\underline{2}} - 2\left[r\Big|\sum \alpha_{\mathfrak{a}} a_{\mathfrak{a}}\right] + \sum \alpha_{\mathfrak{a}}(a_{\mathfrak{a}}^{\underline{2}} - \beta_{\mathfrak{a}}^2), \end{aligned}$$

since $\sum \alpha_{\mathfrak{a}} = \alpha$. Now since the point O is arbitrary, we can, *if α is not zero*, choose it so that it is the center of gravity of the centers of the circles endowed with the coefficients $\alpha_1, \ldots, \alpha_n$. Then {according to 222} $\sum \alpha_{\mathfrak{a}} a_{\mathfrak{a}} = 0$ and the second term drops out. Thus

$$f = \alpha r^{\underline{2}} + \sum \alpha_{\mathfrak{a}}(a_{\mathfrak{a}}^{\underline{2}} - \beta_{\mathfrak{a}}^2).$$

If we set

$$\sum \alpha_{\mathfrak{a}}(\beta_{\mathfrak{a}}^2 - a_{\mathfrak{a}}^{\underline{2}}) = \alpha\beta^2,$$

then

$$f = \alpha(r^{\underline{2}} - \beta^2),$$

that is: *The center of the circle sought is, if the sum of the n given numbers is not zero, the point derivable from the n centers of the given circles by the n given numbers; and one obtains the radius (β) of the circle when one multiplies the n double distances of the center found with the n given numbers respectively, divides the sum of these products by the sum of the n given numbers, multiplies the quotient by -1 and extracts the root of this product.*

REMARK. I leave to the reader the treatment of the case *where* $\alpha = 0$. {More on circle geometry is found in Nos. 405–409.}

6. Relations considered from the standpoint of functional conjunctions

401. DEFINITION. I call two assemblies of magnitudes *related* if every numerical relation that exists between the magnitudes of the first or second assembly also holds between the corresponding magnitudes of the other, that is, if the magnitude

$$p = \alpha a + \beta b + \cdots$$

corresponds to the magnitude

$$p_1 = \alpha a_1 + \beta b_1 + \cdots$$

and conversely, where a, b, ... are arbitrary magnitudes of the first assembly and a_1, b_1, ... are the corresponding members of the other, and α, β, ... are arbitrary numbers.

402. *If two assemblies of magnitudes, in which the magnitudes of each of the assemblies can be numerically derived from n of its magnitudes standing in no numerical relation to one another, are to be related to one another, then one can set any n of the magnitudes, standing in no numerical relation to one another, of one assembly in correspondence with any n of the other, subject to the same condition; then for each magnitude of each of the two assemblies its correspondent on the other is precisely specified.*

PROOF. Let a, b, ... be n arbitrary magnitudes standing in no numerical relation to one another of one assembly, and a_1, b_1, ... be magnitudes subject to the same condition of the other assembly; then by assumption every magnitude p of the first assembly can be numerically derived from a, b, ... Let

$$p = \alpha a + \beta b + \cdots$$

be the expression for this derivation; then (according to 29) the numbers α, β, ... are precisely defined as soon as p is a specific magnitude. Now if the two assemblies are to be related, then (according to 401) the magnitude p must correspond to a magnitude

$$p_1 = \alpha a_1 + \beta b_1 + \cdots .$$

Thus to each magnitude of the one domain its correspondent in the other is precisely specified.

It is now to be shown that the assemblies constructed thus are in fact related to one another, that is, that every numerical relation that holds between the magnitudes of the first assembly also holds between the magnitudes of the second and conversely.

Let

$$(a) \qquad \rho r + \sigma s + \cdots = 0$$

be a numerical relation holding between the magnitudes r, s, ... of the first assembly, and let r_1, s_1, ... be the magnitudes of the second assembly corresponding to r, s, ... ; then it is to be shown that

$$\rho r_1 + \sigma s_1 + \cdots = 0$$

as well. If in (a) one replaces r, s, ... with the expressions for their numerical derivations from a, b, ... , removes the parentheses, and combines the terms that include a into one term, and so forth, one obtains an expression of the form

$$\alpha a + \beta b + \cdots = 0,$$

where α, β, ... are functions of the numerical magnitudes ρ, σ, ... and are the derivation numbers of r, s, ... Thus it follows, since a, b, ... stand in no numerical relation to one another, that (according to 29)

$$\alpha = 0, \ \beta = 0, \ \ldots .$$

Now if one applies the same procedure to the expression $\rho r_1 + \sigma s_1 + \cdots$, then, since the derivation numbers of r_1, s_1, ... are the same as those of r, s, ... , one obtains

$$\rho r_1 + \sigma s_1 + \cdots = \alpha a_1 + \beta b_1 + \cdots,$$

where α, β, ... have the same meaning as above. Since however α, β, ... are zero, one obtains

$$\rho r_1 + \sigma s_1 + \cdots = 0,$$

that is, every numerical relation that holds between the magnitudes of the first assembly also holds between their correspondents in the second and conversely, that is, the two assemblies are related.

403. *If from two related assemblies one derives two new assemblies in such a way that one sets every linear product P formed from magnitudes of the first assembly in correspondence with the product formed in the same way from the corresponding magnitudes of the second assembly, then these two new assemblies are likewise related to one another; that is, if r, s, ... are arbitrary magnitudes of the first assembly and r_1, s_1, ... are their correspondents in the other, and the linear products $P(r, s, \ldots)$ and $P(r_1, s_1, \ldots)$ are set in correspondence with one another, however r, s, ... may be chosen, then the two assemblies so obtained are likewise related.*

PROOF. Let a_1, a_2, ... , a_n be magnitudes of the first assembly that stand in no numerical relation to one another, and from which all the magnitudes of the first assembly can be numerically derived, and b_1, b_2, ... , b_n be their correspondents of the other, which are therefore subject to the same condition, and let

$$r = \sum \rho_{\mathfrak{a}} a_{\mathfrak{a}} = \rho_1 a_1 + \cdots + \rho_n a_n, \quad s = \sum \sigma_{\mathfrak{a}} a_{\mathfrak{a}}, \quad \ldots,$$

then (according to 401)

$$r_1 = \sum \rho_{\mathfrak{a}} b_{\mathfrak{a}}, \quad s_1 = \sum \sigma_{\mathfrak{a}} b_{\mathfrak{a}}, \quad \ldots,$$

whence

$$\left.\begin{aligned} P(r, s, \ldots) &= \sum \rho_{\mathfrak{a}} \sigma_{\mathfrak{b}} \cdots P(a_{\mathfrak{a}}, a_{\mathfrak{b}}, \ldots), \\ P(r_1, s_1, \ldots) &= \sum \rho_{\mathfrak{a}} \sigma_{\mathfrak{b}} \cdots P(b_{\mathfrak{a}}, b_{\mathfrak{b}}, \ldots). \end{aligned}\right\} \qquad [45]$$

Now since the products are linear, every defining equation that holds between the products $P(a_{\mathfrak{a}}, a_{\mathfrak{b}}, \ldots)$ must (according to 50) continue to hold if one replaces a_1, a_2, ... , a_n by the magnitudes b_1, b_2, ... , b_n. Now (according to 49) if p is the number of different products of the form $P(a_{\mathfrak{a}}, a_{\mathfrak{b}}, \ldots)$ and q is the number of independent defining equations, all the products $P(a_{\mathfrak{a}}, a_{\mathfrak{b}}, \ldots)$ can be numerically derived from $p - q$ of them that stand in no numerical relation to one another, and in fact so that if these $p - q$ products are specified, the derivation numbers for each of the other products are also specified.

The expressions for these derivations are dependent only on the defining equations. If therefore one replaces a_1, a_2, ... everywhere by b_1, b_2, ... , then, since the defining equations remain valid under this substitution, the expression of that derivation must also remain valid, that is, if A_1, A_2, ... are the products standing in no numerical relation to one another, from which all the other products of the form $P(a_{\mathfrak{a}}, a_{\mathfrak{b}}, \ldots)$ can be derived, and

$$P(a_{\mathfrak{a}}, a_{\mathfrak{b}}, \ldots) = \alpha_{1,\mathfrak{a},\mathfrak{b},\ldots} A_1 + \alpha_{2,\mathfrak{a},\mathfrak{b},\ldots} A_2 + \cdots ,$$

and if, further, B_1, B_2, ... are those products that result from the products A_1, A_2, ... when one replaces a_1, a_2, ... in them by b_1, b_2, ... , then

$$P(b_{\mathfrak{a}}, b_{\mathfrak{b}}, \ldots) = \alpha_{1,\mathfrak{a},\mathfrak{b},\ldots} B_1 + \alpha_{2,\mathfrak{a},\mathfrak{b},\ldots} B_2 + \cdots .$$

Thus

$$P(r, s, \ldots) = \sum \rho_{\mathfrak{a}} \sigma_{\mathfrak{b}} \cdots (\alpha_{1,\mathfrak{a},\mathfrak{b},\ldots} A_1 + \alpha_{2,\mathfrak{a},\mathfrak{b},\ldots} A_2 + \cdots),$$
$$P(r_1, s_1, \ldots) = \sum \rho_{\mathfrak{a}} \sigma_{\mathfrak{b}} \cdots (\alpha_{1,\mathfrak{a},\mathfrak{b},\ldots} B_1 + \alpha_{2,\mathfrak{a},\mathfrak{b},\ldots} B_2 + \cdots),$$

that is, $P(r, s, \ldots)$ is derived by the same numbers from A_1, A_2, ... as the corresponding product $P(r_1, s_1, \ldots)$ from the corresponding products B_1, B_2, ... , which is to say (according to 401), the assembly of products $P(r, s, \ldots)$ is related to the assembly of corresponding products $P(r_1, s_1, \ldots)$.

404. *In two assemblies, each of which is derivable from n of its magnitudes, and which are to be related to one another, one can in each take any $n + 1$ magnitudes, no n of which stand in a numerical relation to one another, and arrange that to the $n + 1$ magnitudes of the first assembly will correspond magnitudes that are congruent to the $n + 1$ magnitudes taken in the second; then to every magnitude of one domain its correspondent in the other is precisely specified, except for a numerical coefficient, the same for all of them.*

PROOF. Let a_1, ... , a_{n+1} be the magnitudes of the first assembly subject to the condition stated in the theorem, and b_1, ... , b_{n+1} be those of the second; then, on account of this condition, each of the magnitudes a_1, ... , a_{n+1} can be derived from the others by numbers, all of which are different from zero. For, since the assembly is to be derivable from n magnitudes standing in no numerical relation to one another, it must also be derivable from every n of the magnitudes subject to this condition, and therefore each of the magnitudes a_1, ... , a_{n+1} is derivable from the rest; and should any one of the derivation numbers be zero, there would, contrary to hypothesis, exist a numerical relation among the n others. The same holds for the magnitudes b_1, ... , b_{n+1}. Now let

$$(a) \qquad \begin{cases} a_{n+1} = \alpha_1 a_1 + \cdots + \alpha_n a_n, \\ b_{n+1} = \beta_1 b_1 + \cdots + \beta_n b_n, \end{cases}$$

where therefore α_1, ... , α_n, β_1, ... , β_n are all nonzero.

Further, let c_1, ... , c_{n+1} be the magnitudes that are to correspond respectively to a_1, ... , a_{n+1} and are congruent to the magnitudes b_1, ... , b_n. From this congruence it follows that, for every index r from 1 to $n+1$, c_r must be representable as the product of b_r with a nonzero number x_r, thus

$$(b) \qquad c_r = x_r b_r.$$

Since further c_1, ... , c_{n+1} are so to correspond to the magnitudes a_1, ... , a_{n+1} that the assemblies are related, it is necessary (according to 401) that

$$(c) \qquad c_{n+1} = \alpha_1 c_1 + \cdots + \alpha_n c_n.$$

If one substitutes the values from (b) into (c) and divides by x_{n+1}, one obtains

$$b_{n+1} = \frac{x_1\alpha_1}{x_{n+1}} b_1 + \cdots + \frac{x_n\alpha_n}{x_{n+1}} b_n.$$

But from (a) one has at the same time

$$b_{n+1} = \beta_1 b_1 + \cdots + \beta_n b_n,$$

whence necessarily (according to 29)

$$\frac{x_1\alpha_1}{x_{n+1}} = \beta_1, \ \ldots, \ \frac{x_n\alpha_n}{x_{n+1}}\beta_n.$$

Hereby all unknowns but one are specified. If we set $x_{n+1} = \lambda$, then

$$(d) \qquad x_1 = \frac{\lambda\beta_1}{\alpha_1}, \ \ldots, \ x_n = \frac{\lambda\beta_n}{\alpha_n}, \ x_{n+1} = \lambda.$$

If these conditions (d) are satisfied, then conversely equation (c) is satisfied. Thus the assemblies are related with respect to the $n+1$ magnitudes $a_1, \ldots, a_{n+1}$ and their correspondents $c_1, \ldots, c_{n+1}$, and every magnitude

$$p = u_1 a_1 + \cdots + u_n a_n$$

corresponds to a magnitude

$$q = u_1 c_1 + \cdots + u_n c_n,$$

or, when one replaces $c_1, \ldots, c_n$ with their values from (b) and $x_1, \ldots, x_n$ with their values from (d),

$$q = \lambda\left(\frac{u\beta_1}{\alpha_1} b_1 + \cdots + \frac{u\beta_n}{\alpha_n} b_n\right),$$

that is, q is, apart from a constant factor, precisely specified.[96]

REMARK. The relation of two assemblies without regard to the metric value of the corresponding magnitudes can also be specified in another way, in which one arranges that every three magnitudes of the first assembly standing in a numerical relation are to correspond to three magnitudes standing in a numerical relation in the second and conversely. The proof of the identity of this definition with that given above {in No. 401} (where one neglects the metric values of the corresponding magnitudes) follows easily if one applies the excellent development of collineation given by MÖBIUS in his *Barycentrische Kalkül* in §§200–206, and especially in §203 {*Werke*, Bd. 1}, to the general relation considered here.

405. *Space and the plane can be set as related to one another if to each point in space corresponds a circle in the plane and conversely. Then to the points lying in a single plane correspond such circles as are orthogonally intersected by the same circle. And in fact one can set five arbitrary points of space, no four of which lie in the same plane, in correspondence with five arbitrary circles in the plane, no four of which are orthogonally intersected by the same circle. But then to each point of space the corresponding circle in the plane is completely specified. Every theorem of stereometry can in this sense be applied to circles in the plane, and conversely every theorem on circles in the plane can be applied to points in space.*

PROOF. According to 395 every circular function is numerically derivable from four arbitrary circular functions not standing in a numerical relation to one another, and according to 232 every point in space is numerically derivable from four arbitrary points standing in no numerical relation to one another. Consequently (according to 404), if the points of space and the circles in a plane are to form two related assemblies, one can take five arbitrary points in space, no four of which stand in a numerical relation, that is, no four of which lie in the same plane, and five arbitrary circles in a plane, no four of which stand in a numerical relation, that is, no four of which are orthogonally intersected by the same circle {Remark following 399}, and arrange that those five points in space are to correspond to these five circles. Then to each point in space the corresponding circle in the plane is specified and conversely.

Further, since according to the concept of relationship (401) every numerical relation that holds between the magnitudes of one assembly also holds between the corresponding magnitudes of the related assembly, it follows that if there is a numerical relation between four points of space, such a relation must also hold between the four corresponding circles, that is, if the four points lie in the same plane, the four corresponding circles must be intersected by the same circle.

Finally the applicability of the theorems follows thereby, since every theorem in space can be expressed by means of the four derivation numbers by which each point is representable in an analytic theorem, and this again, when one sets the four derivation numbers as the derivation numbers of the circles corresponding to those points, is transformed into a theorem on circles in the plane, and likewise conversely.

406. *In the plane one can take two related assemblies of circles, and thereby set five arbitrary circles of one assembly in correspondence with five arbitrary circles of the other, provided that no four of the five circles taken in either plane are orthogonally intersected by the same circle. Then to each sixth circle of one assembly its correspondent in the other is specified, and every four circles of one assembly that are not orthogonally intersected by the same circle correspond to four circles of the other that likewise are not orthogonally intersected by the same circle.*

PROOF. Follows automatically from the above.

REMARK. We call the relation just considered syncyclic. Of particular interest is the case where such circles as are contracted into points also correspond to such in the other assembly.[97]

407. *If the circle whose equation is*

$$(a) \qquad \alpha(x^2+y^2)+2\beta x+2\gamma y+\delta=0,$$

where α, β, γ, δ are **real**, *is* **contracted into a point**, *then necessarily*

$$(b) \qquad \alpha\delta=\beta^2+\gamma^2.$$

Conversely, if equation (b) obtains and not all the coefficients are zero, then the circle represented by (a) must either be contracted to a point or cast onto the infinitely distant straight line, this last if α, β, γ are simultaneously zero.

PROOF. From equation (a) it follows that, if α is not zero, for the radius r of the circle belonging to the equation,

$$r^2=\frac{\beta^2+\gamma^2-\alpha\delta}{\alpha^2},$$

from which the rest results. If on the other hand α is zero, then equation (a) is the equation of a straight line; but then it follows from (b) that $\beta^2+\gamma^2$ is zero, that is, that β and γ are zero; thus the {straight} line represented by equation (a) is infinitely distant.

408. *If one takes x and y as (rectangular) coordinates of one assembly of circles and x' and y' as those of another, and sets the four functions*

$$x^2 + y^2, \quad x, \quad y, \quad 1$$

in correspondence, serially, with the functions

$$1, \quad x', \quad y', \quad x'^2 + y'^2,$$

so that each circle whose equation is

$$(a) \qquad \alpha(x^2 + y^2) + 2\beta x + 2\gamma y + \delta = 0$$

corresponds to that circle whose equation is

$$(b) \qquad \alpha + 2\beta x' + 2\gamma y' + \delta(x'^2 + y'^2) = 0,$$

then every point of the first assembly, with the exception of the origin of the abscissa, corresponds to a point of the second and conversely; on the other hand to the origin of the abscissa in each case corresponds the infinitely distant straight line in the other assembly.

PROOF. If the circle whose equation is (a) is contracted to a point, then $\alpha\delta = \beta^2 + \gamma^2$ (407). But if this equation holds, the circle whose equation is (b) is also (according to 407) either a point (if $\delta \neq 0$) or the infinitely distant straight line, the latter if β, γ, δ are zero, that is, if the point of the first assembly is defined by the equation $\alpha(x^2 + y^2) = 0$, that is, if it is the origin of coordinates.

409. {CONVERSE}. *If in two syncyclically related assemblies of circles the infinitely distant straight line of each assembly corresponds to a point of the other, and all other points of each assembly again correspond to points of the other, one can always give the (mutually orthogonal) coordinate axes of each assembly such positions, and the measure of length taken as unity such a value, that the four functions*

$$x^2 + y^2, \quad x, \quad y, \quad 1$$

of the first assembly correspond, serially, to the four functions

$$1, \quad x', \quad y', \quad x'^2 + y'^2$$

of the second.

PROOF. The infinitely distant point is represented by a function that consists of just a constant. Let the point, which in each assembly corresponds to the infinitely distant line in the other, be made the origin of the abscissa. The origin of the abscissa is represented by the circular function $x^2 + y^2$. This function corresponds in the second assembly to the constant a, by which the infinitely distant straight line is represented; likewise the function $x'^2 + y'^2$ of the second assembly corresponds to the constant b in the first.

Now if one changes the measure of length chosen as unity, the coordinates are multiplied by a constant factor λ, and thus there then correspond

$$\begin{matrix} \lambda^2(x^2 + y^2), & b \\ a, & \lambda^2(x'^2 + y'^2). \end{matrix}$$

Let λ^2 be so defined that $a : \lambda^2 = \lambda^2 : b$, that is, $\lambda^4 = ab$. If we then set $a : \lambda^2 = \mu$, there correspond

$$\begin{array}{cc} (x^2+y^2), & b \\ \mu a, & \mu(x'^2+y'^2). \end{array}$$

But the spatial structures defined by functions are not altered if one divides all of them by a constant number, here by μ, and we can therefore set x^2+y^2 and 1 in correspondence with 1 and $x'^2+y'^2$.

Further, the functions x and y of the first assembly may be set in correspondence with the functions f_1 and f_2, namely

$$f_1 = \alpha_1(x'^2+y'^2)+\beta_1 x'+\gamma_1 y'+\delta_1$$
$$f_2 = \alpha_2(x'^2+y'^2)+\beta_2 x'+\gamma_2 y'+\delta_2,$$

so that the functions

$$\begin{array}{cccc} x^2+y^2, & x, & y, & 1 \\ 1, & f_1, & f_2, & x'^2+y'^2 \end{array}$$

correspond. From the former let a function f be derived by the coefficients α, β, γ, δ; then it corresponds to a function f' in the second assembly which is derived by the same coefficients from the four latter functions. Thus the condition that the former function f represents just a point is (according to 407)

$$4\alpha\delta = \beta^2+\gamma^2.$$

For the second function,

$$\begin{aligned} f' &= \alpha+\beta f_1+\gamma f_2+\delta(x'^2+y'^2) \\ &= (\delta+\beta\alpha_1+\gamma\alpha_2)(x'^2+y'^2)+(\beta\beta_1+\gamma\beta_2)x' \\ &\quad+(\beta\gamma_1+\gamma\gamma_2)y'+\alpha+\beta\delta_1+\gamma\delta_2, \end{aligned}$$

this condition reads

$$4(\delta+\beta\alpha_1+\gamma a_2)(\alpha+\beta\delta_1+\gamma\delta_2) = (\beta\beta_1+\gamma\beta_2)^2+(\beta\gamma_1+\gamma\gamma_2)^2.$$

If one replaces δ here by its value from (a), one obtains

$$(\beta^2+\gamma^2+4\alpha\beta\alpha_1+4\alpha\gamma\alpha_2)(\alpha+\beta\delta_1+\gamma\delta_2) = \alpha(\beta\beta_1+\gamma\beta_2)^2+\alpha(\beta\gamma_1+\gamma\gamma_2)^2.$$

This equation must hold for arbitrary values of α, β, γ, and thus the coefficients belonging to the same powers of these magnitudes must be equal on both sides. Since the right side includes α only to the first power, the coefficients of the terms that include α^2, and those that do not include α, must be zero, whence α_1, α_2, δ_1, δ_2 must be zero, that is, f_1 and f_2 represent straight lines that go through the origin of the abscissa. If we position the abscissa axis so that it coincides with the line represented by f_2, then f_2 is reduced to just the term that includes y, that is, β_2 is zero. If we then divide the reduced equation by α, it goes over into

$$\beta^2+\gamma^2 = \beta^2\beta_1^2+(\beta\gamma_1+\gamma\gamma_2)^2.$$

Since in the developed equation $2\gamma_1\gamma_2$ is the coefficient of $\beta\gamma$, it is necessary that $\gamma_1\gamma_2 = 0$; γ_2 cannot be zero, since otherwise f_2 would be equal to zero, so γ_1 must be zero. Thus one obtains

$$\beta^2(1-\beta_1^2)+\gamma^2(1-\gamma_2^2) = 0,$$

whence $\beta_1 = \pm 1$, $\gamma_2 = \pm 1$. Since for each of the two coordinate axes the side toward which the coordinates are taken as positive can be chosen arbitrarily, we can set β_1 and γ_2 both $= 1$, whence $f_1 = x'$, $f_2 = y'$, and thus to the functions

$$x^2 + y^2, \quad x, \quad y, \quad 1$$

of the first assembly correspond the functions

$$1, \quad f_1, \quad f_2, \quad x'^2 + y'^2$$

of the second.

REMARK. The special type of syncyclic relation treated here was first formulated by MÖBIUS and called by him a *circular relation*; cf. MÖBIUS, "Uber eine neue Verwandschaft zwischen ebenen Figuren" (in *Berichten der Königl. Sächs. Gesellsch. der Wiss.*, 5 Feb. 1853 {*Werke*, Bd. 2, p. 205}).

It follows from the above that the point in each of the two circularly related assemblies which corresponds to the infinitely distant straight line in the other assembly plays a prominent role. Let this point be called the *central point* of the assembly. If one now positions the two assemblies upon one another so that the central points, the x-axes and the y-axes are reciprocally covered, then *every straight line going through the central points*, for example the straight line $qx + ry = 0$, *also covers its correspondent.* Now if one scribes a circle around the central point with the length taken as the fundamental measure of the coordinates, called the *principal circle*, then the whole variety of reciprocal correspondents is most plainly represented. For *the periphery of the principal circle consists of all the points that cover their correspondents*; every point in the interior of the principal circle corresponds in the other assembly to a point outside the circle lying on the same radius, and in fact such that *the radius is always the geometric mean between the distances of the two corresponding points from the center.*

The latter follows immediately for a point on the x-axis from the corresponding functions; for the circle equation of a point on the x-axis whose abscissa is $= c$, is $(x - c)^2 + y^2 = 0$, that is

$$x^2 + y^2 - 2cx + c^2 = 0,$$

whence that of the corresponding point is

$$1 - 2cx + c^2(x^2 + y^2) = 0,$$

that is

$$\left(x - \frac{1}{c}\right)^2 + y^2 = 0,$$

so that the distance of this point from the central point $= 1 : c$, while that of the corresponding point was c, whence their product is 1, that is, the length taken as the unit is the geometric mean between the distances of the corresponding points on the x-axis. Now since one can take any straight line going through the central point as the x-axis, the relation holds in general.

If instead of the assumption that the infinitely distant line is to correspond to a point, one makes the assumption that in the two syncyclic assemblies each point of one assembly, without exception, is to correspond to a point of the other, then the infinitely distant line of one assembly also corresponds to the infinitely distant line of the other, and one arrives at similarity, which in this way therefore stands opposed to the circular relation.

7. Normal units of functions, continuity of the latter

410. DEFINITION. *Normal units of real magnitudes.* For *real numbers* I set 1 as the normal unit, *for real extensive magnitudes* {*of first order*} I set as normal units the original units $e_1, e_2, \ldots, e_n$; further, *for real extensive magnitudes of mth order*

I set the well-ordered (multiplicative) combinations *without repetition* of mth class of the original units; finally, *for real algebraic products of magnitudes* of the same order, the (well-ordered) combinations *with repetition* of the normal units of the factors (where each of these combinations is interpreted as the algebraic product of the elements included in it).

REMARK. The definitions introduced previously are just combined here. It is still important to establish the normal units for real open expressions {with interchangeable openings} as well.

We have (in the Remark following 363) seen that the product of the factors that are to fill the openings of {such} an open expression is interpreted as an algebraic product by which the open expression is to be multiplied. Consequently it is important what value the open expression assumes when the normal units of that algebraic product are multiplied by it. Let E_1, E_2, ... be the normal units of this algebraic product, and L be the open expression; then it depends on the values LE_1, LE_2, ... These values can again be extensive magnitudes whose normal units are e_1, e_2, ..., whence the normal units of L are those open expressions which, multiplied by E_1, E_2, ..., yield any one of the units e_1, e_2, ...

411. DEFINITION. *Normal units of real open expressions.* If E_1, E_2, ... are the normal units of those algebraic products whose factors can fill the openings of a real open expression L {whose openings are interchangeable}, and e_1, e_2, ... are the normal units of those magnitudes into which L is transformed after its openings are filled, I set as normal units of L those open expressions $E_{r,s}$ that satisfy the equations

$$E_{r,s}E_r = e_s \quad \text{and} \quad E_{r,s}E_t = 0 \ \ (t \neq r).$$

REMARK. This definition is in agreement with that given in 381 for the units of quotients, that is, of open expressions with a single opening.

412. *Every (real) open expression {with interchangeable openings} can be numerically derived from its normal units as defined in* 411, *and these latter stand in no numerical relation to one another.*

PROOF. As in 381.

413. DEFINITION. *Normal units of a family of magnitudes.* I define as *magnitudes of the same family* all those magnitudes that, according to what has been established so far, can be added to one another. The number of normal units of a family of magnitudes I always assume is even, since half of them are real, and the others result from them through multiplication by $i = \sqrt{-1}$. I always assume the derivation numbers, by which a magnitude is numerically derived from its normal units, are *real*.

REMARK. For the general numerical magnitudes the normal units are therefore 1 and $\sqrt{-1} = i$, for the general magnitudes of first order

$$e_1, e_2, \ldots, e_n, e_1 i, e_2 i, \ldots, e_n i,$$

where e_1, e_2, ..., e_n are the original units, and so forth.

414. DEFINITION. The *numerical value* of a magnitude means the positive square root of the sum of the squares of all numbers by which that magnitude is derivable from its normal units, that is, if E_1, E_2, ... are the normal units of a magnitude P and

$$P = \alpha_1 E_1 + \alpha_2 E_2 + \cdots,$$

then the numerical value of P is equal to

$$\sqrt{\alpha_1^2 + \alpha_2^2 + \cdots}.$$

REMARK. This definition is in agreement with that given in 151. The numerical value of a complex numerical magnitude $p + qi$ is therefore equal to $\sqrt{p^2 + q^2}$.

415. *If the numerical value of a magnitude is zero, then all the numbers by which this magnitude is derived from its normal units are individually zero.*

PROOF. Let E_1, E_2, ... be the normal units of the magnitude P and let

$$P = \alpha_1 E_1 + \alpha_2 E_2 + \cdots .$$

Now if the numerical value of P is zero, this means (according to 414)

$$\sqrt{\alpha_1^2 + \alpha_2^2 + \cdots} = 0,$$

whence

$$\alpha_1^2 + \alpha_2^2 + \cdots = 0.$$

But since (according to 413) α_1, α_2, ... are all real, the sum of their squares can only be zero if they are all individually zero, thus

$$0 = \alpha_1 = \alpha_2 = \cdots .$$

416. DEFINITION. If the numerical value of a magnitude a is smaller than that of a magnitude b, I say that a is *numerically smaller* than b, and write this as

$$a \text{ num.} < b.$$

417. *If α_1, α_2, ... are the numbers by which a is derivable from its normal units, and likewise β_1, β_2, ... are the numbers by which b is derivable from its normal units, then the inequalities*

$$a \text{ num.} < b$$

and

$$\alpha_1^2 + \alpha_2^2 + \cdots < \beta_1^2 + \beta_2^2 + \cdots$$

are equivalent to one another.

PROOF. Follows directly from 416 and 414.

418. *If p, q, ... are positive numerical values and a, b, ... arbitrary magnitudes (derivable from the same normal units) such that*

$$a \text{ num.} < p, \quad b \text{ num.} < q, \quad \ldots,$$

then in addition

$$a + b + \cdots \text{ num.} < p + q + \cdots .$$

PROOF. 1. For *two* magnitudes. Let e_1, e_2, ... be the normal units of a and b and let

$$a = \alpha_1 e_1 + \alpha_2 e_2 + \cdots ,$$
$$b = \beta_1 e_1 + \beta_2 e_2 + \cdots ,$$

and let α be the numerical value of a, β be that of b, and γ be that of $a + b$; then $\alpha < p$, $\beta < q$. It is to be shown that $\gamma < p + q$.

According to 414,

$$a^2 = \alpha_1^2 + \alpha_2^2 + \cdots,$$
$$\beta^2 = \beta_1^2 + \beta_2^2 + \cdots,$$
$$\gamma^2 = (\alpha_1 + \beta_1)^2 + (\alpha_2 + \beta_2)^2 + \cdots,$$

whence

$$(*) \qquad \gamma^2 = \alpha^2 + \beta^2 + 2(\alpha_1\beta_1 + \alpha_2\beta_2 + \cdots).$$

Now we can show that $\alpha_1\beta_1 + \alpha_2\beta_2 + \cdots \leq \alpha\beta$. In fact,

$$\begin{aligned}(\alpha\beta)^2 &- (\alpha_1\beta_1 + \alpha_2\beta_2 + \cdots)^2 \\ &= (\alpha_1^2 + \alpha_2^2 + \cdots)(\beta_1^2 + \beta_2^2 + \cdots) - (\alpha_1\beta_1 + \alpha_2\beta_2 + \cdots)^2 \\ &= (\alpha_1\beta_2 - \alpha_2\beta_1)^2 + (\alpha_1\beta_3 - \alpha_3\beta_1)^2 + \cdots.\end{aligned}$$

The right side is the sum of several squares, and thus is greater than or equal to zero, and therefore the same holds for the left side, that is

$$(\alpha\beta)^2 \geq (\alpha_1\beta_1 + \alpha_2\beta_2 + \cdots)^2,$$

whence also, since α and β are positive,

$$\alpha\beta \geq \alpha_1\beta_1 + \alpha_2\beta_2 + \cdots.$$

If we apply this inequality to equation $(*)$, it follows that

$$\gamma^2 \leq \alpha^2 + \beta^2 + 2\alpha\beta, \quad \text{that is,} \quad \gamma^2 \leq (\alpha + \beta)^2,$$

whence, since γ and $\alpha + \beta$ are positive,

$$\gamma \leq \alpha + \beta;$$

but, since α and β are (positive) numbers that are respectively smaller than the positive numbers p and q,

$$\alpha + \beta < p + q,$$

whence

$$\gamma < p + q,$$

that is, the numerical value of $a + b$ is less than $p + q$.

2. For *more* magnitudes. Since (according to Part 1 of the Proof) $a+b < p+q$, and (by hypothesis) c num. $< r$, it follows (according to Part 1 of the Proof) that

$$a + b + c \text{ num.} < p + q + r,$$

and so forth for arbitrarily many magnitudes.

419. Corollary. *If p and q are positive numerical values and a and b are arbitrary magnitudes (derivable from the same normal units) such that*

$$a \text{ num.} < p, \quad b \text{ num.} < q,$$

then in addition

$$a - b \text{ num.} < p + q.$$

419b.* *If a is an arbitrary magnitude, and b, c, ... are numerical magnitudes, then the numerical value (ρ) of the product $abc\cdots$ of these magnitudes is equal to*

*{Theorems 419b and 419c bore the numbers 457 and 458 in the original edition; yet Grassmann himself said, in a Remark after No. 458: "These two theorems, which systematically stand after 419, are held back to here in order to be able to apply them in the following," whence they are now inserted after No. 419.}

the product of their numerical values $(\alpha, \beta, \gamma, \dots)$*, that is*

$$\rho = \alpha\beta\gamma\cdots .$$

PROOF. 1. For *two* magnitudes. Let $e_1, e_2, \dots$ be the real units of $a, ie_1, ie_2, \dots$ $(i = \sqrt{-1})$ its imaginary units, and let

$$a = \sum \{\alpha_{\mathfrak{a}} e_{\mathfrak{a}} + i\gamma_{\mathfrak{a}} e_{\mathfrak{a}}\},$$

where the $\alpha_{\mathfrak{a}}$ and $\gamma_{\mathfrak{a}}$ are all real, and let $b = \delta + \varepsilon i$; then (according to 414)

$$\alpha^2 = \sum \{\alpha_{\mathfrak{a}}^2 + \gamma_{\mathfrak{a}}^2\}, \quad \beta^2 = \delta^2 + \varepsilon^2.$$

Further,

$$ab = \sum \{(\delta\alpha_{\mathfrak{a}} - \varepsilon\gamma_{\mathfrak{a}})e_{\mathfrak{a}} + i(\delta\gamma_{\mathfrak{a}} + \varepsilon\alpha_{\mathfrak{a}})e_{\mathfrak{a}}\},$$

whence

$$\begin{aligned} \rho^2 &= \sum \{(\delta\alpha_{\mathfrak{a}} - \varepsilon\gamma_{\mathfrak{a}})^2 + (\delta\gamma_{\mathfrak{a}} + \varepsilon\alpha_{\mathfrak{a}})^2\} \\ &= \sum \{\delta^2\alpha_{\mathfrak{a}}^2 - \varepsilon^2\gamma_{\mathfrak{a}}^2 + \delta^2\gamma_{\mathfrak{a}}^2 + \varepsilon^2\alpha_{\mathfrak{a}}^2\} \\ &= (\delta^2 + \varepsilon^2)\sum \{\alpha_{\mathfrak{a}}^2 + \gamma_{\mathfrak{a}}^2\} = \beta^2\alpha^2, \end{aligned}$$

whence, since ρ, α, β are positive,

$$\rho = \alpha\beta,$$

that is

$$ab \text{ num.} = \alpha\beta.$$

2. {For *more* magnitudes.} Now since (according to Part 1 of the Proof) ab num. $= \alpha\beta$, we have (according to Part 1 of the Proof)

$$abc \text{ num.} = \alpha\beta\gamma,$$

and so forth.

419C. *If* a *and* a_1 *are arbitrary magnitudes,* b*,* c*,* $\dots$ *,* b_1*,* c_1*,* $\dots$ *are numerical magnitudes, and*

$$a \text{ num.} < a_1,$$
$$b \text{ num.} \leq b_1, \quad c \text{ num.} \leq c_1, \ \dots,$$

then in addition

$$abc\cdots \text{ num.} < a_1 b_1 c_1 \cdots .$$

PROOF. Let α, β, γ, $\dots$, α_1, β_1, γ_1, $\dots$, respectively, be the numerical values of a, b, c, $\dots$, a_1, b_1, c_1, $\dots$; then (according to 419B) $\alpha\beta\gamma\cdots$ is the numerical value of $abc\cdots$ and $\alpha_1\beta_1\gamma_1\cdots$ that of $a_1b_1c_1\cdots$. But since α, β, γ, $\dots$, α_1, β_1, γ_1, $\dots$ are positive numbers and $\alpha < \alpha_1$, $\beta \leq \beta_1$, $\gamma \leq \gamma_1$, $\dots$, so also $\alpha\beta\gamma\cdots < \alpha_1\beta_1\gamma_1\cdots$, that is

$$abc\cdots \text{ num.} < a_1 b_1 c_1 \cdots .$$

420. DEFINITION. I say a function $f(q)$ of a positive numerical magnitude $\{q\}$ *vanishes with* q if for every positive number p a positive value of q can be given such that

$$f(q) \text{ num.} < p,$$

and remains so if q arbitrarily diminishes, but remains positive.[98] If on the other hand $f(0) = 0$, I say $f(q)$ is *zero with* q.

REMARK. The two expressions "vanishes with (positive) q" and "is zero with q" are therefore not equivalent; rather, only the second includes the first, not conversely. For the condition that $f(q)$ to vanish with q could be fulfilled, and yet for $q = 0$, $f(q)$ could pass abruptly to an isolated nonzero value.

421. *If several functions $f_1(q)$, $f_2(q)$, ... , $f_n(q)$ of a positive numerical magnitude q vanish with q, then*

$$(a) \qquad a_1 f_1(q) + a_2 f_2(q) + \cdots + a_n f_n(q)$$

also vanishes with q, where a_1, a_2, ... , a_n are finite numbers or arbitrary finite open expressions with one opening that can be filled by $f_1(q)$, $f_2(q)$, ... (And likewise, if those functions are zero with q, so also will the last expression (a) be zero with q.)

PROOF. For *numbers.* Should some of the magnitudes a_1, ... , a_n be zero, one can drop the terms in expression (a) in which these coefficients equal to zero appear. We assume this already done, and thus a_1, ... , a_n are all nonzero (finite) numbers.

Now since (by hypothesis) $f_1(q)$ vanishes with q, (according to 420) to every nonzero number, for example to $p : a_1 n$, it must be possible to find a positive value q_1 such that $f_1(q_1)$ is numerically smaller than $p : a_1 n$ and remains so if q_1 arbitrarily diminishes, but remains positive; on the same grounds one can also find a positive value q_2 such that $f_2(q_2)$ is numerically smaller than $p : a_2 n$ and remains so with diminishing q_2, ... Now if q is a positive value that is even smaller than each of the magnitudes q_1, q_2, ... , q_n, then in addition $f_1(q)$ num. $< p : a_1 n$, ... , or

$$a_1 f_1(q) \text{ num.} < \frac{p}{n}, \quad a_2 f_2(q) \text{ num.} < \frac{p}{n}, \quad \ldots, \quad a_n f_n(q) \text{ num.} < \frac{p}{n};$$

consequently (according to 418) the sum of the left sides is numerically smaller than that of the right, that is

$$a_1 f_1(q) + a_2 f_2(q) + \cdots + a_n f_n(q) \text{ num.} < p,$$

an inequality that also remains valid if q arbitrarily diminishes but remains positive, that is, expression (a) vanishes with q when a_1, ... , a_n are {finite} numbers.

2. Let expression (a) be reduced to $A_1 f_1(q)$, where a_1 is a normal unit of an open expression whose opening can be filled by $f_1(q)$; further, let e_1, e_2, ... be the normal units of $a_1 f_1(q)$ and E_1, E_2, ... be those of $f_1(q)$, and let

$$(b) \qquad f_1(q) = E_1\phi_1(q) + E_2\phi_2(q) + \cdots + E_n\phi_n(q);$$

then (according to 411) a_1 has the property that, multiplied by one of the units E_1, E_2, ... , for example E_r, it yields one of the units e_1, e_2, ... , for example the unit e_s, while multiplied by any of the other units E_1, E_2, ... it gives zero, so that

$$(c) \qquad a_1 E_r = e_s, \quad a_1 E_t = 0, \quad \text{for } t \neq r.$$

Thus we obtain

$$a_1 f_1(q) = a_1 \sum E_{\mathfrak{a}}\phi_{\mathfrak{a}}(q) = \sum a_1 E_{\mathfrak{a}}\phi_{\mathfrak{a}}(q) \qquad [44]$$
$$= a_1 E_r \phi_r(q) = e_s \phi_r(q). \qquad [(c)]$$

Thus (according to 414) the numerical value of $a_1 f_1(q)$ is equal to $\sqrt{(\phi_r(q))^2}$. Now since (by hypothesis) $f_1(q)$ vanishes with q, (according to 420) for each positive number p a value of q can be given such that $f_1(q)$ num. $< p$, and this remains so with diminishing q. But since (according to 413) $\phi_1(q)$, $\phi_2(q)$, ... are real, $(\phi_1(q))^2$, $(\phi_2(q))^2$, ... are positive, and thus each of these squares must be smaller than p^2, thus $(\phi_r(q))^2 < p^2$, that is $a_1 f_1(q)$ num. $< p$, and thus $a_1 f_1(q)$ vanishes with q.

3. Finally, let a_1, a_2, ... be arbitrary open expressions (each with one opening), and let

$$a_r = \sum \alpha_{r,\mathfrak{b}} E_{r,\mathfrak{b}},$$

where $E_{r,1}$, $E_{r,2}$, ... represent the normal units of a_r; then

$$a_1 f_1(q) + a_2 f_2(q) + \cdots = \sum \alpha_{\mathfrak{a},\mathfrak{b}} E_{\mathfrak{a},\mathfrak{b}} f_{\mathfrak{a}}(q).$$

Now since (according to Part 2 of the Proof) $E_{\mathfrak{a},\mathfrak{b}} f_{\mathfrak{a}}(q)$ vanishes with q, so also (according to Part 1 of the Proof) the multiple sum of these expressions vanishes with q; and thus so also does $a_1 f_1(q) + a_2 f_2(q) + \cdots$. If in addition $f_1(q), f_2(q), \ldots$ are zero for $q = 0$, the same is also true for $a_1 f_1(q) + a_2 f_2(q) + \cdots$, whence, since in addition the last expression vanishes with q, it is now zero with q.

422. *If each of two functions $f_1(q)$ and $f_2(q)$ of a positive numerical magnitude q vanishes (or is zero) with it, then the difference $f_1(q) - f_2(q)$ must also vanish (or be zero) with it.*

PROOF. In 421 {cf. the Corollary to 419}.

423. DEFINITION. If, for a specific value x, $f(x)$ has the property that there is always a constant value c such that $f(x+qdx)-c$ always vanishes with the positive value q, whatever finite magnitude, of the same family as x, may be meant by dx, then I say that the function $f(x)$ *converges to c about x*.

REMARK. Thus for the present dx will mean nothing but an arbitrary finite magnitude of the same family as x. But I have already chosen this notation here, since it is most convenient for the following.

424. *If $f(x)$ converges to c about x, then it cannot simultaneously converge to a value c_1 different from c about x.*

PROOF. For should both simultaneously be the case, then (according to 423) necessarily $f(x + qdx) - c$ and $f(x + qdx) - c_1$ must both vanish with q, and thus (according to 422) also their difference, that is $c - c_1$, which is impossible, since c and c_1 are two different constants.

425. DEFINITION. A function $f(x)$ is continuous in x if $f(x)$ converges about x to that particular value that $f(x)$ has at x.

426. *If $f(x)$ is continuous in x, then for every finite dx {of the same family as x} the difference $f(x + qdx) - f(x)$ vanishes with q {and is also simultaneously zero with q}.*

PROOF. Directly from 425, 423.

REMARK. If $f(x)$ is not continuous in x, then the difference $f(x+qdx)-f(x)$ does not vanish with q for every finite value of dx; rather, $f(x+qdx)$ could converge to different limits for different magnitudes dx, or, if it converges to the same value c for all finite values dx, and thus (according to 423) the function $f(x)$ itself converges *about* x to this value, even so, if $f(x)$ is not continuous in x, it could pass abruptly to a value different from c there.

427. DEFINITION. If $f_1(x)$ is a numerical function and $f(x)$ is an arbitrary function and both are zero for the same value of $x=a$, but so that the quotient $f(x):f_1(x)$ converges to a constant value c, then by the fraction

$$\frac{f(x)}{f_1(x)}$$

I mean that function which otherwise agrees with that quotient, but for $x=a$ assumes the value c, and denote the value c that this fraction assumes for $x=a$, as

$$c=\left[\frac{f(x)}{f_1(x)}\right]_{(x=a)}.$$

REMARK. This definition, just as the previous one, is necessary not only for functions of extensive magnitudes, but also for the ordinary theory of functions.

In fact, whether x is an extensive magnitude or a numerical magnitude, $f(x)$ an extensive function or a numerical function, if the fraction

$$\frac{f(x)}{f_1(x)}$$

is to be treated as a function, then for every specific value of x it must likewise assume a specific value (348). This value is in general found by division of the particular values that $f(x)$ and $f_1(x)$ then assume. But if both $f(x)$ and $f_1(x)$ are zero for $x=a$, the quotient of these two special values is completely undefined; a lack of definition that must be completely removed for the fraction $f(x):f_1(x)$ in case one should subject the fraction to conjunctions that are to hold for the case that $x=a$.

To this end it is possible in this case to fix an arbitrary specific value b for that fraction, except that then this specification, that for $x=a$ it is to assume the specific value b, must be incorporated into the notation for the fraction. This arbitrary definition is superfluous if the concept advanced in the above definition is established, according to which that fraction is set as continuous in x. But this concept assumes that the fraction converges to a definite value c about x. If this is not the case, but rather the fraction

$$\frac{f(a+qb)}{f_1(a+qb)}$$

converges to different limits with the vanishing of the positive numerical magnitude q, according as b assumes other values, for example with numerical magnitudes, if b assumes the values $+1,-1$, or $\cos p+i\sin p$, then the concept given above is not applicable, and nothing remains but to introduce an arbitrary definition, and incorporate it into the notation.

The misunderstanding of all these relations has introduced an unholy confusion into higher analysis, which often enough is betrayed by contradictions and defective results. In order to escape these errors, attempts have been made here and there to improve the method; in particular it is to CAUCHY's credit that, with an inexhaustible wealth of the most genial tricks, he has sought to make the method secure wherever it seemed it could lead into error. But even he could not thereby attain the goal, since he did not grasp the evil at the root, and did not add the essential definitions from whose lack all those confusions result. I have therefore found it requisite, insofar as seems necessary for the following, to append these definitions here, and, instead of being able to depend on earlier works on differential calculus, power series, and integral calculus, I must reconstruct these disciplines from the beginning in order to be able to apply them with confidence to extensive magnitudes as well. It is therefore the more advisable that I limit myself to only what is most necessary.

Still I note here what easily follows from the above remarks, that similar definitions are established for all cases where, for certain values of the variables, the functions to be conjoined go over into such expressions as are subject to no conjunctions (or at least not those by which those functions are bound as such), thus in particular, if one or more of them are infinite or ambiguous. In all these cases the definition can be effected in complete analogy with that just given.

The notation that I introduced above, where after the function I annex the specific value of the variable in parentheses, in order that the whole express the value that the function assumes for this specific value of the variable, may in many other cases as well be applied to advantage, and is in part inescapable.

CHAPTER 2

Differential Calculus

1. Differential of first order

428. DEFINITION. If q denotes a real numerical magnitude, dx an arbitrary finite magnitude *of the same family as* x, then by the *difference of the function* $f(x)$ (with respect to the variable x and the numerical factor q), written $d_{x,q}f(x)$, I mean that function satisfying the equation

$$d_{x,q}f(x) = \frac{f(x+qdx) - f(x)}{q}$$

(where the division by q is to be interpreted in the sense defined in 427).

429. DEFINITION. If the expression $d_{x,q}f(x)$ is continuous in x at $q = 0$ (425) {cf. also 427}, then I denote $d_{x,0}f(x)$ by $d_x f(x)$ and call $d_x f(x)$ the *differential of* $f(x)$ with respect to x, that is I set

$$d_x f(x) = d_{x,0}f(x) = \left[\frac{f(x+qdx) - f(x)}{q}\right]_{(q=0)}.$$

If $d_{x,q}f(x)$ does not have the property that it is continuous in x at $q = 0$, then I say that $d_x f(x)$ is also discontinuous in x.

If in a formula the differential symbol d in front of a function is written without any index, this will mean that the formula holds generally, with respect to whatever variable the differentiation so expressed is taken, that is, whatever index one may append to the d, assuming only that one then appends the same index to every differential symbol d (appearing before a magnitude) in this formula.

REMARK. The concept of the differential can also be established for the case that it is discontinuous, and under certain circumstances valid conjunctions can still be undertaken with such differentials. But in any treatment of differential calculus it is still most appropriate at the beginning to exclude this case completely, and to cover in particular the case where the differential is infinite together with the general treatment of unboundedly increasing functions in a separate section treating the entire analysis of infinities. In the present work however we exclude this treatment, and *in the following we assume that every differential is continuous.*

But I remark that the continuity of $d_x f(x)$ assumes that

$$f(x+qdx) - f(x)$$

is likewise zero about $q = 0$, thus $f(x)$ is also continuous.

430. *If $d_x f(x)$ is continuous and one sets $f(x) = y$, then*

$$\begin{aligned} f(x+qdx) &= f(x) + q(d_x f(x) + N) \\ &= y + q(d_x y(x) + N), \end{aligned}$$

where N is zero simultaneously with the real numerical magnitude q.

PROOF. One sets

$$\frac{f(x+qdx)-f(x)}{q}-d_x f(x)=N.$$

Since $d_x f(x)$ is continuous (by hypothesis), so also (according to 429) the quotient

$$\frac{f(x+qdx)-f(x)}{q}$$

is continuous at $q=0$, and thus $=d_x f(x)$, whence, as the difference of these two expressions, N is zero simultaneously with q. But then we obtain

$$f(x+qdx)=f(x)+q(d_x f(x)+N)=y+q(d_x y+N).$$

431. *If* A *is a constant open expression with* n *{interchangeable} openings (in each term), in which magnitudes of the species of* x *are to be entered, then*

(*a*) $$d_x(\mathsf{A}x^n)=n\mathsf{A}x^{n-1}dx;$$

in particular,

(*b*) $$d_x(\mathsf{A}x)=\mathsf{A}dx,$$

(*c*) $$d_x\mathsf{A}=0.$$

PROOF. Since the usual laws of algebra hold for the products whose factors are to be entered into the openings (according to {the Remark following} 363), it follows, as in algebra, that

$$\mathsf{A}(x+qdx)^n=\mathsf{A}x^n+nq\mathsf{A}x^{n-1}dx+q^2\mathsf{B},$$

where B is a continuous power series in q. From this it follows directly that

$$\frac{\mathsf{A}(x+qdx)^n-\mathsf{A}x^n}{q}=n\mathsf{A}x^{n-1}dx+q\mathsf{B}$$

is continuous at $q=0$, whence (according to 429)

$$d_x(\mathsf{A}x^n)=n\mathsf{A}x^{n-1}dx.$$

From this formulas (*b*) and (*c*) follow for $n=1$ and 0.

432. *If* $u_1, u_2, \ldots$ *are arbitrary functions of an arbitrary variable* x, *then (if* $du_1, du_2, \ldots$ *are continuous)*

$$d(u_1+u_2+\cdots)=du_1+du_2+\cdots.$$

PROOF. Let $u_1=f_1(x)$, $u_2=f_2(x)$, $\ldots$; then (according to 430)

$$f_1(x+qdx)=u_1+q(du_1+N_1),$$

where N_1 is zero simultaneously with q, and thus with every other index. Therefore

$$\sum f_{\mathfrak{a}}(x+qdx)=\sum\{u_{\mathfrak{a}}+q(d_x u_{\mathfrak{a}}+N_{\mathfrak{a}})\}.$$

Consequently

$$\begin{aligned}\frac{\{\sum f_{\mathfrak{a}}(x+qdx)-u_{\mathfrak{a}}\}}{q}&=\sum\{d_x u_{\mathfrak{a}}+N_{\mathfrak{a}}\}\\&=\sum d_x u_{\mathfrak{a}}+\sum N_{\mathfrak{a}}.\end{aligned}$$

Now since N_1, N_2, ... are zero with q, as shown, so also (according to 421) their sum $\sum N_{\mathfrak{a}}$ is zero with q, whence

$$\left[\frac{\{\sum f_{\mathfrak{a}}(x+qdx)-u_{\mathfrak{a}}\}}{q}\right]_{(q=0)}=\sum d_x u_{\mathfrak{a}}.$$

But the left side is

$$d_x\sum f_{\mathfrak{a}}(x)=d_x\sum u_{\mathfrak{a}},$$

whence

$$d_x\sum u_{\mathfrak{a}}=\sum d_x u_{\mathfrak{a}},$$

or, since the formula holds for every index x,

$$d\sum u_{\mathfrak{a}}=\sum du_{\mathfrak{a}}.$$

433. *If y and z are arbitrary functions of x, and $[yz]$ is any product of them, then (assuming dy and dz are continuous),*

$$d[yz]=[dy\cdot z]+[y\cdot dz].$$

PROOF. If $y=f(x)$, $z=F(x)$, then (according to 430)

$$f(x+qdx)=y+q(d_x y+N),$$
$$F(x+qdx)=z+q(d_x z+N'),$$

where N and N' are zero simultaneously with q. Thus

$$\begin{aligned}d_x[yz]&=\left[\frac{[f(x+qdx)\cdot F(x+qdx)]-[yz]}{q}\right]_{(q=0)}\\&=[y(d_x z+N')]+[(d_x y+N)z]\quad\text{for }q=0,\end{aligned}$$

or (according to 429) upon dropping the index

$$d[yz]=[y\cdot dz]+[dy\cdot z]+[yN']+[nz]\quad\text{for }q=0.$$

Now since N and N' are zero with q, (according to 421)

$$[yl]N'+[lz]N,$$

where l denotes an opening in which N {or N'} is to enter, is also zero with q, that is

$$d[yz]=[y\cdot dz]+[dy\cdot z].$$

434. *If y is derivable from its normal units e_1, e_2, ... by the numerical magnitudes y_1, y_2, ... , and y_1, y_2, ... are functions of an arbitrary variable x, then (assuming dy_1, dy_2, ... are continuous),*

$$dy=e_1dy_1+e_2dy_2+\cdots.$$

PROOF. Since

$$y=e_1y_1+e_2y_2+\cdots$$

(by hypothesis), it follows that

$$\begin{aligned}dy&=d(e_1y_1)+d(e_2y_2)+\cdots &&[432]\\&=e_1dy_1+e_2dy_2+\cdots. &&[433,\ 431(c)]\end{aligned}$$

REMARK. With this the differential of an extensive magnitude can be reduced to the differentials of numerical magnitudes.

2. Differential quotient of first order

435. DEFINITION. By $\frac{d}{dx}f(x)$ or $f'(x)$ I mean (assuming $d_x f(x)$ is continuous) the expression which, multiplied by any magnitude dx (which is of the same family as x), yields $d_x f(x)$, that is, which satisfies the equation

$$\frac{d}{dx}f(x)\cdot dx = f'(x)dx = \left[\frac{f(x+qdx)-f(x)}{q}\right]_{(q=0)}.$$

I call $\frac{d}{dx}f(x)$ the *differential quotient of first order of* $f(x)$ with respect to x, and $f'(x)$ the first derived function of $f(x)$.

436. DEFINITION. If one forms the differential quotients of a function $u = f(x, y, \dots)$ of several variables x, y, ... in such a way that in each case one takes the differential quotient with respect to one of these variables, while the remaining variables are treated as constants, then I call the resulting differential quotients the *partial differential quotients* belonging to the assembly of variables x, y, ... , and then denote the differential quotients taken in this sense with respect to x, y, ... by

$$\frac{d}{dx}u, \quad \text{or} \quad \frac{d}{dx}f(x,y,\dots),\dots,$$

{and the corresponding differences and differentials by $d_{x,q}u$ and $d_x u$ or by $d_{x,q}f(x,y,\dots)$ and $d_x f(x,y,\dots 0)$, ... }.

REMARK. For partial differential quotients it is absolutely necessary (as JACOBI has already pointed out in *Crelles Journal* **22**, 321 {*Werke*, Bd. 3, p. 397}) to state the pertinent assembly of variables, and thus not just to name that variable with respect to which the differential quotient is to be taken, but also those which are to be treated as constants in its formation. For if, for example, there is an equation between x and y, the number of variables can be diminished by one; if for example one removes x, there remain only y, z, ... ; and if one now treats these as forming the assembly of variables, $\frac{d}{dy}u$ takes on a completely different meaning, and in general a completely different value, from before.

But it would be very inconvenient if one were to include the whole assembly of variables to which the partial differential quotient belongs in its notation. One forestalls any confusion if one gives the assembly of variables each time, and whenever, in a connected presentation, under differentiation with respect to the same variable, for example x, different magnitudes are to be treated as constant at one time, then at another time, a new, otherwise arbitrary symbol replaces $\frac{d}{dx}$; if one has then provided the meaning of this symbol, no confusion is possible.

The general notation by

$$\frac{d}{dx}u$$

that I have chosen for the partial differential quotient with respect to x, although unusual, scarcely needs justification, since, without being arbitrary, it is extremely convenient, and allows a more unhindered application than any other.

437. *If* e_1, e_2, ... *are the normal units of* $x = x_1e_1 + x_2e_2 + \cdots$, *and* $\delta_1 f(x)$, $\delta_2 f(x)$, ... *are the {partial} differential quotients of* $f(x)$ *with respect to* $x_1, x_2, \dots$

that belong to the assembly of variables $x_1, x_2, \ldots$, *then (assuming* $d_x f(x)$ *is continuous)*

$$d_x f(x) = \delta_1 f(x) \cdot dx_1 + \delta_2 f(x) \cdot dx_2 + \cdots.$$

PROOF. 1. Let the normal units $e_1, e_2, \ldots$ be split into two groups, and let y be numerically derived from one group, z from the other, so that $x = y + z$; then I say that $d_x f(x) = d_y f(x) + d_z f(x)$, where in the differentiations symbolized by d_y and d_z, y and z are thought of as forming the assembly of variables.

Thus let dy be derivable from the same units as y, and dz from the same as z, and let

$$dy + dz = dx = e_1 dx_1 + e_2 dx_2 + \cdots.$$

Now (by hypothesis) $d_x f(x)$ is continuous, that is,

$$\frac{f(x+qdx) - f(x)}{q}$$

is $= 0$ with q and continuous in x for any value of dx (derivable from $e_1, e_2, \ldots$), and thus also, if one substitutes dy for dx; that is

$$\frac{f(x+qdy) - f(x)}{q}$$

is continuous in x at $q = 0$. Furthermore,

$$\frac{f(x+qdy) - f(x)}{q} = \frac{f(y+qdy+z) - f(y+z)}{q} = d_{y,q} f(x),$$

whence $d_y f(x)$ is different from $d_{y,q} f(x)$ by a magnitude N that is zero with q, whence

$$d_y f(x) = \frac{f(x+qdy) - f(x)}{q} + N$$

and likewise

$$d_z f(x) = \frac{f(x+qdz) - f(x)}{q} + N_1,$$

where N and N_1 are zero with q, and the first terms are $= 0$ with q and continuous in x.

Now if a function $\phi(x)$ is continuous in x, then (according to 425) one says that $\phi(x+qdx)$, where dx is an arbitrary magnitude of the same species as x and q denotes a positive number, converges about $q = 0$ to a value that it attains at $q = 0$, that is, $\phi(x+qdx)$ can be represented in the form $\phi(x) + N_2$, where N_2 is zero with q. Thus $\phi(x) = \phi(x+qdx) - N_2$, or, if we should replace dx, which was arbitrary, by the above dz, $\phi(x) = \phi(x+qdz) - N_2$. If we apply this transformation to the first term of $d_y f(x)$, and thus to the function

$$\phi(x) = \frac{f(x+qdy) - f(x)}{q},$$

we obtain

$$d_y f(x) = \frac{f(x+qdz+qdy) - f(x+qdz)}{q} + N - N_2.$$

Here

$$qdz + qdy = q(dz+dy) = qdx,$$

since we set $dy + dz = dx$ above, whence

$$d_y f(x) + d_z f(x) = \frac{f(x+qdx) - f(x+qdz) + f(x+qdz) - f(x)}{q} + N + N_1 - N_2.$$

Here the second and third terms in the numerator cancel, and since (according to 421) $N + N_1 - N_2$ is zero with q, we obtain

$$d_y f(x) + d_z f(x) = \frac{f(x + qdx) - f(x)}{q} + \mathsf{N},$$

where N is zero with q, whence

$$d_y f(x) + d_z f(x) = d_x f(x).$$

2. Now since just as one split x into y and z, one can split y or z, the theorem holds for as many pieces as one can split x into, provided each piece is numerically derivable from a group of units $e_1, e_2, \ldots$, and the different groups do not contain the same units; thus, if $x_1 e_1 = y_1$, $x_2 e_2 = y_2$, $\ldots$ and accordingly $dy_1 = e_1 dx_1$, $dy_2 = e_2 dx_2, \ldots$, then

$$d_x f(x) = d_{y_1} f(x) + d_{y_2} f(x) + \cdots,$$

where the partial differentials denoted by $d_{y_1}, \ldots$ refer to the assembly of variables $y_1, y_2, \ldots$.

3. Now

$$d_{y_1} f(x) = \left[\frac{f(x + qdy_1) - f(x)}{q}\right]_{(q=0)}.$$

But

$$f(x + qdy_1) = f(x + qe_1 dx_1) = f(x_1 e_1 + z + qe_1 dx_1),$$

if the $x_2 e_2 + x_3 e_3 + \cdots$ is denoted z for short, whence

$$f(x + qdy_1) = f(e_1(x_1 + qdx_1) + z),$$

and thus

$$\begin{aligned} dy_1 f(x) &= \left[\frac{f(e_1(x_1 + qdx_1) + z) - f(e_1 x_1 + z)}{q}\right]_{(q=0)} \\ &= dx_1 f(x) = \frac{d}{dx_1} f(x) \cdot dx_1 \qquad \text{[according to 436, \{435\}]} \\ &= \delta_1 f(x) \cdot dx_1, \end{aligned}$$

and similarly for the other indices. If one inserts these values into the equation previously found, one obtains

$$d_x f(x) = \delta_1 f(x) \cdot dx_1 + \delta_2 f(x) \cdot dx_2 + \cdots.$$

438. *If $d_x f(x)$ is continuous, then $\frac{d}{dx} f(x)$ or $f'(x)$ is a quotient independent of dx, and indeed, if $e_1, e_2, \ldots$ are the normal units of $x = e_1 x_1 + e_2 x_2 + \cdots$, then*

$$f'(x) e_r = \frac{d}{dx} f(x) = \delta_r f(x)$$

and

$$f'(x) = \frac{(\delta_1 f(x), \delta_2 f(x), \ldots)}{(e_1, e_2, \ldots)},$$

where $\delta_1, \delta_2, \ldots$ or $\frac{d}{dx_1}, \frac{d}{dx_2}, \ldots$ denote the {partial} differential quotients with respect to $x_1, x_2, \ldots$ belonging to the assembly of variables $x_1, x_2, \ldots$.

PROOF. 1. If x is a numerical magnitude, then (according to 428) dx is also a numerical magnitude, and

$$\frac{d_{xq}F(x)}{dx} = \frac{f(x+qdx)-f(x)}{qdx} = \frac{f(x+q')-f(x)}{q},$$

if one denotes qdx by q'. Further (by hypothesis) $d_x f(x)$ is continuous, and thus, since $dx \neq 0$, so is $\frac{d_x f(x)}{dx}$, and consequently

$$\frac{f(x+q')-f(x)}{q}$$

is also continuous at $q=0$, that is (according to 425), if x is constant, this expression converges about $q'=0$ to a constant value (independent of q') that it attains at $q'=0$; with a variable x this value is therefore a function only of x, independent of q', that is, of qdx. If this function is $\phi(x)$, then

$$d_x f(x) = \phi(x)\cdot dx,$$

whence $\phi(x)$ is the magnitude which, multiplied by dx, yields $d_x f(x)$, that is (according to 435),

$$\phi(x) = \frac{d}{dx}f(x),$$

whence $\frac{d}{dx}f(x)$ is independent of dx.

2. Let $x = x_1e_1 + x_2e_2 + \cdots$; then (according to 437)

$$d_x f(x) = \delta_1 f(x)\cdot dx_1 + \delta_2 f(x)\cdot dx_2 + \cdots,$$

where δ_1, δ_2, ... have the meaning given in that theorem. Now (according to 435) $f'(x)$ is the expression which, multiplied by each

$$dx = e_1dx_1 + e_2dx_2 + \cdots,$$

gives $d_x f(x)$, whence one has

$$f'(x)(e_1dx_1 + e_2dx_2 + \cdots) = \delta_1 f(x)\cdot dx_1 + \delta_2 f(x)\cdot dx_2 + \cdots,$$

and thus

$$f'(x)e_1\cdot dx_1 + f'(x)e_2\cdot dx_2 + \cdots = \delta_1 f(x)\cdot dx_1 + \delta_2 f(x)\cdot dx_2 + \cdots.$$

But according to Part 1 of the Proof and according to 350, the magnitudes $\delta_1 f(x)$, $\delta_2 f(x)$, ... are just functions of x, and thus independent of dx_1, dx_2, ... ; that is, for each value of x the right side of the above equation is a sum of products of the variable numerical magnitudes dx_1, dx_2, ... with magnitudes that are not changed with unchanging x; thus the left side must be of the same form, and the corresponding coefficients must be equal, that is

$$f'(x)e_1 = \delta_1 f(x),\ f'(x) = \delta_2 f(x),\ \ldots.$$

Thus $f'(x)$ is defined as that expression which, multiplied individually by $e_1, e_2, \ldots$, yields the values $\delta_1 f(x)$, $\delta_2 f(x)$, ... respectively, that is (according to 377),

$$f'(x) = \frac{(\delta_1 f(x), \delta_2 f(x), \ldots)}{(e_1, e_2, \ldots)}.$$

REMARK. With this, *differentiation with respect to an extensive magnitude* x is reduced to the partial differential quotients with respect to numerical magnitudes, while in 434 the *differential of the extensive function* was reduced to the differentials of numerical functions, whereby the reduction of the differential quotients of an extensive function with respect to an extensive magnitude to the differential quotients of numerical functions with respect to numerical magnitudes is completed.

439. *If z denotes an arbitrary finite magnitude of the same family as x, and q as before denotes a positive numerical magnitude, then*

$$f(x+qz) = f(x) + q(f'(x)z + N),$$

where N is zero with q (and it is assumed that $d_x f(x)$ is continuous).

PROOF. For every finite magnitude dx of the same family as x, one has (according to 430)

$$f(x+qdx) = f(x) + q(d_x f(x) + N),$$

where N is zero with q. But then $d_x f(x) = f'(x)dx$, whence

$$f(x+qdx) = f(x) + q(f'(x)dx + N);$$

but since by assumption z has the same meaning as dx, we can also replace the one by the other, and thus obtain the required equation.

440. *One also has*

$$df(y) = f'(y) = \frac{d}{dy}f(y) = d_y f(y),$$

if y is again a function of an arbitrary magnitude, to which the differentiation represented by the symbol d refers (assuming that dy and $df(y)$ are continuous).

PROOF. If the differentiation refers to x and $y = \phi(x)$, then (according to 430)

$$(*) \qquad f(x+qdx) = y + q(dy + N),$$

where N is zero with q, and

$$df(y) = df(\phi(x)) = \left[\frac{f[\phi(x+qdx)] - f(y)}{q}\right]_{(q=0)} \qquad [429]$$

$$= \left[\frac{f[y+q(dy+N)] - f(y)}{q}\right]_{(q=0)} \qquad [(*)]$$

$$= \left[\frac{f(y) + q[f'(y)(dy+N) + N'] - f(y)}{q}\right]_{(q=0)}, \qquad [439]$$

where N' is zero with q,

$$= f'(y)(dy+N) + N' \qquad \text{for } q = 0.$$

Now since N and N' are zero with q, (according to 421) $f'(y)N + N'$ is also zero with q, and thus

$$df(y) = f'(y)dy.$$

441. *If x and $y = f(x)$ are derivable from the n original units $e_1, e_2, \ldots$, and*

$$x = x_1e_1 + x_2e_2 + \cdots + x_ne_n,$$
$$y = y_1e_1 + y_2e_2 + \cdots + y_ne_n,$$

and d_xy is continuous, the power of the quotient $\frac{d}{dx}y$ is equal to the ***functional determinant*** *of $y_1, y_2, \ldots$ with respect to $x_1, x_2, \ldots$, that is, it is equal to the determinant formed from the partial differential quotients of the functions $y_1, y_2, \ldots$ with respect to the variables $x_1, x_2, \ldots$, thus*

$$[f'(x)^n] = \left[\left(\frac{d}{dx}y\right)^n\right] = \sum \pm \frac{d}{dx_1}y_1 \cdot \frac{d}{dx_2}y_2 \cdots \frac{d}{dx_n}y_n,$$

where for each index r from 1 to n the symbol $\frac{d}{dx_r}$ denotes the partial differential quotient taken with respect to x_r so that all the rest of the variables $x_1, \ldots, x_n$ (except for x_r) are constants in the differentiation.

PROOF. If the symbol $\frac{d}{dx_r}$ is replaced by δ_r for short, then (according to 438)

$$f'(x) = \frac{(\delta_1 y, \delta_2 y, \ldots, \delta_n y)}{(e_1, e_2, \ldots, e_n)},$$

whence (according to 383) the power

$$[f'(x)^n] = [\delta_1 y \cdot \delta_2 y \cdots \delta_n y].$$

But since $y = e_1y_1 + e_2y_2 + \cdots$, it follows (according to 434) that

$$\delta_1 y = e_1\delta_1 y_1 + e_2\delta_1 y_2 + \cdots,$$

whence

$$\begin{aligned}&[f'(x)^n]\\ &= [(e_1\delta_1y_1 + e_2\delta_1y_2 + \cdots)(e_1\delta_2y_1 + e_2\delta_2y_2 + \cdots)\cdots(e_1\delta_ny_1 + e_2\delta_ny_2 + \cdots)]\\ &= \sum \pm \delta_1y_1 \cdot \delta_2y_2 \cdots \delta_ny_n, \qquad [63]\end{aligned}$$

since (according to 94) $[e_1e_2\cdots e_n]$ is equal to 1.

REMARK. The concept of the functional determinant, as it was first formulated by JACOBI in *Crelles Journal* **22**, p. 319ff {*Werke*, Bd. 3, p. 393ff}, appears here in its true meaning as the power of the derived function, and the theorem proved there follows most easily from this meaning; I therefore leave this proof to the reader.

442. *If u is an arbitrary function of the variable magnitudes $x, y, \ldots$, then*

$$\begin{aligned}du &= \frac{d}{dx}u \cdot dx + \frac{d}{dy}u \cdot dy + \cdots\\ &= d_xu + d_yu + \cdots,\end{aligned}$$

where

$$\frac{d}{dx}, \frac{d}{dy}, \ldots$$

denote the partial differential quotients belonging to the assembly of variables x, $y, \ldots$, and dx, dy, $\ldots$ the partial differentials in the same sense (and the latter are assumed to be continuous).

PROOF. Let x be derivable from its normal units by the variable numerical magnitudes $x_1, x_2, \ldots$, y from its normal units by the variable numerical magnitudes $y_1, y_2, \ldots$, and so on. Now if one forms a new system of normal units $e_1, e_2, \ldots, f_1, f_2, \ldots$ and sets

$$\begin{aligned} v = x_1e_1 + x_2e_2 + \cdots \\ + y_1f_1 + y_2f_2 + \cdots, \end{aligned}$$

then (according to 352) u is a function of the single variable v and (according to 437) one obtains

$$\begin{aligned} d_vu = \frac{d}{dx_1}u \cdot dx_1 + \frac{d}{dx_2}u \cdot dx_2 + \cdots \\ + \frac{d}{dy_1}u \cdot dy_1 + \frac{d}{dy_2}u \cdot dy_2 + \cdots, \end{aligned}$$

where $\frac{d}{dx_1}, \ldots$ refer to the assembly of variables $x_1, x_2, \ldots, y_1, y_2, \ldots$. Since u is a function of v, we can (according to 440) write du instead of d_vu.

Further, if one sets $y, z, \ldots$, that is $y_1, y_2, \ldots, z_1, z_2, \ldots$ to constants, then (according to 437)

$$\frac{d}{dx}u \cdot dx = \frac{d}{dx_1}u \cdot dx_1 + \frac{d}{dx_2}u \cdot dx_2 + \cdots,$$

and likewise

$$\frac{d}{dy}u \cdot dy = \frac{d}{dy_1}u \cdot dy_1 + \frac{d}{dy_2}u \cdot dy_2 + \cdots.$$

Thus

$$du = \frac{d}{dx}u \cdot dx + \frac{d}{dy}u \cdot dy + \cdots.$$

3. Differentials of higher order

443. DEFINITION. If u is an arbitrary function and δ and δ_1 are two arbitrary difference symbols ($d_{x,q}$, $d_{y,q}$) or differential symbols (d_x, d_y), which however refer to the same assembly of variables, *whose differentials are set to constants in each differentiation*, then I mean by $\delta\delta_1 u$ the expression $\delta(\delta_1 u)$, and call $\delta\delta_1$ in this sense a product of difference symbols, and retain this definition when δ_1 is a product of difference symbols, that is I set

$$\begin{aligned} \delta\delta_1 u = \delta(\delta_1 u), \\ \delta\delta_1\delta_2 u = \delta(\delta_1\delta_2 u) = \delta(\delta_1(\delta_2 u)), \end{aligned}$$

and so on, where $\delta, \delta_1, \delta_2, \ldots$ refer to the same assembly of variables whose differentials are set to constants.

444. DEFINITION. If δ is an arbitrary difference symbol ($d_{x,q}$) {or a differential symbol (d_x)} I set

$$\begin{aligned} \delta^0 u = u, \\ \delta^{n+1} u = \delta\delta^n u, \end{aligned}$$

the latter however, only when $n + 1$ is a positive integer.

REMARK. A meaning can also be attributed to negative exponents, which however can only be made clear in integral calculus.

445. *One also has*

(*a*)

$$d_{x,q}d_{y,q}f(x,y) = \frac{f(x+qdx, y+qdy) - f(x+qdx, y) - f(x, y+qdy) + f(x,y)}{q^2},$$

when $f(x,y)$ includes other variables, which however are all to be set constant under the differentiation, and likewise

(*b*)
$$d_{x,q}d_{y,q}u = d_{y,q}d_{x,q}u.$$

PROOF. One has (according to 443)

$$d_{x,q}d_{y,q}f(x,y) = d_{x,q}[d_{y,q}f(x,y)]$$
$$= d_{x,q}\frac{f(x, y+qdy) - f(x,y)}{q} \qquad [436, 428]$$

and on the same grounds this

$$= \frac{f(x+qdx, y+qdy) - f(x+qdx, y) - f(x, y+qdy) + f(x,y)}{q^2}.$$

Thus formula (*a*) is proven. But this formula shows at once that $d_{x,q}d_{y,q}f(x,y)$ yields the same expression as $d_{y,q}d_{x,q}f(x,y)$, whence formula (*b*) is also proven.

REMARK. One could also have set the q belonging to y as different from that belonging to x and denoted it by q_1, say, and the formulas would still hold, a generalization which however is without any particular use.

446. *The order of the sequential difference symbols ($d_{x,q,\ldots}$), among which the differential symbols are included, is of no consequence for the resultant.*

PROOF. For according to 445(*b*) two sequential difference symbols can be interchanged.

447. *If a higher differential is continuous, then the lower differentials by whose progressive differentiations the former results, are also continuous.*

PROOF. Let u be an arbitrary differential, and let $d_x u$ be continuous. In general u will be a function of the variables x, y, ... and their differentials. However, since under differentiation with respect to x all other variables and all the differentials are treated as constants, for this differentiation it suffices to treat u as just a function of x. In this sense let $u = f(x)$; then

$$d_x u = \left[\frac{f(x+qdx) - f(x)}{q}\right]_{(q=0)}.$$

Now since $d_x u$ is continuous, the fraction enclosed in brackets must converge about $q = 0$ to a definite finite limit, which it achieves at $q = 0$, and thus the numerator must also be zero with the denominator (q), that is $f(x+qdx) - f(x)$ must be zero with q, or (according to 425) $f(x)$ is continuous in x, and thus also u.

448. *If* A *denotes an expression with n {interchangeable} openings of the x species, then*

$$d_x^m(\mathsf{A}x^n) = \frac{n!}{(n-m)!}\mathsf{A}x^{n-m}dx^m.$$

PROOF. 1. The theorem holds for $m = 1$ (according to 431).

2. Now if the theorem holds for any given value m, then it also holds for $m+1$; for then

$$d_x^{m+1}(\mathsf{A}x^n) = d_x(d_x^m(\mathsf{A}x^n)) \qquad [444]$$

$$= d_x\left(\frac{n!}{(n-m)!}\mathsf{A}x^{n-m}dx^m\right),$$

since by assumption the theorem holds for the value m,

$$= d_x\left(\frac{n!}{(n-m)!}\mathsf{A}dx^m x^{n-m}\right). \qquad [362]$$

Now since (according to 443) dx is to be treated as constant under the differentiation, and is of the same species as x,

$$\frac{n!}{(n-m)!}\mathsf{A}dx^m$$

is an expression with $n-m$ openings, as a consequence of which we obtain (according to 431) for the last expression found

$$= \frac{(n-m)n!}{(n-m)!}\mathsf{A}dx^m x^{n-m-1}dx$$

$$= \frac{n!}{(n-m-1)!}\mathsf{A}x^{n-m-1}dx^{m+1}, \qquad [362]$$

that is, the theorem also holds if one replaces m by $m+1$; now since (according to Part 1 of the Proof) it holds for $m = 1$, it also holds for $m = 2$, and since it holds for $m = 2$, also for $m = 3$, and thus for all positive values.

449. *If* $e_1, e_2, \ldots$ *are the normal units of* $x = x_1e_1 + x_2e_2 + \cdots$, *and* $\delta_1, \delta_2, \ldots$ *denote the partial differential quotients with respect to* $x_1, x_2, \ldots$ *belonging to the assembly of variables* $x_1, x_2, \ldots$ *, then (assuming* $d_x^n u$ *is continuous)*

$$d_x^n u = \sum\{\delta_{\mathfrak{a}}\delta_{\mathfrak{b}}\cdots u \cdot dx_{\mathfrak{a}}dx_{\mathfrak{b}}\cdots\},$$

where the number of factors $dx_{\mathfrak{a}}, dx_{\mathfrak{b}}, \ldots$ *in each term is* n*, and the sum refers to all positive integer values of* $\mathfrak{a}, \mathfrak{b}, \ldots$ *possible under this condition.*

PROOF. According to 437,

$$d_x u = \sum\{\delta_{\mathfrak{a}}u \cdot dx_{\mathfrak{a}}\}.$$

If one differentiates again with respect to x, then since (according to 443) under this differentiation dx, and thus also $dx_1, dx_2, \ldots$ are to be set to constants, (according to 437)

$$d_x^2 u = \sum\{\delta_{\mathfrak{b}}\delta_{\mathfrak{a}}u \cdot dx_{\mathfrak{a}}dx_{\mathfrak{b}}\} = \sum\{\delta_{\mathfrak{a}}\delta_{\mathfrak{b}}u \cdot dx_{\mathfrak{a}}dx_{\mathfrak{b}}\}, \qquad [446]$$

and so on.

450. DEFINITION. By

$$\frac{d^n}{dx^n}f(x) \quad \text{or} \quad f^{(n)}(x)$$

I mean (assuming that $d_x^n f(x)$ is continuous) the expression which, multiplied by dx^n, yields $d_x^n f(x)$, whatever magnitude, of the same species as x, dx may be.

451. *If $d_x^n f(x)$ is continuous, then $f^{(n)}(x)$ is that expression, always with n {interchangeable} openings in every term, which has the property that*

$$(a) \qquad f^{(n)}(x)(e_r e_s \cdots) = \delta_r \delta_s \cdots f(x),$$

for every series of n indices r, s, ..., where the meaning of e_1, e_2, ..., δ_1, δ_2, ... is the same as in 449.

PROOF. According to 450 it is to be shown that always

$$f^{(n)}(x)dx^n = d_x^n f(x),$$

if $f^{(n)}(x)$ satisfies equation (a). Then

$$\begin{aligned} f^{(n)}(x)dx^n &= f^{(n)}(x)(e_1 dx_1 + e_2 dx_2 + \cdots)^n \\ &= \sum f^{(n)}(x)(e_{\mathfrak{a}} e_{\mathfrak{b}} \cdots) dx_{\mathfrak{a}} dx_{\mathfrak{b}} \cdots, \end{aligned}$$

according to the general polynomial theorem (or alternatively according to 45)

$$\begin{aligned} &= \sum \{\delta_{\mathfrak{a}} \delta_{\mathfrak{b}} \cdots f(x) \cdot dx_{\mathfrak{a}} dx_{\mathfrak{b}} \cdots\} && [(a)] \\ &= d_x^n f(x). && [449] \end{aligned}$$

452. DEFINITION. If x, y, ... are numerical magnitudes and u is a function of them, then if $a + b + \cdots = n$, by

$$\frac{d^n}{dx^a dy^b \cdots} u$$

I mean the expression

$$\frac{d^n}{dx^a dy^b \cdots} u = \frac{d_x^a d_y^b \cdots}{dx^a dy^b \cdots} u,$$

where the differentiations refer to the assembly of variables x, y, ..., and $dx, dy, \ldots$ are assumed to be different from zero.

REMARK. The partial differential quotients with respect to different extensive variables can almost always be dispensed with, since one can always reduce several extensive variables to a single one (according to 352).

453. *If y is again a function of an arbitrary variable, then (assuming the continuity of the relevant differentials)*

$$\frac{d^n f(y)}{n!} = \sum \frac{f^{(\mathfrak{r})}(y)}{\mathfrak{r}!} \left(\frac{d^{\mathfrak{a}} y}{\mathfrak{a}!} \cdot \frac{d^{\mathfrak{b}} y}{\mathfrak{b}!} \cdots \right) \quad (\mathfrak{a} + \mathfrak{b} + \cdots = n).$$

PROOF. As in ordinary analysis.

CHAPTER 3

Infinite Series

1. Infinite series in general

454. DEFINITION. A series

$$u_0 + u_1 + u_2 + \cdots$$

is called *true* {echt} if a positive number $T > 1$ can be found such that u_0, u_1T, $u_2T^2, \ldots$ out to infinity remain finite, that is, that they remain numerically smaller than a certain magnitude M, that is that

$$u_r T^r \text{ num.} < M$$

for every index r.[99]

455. COROLLARY. If we set $1 : T = t$, then we can express the condition for trueness as that two positive numbers, t and M, with $t < 1$, can be found such that

$$u_r : t^r \text{ num.} < M$$

is always satisfied.

456. *Every true series is convergent.*

PROOF. If

$$R = u_0 + u_1 + u_2 + \cdots$$

is a true series, then (according to 455) there is a positive number $t < 1$ such that for each index r, the quotient $u_r : t^r$, which we will denote by a_r, is numerically smaller than a certain finite (positive) number M; then

$$R = a_0 + a_1 t + a_2 t^2 + \cdots ,$$

where a_r num. $< M$.

The remainder ρ_n of this series from the term $a_n t^n$ on is

$$\rho_n = a_n t^n + a_{n+1} t^{n+1} + \cdots .$$

Now

$$a_r \text{ num.} < M, \quad a_r t^r \text{ num.} < M t^r,$$

whence (according to 418)

$$\begin{aligned} \rho_n \text{ num.} &< M t^n + M t^{n+1} + \cdots \\ \text{num.} &< M t^n (1 + t + t^2 + \cdots) \\ \text{num.} &< M \frac{t^n}{1 - t}. \end{aligned}$$

Now here n can be chosen so large that ρ_n is numerically smaller than an arbitrarily given positive magnitude k, and remains so if n increases even more; thus this will be satisfied if

$$n > \log \frac{M}{k(1-t)} : \log\left(\frac{1}{t}\right).$$

Therefore, since the remainder ρ_n converges to zero with increasing n, the series is convergent.

REMARK. In order for the relation between true, false {unecht}, convergent, and divergent series to appear even more clearly, I will touch upon false series.

If all the terms of a *false* series remain finite, that is, remain numerically smaller than a finite positive number M, I will call this series a *transitional series*, whereas if some terms of a series are infinite, that is, if for every positive number M there are terms of the series that are even greater than M, such a series may be called *absurd*. Thus for example the series $t + \frac{1}{2}t^2 + \frac{1}{3}t^3 + \cdots$ is true if the numerical value of the numerical magnitude t is less than 1; it is a transitional series if t is numerically equal to 1, and in fact a divergent transitional series if $t = 1$, convergent if $t = -1$; it is absurd if t num. > 1.

Such an absurd series is always to be rejected. On the other hand the transitional series has in common with the true series that it really expresses the value of the function that is to be represented by the series, whether it is convergent or divergent. In the latter case, *should it approach infinity*, it shows that for this case the function is in fact infinite; thus for example the above series is well known to be the series for $-\log(1-t)$; this function is infinite for $t = 1$, just as is the series $t + \frac{1}{2}t^2 + \frac{1}{3}t^3 + \cdots$, and this therefore also represents the value of that function for this case. Or, in order to choose a simpler example, the series $1 + t + t^2 + \cdots$ is a transitional series for $t = \pm 1$; and for $t = 1$, corresponding to the function

$$\frac{1}{1-t},$$

whose development it represents, it assumes an infinite value. — If on the other hand the divergent transitional series *approaches no infinite value*, but, however far one follows it, it continually oscillates between different values, as for example the series $1+t+t^2+\cdots$ for the value $t = -1$, one can still define its value from the limit to which the series converges if one first sets t as less than 1 and then lets t approach infinitely close to 1. But all these transitional series, even when they converge, can only be used with caution, since the calculational theorems of true series are no longer applicable to them.

{457 and 458 of the original edition have been moved and renumbered 419B and 419C, respectively.}

459. *If several series are true, their multiple sum is also true, that is, if*

$$\begin{aligned} R_1 &= u_1 + u_1' + u_1^{(2)} + u_1^{(3)} + \cdots, \\ R_2 &= u_2 + u_2' + u_2^{(2)} + u_2^{(3)} + \cdots, \\ &\cdots\cdots\cdots\cdots\cdots\cdots \\ R_n &= u_n + u_n' + u_n^{(2)} + u_n^{(3)} + \cdots \end{aligned}$$

are true series, and $\alpha_1, \alpha_2, \ldots, \alpha_n$ *are arbitrary finite numerical magnitudes, then also the series*

$$R = u + u' + u^{(2)} + u^{(3)} + \cdots$$

is a true series, where for each index s

$$u^{(s)} = \alpha_1 u_1^{(s)} + \alpha_2 u_2^{(s)} + \cdots + \alpha_n u_n^{(s)}.$$

PROOF. Since R_1 is a true series, (according to 454) there are two positive numbers T_1 and M, the first of which is > 1, such that for each index s

$$u_1^{(s)} T_1^s \text{ num.} < M_1.$$

Likewise such pairs of positive numbers T_2, M_2, ..., T_n, M_n can be found for the other series R_2, ..., R_n, of which the first member of each pair is > 1, and such that

$$u_2^{(s)} T_2^s \text{ num.} < M_2, \ \dots, \ u_n^{(s)} T_n^s \text{ num.} < M_n.$$

Let T be the smallest of the numbers T_1, ..., T_n; then $T > 1$ still, and thus

$$u_1^{(s)} T^s \text{ num.} < M_1, \ u_2^{(s)} T^s \text{ num.} < M_2, \ \dots, \ u_n^{(s)} T^s \text{ num.} < M_n,$$

and thus also (according to 419C), if β_1, ..., β_n are the numerical values of α_1, ..., α_n,

$$\alpha_1 u_1^{(s)} T^s \text{ num.} < \beta_1 M_1, \ \dots, \ \alpha_1 u_1^{(s)} T^s \text{ num.} < \beta_n M_n;$$

consequently, since the right sides of these inequalities are positive, (according to 418)

$$\alpha_1 u_1^{(s)} T^s + \cdots + \alpha_n u_n^{(s)} T^s \text{ num.} < \beta_1 M_1 + \cdots + \beta_n M_n,$$

that is

$$u^{(s)} T^s \text{ num.} < M,$$

if $\beta_1 M_1 + \cdots + \beta_n M_n$ is denoted by M; and consequently (according to 454) the series R is true.

2. Series as functions of a numerical magnitude

460. *The differential quotients with respect to the numerical magnitude x of a true series*

$$R = a_0 + a_1 x + a_2 x^2 + \cdots = \sum a_{\mathfrak{a}} x^{\mathfrak{a}}$$

are again true series.[100]

PROOF. 1. Since R is a true series, it must be possible (according to 455) to find two positive magnitudes t and M, the first of which is < 1, such that for each r

$$\frac{a_r}{x^r t^r} \text{ num.} < M.$$

Now let τ be a positive number between t and 1, that is $\tau > t$ but < 1; then I show that all the terms of the series

$$\frac{d}{dx} R = \sum \mathfrak{a} a_{\mathfrak{a}} x^{\mathfrak{a}-1}$$

have the property that for each index r out to infinity the expression $r a_r x^{r-1} : \tau^r$ is finite. In fact

$$\frac{r a_r x^{r-1}}{\tau^r} = \frac{r t^r}{x \tau^r} \cdot \frac{a_r x^r}{t^r}.$$

The second factor is (by hypothesis) numerically smaller than M, and thus (according to 419C) the whole expression

$$\text{num.} < \frac{r t^r}{\tau^r} M_1,$$

if for short we denote the numerical value of $M : x$ by M_1. Now let

$$n > \frac{t}{\tau - t},$$

which is always possible, since τ is greater than t, whence $\tau - t$ is not equal to zero. Then

$$n > (n+1)\frac{t}{\tau}$$

or, when we multiply by $t^n : \tau^n$,

$$\frac{nt^n}{\tau^n} > \frac{(n+1)t^{n+1}}{\tau^{n+1}},$$

and on the same grounds

$$\frac{(n+1)t^{n+1}}{\tau^{n+1}} > \frac{(n+2)t^{n+2}}{\tau^{n+2}} > \cdots.$$

But the expressions

$$\frac{t}{\tau}, \frac{2t^2}{\tau^2}, \ldots, \frac{nt^n}{\tau^n},$$

the number of which is finite, and all of which have finite values, are {all} smaller than a certain finite positive number; call this m. Now since, as just proved, for every r greater than n, all expressions $rt^r : \tau^r$ are smaller than $nt^n : \tau^n$, and this latter is $< m$, all those expressions are less than m for every value of r, and thus also

$$\frac{rt^r}{\tau^r} M_1 < mM_1.$$

Here m is a finite magnitude, but so is M_1, unless x is zero, and thus also mM_1 is finite, and therefore $ra_r x^{r-1} : \tau^r$ is also smaller than a finite magnitude, that is, the series $\frac{d}{dx}R$ is true, assuming however that $x \neq 1$. But if $x = 0$, then

$$\frac{d}{dx}R = a_1,$$

and thus is certainly a true series.

2. Now since $\frac{d}{dx}R$ is a true series, so also (according to Part 1 of the Proof) is its differential quotient with respect to x, that is

$$\frac{d^2}{dx^2}R$$

is a true series, and so on.

461. *If a series*

$$a_0 + a_1 x + a_2 x^2 + \cdots$$

is true for any given value x' of the numerical magnitude x, it is also true for every value numerically equal to or smaller than x'.

PROOF. For, if the series is true for $x = x'$, it must (according to 454) be possible to state two positive values T and M, the first of which is > 1, such that for every r

$$a_r x'^r T^r \text{ num.} < M.$$

But then, if x num. $\leq x'$,

$$a_r x^r T^r \text{ num.} \leq a_r x'^r T^r \qquad [419c]$$
$$\text{num.} < M,$$

that is, the series $a_0 + a_1 x + a_2 x^2 + \cdots$ is then also true (according to 454).

REMARK. From this it follows at once (according to 460) that the differential quotients of that series with respect to x are true series for every x numerically equal to or less than x', and thus must also be continuous. Thereby it also follows conversely that, if a function $f(x)$ can be developed in a true series for any given value of x numerically equal to x', then necessarily $f(x)$ and its differential must also be continuous for every value numerically equal to or less than x'. But it remains to investigate whether this condition of continuity is therefore sufficient for $f(x)$ to be developable in a true series.

To this end it is important to consider $f(x)$ for different numerically equal values, namely for a series of such values, each of which results from its predecessor by the same circular evolution. Now CAUCHY has proved that, if $f(x)$ is continuous, the arithmetic mean of all values that $f(x)$ attains, when x progressively undergoes a constant circular evolution until x returns to its original value, is an expression that always converges to a constant limit (independent of x), so long as the angle of the circular evolution is vanishingly small. From this theorem he has very ingeniously derived the condition under which a function $f(x)$ can be developed in a convergent (or more precisely a true) series, to which MOIGNO, *Leçons de Calcul Différentiel*, Tome 1, p. 150ff {Paris 1840} should be compared. The procedure in the following development is essentially the same as that chosen in the work just cited; but here the treatment is generalized, insofar as $f(x)$ is considered as an extensive magnitude, while x itself remains a numerical magnitude.

462. THEOREM AND DEFINITION. *If $f'(x)$ is continuous for every numerical magnitude x whose numerical value lies between the limits a and b, and Θ is an nth root of the absolute unit, specifically*

$$\Theta = \cos\frac{2\pi}{n} + i\sin\frac{2\pi}{n},$$

then with infinitely increasing n the expression

$$\frac{1}{n}\sum f(x\Theta^{\mathfrak{a}}) = \frac{f(x\Theta) + f(x\Theta^2) + \cdots + f(x\Theta^n)}{n}$$

converges to a constant limit (independent of x). Let this constant limit be called the constant term of the function $f(x)$ belonging to that domain of continuity, and denote it by

$$\mathsf{C}[f(x)].$$

PROOF OF THE THEOREM. Since $f'(x)$ is continuous, (according to 439)

$$f(x+q) - f(x) = q(f'(x) + N),$$

where N is zero with q. Now since $\Theta^{\mathfrak{a}}$ is numerically equal to 1, $x\Theta^{\mathfrak{a}}$ is numerically $= x$, and thus also $f'(x\Theta^{\mathfrak{a}})$ is continuous. Now in the above equation one replaces x by $x\Theta^{\mathfrak{a}}$, and $q = x\Theta^{\mathfrak{a}}(\Theta - 1)$; then $x + q$ is transformed into $x\Theta^{\mathfrak{a}+1}$, and, if we append the index $\mathfrak{a}$ to N, we obtain

$$f(x\Theta^{\mathfrak{a}+1}) - f(x\Theta^{\mathfrak{a}}) = x\Theta^{\mathfrak{a}}(\Theta - 1)[f'(x\Theta^{\mathfrak{a}}) + N_{\mathfrak{a}}].$$

Now, if we denote $\frac{d}{dx}$ by δ,

$$\delta f(x\Theta^{\mathfrak{a}}) = \Theta^{\mathfrak{a}} f'(x\Theta^{\mathfrak{a}})$$

(according to 440); thus

$$\frac{f(x\Theta^{\mathfrak{a}+1}) - f(x\Theta^{\mathfrak{a}})}{x(\Theta - 1)} = \delta f(x\Theta^{\mathfrak{a}}) + N'_{\mathfrak{a}},$$

where instead of $\Theta^{\mathfrak{a}} N_{\mathfrak{a}}$, which is numerically equal to $N_{\mathfrak{a}}$ and thus, like it, is zero with q, we have written $N'_{\mathfrak{a}}$.

If we now go over to the arithmetic mean, then, if the following sum from $\mathfrak{a} = 1$ to n is taken,

$$\frac{1}{nx(\Theta - 1)} \sum \{f(x\Theta^{\mathfrak{a}+1}) - f(x\Theta^{\mathfrak{a}})\} = \delta \sum \frac{f(x\Theta^{\mathfrak{a}})}{n} + \frac{1}{n} \sum N'_{\mathfrak{a}}.$$

The left side is zero;[101] for the sum appearing there is equal to

$$f(x\Theta^{n+1}) + f(x\Theta^n) + \cdots + f(x\Theta^2) - f(x\Theta^n) - \cdots - f(x\Theta^2) - f(x\Theta),$$

whence

$$= f(x\Theta^{n+1}) - f(x\Theta) = 0,$$

since $\Theta^n = 1$, and thus $x\Theta^{n+1} = x\Theta$. Consequently we obtain

$$\delta \sum \frac{f(x\Theta^{\mathfrak{a}})}{n} = -\frac{1}{n} \sum N'_{\mathfrak{a}}.$$

But $\sum N'_{\mathfrak{a}} : n$ is the arithmetic mean of the magnitudes N'_1, N'_2, ... , and thus, however large n may be, is numerically smaller than the greatest numerical value of these magnitudes, which we will denote by N'. Now if n becomes infinite,

$$\Theta = \cos \frac{2\pi}{n} + i \sin \frac{2\pi}{n}$$

converges to 1, and thus $\Theta - 1$ to 0, whence $q = x\Theta^{\mathfrak{a}}(\Theta - 1)$, which is numerically equal to $x(\Theta - 1)$, also converges to 0, and thus also N'_1, N'_2, ... , and thus also N'; and finally also

$$\delta \sum \frac{f(x\Theta^{\mathfrak{a}})}{n},$$

since its numerical value is even smaller that that of N'. If we set the limit to which

$$\sum \frac{f(x\Theta^{\mathfrak{a}})}{n}$$

converges $= \phi(x)$, then we have

$$\delta\phi(x) = 0,$$

that is

$$\phi(x) = \text{const.}$$

REMARK. Here I have assumed known the theorem that, if the differential of a function remains zero, the function is constant, in order not to break into the development. The proof of this theorem is recovered at the beginning of the following chapter (on integral calculus), and specifically without referring back to any theorem of the previous chapter in that proof.

463. *The constant term of a multiple sum of functions (whose first derived functions are continuous) is the corresponding multiple sum of the constant terms of the functions, that is (if $f'_1(x)$, $f'_2(x)$, ... are constant)*

$$\mathsf{C}[\alpha_1 f_1(x) + \alpha_2 f_2(x) + \cdots] = \alpha_1 \mathsf{C}[f_1(x)] + \alpha_2 \mathsf{C}[f_2(x)] + \cdots.$$

PROOF. If Θ has the same meaning as in 462, then $\mathsf{C}[f(x)]$ is the limit to which

$$\frac{1}{n} \sum f_1(x\Theta^{\mathfrak{a}})$$

converges with infinite n, that is

$$\frac{1}{n} \sum f_1(x\Theta^{\mathfrak{a}}) - \mathsf{C}[f_1(x)]$$

vanishes with $\frac{1}{n}$, and likewise

$$\frac{1}{n}\sum f_2(x\Theta^{\mathfrak{a}}) - \mathsf{C}[f_2(x)]\,,\ \ldots\,,$$

and therefore (according to 421) also their multiple sum, that is

$$\frac{1}{n}\sum \{\alpha_1 f_1(x\Theta^{\mathfrak{a}}) + \alpha_2 f_2(x\Theta^{\mathfrak{a}}) + \cdots\} - \alpha_1\mathsf{C}[f_1(x)] - \alpha_2\mathsf{C}[f_2(x)] - \cdots\,,$$

that is,

$$\frac{1}{n}\sum \{\alpha_1 f_1(x\Theta^{\mathfrak{a}}) + \alpha_2 f_2(x\Theta^{\mathfrak{a}}) + \cdots\}$$

converges to

$$\alpha_1\mathsf{C}[f_1(x)] + \alpha_2\mathsf{C}[f_2(x)] + \cdots.$$

But the limit to which that expression converges is (according to 462) denoted by

$$\mathsf{C}[\alpha_1 f_1(x) + \alpha_2 f_2(x)] + \cdots],$$

whence

$$\mathsf{C}[\alpha_1 f_1(x) + \alpha_2 f_2(x) + \cdots] = \alpha_1\mathsf{C}[f_1(x)] + \alpha_2\mathsf{C}[f_2(x)] + \cdots.$$

464. *If m is a {positive or negative} whole number, but not zero, then*

$$\mathsf{C}[x^m] = 0.$$

PROOF. $\mathsf{C}[x^m]$ is (according to 462) the limit to which

$$\frac{1}{n}\sum (x\Theta^{\mathfrak{a}})^m$$

converges with infinitely increasing n. But

$$\frac{1}{n}\sum (x\Theta^{\mathfrak{a}})^m = \frac{x^m}{n}\sum \Theta^{\mathfrak{a}m}.$$

If we take n to be so large that m num. $< n$, and set $\sum \Theta^{\mathfrak{a}m} = s$, then

$$s = \Theta^m + \Theta^{2m} + \cdots + \Theta^{nm},$$

$$s = 1 + \Theta^m + \Theta^{2m} + \cdots + \Theta^{(n-1)m},$$

since $\Theta^{nm} = 1$. But the upper expression results from the lower upon multiplication by Θ^m; thus we have

$$s\Theta^m = s, \quad \text{that is} \quad s(1 - \Theta^m) = 0.$$

But (according to 462)

$$\Theta = \cos\frac{2\pi}{n} + i\sin\frac{2\pi}{n},$$

whence

$$\Theta^m = \cos\frac{2m\pi}{n} + i\sin\frac{2m\pi}{n};$$

thus, since $m : n$ is a true fraction, $\Theta^m \neq 1$, whence it follows from the equation $s(1 - \Theta^m) = 0$ that s is equal to zero, and thus also

$$\frac{x^m}{n}\sum \Theta^{\mathfrak{a}m} = 0,$$

that is

$$\frac{1}{n}\sum (x\Theta^{\mathfrak{a}})^m = 0.$$

REMARK. The grounds for the above nomenclature lie in this theorem, since if $f(x)$ is an arbitrary (bounded) power series of x with whole positive or negative exponents and constant term a, $\mathsf{C}[f(x)]$ is equal to the constant term a.

465. *If x num. $> a$, then*

$$\mathsf{C}\left[\frac{x}{x-a}\right] = 1.$$

PROOF. One has

$$\frac{x}{x-a} = 1 + \frac{a}{x} + \frac{a^2}{x^2} + \cdots + \frac{a^{r-1}}{x^{r-1}} + \frac{a^r}{x^{r-1}(x-a)}.$$

Thus (according to 463, {464})

$$\mathsf{C}\left[\frac{x}{x-a}\right] = 1 + \mathsf{C}\left[\frac{a^r}{x^{r-1}(x-a)}\right].$$

Now the last term on the right side is (according to {the method of} 462) numerically smaller than the largest of the expressions that result from

$$\frac{a^r}{x^{r-1}(x-a)}$$

when one replaces x by values numerically equal to x. If A and X are the numerical values of a and x, the greatest of these expressions is

$$= \frac{A^r}{X^{r-1}(X-A)}.$$

Now if p is an arbitrary positive magnitude, then one can always choose r so large that

$$\frac{A^r}{X^{r-1}(X-A)} \text{ num.} < p,$$

and that it remains so as r increases; thus therefore

$$\mathsf{C}\left[\frac{x}{x-a}\right] - 1 \text{ num.} < p,$$

that is, numerically smaller than any positive number, i.e. $= 0$, whence

$$\mathsf{C}\left[\frac{x}{x-a}\right] = 1.$$

466. *If the second derived function of $f(x)$ is continuous for every numerical value x numerically smaller than x', then $f(x)$ can be developed in a true ascending power series in x. And in particular, if z* num. $> x$, *but* num. $< x'$, *and the symbol* C *refers to the variable z, while x is set to a constant, then*

$$f(x) = \mathsf{C}\left[\frac{zf(z)}{z-x}\right] = \sum x^{\mathfrak{a}}\mathsf{C}\left[\frac{f(z)}{x^{\mathfrak{a}}}\right],$$

and if X and Z are respectively the numerical values of x and z, and F the largest numerical value that $f(z)$ assumes for the different values of z that are numerically $= Z$, then every term of the above developed series for $f(x)$ and also the remainder of the series, is numerically smaller than the corresponding term and the corresponding remainder of the series

$$\frac{FZ}{Z-X} = F\sum\frac{X^{\mathfrak{a}}}{Z^{\mathfrak{a}}}.$$

PROOF. First of all, if it is assumed that z is numerically smaller than x', then (by hypothesis) $f''(z)$ is continuous, whence (according to 447) so are $f'(x)$ and $f(x)$. Now let x be regarded as a constant, and z as a variable, and consider the constant term of the function

$$(*) \qquad \phi(z) = \frac{z(f(z) - f(x))}{z - x};$$

thus, investigate the continuity of $q'(z)$.

First, for $z = x$ the expression

$$\frac{(f(z) - f(x))}{z - x} = f'(x) = f'(z)$$

(according to 429, where one has only to set $dx = 1$, and $x + q = z$, {and 435, 438}), whence in this case $\phi(z) = zf'(z)$, and thus in this case $\phi'(z) = f'(z) + zf''(z)$ is also continuous, since $f'(z)$ and $f''(z)$ are.

Further, if $z \neq x$ and thus $z - x \neq 0$, then

$$\phi'(z) = \frac{(f(z) - f(x))}{z - x} + \frac{zf'(x)}{z - x} + \frac{z(f(z) - f(x))}{(z - x)^2}.$$

Now since $f(z)$, $f'(z)$ are continuous and $z - x \neq 0$, $\phi'(z)$ is also continuous in this case; thus $\phi'(z)$ is continuous so long as z num. $< x'$.

Consequently (according to 462) $\mathsf{C}[\phi(z)]$ remains of unchanging value so long as z num. $< x'$, but for $z = 0$ (according to $*$) $\phi(z)$ is likewise zero, thus $\mathsf{C}[\phi(0)] = 0$, whence also for $\mathsf{C}[\phi(z)]$, so that we obtain the equation

$$\mathsf{C}\left[\frac{z(f(z) - f(x))}{(z - x)}\right] = 0.$$

Now if we assume z is numerically $> x$ but still always num. $< x'$, then $z - x \neq 0$ and therefore

$$\frac{zf(z)}{z - x} \quad \text{and} \quad \frac{zf(x)}{z - x}$$

as well as their differentials are continuous, and thus (according to 463) {also}

$$\mathsf{C}\left[\frac{zf(z)}{z - x}\right] - f(x)\mathsf{C}\left[\frac{z}{z - x}\right] = 0.$$

But

$$\mathsf{C}\left[\frac{z}{z - x}\right] = 1$$

(according to 465, where one has only to write z instead of x, and x instead of a), whence one has

$$f(x) = \mathsf{C}\left[\frac{zf(z)}{z - x}\right].$$

Now

$$\frac{z}{z - x} = 1 + \frac{x}{z} + \frac{x^2}{z^2} + \cdots + \frac{x^{r-1}}{z^{r-1}} + \frac{x^r}{z^{r-1}(z - x)},$$

whence (according to 463)

$$f(x) = \mathsf{C}\left[f(z)\right] + x\mathsf{C}\left[\frac{f(z)}{z}\right] + x^2\mathsf{C}\left[\frac{f(z)}{z^2}\right] + \cdots + x^{r-1}\mathsf{C}\left[\frac{f(z)}{z^{r-1}}\right] + \mathsf{C}\left[\frac{x^r f(z)}{z^{r-1}(z - x)}\right].$$

Here (according to {the method of} 462) the last term is numerically smaller than the numerically largest of the expressions that one obtains when in

$$\frac{x^r f(z)}{z^{r-1}(z-x)}$$

one replaces z by all possible values numerically equal to it. The largest of the numerical values that $f(z)$ assumes thereby is denoted by F above, the numerical values of z and x by Z and X; the numerically largest value that $1:(z-x)$ can assume is $1:(Z-X)$, whence

$$\mathsf{C}\left[\frac{x^r f(z)}{z^{r-1}(z-x)}\right] \text{num.} < \frac{X^r F}{Z^{r-1}(Z-X)};$$

and on the same grounds the other terms, starting from the beginning, are numerically smaller than

$$F, \frac{XF}{Z}, \frac{X^2F}{Z^2}, \ldots, \frac{X^rF}{Z^{r-1}(Z-X)}.$$

These are however the corresponding terms and the former the corresponding remainder of the series

$$\frac{FZ}{Z-X} = F\sum\frac{X^{\mathfrak{a}}}{Z^{\mathfrak{a}}}.$$

Now since, finally, this last series is true, so also is the series for $f(x)$, since its terms are numerically even smaller than the terms of this, true, series.

467. TAYLOR'S AND MACLAURIN'S THEOREMS. *If $f''(x)$ is continuous for every x numerically smaller than x', then in the same range*

$$f(x) = f(0) + xf'(0) + \frac{x^2}{2!}f''(0) + \frac{x^3}{3!}f^{(3)}(0) + \cdots = \sum f^{(\mathfrak{a})}(0)\frac{x^{\mathfrak{a}}}{\mathfrak{a}!}.$$

PROOF. For then (according to 466) $f(x)$ can be developed in a series. Let this series be

$$(*) \qquad f(x) = \sum a_{\mathfrak{a}} x^{\mathfrak{a}};$$

then

$$f^{(n)} = \sum \frac{\mathfrak{a}!}{(\mathfrak{a}-n)!} a_{\mathfrak{a}} x^{\mathfrak{a}-n} \qquad [448, \{450\}]$$

whence

$$f^{(n)}(0) = n! a_n,$$

since all the other terms are zero, and thus

$$a_n = \frac{f^{(n)}(0)}{n!}.$$

This substituted into $(*)$ gives the equation to be proved.

REMARK. Since $f(a+x)$ can be treated as a function of x, it is superfluous to separate the theorem into two theorems TAYLOR's and MACLAURIN's).

3. Development of functions of several numerical magnitudes or a single extensive magnitude in series

468. THEOREM AND DEFINITION. (Enlargement of 462.) *If $f(x_1, x_2, \dots)$ is a function of several variable numerical magnitudes x_1, x_2, $\dots$, and the first partial differential quotients*

$$\frac{d}{dx_1} f(x_1, x_2, \dots), \quad \frac{d}{dx_2} f(x_1, x_2, \dots), \ \dots$$

belonging to this assembly are always continuous, provided the numerical value of x_1 lies between the limits a_1 and b_1, that of x_2 between a_2 and b_2, and so forth; and finally if

$$\Theta_1 = \cos\frac{2\pi}{n_1} + i\sin\frac{2\pi}{n_1}, \quad \Theta_2 = \cos\frac{2\pi}{n_2} + i\sin\frac{2\pi}{n_2}, \ \dots,$$

then with infinitely increasing whole numbers n_1, n_2, $\dots$, the expression

$$\frac{1}{n_1 n_2 \cdots} \sum f(x_1\Theta_1^{\mathfrak{a}}, x_2\Theta_2^{\mathfrak{b}}, \dots)$$

converges to a constant limit (independent of x_1, x_2, $\dots$). Let this constant limit be called the **series constant term** *of the function $f(x_1, x_2, \dots)$ belonging to that domain of continuity, and denote it by*

$$\mathsf{C}[f(x_1, x_2, \dots)].$$

Then for two variables

$$\mathsf{C}[f(x_1, x_2)] = \mathsf{C}_1(\mathsf{C}_2[f(x_1, x_2)]),$$

where C_2 refers only to the variable x_2 (x_1 set to a constant) and C_1 refers only to the variable x_1 (x_2 set to a constant); and correspondingly for several variables.

PROOF. 1. For *two* variables. According to the meaning of the summation symbol

$$\frac{1}{n_1 n_2} \sum f(x_1\Theta_1^{\mathfrak{a}}, x_2\Theta_2^{\mathfrak{b}}) = \sum \frac{1}{n_1} \sum \frac{1}{n_2} f(x_1\Theta_1^{\mathfrak{a}}, x_2\Theta_2^{\mathfrak{b}}),$$

where the inner sum refers only to the index $\mathfrak{b}$, the outer only to the index $\mathfrak{a}$. Now if to begin with we let n_2 infinitely increase, the inner sum converges (according to 462) to a limit independent of x_2, which we have to denote by

$$\mathsf{C}_2[f(x_1\Theta_1^{\mathfrak{a}}, x_2)].$$

This limit will therefore be only a function of $x_1\Theta_1^{\mathfrak{a}}$; let it be denoted by $\phi(x_1\Theta_1^{\mathfrak{a}})$; then the limit to which the above expression converges with infinitely increasing n_2 is

$$= \frac{1}{n_1} \sum \phi(x_1\Theta_1^{\mathfrak{a}}).$$

Now if n_1 also increases infinitely, then (according to 462) this expression converges to the limit $\mathsf{C}_1[\phi(x_1)]$, also independent of x_1. Thus the original expression converges to this limit, if in it both n_1 and n_2 infinitely increase; that is,

$$\mathsf{C}[f(x_1, x_2)] = \mathsf{C}_1[\phi(x_1)].$$

But we had set

$$\phi(x_1\Theta_1^{\mathfrak{a}}) = \mathsf{C}_2[f(x_1\Theta_1^{\mathfrak{a}}, x_2)],$$

whence (for $\mathfrak{a} = 0$)

$$\phi(x_1) = \mathsf{C}_2[f(x_1, x_2)];$$

and thus

$$\mathsf{C}[f(x_1, x_2)] = \mathsf{C}_1(\mathsf{C}_2[f(x_1, x_2)]).$$

2. The same series of conclusions can be taken over for *arbitrarily many* variables.

REMARK. It is self-evident that one can also set $n_1 = n_2 = \cdots$, and thus also $\Theta_1 = \Theta_2 = \cdots$, without the theorem ceasing to be correct.

469. (Enlargement of 466.) *If $f(x_1, x_2, \ldots)$ is a function of several variable magnitudes x_1, x_2, ..., and the second partial differential quotients*

$$\frac{d^2}{dx_1^2} f(x_1, x_2, \ldots), \quad \frac{d^2}{dx_2^2} f(x_1, x_2, \ldots), \ \ldots$$

belonging to the assembly of these variables are always continuous so long as simultaneously x_1 is numerically smaller than x_1', x_2 is numerically smaller than x_2', etc., then $f(x_1, x_2, \ldots)$ can be developed in a true series in ascending completely homogeneous functions of x_1, x_2, ... And in particular, if the symbol C refers to the variables z_1, z_2, ..., while x_1, x_2, ... are set to constants, then

$$(a) \qquad f(x_1, x_2, \ldots) = \mathsf{C}\left[f(z_1, z_2, \ldots)\frac{z_1}{z_1 - x_1} \cdot \frac{z_2}{z_2 - x_2} \cdots\right]$$

$$(b) \qquad = \sum x_1^{\mathfrak{a}} x_2^{\mathfrak{b}} \cdots \mathsf{C}\left[\frac{f(z_1, z_2, \ldots)}{z_1^{\mathfrak{a}} z_2^{\mathfrak{b}}}\right];$$

and if X_1, Z_1, X_2, Z_2, ... are the numerical values of x_1, z_1, x_2, z_2, ..., respectively, and F is the greatest of the numerical values that $f(z_1, z_2, \ldots)$ assumes for the different values of z_1, z_2, ... that are numerically $= Z_1$, Z_2, ..., respectively, then every term of the above developed series for $f(x_1, x_2, \ldots)$, as well as every sum of those terms, and in particular the remainder of the series beginning with the homogeneous term of an arbitrary (nth) degree, is numerically smaller than the corresponding term or the corresponding remainder of the series

$$F\frac{Z_1}{Z_1 - X_1} \cdot \frac{Z_2}{Z_2 - X_2} \cdots = \sum\left(F\frac{X_1^{\mathfrak{a}}}{Z_1^{\mathfrak{a}}} \cdot \frac{X_2^{\mathfrak{b}}}{Z_2^{\mathfrak{b}}} \cdots\right).$$

PROOF. 1. For *two* variables. If we first treat x_1 as a constant, then (according to 466)

$$(*) \qquad f(x_1, x_2) = \mathsf{C}_2\left[\frac{z_2 f(x_1, z_2)}{z_2 - x_2}\right].$$

The expression on the right side is still only a function of x_1 and x_2; let this function be denoted by $\phi(x_1, x_2)$; then (according to 466)

$$f(x_1, x_2) = \phi(x_1, x_2) = \mathsf{C}_1\left[\frac{z_1}{z_1 - x_1}\phi(z_1, x_2)\right],$$

where C_1 refers only to the variable z_1. If we now replace $\phi(z_1, x_2)$ by its value that results from the right side of the above equation $(*)$ when one substitutes z_1 for x_1, we obtain

$$f(x_1, x_2) = \mathsf{C}_1\left(\frac{z_1}{z_1 - x_1}\mathsf{C}_2\left[\frac{z_2}{z_2 - x_2} f(z_1, z_2)\right]\right).$$

Since C_2 refers only to the variable z_2, and thus $z_1 : (z_1 - x_1)$ is a constant with respect to C_2, we can (according to 463) also set the symbol C_2 in front of this factor and obtain

$$f(x_1, x_2) = \mathsf{C}_1\left(\mathsf{C}_2\left[\frac{z_1}{z_1 - x_1}\cdot\frac{z_2}{z_2 - x_2}f(z_1, z_2)\right]\right)$$

$$= \mathsf{C}\left[\frac{z_1}{z_1 - x_1}\cdot\frac{z_2}{z_2 - x_2}f(z_1, z_2)\right], \qquad \text{[according to 468]}$$

whence formula (a) is proved.

It is now a question of developing the expression enclosed in brackets, in which we will for short write f for $f(z_1, z_2)$, in an ascending series of completely homogeneous functions of x_1 and x_2, and adding the remainder belonging to it. If we set $u_0, u_1, \ldots, u_{n-1}$ as the first n terms and r_s as the corresponding remainder of this series, then

$$(**) \qquad \frac{z_1}{z_1 - x_1}\cdot\frac{z_2}{z_2 - x_2}f = u_0 + u_1 + \cdots + u_{n-1} + r_n,$$

whence, as is well known,

$$u_0 = f, \quad u_1 = \left(\frac{x_1}{z_1} + \frac{x_2}{z_2}\right)f,$$

and for every index r

$$(***) \qquad u_r = \sum \frac{x_1^{\mathfrak{a}} x_2^{\mathfrak{b}}}{z_1^{\mathfrak{a}} z_2^{\mathfrak{b}}} f \quad (\mathfrak{a} + \mathfrak{b} = r),$$

and

$$r_n = \frac{z_1}{z_1 - x_1}\cdot\frac{z_2}{z_2 - x_2}u_n.$$

Thus then

$$f(x_1, x_2) = u_0 + \mathsf{C}(u_1) + \mathsf{C}(u_2) + \cdots + \mathsf{C}(u_{n-1}) + \mathsf{C}(u_n).$$

Here every term on the right side is (according to {the method of} 462) numerically smaller than the numerically largest of the expressions that one obtains if, in the function enclosed in parentheses, one replaces the variables z_1 and z_2 by all possible values numerically equal to them. The largest of the numerical values that f assumes thereby is denoted by F above, the numerical values of z_1 and z_2, x_1 and x_2 by Z_1 and Z_2, X_1 and X_2. Consequently

$$\frac{x_1^{\mathfrak{a}} x_2^{\mathfrak{b}} f}{z_1^{\mathfrak{a}} z_2^{\mathfrak{b}}} \text{ num.} < \frac{X_1^{\mathfrak{a}} X_2^{\mathfrak{b}} F}{Z_1^{\mathfrak{a}} Z_2^{\mathfrak{b}}};$$

thus, since the expression on the right is positive, in addition (according to 418)

$$\sum \frac{x_1^{\mathfrak{a}} x_2^{\mathfrak{b}} f}{z_1^{\mathfrak{a}} z_2^{\mathfrak{b}}} \text{ num.} < \sum \frac{X_1^{\mathfrak{a}} X_2^{\mathfrak{b}} F}{Z_1^{\mathfrak{a}} Z_2^{\mathfrak{b}}},$$

and thus also (according to $(***)$) for every index r,

$$u_r \text{ num.} < U_r,$$

where U_r denotes what results from u_r when one substitutes X_1, Z_1, X_2, Z_2 for x_1, z_1, x_2, z_2 in it, so that

$$(****) \qquad \frac{Z_1}{Z_1 - X_1}\cdot\frac{Z_2}{Z_2 - X_2}F = U_0 + U_1 + \cdots + U_{n-1} + R_n,$$

where the remainder R_n also results by the same substitutions. This remainder is yet to be analyzed.

One has, as just shown,

$$u_r \text{ num.} < U_r.$$

But furthermore, since among all the values that $z_1 - x_1$ can assume when x_1 and z_1 are replaced by all possible values numerically equal to them, $Z_1 - X_1$ is numerically the smallest,

$$\frac{z_1}{z_1 - x_1} \text{ num.} < \frac{Z_1}{Z_1 - X_1},$$

and on the same grounds

$$\frac{z_2}{z_2 - x_2} \text{ num.} < \frac{Z_2}{Z_2 - X_2}.$$

Thus, since the last two inequalities only include numerical magnitudes, (according to 419C)

$$\frac{z_1}{z_1 - x_1} \cdot \frac{z_2}{z_2 - x_2} u_n \text{ num.} < \frac{Z_1}{Z_1 - X_1} \cdot \frac{Z_2}{Z_2 - X_2} U_n,$$

that is

$$r_n \text{ num.} < R_n.$$

Thus (according to 468)

$$\mathsf{C}(u_r) \text{ num.} < U_r, \quad \mathsf{C}(r_n) \text{ num.} < R_n,$$

that is, every developed term in the series for $f(x_1, x_2)$ and its remainder is numerically smaller than the corresponding term and than the corresponding remainder of the developed series (****). This latter series is however known to be convergent, that is, with infinitely increasing n its remainder R_n converges to zero; thus so also does the remainder $\mathsf{C}(r_n)$, since it is numerically smaller than R_n, that is the developed series for $f(x_1, x_2)$ is also convergent provided the conditions that x_1 num. $< x_1'$ and x_2 num. $< x_2'$ remain satisfied.

But the series for $f(x_1, x_2)$ was, if we drop the remainder, as is done for convergent series,

$$f(x_1, x_2) = u_0 + \mathsf{C}(u_1) + \mathsf{C}(u_2) + \cdots,$$

where

$$\mathsf{C}(u_r) = \mathsf{C}\left[\sum \frac{x_1^{\mathfrak{a}} x_2^{\mathfrak{b}} f}{z_1^{\mathfrak{a}} z_2^{\mathfrak{b}}}\right] \quad (\mathfrak{a} + \mathfrak{b} = r),$$

that is

$$= \sum x_1^{\mathfrak{a}} x_2^{\mathfrak{b}} \mathsf{C}\left[\frac{f(z_1, z_2)}{z_1^{\mathfrak{a}} z_2^{\mathfrak{b}}}\right] \quad (\mathfrak{a} + \mathfrak{b} = r),$$

wherewith formula (b) is proved.

It still remains to show that the series

$$f(x_1, x_2) = u_0 + \mathsf{C}(u_1) + \mathsf{C}(u_2) + \cdots$$

is not just convergent, but also true.

Since x_1, x_2 num. $< x_1'$, x_2', one has $x_1' : x_1$ and $x_2' : x_2$ num. > 1; consequently there must be a positive number T, which is > 1, but numerically smaller than $x_1' : x_1$ and $x_2' : x_2$. Then one has $x_1 T$ num. $< x_1'$ and $x_2 T$ num. $< x_2'$, and consequently {according to 461} the series for $f(x_1, x_2)$ must still remain convergent when one substitutes $x_1 T$ for x_1 and $x_2 T$ for x_2; but then $\mathsf{C}(u_r)$, being a homogeneous function of rth degree in x_1, x_2, is transformed into $T^r\mathsf{C}(u_r)$, and consequently the series

$$u_0 + T\mathsf{C}(u_1) + T^2\mathsf{C}(u_2) + \cdots$$

remains convergent, and thus also its terms out to infinity are finite, whence (according to 454) the series

$$u_0 + \mathsf{C}(u_1) + \mathsf{C}(u_2) + \cdots$$

is true.

2. Part 1 of the Proof is treated throughout so that it can be carried over directly to *arbitrarily many* variables.

470. ***Taylor's*** *theorem* (467) *still holds if x is an arbitrary extensive magnitude; that is, in this case as well* {*for every x numerically smaller than x'*},

$$f(x) = f(0) + xf'(0) + \frac{x^2}{2!}f''(0) + \cdots = \sum \frac{x^{\mathfrak{a}}}{\mathfrak{a}!}f^{(\mathfrak{a})}(0),$$

assuming that $d^2f(x)$ is continuous for every value x numerically smaller than x'.

PROOF. If

$$x = x_1e_1 + x_2e_2 + \cdots ,$$

where e_1, e_2, ... are the normal units of x, then (by hypothesis) $d^2f(x)$ is continuous; but (according to 449)

$$d_x^2f(x) = \delta_1^2f(x)\cdot dx_1^2 + \delta_2^2f(x)\cdot dx_2^2 + \cdots + 2\delta_1\delta_2f(x)\cdot dx_1dx_2 + \cdots ,$$

where δ_1, δ_2, ... are the partial differential quotients belonging to the assembly of variables x_1, x_2, ... This equation holds for every series of values of $dx_1, dx_2, \ldots$, and thus in particular if one sets dx_2, dx_3, ... to zero. But then $d_x^2f(x) = \delta_1^2f(x)\cdot dx_1^2$, whence $\delta_1^2f(x)$ is continuous, and on the same grounds so is $\delta_2^2f(x), \ldots$; thus (according to 469) $f(x)$, as a function of x_1, x_2, ... , can be developed in a true series, whose terms progress in completely homogeneous functions of $x_1, x_2, \ldots$; let

$$f(x) = u_0 + u_1 + u_2 + \cdots$$

be this series, where

$$u_r = \sum a_{\mathfrak{a},\mathfrak{b},\ldots}x_1^{\mathfrak{a}}x_2^{\mathfrak{b}}\cdots \quad (\mathfrak{a} + \mathfrak{b} + \cdots = r).$$

If here we set

$$\sum a_{\mathfrak{a},\mathfrak{b},\ldots}[l|e_1]^{\mathfrak{a}}[l|e_2]^{\mathfrak{b}}\cdots(\mathfrak{a} + \mathfrak{b} + \cdots = r) = a_r,$$

where l denotes an opening fillable by x, then $u_r = a_rx^r$, and thus

$$(*) \qquad f(x) = a_0 + a_1x + a_2x^2 + \cdots .$$

If we set $x = yz$ here, where z is a number, then

$$f(x) = f(yz) = a_0 + a_1y\cdot z + a_2y^2\cdot z^2 + \cdots = \sum\{a_{\mathfrak{a}}y^{\mathfrak{a}}\cdot z^{\mathfrak{a}}\}.$$

Thus (according to 460) the differential quotients of this series with respect to z are likewise true series, and they are therefore

$$\frac{d^n}{dz^n}f(yz) = \sum\left(\frac{\mathfrak{a}!}{(\mathfrak{a}-n)!}a_{\mathfrak{a}}y^{\mathfrak{a}}\cdot z^{\mathfrak{a}-n}\right).$$

But (according to 440)

$$\frac{d}{dz}f(yz) = f'(yz)\frac{d}{dz}(yz) = f'(x)y,$$

and likewise

$$\frac{d^2}{dz^2}f(yz) = f''(x)y^2, \quad \dots, \quad \frac{d^n}{dz^n}f(yz) = f^{(n)}(x)y^n.$$

Thus

$$f^{(n)}(x)y^n = \sum\left(\frac{\mathfrak{a}!}{(\mathfrak{a}-n)!}a_{\mathfrak{a}}y^{\mathfrak{a}} \cdot z^{\mathfrak{a}-n}\right).$$

Now if one sets $z = 0$, then $x = yz = 0$ as well, whence

$$f^{(n)}(0)y^n = n!a_n y^n,$$

since all the other terms on the right side vanish. Consequently, since this equation holds for every value of y, it follows easily from 357 that[102]

$$f^{(n)}(0) = n!a_n, \quad \text{whence} \quad a_n = \frac{f^{(n)}(0)}{n!},$$

which, substituted into equation $(*)$, yields the required formula.

CHAPTER 4

Integral Calculus

1. Integration of differential expressions

471. *If $f(t)$ is a* ***real*** *numerical function of the* ***real*** *numerical magnitude t, and the derived function $f'(t)$ is continuous and positive between $t = t_1$ and $t = t_2$, then $f(t)$ increases with t between the same limits; on the other hand, if $f'(t)$ is continuous and negative, then $f(t)$ decreases as t increases.*

PROOF. One has (according to 439, when one replaces x by t, and sets $z = 1$)

$$f(t+q) = f(t) + q(f'(t) + N),$$

where N is zero with q, whence

$$f(t+q) - f(t) = q(f'(t) + N).$$

Since N is zero with q, for appropriate small values of q, $f'(t) + N$ must have the same sign as $f'(t)$; thus, if q and $f'(t)$ are magnitudes with the same sign, $q(f'(t)+N)$ is positive, and thus also $f(t+q) > f(t)$, that is, $f(t)$ increases with t; but if q and $f'(t)$ are magnitudes of different signs, then $q(f'(t) + N)$ is negative, whence $f(t + N) + N < f(t)$, that is, $f(t)$ decreases if t increases.

472. *If the* ***real*** *numerical function $f(t)$ of the* ***real*** *numerical magnitude t assumes the same value for $t = t_1$ as for $t = t_2$, where $t_2 > t_1$, and $f'(t)$ is continuous for every value of t lying between t_1 and t_2, then necessarily $f'(t) = 0$ for some value t that lies between t_1 and t_2.*

PROOF. If $f'(t)$ were different from zero for every t lying between t_1 and t_2, it would have to be continuously positive or continuously negative. For if $f'(t)$ were positive for some values, negative for others, there would have to be at least one value where $f'(t)$ ceases to be positive and begins to be negative, or conversely; since however $f'(t)$ is continuous (by hypothesis), it must necessarily be zero for this value of t. But if $f'(t)$ were continuously positive, then (according to 471) $f(t_2) > f(t_1)$, which conflicts with the assumption that $f(t_1) = f(t_2)$. If on the other hand $f'(t)$ were continuously negative, then (according to 471) $f(t_2) < f(t_1)$, which likewise conflicts with the assumption. Thus the assumption that $f'(t)$ is different from zero for every t between t_1 and t_2 is impossible, that is, $f'(t)$ is zero for some t lying between t_1 and t_2.

473. *If $f(t)$ is a* ***real*** *numerical function of a* ***real*** *numerical magnitude t, and $f'(t)$ is continuous for every t lying between t_1 and t_2, then for some t lying between these limits, it is necessary that*

$$\frac{f(t_2) - f(t_1)}{t_2 - t_1} = f'(t).$$

PROOF. The function

$$\phi(t) = f(t) - \frac{f(t_2) - f(t_1)}{t_2 - t_1}(t_2 - t_1)$$

assumes the value $f(t_1)$ for $t = t_1$, and the same value $f(t_1)$ for $t = t_2$; now since

$$\phi'(t) = f'(t) - (f(t_2) - f(t_1)) : (t_2 - t_1),$$

$\phi'(t)$ is also continuous between those limits, whence (according to 472) there is a value t, lying between the same limits, for which $\phi'(t) = 0$, that is

$$f'(t) = \frac{f(t_2) - f(t_1)}{t_2 - t_1}.$$

474. *If $f(t)$ is an* ***arbitrary*** *function of the* ***real*** *numerical magnitude t, then as long as $f'(t) = 0$, $f(t)$ is necessarily constant.*

PROOF. 1. Let $f(t)$ be a *real numerical function.* If it is supposed that $f(t)$ has unequal values for two different values t_1 and t_2, that is $f(t_1) \neq f(t_2)$, while $f'(t)$ is zero between t_1 and t_2, then (according to 473) for some t lying between t_1 and t_2 one has

$$f'(t) = \frac{f(t_2) - f(t_1)}{t_2 - t_1} = f'(t),$$

and it is therefore not equal to zero, which conflicts with the assumption; thus the supposition, that $f(t)$ assumes different values for some two values lying within the limits between which $f'(t) = 0$, is impossible, that is, $f(t)$ is constant within these limits.

2. If $f(t)$ is an arbitrary function, and $e_1, e_2, \dots$ are its normal units and $f_1(t)$, $f_2(t), \dots$ the derivation numbers belonging to them, thus

$$f(t) = e_1 f_1(t) + e_2 f_2(t) + \cdots,$$

then (according to 434)

$$df(t) = e_1 df_1(t) + e_2 df_2(t) + \cdots,$$

that is

$$f'(t) = e_1 f_1'(t) + e_2 f_2'(t) + \cdots.$$

Now since it was assumed that $f'(t) = 0$, it follows (according to 28) that

$$f_1'(t) = f_2'(t) = \cdots = 0,$$

whence (according to Part 1 of the Proof, since $f_1(t), \dots$ are numerical functions) $f_1(t), f_2(t), \dots$ are constants, and thus $e_1 f_1(t) + e_2 f_2(t) + \cdots$ is also constant, that is $f(t)$ is a constant.

475. *If $d_x f(x)$ is zero between certain limits for every dx, then $f(x)$ is constant between the same limits.*

PROOF. Let $e_1, e_2, \dots$ be the normal units, and $x_1, x_2, \dots$ the derivation numbers of x belonging to them, whence $x = x_1 e_1 + x_2 e_2 + \cdots$, and let the partial differential quotients with respect to $x_1, x_2, \dots$ belonging to the assembly of variables $x_1, x_2, \dots$ be denoted respectively by $\delta_1, \delta_2, \dots$; then (according to 437)

$$d_x f(x) = \delta_1 f(x) \cdot dx_1 + \delta_2 f(x) \cdot dx_2 + \cdots.$$

Now since (by hypothesis) $d_x f(x)$ is zero for every dx, and thus if $dx_1 \neq 0$, dx_2, dx_3, ... are zero, one has $\delta_1 f(x) = 0$, whence (according to 474) $f(x)$ is independent of x_1, and on the same grounds $f(x)$ is also independent of $x_2, x_3, \ldots$, that is, it is independent of x, and thus constant.

476. *If between certain limits the differentials of the functions $f(x)$ and $\phi(x)$ are continuously equal, and for some value x between those limits the functions themselves are equal to one another, then this equality also obtains for every other value lying between these limits.*

PROOF. Let $F(x) = f(x) - \phi(x)$; then one has

$$dF(x) = df(x) - d\phi(x),$$

whence $dF(x)$ is zero between those limits for which the assumption that $df(x) = \phi(x)$ obtains. Thus (according to 475) between the same limits $F(x)$ is constant, that is, $f(x) - \phi(x) = \text{const}$. Now since, by assumption, $f(x) = \phi(x)$ for some value of x, the above constant is zero, whence for every value x between those limits $f(x) - \phi(x) = 0$, that is, $f(x) = \phi(x)$.

477. DEFINITION. If t is a *positive* number and the function $f(t)$ is continuous between $t = 0$ and $t = t_1$, then by the *integral* of $f(t)$ I mean that function $F(t)$ that is zero with t and whose differential taken with respect to t is equal to $f(t)dt$ for every t lying between those limits. I denote this integral by $d^{-1}f(t)dt$; that is

$$d^{-1}f(t)dt = F(t),$$

if

$$d_t F(t) = f(t)dt \quad \text{and} \quad F(0) = 0.$$

REMARK. The notation chosen provides the advantage over the usual one that it appears as just an enlargement of that holding for differential calculus; but such a new notation seems desirable, since the concept of the integral, as formulated above, is not identical to the ordinary concept. If, with the notation chosen, we insist that the differential to which the integration refers (here dt) is always to be placed at the end of the expression to be integrated, we can dispense with the parentheses that must properly enclose the expression to be integrated. Likewise one does not need to append the integration symbol of the magnitude with respect to which it is to be integrated, since this too is already denoted by the differential placed at the end. But one must then require that one may not replace this differential by an expression equal to it, if one has not previously proven that the integral entails the same value if it is denoted by this new differential.

The change in the concept of the integral as manifested by the above definition consists in that the vagueness that the so-called *general* {*indefinite*} *integral* entails, by virtue of the arbitrarily introduced constant, is removed, since according to the concept as formulated the integral is always taken between two precisely established limits, as 0 is set as the initial limit, and as final limit the value of the variable itself. The general integral as consisting of a magnitude is to be completely banished from mathematics on the same grounds as are all other multivalued magnitudes and conjunctions of magnitudes, that is since no algebraic law maintains general validity for such multivalued expressions. Thus only the definite integral has any scientific justification.

But the notation chosen suffices to label every *definite integral*. Thus, for example, should the integral of $f(z)dz$ be taken between the limits a and $a + b$, one has only to set $z = a + t$ and take $d^{-1}f(a + t)dt$ and after the integration set $t = b$.

Still, I remark that the continuity of the function to be integrated will be assumed throughout the following, even when this condition is not expressly stated.

478. COROLLARY. *If t is a positive numerical magnitude, then*

$$d_t(d^{-1}f(t)dt) = f(t)dt,$$

and

$$[d^{-1}f(t)dt]_{(t=0)} = 0.$$

479. *If $f(0) = 0$, then for every positive numerical magnitude t lying between the limits 0 and t', between which $df(t)$ is continuous,*

$$d^{-1}df(t) = f(t).$$

PROOF. According to 478, if all differentials are taken with respect to t,

$$d(d^{-1}df(t)) = df(t)$$

and

$$[d^{-1}df(t)]_{(t=0)} = 0;$$

thus the two functions $d^{-1}df(t)$ and $f(t)$ have the properties that for every t lying between 0 and t' their differentials are equal, and for $t = 0$ the two functions are equal, that is equal to zero, which we have just proved for $d^{-1}df(t)$, and (by hypothesis) is the case for $f(t)$. Thus (according to 476) the two functions are equal.

480. *One integrates a sum {with respect to a positive numerical magnitude} when one integrates its terms, and a product, one of whose factors is a constant, when one integrates the variable factor and leaves the constant unchanged; or, combining both in a general formula,*

$$d^{-1}\sum a_{\mathfrak{a}}f_{\mathfrak{a}}(t)\cdot dt = \sum a_{\mathfrak{a}}d^{-1}f_{\mathfrak{a}}(t)dt.$$

PROOF. According to 478 the function $d^{-1}f_{\mathfrak{a}}(t)dt$ is zero with t, and thus so is $a_{\mathfrak{a}}d^{-1}f_{\mathfrak{a}}(t)dt$, and therefore {according to 421 so is the sum of these expressions as well; consequently (according to 479)}

$$\begin{aligned}\sum a_{\mathfrak{a}}d^{-1}f_{\mathfrak{a}}(t)dt &= d^{-1}d\sum a_{\mathfrak{a}}d^{-1}f_{\mathfrak{a}}(t)dt\\ &= d^{-1}\sum a_{\mathfrak{a}}dd^{-1}f_{\mathfrak{a}}(t)dt && [432, 433]\\ &= d^{-1}\sum a_{\mathfrak{a}}f_{\mathfrak{a}}(t)dt. && [478]\end{aligned}$$

481. *If $f(t)$ is continuous for every positive numerical magnitude t lying between the limits 0 and t, then the integration of $f(t)dt$ is feasible for each such t.*

PROOF. If $f(t)$ is a real numerical function the proof is known (cf. for example MOIGNO, *Calcul Integral*, p. 1ff {Paris 1844}). But if $f(t)$ is an arbitrary function, and $e_1, e_2, \ldots$ are its normal units, and $f_1(t), f_2(t), \ldots$ its derivation numbers, whence

$$f(t) = e_1f_1(t) + e_2f_2(t) + \cdots,$$

then (according to 480)

$$d^{-1}f(t)dt = e_1d^{-1}f_1(t) + e_2f_2(t) + \cdots.$$

Now since (according to 413) $f_1(t), f_2(t), \ldots$ are real, then, as just shown, the integrations $d^{-1}f_1(t), d^{-1}f_2(t), \ldots$ are feasible, and thus so also is $d^{-1}f(t)$.

482. *If the positive numerical magnitude* $t = \phi(u)$ *is a function of another positive numerical magnitude* u, *and at the same time* $\phi(u)$ *is zero with* u, *then*

$$d^{-1}f(t)dt = d^{-1}f(t)\phi'(u)du.$$

PROOF. Let $d^{-1}f(t)dt = F(t)$, that is, $d_tF(t) = f(t)dt$ and $F(0) = 0$. Now since $d_tF(t) = F'(t)dt$, it follows from the first equation that $F'(t) = f(t)$. Now (according to 440)

$$\begin{aligned} d_uF(t) &= F'(t)d_ut = F'(t)d_u\phi(u) \\ &= F'(t)\phi'(u)du. \end{aligned} \qquad [435]$$

Further, as shown above, $F'(t) = f(t)$ and $F(t) = F(\phi(u))$, whence, since $\phi(u)$ is zero simultaneously with u, $F(t)$ is not just zero with t, but also with u, and thus we have

$$d_uF(\phi(u)) = f(t)\phi'(u)du$$

and

$$F(0) = 0,$$

whence (according to 477)

$$F(\phi(u)) = d^{-1}f(t)\phi'(u)du;$$

but previously we had

$$F(\phi(u)) = F(t) = d^{-1}f(t)dt,$$

whence

$$d^{-1}f(t)dt = d^{-1}f(t)\phi'(u)du.$$

483. DEFINITION. If x is an arbitrary magnitude whose numerical value is t, and $x : t$ is denoted by e (and thus if the numerical value of e is equal to 1 and $x = et$), I set

$$d^{-1}f(x)dx = d^{-1}f(et)edt,$$

where e is a constant for the integration and it is assumed that $f(et)$ is continuous in t, and remains so if t decreases to zero. If such a *function of* x *approaching zero with* x can be found, whose differential with respect to x is $f(x)dx$, we say that in this case $f(x)dx$ is *completely integrable.*[103]

REMARK. Later {cf. No. 486} we will show that whenever there is a function $F(x)$ such that $d_xF(x) = f(x)dx$, and $F(0) = 0$, then for every x that function $F(x) = d^{-1}f(x)dx$, where $d^{-1}f(x)dx$ is interpreted in the sense given above. On the other hand it will be shown that there is not a function $F(t)$ with the properties stated for every $f(x)dx$, whereas on the other hand $d^{-1}f(x)dx = d^{-1}f(et)edt$ can (according to 481) always be found. Thus $d^{-1}f(x)dx$, in the way we defined it above, is to be interpreted as the general, always feasible integral of $f(x)dx$, which only in special cases can be represented as a function of x such that the differential of this function with respect to x is equal to $f(x)dx$.

484. *Instead of integrating a sum, one can integrate the terms individually, that is*

$$d^{-1}(f_1(x) + f_2(x) + \cdots)dx = d^{-1}f_1(x)dx + d^{-1}f_2(x)dx + \cdots,$$

or

$$d^{-1}\sum f_{\mathfrak{a}}(x) \cdot dx = \sum d^{-1}f_{\mathfrak{a}}(x)dx,$$

{*even when the independent variable* x *is an arbitrary* ***extensive*** *magnitude*}.

PROOF. Let $x = et$, where t is a positive numerical magnitude and e is numerically equal to 1; then

$$d^{-1}\sum f_{\mathfrak{a}}(x)\cdot dx = d^{-1}\left\{\sum f_{\mathfrak{a}}(et)\right\} edt \qquad [83]$$

$$= d^{-1}\sum f_{\mathfrak{a}}(et)e\cdot dt \qquad [39]$$

$$= \sum d^{-1} f_{\mathfrak{a}}(et)edt, \qquad [480]$$

that is, since t is a positive numerical magnitude,

$$= \sum d^{-1} f_{\mathfrak{a}}(x)dx. \qquad [483]$$

485. *Instead of integrating a product of two factors, one of which is constant, one can integrate the other factor and leave the constant factor unchanged, that is*

$$d^{-1}af(x)dx = ad^{-1}f(x)dx,$$

where in general $f(x)$ represents an expression with two openings, one of which is to be filled by a, the other by dx. If we denote the former opening by l, the latter by l_1, and instead of a and dx write respectively

$$\frac{a}{l} \quad \text{and} \quad \frac{dx}{l_1},$$

in order to express symbolically that a is to enter into the opening l, and dx into the opening l_1, we can write the above formula more significantly as

$$d^{-1}\frac{a}{l}f(x)\frac{dx}{l_1} = \frac{a}{l}d^{-1}f(x)\frac{dx}{l_1}.$$

PROOF. If we set $x = et$ (in the sense of 483), then

$$\frac{a}{l}d^{-1}f(x)\frac{dx}{l_1} = \frac{a}{l}d^{-1}f(et)\frac{e}{l_1}dt \qquad [483]$$

$$= d^{-1}d\left[\frac{a}{l}d^{-1}f(et)\frac{e}{l_1}dt\right] \qquad [479, \{477\}]$$

$$= d^{-1}\frac{a}{l}dd^{-1}f(et)\frac{e}{l_1}dt \qquad [433][104]$$

$$= d^{-1}\frac{a}{l}f(et)\frac{e}{l_1}dt \qquad [478]$$

$$= d^{-1}\frac{a}{l}f(x)\frac{dx}{l_1}. \qquad [483]$$

REMARK. {It is self-evident that, if one of the magnitudes a or x (and thus also dx) is a numerical magnitude, the opening belonging to it drops out, and thus distinguishing the openings is superfluous. Likewise if the two openings are interchangeable, i.e. if the same resultant always arises whenever, for two arbitrary magnitudes (here a and dx), the one enters the first opening, the other in the second, or conversely, the former in the second, the latter in the first. Yet I observe in addition that throughout the present Part, where an open expression with n openings is spoken of, unless a more specific condition is appended, the n openings are always taken to be interchangeable.}[105]

486. *$f(x)dx$ is completely integrable if and only if the derived function $f'(x)$ is either an expression without openings (that is, x is a real numerical magnitude)*

or an expression with two interchangeable {bound} openings, such that it is unimportant for the resultant in which apportioning two magnitudes enter in the two openings. If this condition is satisfied, and $F(x)$ is the function of x becoming zero with x whose differential is $f(x)dx$, then always

$$F(x) = d^{-1}f(x)dx.$$

PROOF. 1. If there is a function $F(x)$ approaching zero with x such that

$$d_xF(x) = f(x)dx,$$

then $F'(x) = f(x)$ (according to 435), whence $F''(x) = f'(x)$ (according to 450). But $F''(x)$ is (according to 451) an expression with two interchangeable openings {and these openings also remain interchangeable (by virtue of 446) if one interprets them as no longer free, but bound openings}, whence the same also holds for the expression $f'(x)$, which indeed is equal to $F''(x)$. If the openings are of zeroth order (that is, x is a numerical magnitude), the openings can be dropped, and then $f'(x)$ is an expression without openings.

2. Let $x = yt$, where y is numerically equal to 1 and t is a positive number, and let it be assumed that $f'(x)$ is an expression with two interchangeable openings; then (according to 483) one has

$$d^{-1}f(x)dx = d^{-1}f(yt)ydt,$$

where in the integration y is treated as a constant. Let this integral be set equal to $F(y,t)$, that is, let

$$d_tF(y,t) = f(yt)ydt,$$

and $F(y,0) = 0$; then I prove that

$$d_xF(y,t) = f(x)dx.$$

Since y and t are independent of one another and in addition are functions of x (thus if $t = \sqrt{x^2}$ and $y = x : \sqrt{x^2}$), $F(y,t)$ can also be interpreted as a function of x, and, if d_y and d_t are the {partial} differentials with respect to the assembly of the two variables y and t, (according to 442)

$$d_xF(y,t) = d_yF(y,t) + d_tF(y,t) = d_yF(y,t) + f(yt)ydt.$$

Further (according to 446),

$$\begin{aligned} d_t[d_yF(y,t)] &= d_y[d_tF(y,t)] \\ &= d_y[f(yt)\cdot ydt] \\ &= f'(yt)\cdot tdy\cdot ydt + f(yt)dydt, && [433] \end{aligned}$$

{whereby in the expression $f'(yt)$ the opening resulting from differentiation is interpreted as the first, the opening already present in $f(x)$ as the second. Since however $f'(x)$ is by assumption an expression with two *interchangeable* openings, in the first term the factors tdy and ydt can be interchanged, and one finds that the last expression

$$\begin{aligned} &= f'(y)dt\cdot tdy + f(yt)dydt\} \\ &= d_tf(yt)\cdot tdy + f(yt)dtdy && [440] \\ &= d_t[f(yt)\cdot tdy]. && [433] \end{aligned}$$

Now since, as shown above, $F(y, 0) = 0$ for every y, $d_y F(y, 0) = 0$ as well, and likewise $f(yt) \cdot tdy$ is zero with t, whence (according to 476)

$$d_y F(y, t) = f(yt) \cdot tdy.$$

Now if we insert this value in the expression found above for $d_x F(y, t)$, we obtain

$$\begin{aligned} d_x F(y, t) &= f(yt) \cdot tdy + f(yt) \cdot ydt \\ &= f(yt)d(yt) = f(x)dx. \end{aligned}$$

Finally, if we denote even the function $F(y, t)$, interpreted as a function of x, by $F(x)$, then in every case where $f'(x)$ is an expression with two interchangeable openings (within which the case that $f'(x)$ is an expression without openings can be reckoned), we have found *one* function $F(x)$ approaching zero with $x = yt$, whose differential with respect to x is equal to $f(x)dx$; and in fact this function is equal to $d^{-1}f(x)dx$.

3. Aside from the function $F(x) = d^{-1}f(x)dx$, there is *no other function* $\phi(x)$ whose differential taken with respect to x in the same range as that of $F(x)$ is equal to $f(x)dx$, and which approaches zero with x; for, if $d_x F(x) = d_x \phi(x)$, and for some value (here for $x = 0$) $F(x) = \phi(x)$, this equation obtains generally (according to 476). Consequently, so long as $d_x F(x) = f(x)dx$ and $F(0) = 0$, we must have $F(x) = d^{-1}f(x)dx$.

487. *If x is derivable from its normal units $e_1, e_2, \ldots$ by the numbers $x_1, x_2, \ldots$, thus $x = x_1 e_1 + x_2 e_2 + \cdots$, and if, at the same time, $f(x)dx = A_1 dx_1 + A_2 dx_2 + \cdots$, where A_1, A_2, $\ldots$ are functions of x_1, x_2, $\ldots$; then the condition (complete integrability) that $f'(x)$ be an expression with two interchangeable {bound} openings is identical to the the condition that for every two indices r and s,*

$$\delta_r A_s = \delta_s A_r,$$

where δ_1, δ_2, $\ldots$ denote the partial differential quotients with respect to x_1, x_2, $\ldots$ belonging to the assembly of variables x_1, x_2, $\ldots$.

PROOF. Instead of $A_1 dx_1 + A_2 dx_2 + \cdots$ we can, since (according to 142) $[e_r|e_r] = 1$ and $[e_r|e_s] = 0$ if $r \neq s$, write

$$(A_1[l|e_1] + A_2[l|e_2] + \cdots)dx.$$

Thus, since for every dx

$$f(x)dx = (A_1[l|e_1] + A_2[l|e_2] + \cdots)dx,$$

it follows (according to 357) that

$$\{*\} \qquad f(x) = A_1[l|e_1] + A_2[l|e_2] + \cdots = \sum A_{\mathfrak{a}}[l|e_{\mathfrak{a}}].$$

Now (according to 437)

$$d_x f(x) = \sum \{\delta_{\mathfrak{b}} f(x) \cdot dx_{\mathfrak{b}}\},$$

{and thus also, according to {*}, 433, and 431c}

$$d_x f(x) = \sum \{\delta_{\mathfrak{b}} A_{\mathfrak{a}} \cdot [l|e_{\mathfrak{a}}] dx_{\mathfrak{b}}\}.$$

{If on the left side of this equation one introduces the differential quotient $f'(x)$, according to 435, and on the right side applies the previously utilized method once more, one obtains}

$$f'(x)dx = \sum \{\delta_{\mathfrak{b}} A_{\mathfrak{a}} \cdot [l|e_{\mathfrak{a}}][l_1|e_{\mathfrak{b}}]\} \cdot \frac{dx}{l_1},$$

where l_1 is an opening in which dx is to enter. Consequently (according to 357)

$$f'(x) = \sum \{\delta_{\mathfrak{b}} A_{\mathfrak{a}} \cdot [l|e_{\mathfrak{a}}][l_1|e_{\mathfrak{b}}]\}.$$

Now if l and l_1 are *interchangeable* openings, then one has for every two indices r and s,

$$\sum\{\delta_{\mathfrak{b}} A_{\mathfrak{a}} \cdot [e_s|e_{\mathfrak{a}}][e_r|e_{\mathfrak{b}}]\} = \sum\{\delta_{\mathfrak{b}} A_{\mathfrak{a}} \cdot [e_r|e_{\mathfrak{a}}][e_s|e_{\mathfrak{b}}]\}.$$

But since $[e_r|e_s]$ is zero for every two different indices r and s, and equal to 1 for every two equal ones, one obtains

$$\delta_r A_s = \delta_s A_r,$$

and likewise, from this last equation there follows conversely the next to last, which expresses the interchangeability of the openings.

488. *If $F(x)$ is continuous {and completely integrable} between definite limits that include both $x = 0$ and $x = a$, and*

$$F(x) = d^{-1} f(x)dx,$$

then, if $x = a + y$, also

$$F(a + y) - F(a) = d^{-1} f(a + y)dy.$$

PROOF. If $F(x) = d^{-1} f(x)dx$, then $d_x F(x) = f(x)dx$ {according to 483}, that is, $F'(x) = f(x)$; thus

$$\begin{aligned} d_y[F(a + y) - F(y)] &= F'(a + y)dy \qquad \text{[according to 440]} \\ &= f(a + y)dy. \end{aligned}$$

489. *If a denotes an expression with n openings l and one opening l_1, one has*

$$d^{-1} a \left(\frac{x}{l}\right)^n \frac{dx}{l_1} = \frac{1}{n+1} a x^{n+1}.$$

PROOF. Let $x = et$, where t is the numerical value of x, and e is numerically equal to 1; then

$$d^{-1} a \left(\frac{x}{l}\right)^n \frac{dx}{l_1} = d^{-1} a \left(\frac{e}{l}\right)^n \frac{e}{l_1} t^n dt.$$

Let ae^{n+1}, which, since the same magnitude enters in the openings l and l_1 in the expression

$$a \left(\frac{e}{l}\right)^n \frac{e}{l_1},$$

can replace this expression, be denoted by b; then we obtain

$$d^{-1} a \left(\frac{x}{l}\right)^n \frac{dx}{l_1} = d^{-1} b t^n dt = \frac{1}{n+1} b t^{n+1},$$

since the last expression vanishes with t and differentiated with respect to t yields $bt^n dt$, whence

$$= \frac{1}{n+1} ae^{n+1}t^{n+1}$$
$$= \frac{1}{n+1} ax^{n+1}.$$

490. *If the series*

$$\sum a_{\mathfrak{a}} \left(\frac{x}{l}\right)^{\mathfrak{a}} \frac{dx}{l_1},$$

in which $a_{\mathfrak{a}}$ represents an expression with $\mathfrak{a}$ openings l and one opening l_1, is true, then

$$d^{-1} \sum a_{\mathfrak{a}} \left(\frac{x}{l}\right)^{\mathfrak{a}} \frac{dx}{l_1} = \sum \frac{a_{\mathfrak{a}}}{\mathfrak{a}+1} x^{\mathfrak{a}+1}.$$

PROOF. One has (according to 484)

$$d^{-1} \sum a_{\mathfrak{a}} \left(\frac{x}{l}\right)^{\mathfrak{a}} \frac{dx}{l_1} = \sum d^{-1} a_{\mathfrak{a}} \left(\frac{x}{l}\right)^{\mathfrak{a}} \frac{dx}{l_1}$$
$$= \sum \frac{a_{\mathfrak{a}}}{\mathfrak{a}+1} x^{\mathfrak{a}+1}. \qquad \text{[according to 489]}$$

REMARK. By this formula, which only represents the complete integral if the condition of complete integrability (486) is satisfied, the problem of the integration of differential expressions is solved in general. Since we are here concerned only with the representation of the essential features of integral calculus, we will content ourselves with this solution of the problem.

2. Integration of differential equations when the independent variable is a numerical magnitude

491. DEFINITION. To *completely integrate* a given assembly of differential equations (which however may consist of a single equation) means to find all the assemblies of equations that include no more differentials, and each assembly of which has the property that, when it is satisfied, the given assembly is also satisfied; each such assembly is called an *integrating assembly* (of the given assembly). Thus if A is an assembly of differential equations and B is an integrating assembly of it, this means that: 1) B no longer includes differentials and 2) So long as the equations of the assembly B are assumed to be correct, the equations of the assembly A can thereby be shown to be correct.

REMARK. It will be assumed in these paragraphs that all the variables are interpreted as variable numerical magnitudes, on one of which (t) all the others ($x_1, \ldots, x_n$) depend. Should this dependence be as precisely defined by the given differential equations as is in general possible by differential equations, then there must be given as many (n) differential equations independent of one another as there are dependent variables. If t is the independent {variable} numerical magnitude and $x_1, \ldots, x_n$ the dependent numerical magnitudes, we can take a system of n units $e_1, e_2, \ldots, e_n$ and set $x_1e_1 + \cdots + x_ne_n = x$. Since t is assumed to be the independent variable, all the differential quotients appearing in the given differential equations must be taken with respect to t. If these differential quotients ascend to mth order, each of the n equations will have the form that a numerical function of t, x and the differential quotients of x up to mth order, is set equal to zero. Now if $f_1 = 0$, $f_2 = 0$, ... , $f_n = 0$ are these equations, then if one sets $e_1 f_1 + \cdots + e_n f_n = f$, we have one equation

$$f\left(t, x, \frac{d}{dt}x, \ldots, \frac{d^m}{dt^m}x\right) = 0$$

to solve, in which t is a numerical magnitude, whereas x and the function f are numerically derivable from n units.

The solution of this equation thus forms the subject matter of these paragraphs. First we consider the differential equation of first order, that is, the case where $m = 1$.

492. PROBLEM. *To integrate the equation $f(t, x, \delta x) = 0$, in which t is a numerical magnitude, x and $f(t, x, \delta x)$ are magnitudes that are derivable from n units, and δx denotes the differential quotient of x with respect to t; in which it is assumed that $f(t, x, \delta x)$ cannot be derived from less than n units.*

SOLUTION. Let e_1, ... , e_n be the units from which $x = x_1 e_1 + \cdots + x_n e_n$ and $f = e_1 f_1 + \cdots + e_n f_n$ are numerically derived. Let it be assumed that the equations

$$f_1 = 0\,, \ \ldots\,, \ f_n = 0,$$

that are included in $f = 0$, are independent of one another, that is that none of them results necessarily from the rest. For then in the equation $f = 0$, f would be numerically derivable from less than n units, which was excluded above.

Here f_1, ... , f_n are functions of the numerical magnitudes t, x_1, ... , x_n, δx_1, ... , δx_n. From one of the equations $f_1 = 0$, ... , $f_n = 0$ one specifies one of the unknowns δx_1, ... , δx_n and inserts the value found in the remaining equations; and if in general one proceeds in the same way with the equations still remaining, insofar as they include one of the variables δx_1, ... , δx_n, one finally obtains *either* from the last remaining equation the last of those unknowns, and thus one by one all of those unknowns δx_1, ... , δx_n, that is, δx as a function of t and x, or all the magnitudes δx_1, ... , δx_n disappear from the last, or even from the last m equations. In this case there remain m equations that only express relations between t, x_1, ... , x_n, and in fact these equations must all be independent of one another, since otherwise the n original equations would also be dependent on one another. These m equations therefore form a part of the integral equations sought. With them one can express m of the values t, x_1, ... , x_n by the remaining $n - m + 1$, and thereby the first $n - m$ differential equations (by means of which one expressed $n - m$ of the unknowns δx_1, ... , δx_n) to $n - m$ equations, in which, besides t, only $n - m$ of the magnitudes x_1, ... , x_n and the corresponding $n - m$ of the magnitudes δx_1, ... , δx_n appear; and by which these latter can be represented as functions of the former.

Thus it just comes down to the integration of the equation of the form $\delta x = f(t, x)$, that is $dx = f(t, x)dt$. This integration will be considered in the following Numbers.

493. *If*

$$dx = f(t)dt,$$

where t is a numerical magnitude and x is a magnitude derivable from a system of n units, then

$$x = d^{-1}f(t)dt + c,$$

where c is an arbitrary constant (derivable from n units).

PROOF. If $d^{-1}f(t)dt$ is set equal to y, then (according to 478) $dy = f(t)dt$, whence $dx - dy = 0$, that is (according to 432, 433) $d(x - y) = 0$, so that (according to 475) $x-y$ is a constant. Let this constant, which is of the same family as x, and is therefore derivable from n units, be c; then one has $x=y+c=d^{-1}f(t)dt+c$.

494. *If*

$$(a) \qquad \delta x = f(x,t),$$

and throughout one means by δ the total differential quotient with respect to t (with x regarded as dependent on t), and by $\frac{d}{dx}$, $\frac{d}{dt}$, the partial differential quotients with respect to the assembly of variables x, t, the first of which represents an extensive magnitude, the second a numerical magnitude, then

$$(b) \qquad \begin{cases} \delta^2 x = \frac{d}{dt}f + f\frac{d}{dx}f, \\ \delta^3 x = \frac{d}{dt}(\delta^2 x) + f\frac{d}{dx}(\delta^2 x), \\ \dots\dots\dots\dots\dots\dots \\ \delta^{r+1} x = \frac{d}{dt}(\delta^r x) + f\frac{d}{dx}(\delta^r x), \end{cases}$$

and if one sets

$$\delta^{r+1}x = f_r(x,t),$$

then

$$(c) \qquad x = c + f(c,0)t + f_1(c,0)\frac{t^2}{2!} + f_2(c,0)\frac{t^3}{3!} + \cdots,$$

where c is an arbitrary constant, namely the value that x assumes when t is zero, and where it is assumed that the series on the right side is true. From equation (c) one also finds c as a function of x and t, to wit:

$$(d) \qquad c = x - f(x,t)t + f_1(x,t)\frac{t^2}{2!} - f_2(x,t)\frac{t^3}{3!} + \cdots.$$

PROOF. Formulas (b) follow directly from (a), since if ϕ is an arbitrary function of x and t, and x is regarded as dependent on t,

$$d\phi = \frac{d}{dx}\phi \cdot dx + \frac{d}{dt}\phi \cdot dt,$$

whence $\frac{d\phi}{dt}$, that is

$$\delta\phi = \frac{d}{dx}\phi \cdot \delta x + \frac{d}{dt}\phi;$$

but according to (a), $\delta x = f$, whence one obtains

$$\delta\phi = f\frac{d}{dx}\phi + \frac{d}{dt}\phi,$$

from which formulas (b) arise when one replaces ϕ, one by one, with δx, $\delta^2 x$, ... , $\delta^r x$. But then formula (c) follows immediately from TAYLOR's (MACLAURIN's) theorem (470).

If we set $x = F(t)$, we can also represent TAYLOR's theorem in the form

$$F(t+\tau) = x + f(x,t)\tau + f_1(x,t)\frac{\tau^2}{2!} + \cdots,$$

or, if we set $\tau = -t$,

$$F(0) = x - f(x,t)t + f_1(x,t)\frac{t^2}{2!} - f_2(x,t)\frac{t^3}{3!} + \cdots.$$

But $F(0)$ is the value of $x = F(t)$ for $t = 0$, that is, $F(0)$ is equal to c, whence equation (d) is also proved.

REMARK. It is self-evident that the arbitrary constant c is of the same family as x, and therefore includes n numerical constants, if x is derivable from a system of n units. The integrating equation in the form (d) is of particular interest, insofar as in it a function of x and t is set equal to a constant, and indeed that constant that x equals when $t = 0$, to which we will return in the following paragraphs {cf. No. 517}.

In the above we have derived the differential quotients $\delta^2 x$, $\delta^3 x$, ... progressively, each from the one previous to it. It is of interest to seek as well a direct representation of these differential quotients as functions of x and t, which is done in the following theorem, whose easy but somewhat lengthy proof I leave to the reader.

495.[106] *If in the sense of* 494

$$\delta x = f(x,t),$$

then

$$\delta^{r+1} x = \sum \left\{ \alpha_{a\mathfrak{a}b\mathfrak{b}\ldots} f^{\mathfrak{k}} \cdot \frac{d^{a+\mathfrak{a}}}{dt^a dx^{\mathfrak{a}}} f \cdot \frac{d^{b+\mathfrak{b}}}{dt^b dx^{\mathfrak{b}}} f \cdots \right\},$$

where the sum refers to all possible integer, but not negative, values of a, $\mathfrak{a}$, b, $\mathfrak{b}$, ..., $\mathfrak{k}$, *which are subject to the conditions that*

$$\mathfrak{k} = 1 + (\mathfrak{a} - 1) + (\mathfrak{b} - 1) + \cdots,$$

and further

$$a + \mathfrak{a} + b + \mathfrak{b} + \cdots = r,$$

and the sums $a + \mathfrak{a}$, $b + \mathfrak{b}$, ... *are all greater than zero, and where*

$$\alpha_{a\mathfrak{a}b\mathfrak{b}\ldots} = \frac{\mathfrak{a}(\mathfrak{a}+\mathfrak{b}-1)(\mathfrak{a}+\mathfrak{b}+\mathfrak{c}-2)\cdots}{(r-a)(r-a-b-1)(r-a-b-c-2)\cdots} \cdot \frac{r!}{a!\mathfrak{a}!b!\mathfrak{b}!\cdots}.$$

496. PROBLEM. *To integrate the differential equation*

$$f(\delta^m x, \delta^{m-1} x, \ldots, \delta^0 x, t) = 0,$$

where x *as well as* f *is derivable from a system of* n *units* $\{e_1, e_2, \ldots, e_n\}$, t *represents a numerical magnitude,* δ *the total differential quotient with respect to* t (x *regarded as dependent on* t), *and* $\delta^0 x$ *is written instead of* x.

SOLUTION. If one sets

$$\delta^0 x = p_0, \ \delta x = p_1, \ \ldots, \ \delta^{m-1} x = p_{m-1},$$

then $\delta^m x = \delta p_{m-1}$, and one has the m equations

$$\{a\} \qquad \delta p_0 = p_1, \ \delta p_1 = p_2, \ \ldots, \ \delta p_{m-2} = p_{m-1}$$

and the equation

$$f(\delta p_{m-1}, p_{m-1}, \ldots, p_1, p_0, t) = 0.$$

From this last equation one specifies δp_{m-1}; let it be

$$\{b\} \qquad \delta p_{m-1} = \phi(p_0, p_1, \ldots, p_{m-1}, t).$$

Now, in addition to the n units e_1, ..., e_n from which x is derived, one assumes m new units $e^{(0)}$, $e^{(1)}$, ..., $e^{(m-1)}$ and multiplies the above m equations {{a} and {b}} respectively by $e^{(0)}$, $e^{(1)}$, ..., $e^{(m-1)}$ and adds. If further one sets

$$p_0 e^{(0)} + p_1 e^{(1)} + \cdots + p_{m-1} e^{(m-1)} = p$$

and

$$p_1 e^{(0)} + p_2 e^{(1)} + \cdots + p_{m-1} e^{(m-2)} + \phi(p_0, p_1, \ldots, p_{m-1}, t) e^{(m-1)} = F(p,t),$$

then p and F are derivable from the nm units $e^{(r)}e_s$ (where r can take each of the m values from 0 to $m-1$ and s each of the n values from 1 to n), and one obtains the equation

$$\delta p = F(p, t).$$

This equation is to be integrated by the methods of 494, and yields an arbitrary constant derivable from the nm units $(e^{(r)}e_s)$.

REMARK. With this the problem posed in these paragraphs is solved in complete generality with the use of infinite series, which also includes the so-called particular solutions, as these are completely included in the general method of solution given above, as is already evident from the generality of the method applied. Since however those differential equations, which are of first degree with respect to the dependent variable (x) and its differential, and which include the independent numerical variable (t) only in terms in which that variable and its differential do not appear, are integrable in equations of finite form, I will consider this case here.

{No Number 497 exists in the original edition.}

498. *The differential equation*

$$\delta x + \boldsymbol{A}x = 0, \tag{a}$$

in which δ and x have the same meaning as in the previous Number, but $\boldsymbol{A}$ represents a fraction with n denominators (377ff.), *is integrated by the equation*

$$x = \sum \alpha_{\mathfrak{a}} a_{\mathfrak{a}} e^{-m_{\mathfrak{a}} t}, \tag{b}$$

where $m_1, \ldots, m_n$, which we assume are all different from one another, are the n principal numbers of the fraction $\boldsymbol{A}$, and $a_1, \ldots, a_m$ are the principal domains of first order belonging to them ({387}, 388, 389), *where e is the basis of the natural logarithms, and $\alpha_1, \ldots, \alpha_n$ denote arbitrary constant numbers, and the sum refers to $\mathfrak{a} = 1$ to n. The n values $m = m_1, \ldots, m_n$ are specified by the equation of nth degree*

(c)[107]
$$[(m - \boldsymbol{A})^n] = 0$$

and the n magnitudes $a_1, \ldots, a_n$ by the n equations

$$(m_r - \boldsymbol{A})a_r = 0. \tag{d}$$

PROOF. That the n principal numbers $m = m_1, \ldots, m_n$ are the n roots of equation (c) and the n principal domains $a_1, \ldots, a_n$ belonging to them are defined by equations (d) follow immediately from 388, with which the Remark following 383 {and that following 506} is to be compared. The principal domains $a_1, \ldots, a_n$ have (according to 389) the properties that they stand in no numerical relation to one another and that $\boldsymbol{A}a_r = m_r a_r$. Thus x must be numerically derived from $a_1, \ldots, a_n$. Let

$$x = \sum x_{\mathfrak{a}} a_{\mathfrak{a}} \tag{$*$}$$

be the expression for this derivation; then equation (a) is transformed into

$$0 = \sum a_{\mathfrak{a}} \delta x_{\mathfrak{a}} + \sum \{\boldsymbol{A} a_{\mathfrak{a}} x_{\mathfrak{a}}\},$$

whence, since $\boldsymbol{A}a_r = m_r a_r$, we obtain

$$0 = \sum a_{\mathfrak{a}}(\delta x_{\mathfrak{a}} + m_{\mathfrak{a}} x_{\mathfrak{a}}).$$

Since here $\delta x_{\mathfrak{a}} + m_{\mathfrak{a}} x_{\mathfrak{a}}$ is a numerical magnitude, and $a_1, \ldots, a_n$ stand in no numerical relation to one another, one has (according to 28) $\delta x_{\mathfrak{a}} + m_{\mathfrak{a}} x_{\mathfrak{a}} = 0$, that is $dx_{\mathfrak{a}} = -m_{\mathfrak{a}} x_{\mathfrak{a}} dt$, whence

$$x_{\mathfrak{a}} = \alpha_{\mathfrak{a}} e^{-m_{\mathfrak{a}} t},$$

where $\alpha_{\mathfrak{a}}$ is an arbitrary constant number. If we insert this value into the above equation $(*)$, we obtain

$$x = \sum \alpha_{\mathfrak{a}} e^{-m_{\mathfrak{a}} t} a_{\mathfrak{a}}.$$

REMARK. Here the principal numbers of the fraction $\boldsymbol{A}$ are assumed to be different from one another. If some of them are equal, one easily arrives at the result if, in the well-known way, one first sets those among them that are to be equal as infinitesimally different from one another, then develops x according to the above theorem, and finally, after one has removed the infinitely small differences from the denominators, lets these differences vanish completely. Whether the roots are imaginary or not is of no importance for the whole treatment; one can also easily transform the imaginary forms in the final result into real forms.

499. *The differential equation*

$$\delta x + \boldsymbol{A} x = f(t), \tag{a}$$

in which δ, x, $\boldsymbol{A}$, t *have the same meaning as in* 498, *is, if one also gives the magnitudes* $m_1, \ldots, m_n, a_1, \ldots, a_n$ *the same meaning as there, and*

$$f(t) = a_1 f_1 + a_2 f_2 + \cdots + a_n f_n, \tag{b}$$

and in addition sets $d^{-1} f_r e^{m_r t} dt = y_r$, *integrated by the equation*

$$x = \sum (y_{\mathfrak{a}} + \alpha_{\mathfrak{a}}) e^{-m_{\mathfrak{a}} t} a_{\mathfrak{a}}, \tag{c}$$

in which $\alpha_1, \ldots, \alpha_n$ *are arbitrary constant {numerical magnitudes}.*

PROOF. Since (according to 389) $a_1, \ldots, a_n$ stand in no numerical relation to one another, f as well as x can be numerically derived from them (as shown above). Let

$$x = \sum a_{\mathfrak{a}} x_{\mathfrak{a}}; \tag{$*$}$$

then, since (according to 387) $\boldsymbol{A} a_r = m_r a_r$, one has from equation (a)

$$0 = \sum a_{\mathfrak{a}} (\delta x_{\mathfrak{a}} + m_{\mathfrak{a}} a_{\mathfrak{a}} - f_{\mathfrak{a}}),$$

whence (according to 28) $\delta x_{\mathfrak{a}} + m_{\mathfrak{a}} a_{\mathfrak{a}} - f_{\mathfrak{a}} = 0$, where all the magnitudes are numerical magnitudes, that is $dx_{\mathfrak{a}} + m_{\mathfrak{a}} a_{\mathfrak{a}} dt = f_{\mathfrak{a}} dt$. If one sets

$$x_{\mathfrak{a}} = (y_{\mathfrak{a}} + \alpha_{\mathfrak{a}}) e^{-m_{\mathfrak{a}} t}$$

here, where $y_{\mathfrak{a}}$ is a function of t that is zero with t, and $\alpha_{\mathfrak{a}}$ is a constant, then if one inserts this into the previous equation one obtains

$$dy_{\mathfrak{a}} = f_{\mathfrak{a}} e^{m_{\mathfrak{a}} t} dt,$$

whence (according to 477)

$$y_{\mathfrak{a}} = d^{-1} f_{\mathfrak{a}} e^{m_{\mathfrak{a}} t} dt,$$

as above. If one then replaces $x_{\mathfrak{a}}$ by the value found in equation $(*)$, one obtains the equation to be proved.

REMARK. The integration of an equation that includes differential quotients of higher order than the first with respect to t, but otherwise has the form of equations 498 and 499, is reduced by the method of 496 to equations, all of which have the form of equations 498 and 499, except that instead of the n units $e_1, \ldots, e_n$, here, if the differential equation is of mth order, mn units enter.

3. Integration of differential equations when the independent variable is an extensive magnitude

500. *The integration of any arbitrary partial differential equation of first order can be reduced to the integration of a differential equation of the form $X dx = 0$, in which x represents an extensive magnitude, $X dx$ a numerical magnitude.*

PROOF. If $x_1, \ldots, x_n$ are the independent variables and x_0 the variable dependent on them, and $x_0, x_1, \ldots, x_n$ are numerical magnitudes, then every partial differential equation of first order between these magnitudes can be represented in the form of an equation that obtains between the magnitudes

$$x_0, x_1, \ldots, x_n, \frac{d}{dx_1}x_0, \ldots, \frac{d}{dx_n}x_0.$$

If we denote the magnitudes

$$\frac{d}{dx_1}x_0, \ldots, \frac{d}{dx_n}x_0$$

by $p_1, \ldots, p_n$, we can, by means of that equation, represent one of the magnitudes $p_1, \ldots, p_n$, for example p_n, as a function of all the magnitudes $x_0, \ldots, x_1, p_1, \ldots, p_{n-1}$, and therefore cast the partial differential equation to be integrated into the form

$$(*) \qquad p_n = f(x_0, x_1, \ldots, p_1, \ldots, p_{n-1}) = f.$$

Now $dx_0 = p_1 dx_1 + p_2 dx_2 + \cdots + p_n dx_n$. And conversely, if this equation is satisfied, $p_1, \ldots, p_n$ are the partial differential quotients of x_0 with respect to $x_1, \ldots, x_n$. Thus if in this equation one replaces p_n by its value from the foregoing, then, provided the equation

$$(**) \qquad dx_0 = p_1 dx_1 + \cdots + p_{n-1} dx_{n-1} + f dx_n$$

is satisfied, so also is the given equation. Thus every assembly of equations that integrates the latter also satisfies the former and therefore it is only a matter of the integration of the latter. Now if we set $e_0, e_1, \ldots, e_n$ as a system of units and

$$x = x_0 e_0 + x_1 e_1 + \cdots + x_n e_n,$$

whence

$$dx = e_0 dx_0 + e_1 dx_1 + \cdots + e_n dx_n,$$

and in addition, if l represents an opening, we set

$$X = [l|e_0] - p_1[l|e_1] - p_2[l|e_2] - \cdots - p_{n-1}[l|e_{n-1}] - f[l|e_n],$$

then equation $(**)$ is transformed into

$$X dx = 0,$$

to whose integration the problem reduces.

REMARK. The integration of the equation $Xdx = 0$, at which we arrive here, is completely solved by the celebrated method of PFAFF, especially as it has been simplified by JACOBI (*Crelles Journal* **2**, p. 347, and **17**, p. 138 {JACOBI's *Gesammelte Werke*, Bd. 4, pp. 19 and 101}), or more precisely, it is reduced to the integrations treated in the previous paragraphs. The presentation and completion of this method through the application of extensive magnitudes, by which the solving formulas are for the most part presented in an amazing simplicity, will form the main subject of the following development. But we will first extend the theorem formulated to partial differential equations of higher order as well.[107]

501. *The integration of any arbitrary partial differential equation of higher than first order can be reduced to the integration of a differential equation of the form $Xdx = 0$, in which both x and Xdx represent extensive magnitudes.*

PROOF. Let z be the dependent variable and $y_1, \ldots, y_n$ the independent variables, where $z, y_1, \ldots, y_n$ represent numerical magnitudes.

To be able to label the partial differential quotients of higher order conveniently, we first assume a system of n units $e_1, \ldots, e_n$, and set

$$y_1e_1 + y_2e_2 + \cdots + y_ne_n = y;$$

then the different {partial} differential quotients {of z} out to mth order can be expressed by the differential quotients

$$\frac{d}{dy}z, \ \frac{d^2}{dy^2}z, \ \ldots, \ \frac{d^m}{dy^m}z.$$

Here each of these differential quotients represents an expression with as many (mutually interchangeable {bound}) openings as the order of the differential quotient, and in fact is such that, after filling these openings with the *units* of y, the expression yields a numerical expression, and indeed *one of the ordinary (numerical) {partial} differential quotients*; for example $\frac{d^2}{dy^2}z$ always represents an expression with two interchangeable {bound} openings and in fact such that

$$\frac{d^2}{dy^2}z \cdot e_1e_2 = \frac{d^2}{dy_1dy_2}z,$$

and so forth.

Now let

$$\frac{d}{dy}z = p_1, \ \frac{d^2}{dy^2}z = p_2, \ \ldots, \ \frac{d^m}{dy^m}z = p_m;$$

then the partial differential equation in its full generality assumes the form

$$(*) \qquad f(y, z, p_1, p_2, \ldots, p_n) = 0.$$

Of the variables appearing here only z is a numerical magnitude; all the rest are extensive magnitudes, and in fact y, by virtue of the meaning attributed to it, includes n variable numerical magnitudes, and each of the magnitudes $p_1, \ldots, p_m$ as many variable numerical magnitudes as there are combinations with repetition of n elements of the class equaling the index of that magnitude. If the number of variable numerical magnitudes included in all of the variables appearing in the above equation $(*)$ is r, then by virtue of equation $(*)$ one can express one of these variables by the remaining $r - 1$. Thus there remain $r - 1$ variables. Now if one enlarges the system of n units $e_1, \ldots, e_n$ so that it includes $r - 1$ units, and multiplies each of them by one of the $r - 1$ variable numerical magnitudes, and sets the sum of these products $= x$, then x includes all the $r-1$ variable numerical magnitudes.

Now by virtue of the meaning given above for the magnitudes $p_1, \ldots, p_n$ one has in addition

$$(**) \qquad dz = p_1 dy,\ dp_1 = p_2 dy,\ \ldots,\ dp_{n-1} = p_n dy,$$

and if these equations are satisfied, and at the same time, by means of equation $(*)$, one of the r variable numerical magnitudes that are included in equations $(**)$ is expressed by the $(r-1)$ others, the given partial differential equation $(*)$ is thereby satisfied. Consequently it comes down to integrating equations $(**)$.

Of these only the first is a numerical equation; the rest, since dp_1 is of the same species of magnitude as p_1, and so on, each include as many numerical equations as there are variable numerical magnitudes in the magnitudes $p_1, \ldots, p_{n-1}$. Let the number of all the numerical equations included in the above equations $(**)$ be s; then s is smaller than r (that is, equal to the number of variable numerical magnitudes that are included in y and p_m together). If one casts these numerical equations into a form in which the right side is zero, and multiplies them serially by the units $e_1, \ldots, e_n$, they will have the forms

$$e_1 X_1 dx = 0,\ e_2 X_2 dx = 0,\ \ldots,\ e_s X_s dx = 0.$$

Then these equations are equivalent to the single equation

$$e_1 X_1 dx + e_2 X_2 dx + \cdots + e_s X_s dx = 0,$$

that is

$$(e_1 X_1 + e_2 X_2 + \cdots + e_n X_n) dx = 0;$$

thus if one sets

$$e_1 X_1 + e_2 X_2 + \cdots + e_n X_n = X,$$

equation $(**)$ is equivalent to the equation

$$X dx = 0,$$

to whose integration the problem therefore reduces.

REMARK 1. It follows easily that the numbers r and s depend on n and m in such a way that

$$r = n + \frac{(n+1)(n+2)\cdots(n+m)}{1\cdot 2\cdots m}, \qquad s = \frac{(n+1)(n+2)\cdots(n+m-1)}{1\cdots(m-1)},$$

and further that dx, as it appears in the equation $X dx = 0$, does not include the differentials of all r unknowns; rather, the differentials of the variable numerical magnitudes belonging to p_m do not appear in dx. The number of numerical differentials entering in dx is $s+n$.

As an example, let the partial differential equation of second order with two independent variables be chosen. With this, if one changes the notation for the unknowns, one obtains three equations of the form

$$\begin{aligned} dz &= p dx + q dy, \\ dp &= r dx + s dy, \\ dq &= s dx + t dy, \end{aligned}$$

which include the eight variables z, x, y, p, q, r, s, t, one of which can be expressed in terms of the others by means of the given partial differential equation; furthermore, there appear in it five differentials (dx, dy, dz, dp, dq).[108]

REMARK 2. One perceives that the integration of the equation $X dx$ when dx and $X dx$ represent extensive magnitudes is the most general, indeed one may say the *only* problem of integral calculus, since the problems in integral calculus considered in the earlier part can be reduced to it, and in addition, since every numerical magnitude simultaneously appears as a special species of extensive magnitude, the problem previously considered (in 500) is included in it. With the solution of this problem one thus arrives at the aim of integral calculus.

But PFAFF's method is no longer applicable to the case where $X dx$ is also an extensive magnitude, that is where several numerical differential equations arise, and the methods that one applied for the solution of partial differential equations of higher order, and which would also be of use for the solution of this more general problem, have only an extremely limited range. I will therefore fix my attention only on the case where $X dx$ is a numerical magnitude, and only incidentally indicate the more general case.

502. *If the equation*

$$X dx = 0,$$

(in which, ***as generally in the following****, $X dx$ is a numerical magnitude, X is a function of x, and x is numerically derivable from a system of m units $e_1, \ldots, e_n$) is integrated by an assembly of n numerical equations $u_1 = c_1, \ldots, u_n = c_n$, where $c_1, \ldots, c_n$ are arbitrary*[109] *constants, then $X dx$ can be brought to the form*

$$X dx = U_1 du_1 + \cdots + U_n du_n.$$

PROOF. 1. Let $x = x_1 e_1 + \cdots + x_m e_m$; then $u_1, \ldots, u_n$ are interpreted as functions of $x_1, \ldots, x_m$. Now since the equations $u_1 = c_1, \ldots, u_n = c_n$ form an integrating assembly of the equation $X dx = 0$, as it is called (according to 491), the latter must be derivable from those equations; that is if, from the equations $du_1 = 0, \ldots, du_n = 0$, which with respect to the m differentials $dx_1, \ldots, dx_m$ are homogeneous of first degree, n of these latter magnitudes are expressed in terms of the rest, and these expressions are substituted into $X dx$, then $X dx$ must thereby be *identically* equal to zero, or what is the same thing according to a well-known theorem of the theory of equations, it must be possible to find magnitudes $U_1, U_2, \ldots, U_n$ that satisfy the equation

$$X dx = U_1 du_1 + \cdots + U_n du_n.$$

2. (I append a *second proof* in order at the same time to show how to find the magnitudes $U_1, \ldots, U_n$.) The functions $u_1, \ldots, u_n$ can be regarded as independent of one another, since otherwise the given equation would already be integrated by a part of them. But if $u_1, \ldots, u_n$ are mutually independent functions of $x_1, \ldots, x_m$, then n of these latter magnitudes, for example $x_1, \ldots, x_n$, can be represented as functions of the rest and of the magnitudes $u_1, \ldots, u_n$. Let

$$x_1 e_1 + \cdots + x_n e_n = y,$$

and

$$x_{n+1} e_{n+1} + \cdots + x_m e_m = z,$$

and let the relative partial differential quotients of first order with respect to the assembly of variables $u_1, \ldots, u_n, z$ be denoted sequentially by $\delta_1, \ldots, \delta_n, \delta$, and thus so that the last is taken with respect to the extensive magnitude z; then

$$dy = \delta_1 y \cdot du_1 + \cdots + \delta_n y \cdot du_n + \delta y \cdot dz,$$

and thus

$$\begin{aligned} Xdx &= X(dy+dz) \\ &= X\delta_1 y \cdot du_1 + \cdots + X\delta_n y \cdot du_n + (X\delta y + X)dz, \end{aligned}$$

or, if we set

$$X\delta_1 y = U_1\,,\ \ldots\,,\ X\delta_n y = U_n,$$

then

$$Xdx = U_1 du_1 + \cdots + U_n du_n + (X\delta y + X)dz.$$

Now since, if one sets $u_1, \ldots, u_n$ as constants, Xdx must be *identically* zero, and since then $du_1, \ldots, du_n$ are equal to zero, one has $(X\delta y + X)dz = 0$, and thus

$$Xdx = U_1 du_1 + \cdots + U_n du_n.$$

REMARK. This theorem also holds if Xdx is an extensive magnitude, and in particular the second of the above proofs holds directly for this case.

503. *If the expression Xdx (in the sense of 502) can be reduced to n terms, that is to $U_1 du_1 + \cdots + U_n du_n$, but not to less than n such terms, the equation*

$$(a) \qquad Xdx = 0$$

is integrated by assemblies, each of n mutually independent equations, and in fact the following assembly of n equations:

$$(b) \qquad \begin{gathered} u_1 = \phi_1(u_{r+1}, \ldots, u_n),\ \ldots,\ u_r = \phi_r(u_{r+1}, \ldots, u_n), \\ U_1 \tfrac{d}{du_{r+1}}\phi_1 + \cdots + U_r \tfrac{d}{du_{r+1}}\phi_r + U_{r+1} = 0, \\ \cdots\cdots\cdots\cdots\cdots\cdots\cdots\cdots\cdots \\ U_1 \tfrac{d}{du_n}\phi_1 + \cdots + U_r \tfrac{d}{du_n}\phi_r + U_n = 0, \end{gathered}$$

where r can assume each of the values $0, 1, 2, \ldots, n$, and $\phi_1, \ldots, \phi_r$ denote arbitrary functions, forms the complete system of integrating equations. If in particular $r = 0$, one has the assembly

$$(c) \qquad U_1 = U_2 = \cdots = U_n = 0,$$

and if $r = n$, the assembly

$$(d) \qquad u_1 = c_1,\ u_2 = c_2\,,\ \ldots\,,\ u_n = c_n,$$

where $c_1, \ldots, c_n$ are arbitrary constants.[110]

PROOF. The equation $Xdx = 0$ cannot be integrated by an assembly of less than n numerical magnitudes, since otherwise (according to 502) Xdx could be reduced to less than n terms of the form Udu, which conflicts with the assumption. It thus comes down to finding an assembly of n equations that integrate the equation $Xdx = 0$, and indeed all the possible assemblies of this type.

If the n mutually independent equations $v_1 = 0, v_2 = 0, \ldots, v_n = 0$ form one of the integrating assemblies of the equation $Xdx = 0$, then the functions $v_1, \ldots, v_n$, which were originally interpreted as functions of the m variables $x_1, \ldots, x_m$, can at the same time be represented as functions of $u_1, \ldots, u_n$ and $m - n$ of the magnitudes $x_1, \ldots, x_m$, for example as functions of $u_1, \ldots, u_n, x_{n+1}, \ldots, x_m$. If all those functions $v_1, \ldots, v_n$ only include $u_1, \ldots, u_n$, but are independent of $x_{n+1}, \ldots, x_n$, the $u_1, \ldots, u_n$ are found to be constant, and special case (d) applies. But if at least one of the functions $v_1, \ldots, v_n$, for example v_n, includes at least one of the variables $x_{n+1}, \ldots, x_m$, for example x_m, then, by means of the equation $v_n = 0$, x_m can be expressed by the rest $(u_1, \ldots, u_n, x_{n+1}, \ldots, x_{m-1})$. If one inserts this expression into the remaining equations $v_1 = 0, \ldots, v_{n-1} = 0$, it is possible that in the equations thus obtained at least one of the variables $x_{n+1}, \ldots, x_{m-1}$ still appears, for example x_{m-1} in $v_{n-1} = 0$; in this case one expresses x_{m-1} in terms of the yet remaining variables $u_1, \ldots, u_n, x_{n+1}, \ldots, x_{m-2}$ by means of this last equation, and inserts this expression into the remaining equations; and thus one proceeds until finally either all the equations $v_1 = 0, \ldots, v_n = 0$ are exhausted, or only such equations are left over as include only the $u_1, \ldots, u_n$.

In the former case (according to 491) the expression

$$U_1 du_1 + \cdots + U_n du_n$$

must be identically zero, whence $U_1 = 0, \ldots, U_n = 0$, which yields special case (c). In the latter case let there remain r equations $v_1 = 0, \ldots, v_r = 0$ finally left over that only include the variables $u_1, \ldots, u_n$; then by means of these equations we can represent r of the magnitudes $u_1, \ldots, u_n$ as functions of the rest, for example $u_1, \ldots, u_r$ as functions of $u_{r+1}, \ldots, u_n$. In this case the assembly will always be an integrating assembly, only if, whatever type of functions those r functions may be, by them the expression $U_1 du_1 + \cdots + U_n du_n$ is made identically zero. Since we have represented $u_1, \ldots, u_r$ as functions of $u_{r+1}, \ldots, u_n$, we obtain

$$U_1 \tfrac{d}{du_{r+1}} \phi_1 + \cdots + U_r \tfrac{d}{du_{r+1}} \phi_r + U_{r+1} = 0,$$
$$\cdots\cdots\cdots\cdots\cdots\cdots\cdots\cdots\cdots\cdots$$
$$U_1 \tfrac{d}{du_n} \phi_1 + \cdots + U_r \tfrac{d}{du_n} \phi_r + U_n = 0,$$

as the necessary, and also the sufficient conditions for $U_1 du_1 + \cdots + U_n du_n$ to be identically equal to zero in this case.

Consequently the assembly (b) (of which (c) and (d) only represent special cases) has been given as the most general of all possible assemblies that integrate the equation $X dx = 0$ under the assumptions of the theorem.

REMARK. One part of this theorem is proved in the article of JACOBI already referred to (*Crelles Journal* **2**, p. 348 {*Gesammelte Werke*, Bd. 4, p. 20}). But the most important part of our theorem, that is that other than the assembly of equations (b) there is no other assembly of integrating equations, is not proved there. In addition it is not shown there in what the different character of the integrating assembly (b) consists for every value of the index r, although JACOBI frequently refers to this difference of character. Obviously JACOBI was perfectly familiar with this aspect of our theorem, and it was only due to the particular aim of the article that he said nothing further about it.

One perceives from this and the previous theorem that it is the same thing to say that $X dx = 0$ can be integrated by assemblies of any n (and not less than n) equations, or to say that $X dx$ (in the sense of 502) can be reduced to a sum of n (and not less than n) terms of the form $U du$. It is therefore a matter of discovering the conditions under which this reduction is possible, and, if these conditions are satisfied, to give the method for the reduction. In order to attain both these aims easily it is appropriate to formulate the following definitions and theorems on open expressions {with *bound openings*}.

But I observe that the above theorem no longer holds if $X dx$ is an extensive magnitude; and that, if the reduction of $X dx$ is carried out to the smallest possible number of terms of the form $U du$, the integration of the equation $X dx = 0$ is still by no means accomplished. But from 502 it follows that, if one knows all the possible reductions of this type, one has at the same time completely integrated the equation $X dx = 0$. And this consideration seems to me to indicate the way, one can hope, to proceed to this final goal of integral calculus.

504. DEFINITION. If $\boldsymbol{L}$ represents an expression with n {bound} openings of the same species, for which however no interchangeability of openings is assumed, then by

$$[\boldsymbol{L}a_1a_2\cdots a_n]$$

I mean the expression that results when one lets $a_1, \ldots, a_n$ enter into the openings of $\boldsymbol{L}$ in all possible orderings, the expressions obtained being prefixed by the symbol $+$ or $-$, according as the combinatorial product of the magnitudes that enter sequentially into the n openings of $\boldsymbol{L}$ is equal or opposite to the product $[a_1a_2\cdots a_n]$, the sum of the terms so obtained then being divided by their number.

If $\boldsymbol{L}$ includes *less than* n, for example only $n-r$, openings, then by

$$[\boldsymbol{L}a_1a_2\cdots a_n]$$

I mean the expression that results when one *precedes* $\boldsymbol{L}$ by r factors 1, each of which one treats as an expression with *one* opening, thus so that, if one imagines this opening as filled by a magnitude a, from 1 the magnitude $1\cdot a$, that is a, results. {Thus in this way the newly appearing openings of the r factors 1 may be treated as the *first* openings of the resulting expression with n openings, and specifically the opening of the first factor 1 as the first opening of the whole expression, that of the second factor as the second opening, and so forth. However the $n-r$ openings of the expression $\boldsymbol{L}$ in their original rank order correspond to the $(r+1)$th, $(r+2)$th, $\ldots$, nth openings in the new expression. Finally, the magnitudes a_i entering into the places of the factors 1 are to be combinatorially multiplied with each other and with the expression resulting from $\boldsymbol{L}$ by the filling of its openings.

With these definitions the case of an expression $\boldsymbol{L}$ with less than n openings is reduced to the previous one.}

In particular, if $\boldsymbol{L}$ is an expression with $n-1$ openings, *which by the filling of these openings is a numerical magnitude*, then

$$[\boldsymbol{L}a_1a_2\cdots a_n] = \frac{1}{n}(a_1[\boldsymbol{L}A_1] + \cdots + a_n[\boldsymbol{L}A_n]),$$

where A_r (for each index r) includes all magnitudes $a_1, \ldots, a_n$ with the exception of a_r as factors, and $[a_rA_r] = [a_1a_2\cdots a_n]$.

{If $\boldsymbol{L}$ includes *more than* n, say m, openings, then by

$$[\boldsymbol{L}a_1a_2\cdots a_n]$$

will be meant the expression with $m-n$ bound openings that one obtains when one develops the product

$$[\boldsymbol{L}a_1a_2\cdots a_na_{n+1}\cdots a_m]$$

according to the same procedure, and replaces the $m-n$ magnitudes a_{n+1}, a_{n+2}, $\ldots$, a_m by the bound openings l_1, l_2, $\ldots$, l_{n-m}, whose indices simultaneously denote the rank of the individual openings in the resulting expression.}

I also call this product (as generally all products that are enclosed by brackets, cf. No. 94) a relative product, and in fact I set as the principal domain to which it is referred the domain of the units from which all the magnitudes that are capable of entering the openings can be derived.[111]

505. *If* $\boldsymbol{L}$ *denotes an expression with* n *{bound} openings, and*

1) $$[a_1\cdots a_n] = [b_1\cdots b_n],$$

then in addition

$$[\boldsymbol{L}a_1 \cdots a_n] = [\boldsymbol{L}b_1 \cdots b_n],$$

and if

2) $$[a_1 \cdots a_n] = 0,$$

then in addition

$$[\boldsymbol{L}a_1 \cdots a_n] = 0.$$

PROOF. {1.} If two of the magnitudes $a_1, \ldots, a_n$, for example a_r and a_s, are equal to one another, one orders the expressions that (according to 504) result upon the development of $[\boldsymbol{L}a_1 \cdots a_n]$ so that one lets $a_1, \ldots, a_n$ enter into the openings of $\boldsymbol{L}$ in all possible sequences pairwise in such a way that every two of them, that are only distinguished in the sequence of those magnitudes by the opposite positions of a_r and a_s, form a pair. Then the expressions for each pair will (according to 504) have opposite signs. Now if a_r and a_s are equal to one another, then these expressions are, aside from their opposite signs, identical; thus their sum will be zero, whence in this case $[\boldsymbol{L}a_1 \cdots a_n] = 0$ as well.

2. If $[a_1 \cdots a_n] = 0$, this means (according to 66) that the factors $a_1, \ldots, a_n$ stand in a numerical relation, that is (according to 2) one of them, for example a_n, can be expressed as a multiple sum of the remaining $a_1, \ldots, a_{n-1}$. If one inserts this expression for a_n into $[\boldsymbol{L}a_1 \cdots a_n]$, and removes the parentheses enclosing this expression, then $[\boldsymbol{L}a_1 \cdots a_{n-1}a_n]$ is represented as a multiple sum of expressions, each of which includes two equal magnitudes among the $a_1, \ldots, a_{n-1}$, and thus (according to Part 1 of the Proof) is zero. Thus that multiple sum, that is $[La_1 \cdots a_n]$, is also equal to zero.

3. If the series of magnitudes $a_1, \ldots, a_n$ undergoes an elementary linear evolution (cf. 71), for example if a_r is transformed into $a_r + \alpha a_s$, where α is a numerical magnitude and r and s are different from one another, then $[\boldsymbol{L}a_1 \cdots a_r a_{r+1} \cdots a_n]$ is transformed into

$$[\boldsymbol{L}a_1 \cdots a_r a_{r+1} \cdots a_n] + \alpha[\boldsymbol{L}a_1 \cdots a_{r-1} a_s a_{r+1} \cdots a_n].$$

Since s is different from r, the second term of this expression contains a_s twice, and therefore (according to Part 1 of the Proof) is zero, whence the expression $[\boldsymbol{L}a_1 \cdots a_n]$ remains of unchanged value if the series of magnitudes $a_1, \ldots, a_n$ undergoes an elementary linear evolution, and thus also when it undergoes repeated elementary linear evolutions, that is (according to 71) in general when that series evolves linearly.

Now if

$$[a_1 \cdots a_n] = [b_1 \cdots b_n] \neq 0,$$

then (according to 76) the series $b_1, \ldots, b_n$ can be derived from the series $a_1, \ldots, a_n$ by a linear evolution, whence, as just shown, the value of $[La_1 \cdots a_n]$ remains unchanged, that is

$$[\boldsymbol{L}a_1 \cdots a_n] = [\boldsymbol{L}b_1 \cdots b_n].$$

{REMARK. One can therefore interpret the expression

$$[\boldsymbol{L}a_1 \cdots a_n]$$

as a product of the open expression $\boldsymbol{L}$ and the combinatorial product $[a_1 \cdots a_n]$ (cf. the Remark following No. 383).}

506. DEFINITION. If the expression $\boldsymbol{L}$ with n {bound} openings has the property that, with any n magnitudes $a_1, \ldots, a_n$ that can enter into the openings, it yields a product $[\boldsymbol{L}a_1 \cdots a_n]$ that is zero, then I set $[\boldsymbol{L}] = 0$. Further, if a product $[a_1 \cdots a_n] = 1$, and L includes n openings, I set

$$[\boldsymbol{L}] = [\boldsymbol{L}a_1 \cdots a_n].$$

REMARK. The notation chosen here has already been used earlier, although only incidentally, in the treatment of the power of a fraction $\boldsymbol{Q}$ (No. 383), since, if $e_1, \ldots, e_n$ are the n denominators whose product $[e_1 \cdots e_n] = 1$, and $a_1, \ldots, a_n$ are the numerators belonging to them, the product $[a_1 \cdots a_n]$ was meant by $[\boldsymbol{Q}^n]$. Now since $\boldsymbol{Q}$ can be interpreted as an expression with one opening, since (for each index r) $\boldsymbol{Q}e_r = a_r$, $\boldsymbol{Q}^n$ is an expression with n openings, and thus $[\boldsymbol{Q}^n]$ belongs among the expressions defined here (in 506). One is easily convinced that in addition according to this definition (506) the expression $[\boldsymbol{Q}^n] = [a_1 \cdots a_n]$, and therefore both definitions stand in complete agreement.

The generality of the concepts formulated here (in 504 and 506), and their essential significance for analysis, only appeared to me during the course of this work. Otherwise I would have treated these concepts and the theorems flowing from them at once in their proper place (in the first chapter of this Part). Since these theorems interrupt the course of development here, I will limit myself to those theorems that seem indispensable for the following presentation.

507. *If $\boldsymbol{L}$ contains two or more interchangeable openings, then*

$$[\boldsymbol{L}] = 0.$$

PROOF. It is to be shown that if even only two of the n openings of $\boldsymbol{L}$ are interchangeable with one another, then $[\boldsymbol{L}a_1 \cdots a_n]$ is always zero, whatever $a_1, \ldots, a_n$ may be for magnitudes. For if the other openings are filled by any of those magnitudes there results an expression with two interchangeable openings; let this expression be $\boldsymbol{P}$. Now if b and c are two arbitrary magnitudes that can enter in these two openings, then, since the openings are interchangeable, {according to the Remark following 485} $\boldsymbol{P}bc = \boldsymbol{P}cb$; but

$$[\boldsymbol{P}bc] = \frac{1}{2}(\boldsymbol{P}bc - \boldsymbol{P}cb),$$

and thus $= 0$, whence also $[\boldsymbol{L}a_1 \cdots a_n] = 0$ for arbitrary magnitudes $a_1, \ldots, a_n$, that is (according to 506) $[\boldsymbol{L}] = 0$.

508. *If $\boldsymbol{A}_1, \ldots, \boldsymbol{A}_n$ are magnitudes, all with an opening of the same species, either* ***all****, or* ***all but one*** *of which are* ***numerical magnitudes after filling these openings****, and $\boldsymbol{P}$ is an expression with arbitrarily many openings, then the product*

$$[\boldsymbol{A}_1 \cdots \boldsymbol{A}_n \boldsymbol{P}]$$

is completely subject to the laws of combinatorial multiplication (No. 52ff.) and indeed in the sense that $\boldsymbol{A}_1, \ldots, \boldsymbol{A}_n$ can be treated as elementary combinatorial factors. In particular, two products that are only distinguished by the opposed placement of two of these factors are opposites of one another, that is

$$(a) \qquad [\boldsymbol{A}_1 \cdots \boldsymbol{A}_r \cdots \boldsymbol{A}_s \cdots \boldsymbol{A}_n \boldsymbol{P}] = -[\boldsymbol{A}_1 \cdots \boldsymbol{A}_s \cdots \boldsymbol{A}_r \cdots \boldsymbol{A}_n \boldsymbol{P}],$$

where the two sides of the equation are only distinguished by the opposed positions of the factors $\boldsymbol{A}_r$ and $\boldsymbol{A}_s$, and if two of those factors are equal the product is zero, that is

$$(b) \qquad [\boldsymbol{A}_1 \cdots \boldsymbol{A}_r \cdots \boldsymbol{A}_r \cdots \boldsymbol{A}_n \boldsymbol{P}] = 0.$$

PROOF. If for example one considers only the first two factors $\boldsymbol{A}_1$ and $\boldsymbol{A}_2$ and calls the product of the rest $\boldsymbol{Q}$, and sets $e_1, \ldots, e_n$ as units whose combinatorial product is 1, then

$$[\boldsymbol{A}_1\boldsymbol{A}_2\boldsymbol{Q}] = [\boldsymbol{A}_1\boldsymbol{A}_2\boldsymbol{Q}e_1e_2e_3\cdots e_n].$$

Here (according to 504) the factors $e_1, e_2, e_3, \ldots, e_n$ are to enter in all possible orderings in the openings of $\boldsymbol{A}_1\boldsymbol{A}_2\boldsymbol{Q}$, and the sign is positive if $e_1, \ldots, e_n$ enter either in this order, thus e_1 into $\boldsymbol{A}_1$, e_2 into $\boldsymbol{A}_2$, and the remainder $e_3, \ldots, e_n$ sequentially into $\boldsymbol{Q}$, or in any other ordering that results from this one by an *even* number of transpositions; but $[\boldsymbol{A}_2\boldsymbol{A}_1\boldsymbol{Q}e_2e_1e_3\cdots e_n]$ has exactly the same meaning, since here too the sign is positive if e_1 enters into $\boldsymbol{A}_1$, e_2 into $\boldsymbol{A}_2$, and $e_3, \ldots, e_n$ in this sequence into $\boldsymbol{Q}$, and so forth, whence

$$[\boldsymbol{A}_1\boldsymbol{A}_2\boldsymbol{Q}e_1e_2e_3\cdots e_n] = [\boldsymbol{A}_2\boldsymbol{A}_1\boldsymbol{Q}e_2e_1e_3\cdots e_n],$$

but since (according to 55) $[e_2e_1e_3\cdots e_n] = -[e_1e_2e_3\cdots e_n]$, this last is (according to {the Remark following} 505 {and according to 46}) equal to $-[\boldsymbol{A}_2\boldsymbol{A}_1\boldsymbol{Q}e_1e_2e_3\cdots e_n]$, whence

$$[\boldsymbol{A}_1\boldsymbol{A}_2\boldsymbol{Q}] = -[\boldsymbol{A}_2\boldsymbol{A}_1\boldsymbol{Q}].$$

If $\boldsymbol{A}_1$ and $\boldsymbol{A}_2$ are equal to one another, then it follows from this last formula that $[\boldsymbol{A}_1\boldsymbol{A}_2\boldsymbol{Q}]$ is zero. The same conclusion holds on the same grounds if instead of the first two factors of the expression $[\boldsymbol{A}_1\boldsymbol{A}_2\cdots\boldsymbol{A}_n\boldsymbol{Q}]$ one considers any two others, $\boldsymbol{A}_r$ and $\boldsymbol{A}_s$.

Consequently formulas (a) and (b) hold, and upon them depend the remaining laws of combinatorial multiplication.

509. *If $\boldsymbol{A}$ is an expression with one opening and $\boldsymbol{B}$ an expression with $m-1$ openings of the same family {which are numerical magnitudes after filling these openings}, and $a_1, \ldots, a_m$ are magnitudes of that species {that is to enter the openings}, then*

$$\begin{aligned}[\boldsymbol{AB}a_1\cdots a_m] &= \boldsymbol{A}[\boldsymbol{B}a_1\cdots a_m]\\ &= \frac{1}{m}\sum \boldsymbol{A}a_{\mathfrak{a}}[\boldsymbol{B}A_{\mathfrak{a}}],\end{aligned}$$

where $A_{\mathfrak{a}}$ contains all the factors $a_1, \ldots, a_m$ with the exception of the factor $a_{\mathfrak{a}}$, in such a way that

$$[a_{\mathfrak{a}}A_{\mathfrak{a}}] = [a_1\cdots a_m],$$

and where the sum refers to the values $\mathfrak{a} = 1, \ldots, m$.

PROOF. 1. In order to develop the expression $[\boldsymbol{AB}a_1\cdots a_n]$, one must let the factors $a_1, \ldots, a_n$ enter into the openings of $\boldsymbol{AB}$ in all possible orderings, one by one; thus each of the magnitudes $a_1, \ldots, a_n$ must enter into $\boldsymbol{A}$ one by one.

Now if, first, the magnitude a_1 enters into $\boldsymbol{A}$, the remaining factors must enter into $\boldsymbol{B}$, and thus the factors of A_1 in all possible orderings, and in fact together with the sign law that two expressions produced in this way have the same or opposite signs, according as the two sequences yield equal or opposite combinatorial products. The terms so produced thus yield $+\boldsymbol{A}a_1[\boldsymbol{B}A_1]$. Here however the + sign is to be chosen, since $[a_1A_1] = [a_1a_2\cdots a_m]$. On the same grounds the sum of terms in which the magnitude a_2 enters $\boldsymbol{A}$ is $= +\boldsymbol{A}a_2[\boldsymbol{B}A_2]$, and so forth; and thus, since one must still divide these sums by the number of their terms,

$$[\boldsymbol{AB}a_1\cdots a_m] = \frac{1}{m}\sum \boldsymbol{A}a_{\mathfrak{a}}[\boldsymbol{B}A_{\mathfrak{a}}].$$

3. Further (according to 504)

$$\begin{aligned} \boldsymbol{A}[\boldsymbol{B}a_1 \cdots a_m] &= \frac{1}{m}\boldsymbol{A}\sum a_{\mathfrak{a}}[\boldsymbol{B}A_{\mathfrak{a}}] \\ &= \frac{1}{m}\sum \boldsymbol{A}a_{\mathfrak{a}}[\boldsymbol{B}A_{\mathfrak{a}}] && [39] \\ &= [\boldsymbol{A}\boldsymbol{B}a_1 \cdots a_m]. && \text{[Part 1 of the Proof]} \end{aligned}$$

{REMARK. If upon filling their openings $\boldsymbol{B}$ and $\boldsymbol{A}$ do not transform into numerical magnitudes, but rather into extensive magnitudes, then in place of the second formula of 509 there appears the formula

$$\{*\} \qquad [\boldsymbol{A}\boldsymbol{B}a_1 \cdots a_m] = \frac{1}{m}\sum [(\boldsymbol{A}a_{\mathfrak{a}})[\boldsymbol{B}A_{\mathfrak{a}}]],$$

while the first formula of 509 no longer holds. Should at least $\boldsymbol{A}$ be a numerical magnitude upon filling its openings, one can drop the brackets adjacent to the summation symbol in {∗}.}

510. *If $\boldsymbol{C}$ is an expression with two openings of the same family, which upon filling its openings is a numerical magnitude, and a is a magnitude of that family, B a product of $2n-1$ such magnitudes, and $2n$ the number of units from which the magnitudes of this family are derivable, then*

$$[\boldsymbol{C}a[\boldsymbol{C}^{n-1}B]] = [\boldsymbol{C}^n aB],$$
$$[\boldsymbol{C}[\boldsymbol{C}^{n-1}B]a] = [\boldsymbol{C}^n Ba].$$

PROOF. Since the n factors $\boldsymbol{C}$ are all equal to one another, and after filling their openings are numerical factors, and are therefore interchangeable with one another, $[\boldsymbol{C}^n aB]$ expresses that a enters an opening of one of the factors, for example the first factor $\boldsymbol{C}$, while the $2n-1$ factors of B enter the other openings of that factor and the remaining factors. But the expression $[\boldsymbol{C}a[\boldsymbol{C}^{n-1}B]]$ has exactly the same meaning, whence these two are equal.

The second formula follows in the same way.[112]

511. *If $X dx$ (in the sense of 502) is reducible to n terms of the form $U du$, that is, if*

$$X dx = U_1 du_1 + \cdots + U_n du_n,$$

then necessarily

$$\left[X\left(\frac{d}{dx}X\right)^n\right] = 0.$$

PROOF. Let $U_1 du_1 + \cdots + U_n du_n$ be denoted by $\sum U_{\mathfrak{a}} du_{\mathfrak{a}}$ for short; then

$$\sum U_{\mathfrak{a}} du_{\mathfrak{a}} = \sum U_{\mathfrak{a}}\left\{\frac{d}{dx}u_{\mathfrak{a}} \cdot dx\right\} = \sum U_{\mathfrak{a}}\frac{d}{dx}u_{\mathfrak{a}} \cdot dx,$$

whence

$$X = \sum U_{\mathfrak{a}}\frac{d}{dx}u_{\mathfrak{a}}.$$

Consequently

$$\begin{aligned} \frac{d}{dx}X &= \frac{d}{dx}\sum U_{\mathfrak{a}}\frac{d}{dx}u_{\mathfrak{a}} \\ &= \sum\left\{\frac{d}{dx}U_{\mathfrak{a}} \cdot \frac{d}{dx}u_{\mathfrak{a}}\right\} + \sum U_{\mathfrak{a}}\frac{d^2}{dx^2}u_{\mathfrak{a}}. \end{aligned}$$

Since $\frac{d^2}{dx^2}u_{\mathfrak{a}}$ is an expression with two interchangeable openings (451) {and since (by virtue of 446) the interchangeability of the openings is not voided if one regards the openings as bound}, we can, upon the substitution of $\frac{d}{dx}X$ into the expression

$$\left[X\left(\frac{d}{dx}X\right)^n\right],$$

(according to 507) drop the terms

$$\sum U_{\mathfrak{a}}\frac{d^2}{dx^2}u_{\mathfrak{a}},$$

and obtain

$$\begin{aligned}\left[X\left(\frac{d}{dx}X\right)^n\right] &= \left[\sum U_{\mathfrak{a}}\frac{d}{dx}u_{\mathfrak{a}}\cdot\left(\sum\left\{\frac{d}{dx}U_{\mathfrak{b}}\cdot\frac{d}{dx}u_{\mathfrak{b}}\right\}\right)^n\right]\\ &= \sum\left[U_{\mathfrak{a}}\frac{d}{dx}u_{\mathfrak{a}}\cdot\frac{d}{dx}U_{\mathfrak{b}}\cdot\frac{d}{dx}u_{\mathfrak{b}}\cdot\frac{d}{dx}U_{\mathfrak{c}}\cdot\frac{d}{dx}u_{\mathfrak{c}}\cdots\right],\end{aligned}$$

where the number of indices $\mathfrak{a}$, $\mathfrak{b}$, $\mathfrak{c}$, ... is equal to $n+1$. But since u only has n different indices, there must among the indices $\mathfrak{a}$, $\mathfrak{b}$, $\mathfrak{c}$, ... be at least two that are equal; thus among the magnitudes

$$\frac{d}{dx}u_{\mathfrak{a}},\ \frac{d}{dx}u_{\mathfrak{b}},\ \frac{d}{dx}u_{\mathfrak{c}},\ \ldots$$

there are also at least two that are equal, whence (according to 508) every term of the above series is zero, and therefore so is the sum itself, that is

$$\left[X\left(\frac{d}{dx}X\right)^n\right] = 0.$$ [113]

REMARK. In the following I will show that this equation of condition is at the same time completely sufficient, so that, if it is satisfied, then the reduction to n terms of the form $U du$, and therefore also (according to 503) the integration by assemblies of n equations is also always possible. This marvellously simple formula is therefore of extremely far-reaching significance.

Its proof is carried out above so that the way it is discovered is immediately apparent throughout. But it is not difficult to develop the corresponding formulas for the case that $X dx$ is an extensive magnitude; and so I have on this account not formulated these latter formulas here, since, as already mentioned above, the treatment of this general case, encompassing all of integral calculus, must be left undone here. On the other hand I will re-express the formula developed above in ordinary analysis in the following Number.[114]

512. PROBLEM. *To replace the equation of condition of* 511, *namely*

$$\left[X\left(\frac{d}{dx}X\right)^n\right] = 0,$$

by numerical equations.

SOLUTION. Let

$$x = x_1 e_1 + \cdots + x_m e_m,$$

where $e_1, \ldots, e_m$ form the system of units. Then

$$dx = e_1 dx_1 + \cdots + e_m dx_m,$$

and thus

$$X dx = X e_1 dx_1 + X e_m dx_m = X_1 dx_1 + \cdots + X_m dx_m,$$

if we denote the numerical magnitudes $Xe_1, \dots, Xe_m$ respectively by $X_1, \dots, X_m$. The above equation of condition then (according to 506) expresses that, since X contains one opening and $\frac{d}{dx}X$ two,

$$(*) \qquad \left[X\left(\frac{d}{dx}X\right)^n e_1e_2\cdots e_{2n+1}\right] = 0,$$

and this continues to hold if one replaces $e_1, e_2, \dots, e_{2n+1}$ by any arbitrary $2n+1$ of the m units.

If $m = 2n+1$, no other numerical equation of condition than equation $(*)$ appears; if $m < 2n+1$, there is none at all, since then every $2n+1$ magnitudes with which

$$X\left(\frac{d}{dx}X\right)^n$$

may be multiplied stand in a numerical relation, whence

$$X\left(\frac{d}{dx}X\right)^n$$

multiplied by them yields zero (according to 505), and thus (according to 506)

$$\left[X\left(\frac{d}{dx}X\right)^n\right]$$

is zero of itself. We therefore now assume that $m > 2n+1$; and under this assumption, we seek to replace X and x by the numerical magnitudes $X_1, \dots, X_m, x_1, \dots, x_m$ in equation $(*)$.

Now according to the above, $Xe_r = X_r$. Thus if in the expression $\frac{d}{dx}X$ one interprets the opening of the function X itself as the first opening and the opening in the expression resulting from differentiation as the second opening, then

$$\frac{d}{dx}X \cdot e_re_s = \frac{d}{dx}X_r \cdot e_s = \frac{d}{dx_s}X_r,$$

(according to 451), whence the above formula $(*)$ is transformed into

$$0 = \sum \pm X_1 \cdot \frac{d}{dx_3}X_2 \cdot \frac{d}{dx_5}X_4 \cdots \frac{d}{dx_{2n+1}}X_{2n},$$

where one has to interchange the indices in all possible ways and to prefix each term with the sign $+$ or $-$, according as the number of interchanges by which it results is even or odd. If, following JACOBI's precedent (*Crelles Journal* **2**, p. 351 {*Werke*, Bd. 4, p. 23}), we denote

$$\frac{d}{dx_3}X_2 - \frac{d}{dx_2}X_3$$

by $(2,3)$, and so forth, or if in general we set

$$\frac{d}{dx_s}X_r - \frac{d}{dx_r}X_s = (r,s),$$

we can also write the foregoing formula as

$$0 = \sum \pm X_1(2,3)(4,5)\cdots(2n,2n+1),$$

in which the interchange of any two indices enclosed in parentheses remains disallowed. Likewise, we can, without altering the meaning of the equation, require that the first of the two indices of a factor enclosed in parentheses only increases in the next one sequentially, and never decreases. For, since the order of the numerical factors $(2,3)$, ... is not important, we can always give them that ordering. If, in this sense, we denote (with JACOBI, op. cit. p. 355f {*Werke*, Bd. 4, p. 27}) the sum

$$\sum \pm(2,3)\cdots(2n,2n+1)$$

by

$$(2,3,4,5,\ldots,2n,2n+1),$$ [115]

then the above equation is transformed into

$$0=\sum \pm X_1(2,3,\ldots,2n+1),$$

that is, into

$$0=X_1(2,3,\ldots,2n+1)-X_2(1,3,\ldots,2n+1)+X_3(1,2,4,\ldots,2n+1)-\cdots,$$

which JACOBI, op. cit. p. 356 {*Werke*, Bd. 4, p. 28} writes

(**) $$0=\sum X_1(2,3,\ldots,2n+1).$$

There are as many such equations as combinations without repetition of m elements for $(2n+1)$th class. But if $m>2n+1$, these equations are not independent of one another. In fact, we can show that if equation $(*)$, whose transform is equation $(**)$, is assumed to hold for all combinations of $e_1, \ldots, e_m$ of $(2n+1)$th class in which there appears a unit e_r whose associated $Xe_r = X_r$ is not zero, it must also hold for all remaining combinations (in which e_r does not appear).

Thus if $X_1 \neq 0$, and equation $(*)$ holds for all combinations in which e_r appears, that is, if it is always the case that

$$\left[X\left(\frac{d}{dx}X\right)^n e_1E_r\right]=0,$$

if E_r is an arbitrary combination without repetition of $e_2, \ldots, e_m$ of $2n$th class, then it is to be shown that it is also always the case that

$$\left[X\left(\frac{d}{dx}X\right)^n e_sE_r\right]=0,$$

even if e_s denotes an arbitrary unit not appearing in E_r. Thus (according to 508)

$$\left[X^2\left(\frac{d}{dx}X\right)^n\right]$$

is always zero, whence also (according to 506)

$$\left[X^2\left(\frac{d}{dx}X\right)^n e_1e_sE_r\right]=0,$$

that is,

$$\sum \pm Xe_1\left[X\left(\frac{d}{dx}X\right)^n e_sE_r\right]=0,$$

where the sum refers to the different terms that result from the one under the summation symbol when one interchanges e_1 with each of the units appearing in e_sE_r one by one (and suitably changes the algebraic sign); but all these terms are zero, since then

$$\left[X\left(\frac{d}{dx}X\right)^n\right]$$

is multiplied by a combination of units in which e_1 appears, and these products are zero by assumption, whence all that is left over is the term standing under the summation symbol, that is,

$$Xe_1\left[X\left(\frac{d}{dx}X\right)^n e_s E_r\right] = 0.$$

Now it is likewise assumed that the numerical magnitude $Xe_1 = X_1$ is different from zero, whence one obtains

$$\left[X\left(\frac{d}{dx}X\right)^n e_s E_r\right] = 0,$$

which was to be shown.

We collect this result into a theorem as follows:

513. *If the expression*

$$X_1 dx_1 + \cdots + X_m dx_m,$$

in which $X_1, \ldots, X_m$ *are functions of the variables* $x_1, \ldots, x_m$, *can be reduced to* n *terms, namely to*

$$U_1 du_1 + \cdots + U_n du_n,$$

or, expressed otherwise, if the equation

$$X_1 dx_1 + \cdots + X_m dx_m = 0$$

can be integrated by assemblies of n *equations, then* ***first****, if* $m = 2n + 1$, *the one equation of condition*

$$\sum X_1(2, 3, \ldots, 2n + 1) = 0,$$

which has the meaning described in 512, *must be satisfied; but if* ***second****,* $m > 2n + 1$, *then there are as many equations as there are combinations of* m *elements of* $(2n+1)$*th class, thus since one can replace the indices* $1, 2, \ldots, 2n+1$ *in the above equation with any other group of just as many indices; but a smaller number of these equations already suffice, since if, for example,* X_1 *is not equal to zero, it suffices if in the above equation one replaces the group of indices* $2, 3, \ldots, 2n + 1$ *with every other combination of the indices* $2, 3, \ldots, m$ *of* $2n$*th class. Thus there remain only as many equations of condition as there are combinations without repetition of* $m - 1$ *elements of* $2n$*th class.*

REMARK. For the simplest case, where $m = 2n + 1$, JACOBI (op. cit. p. 356 {*Werke*, Bd. 4, p. 28}) has formulated the equation of condition. For the case $n = 1$ one obtains the well-known equations of condition for integrability, which (according to 511) appear collected together in the equation

$$\left[X\frac{d}{dx}X\right] = 0.$$

It is now a matter of actually carrying out the reduction of Xdx to n terms of the form Udu, as soon as the equations of condition (511) for the possibility of this reduction are satisfied. To this end we solve the following problem according to PFAFF's process:

514. PROBLEM. *To cast the equation*

$$Xdx = 0, \tag{a}$$

in which X *is a function of* x, *and* $x = x_1e_1 + \cdots + x_me_m$ *is derivable from a system of* m *units, into a form in which the resulting equation only contains* $m - 1$ *variable numerical magnitudes.*

SOLUTION. To this end it is only a matter of representing x as a function of a variable a derivable from $m-1$ units and a variable numerical magnitude t such that, if one introduces this expression for x in the given equation, the coefficient of dt in the developed equation is zero, and the coefficient of da either does not contain t at all, or only in a numerical factor N, so that, if one divides the equation by N, the resulting equation no longer contains t.

If one denotes by δ' the differential with respect to a, in which t is set as a constant, and by δ the differential quotient with respect to t, in which a is set as a constant, one obtains

$$dx = \delta' x + \delta x \cdot dt;$$

consequently, if the required problem is to be solved, the two equations

$$X \delta x = 0 \tag{b}$$

$$\delta \frac{X \delta' x}{\mathsf{N}} = 0, \tag{c}$$

must be satisfied, where the latter expresses that $X\delta' x : \mathsf{N}$ no longer depends on t.

When one sets

$$\frac{d\mathsf{N}}{\mathsf{N}} = \lambda, \tag{d}$$

the last of the above equations yields

$$\lambda X \delta' x = \delta(X\delta' x) = \delta X \cdot \delta' x + X \delta\delta' x. \tag{e}$$

If one also differentiates equation (b) with respect to a, one obtains

$$0 = \delta' X \cdot \delta x + X \delta' \delta x. \tag{f}$$

If one subtracts the second of these equations from the first, one obtains

$$\begin{aligned} \lambda X \delta' x &= \delta X \cdot \delta' x - \delta' X \cdot \delta x \\ &= \frac{d}{dx} X \cdot \delta' x \cdot \delta x - \frac{d}{dx} X \cdot \delta x \cdot \delta' x, \end{aligned} \tag{g}$$

$$\lambda X \delta' x = 2 \left[\frac{d}{dx} X \cdot \delta' x \cdot \delta x \right]. \tag{h}$$

Here $\frac{d}{dx}X$ is an expression with two openings; and in fact the first opening, that is, the opening in which the factor placed first (in the first term $\delta' x$) is to enter, is interpreted as that opening contained in X, and as second that introduced by the differentiation by x and division by dx.

Equation (h) will now obviously be satisfied if, for every magnitude c of the same species as x (and thus also as $\delta' x$),

$$\lambda X c = 2 \left[\frac{d}{dx} X \cdot c \cdot \delta x \right]. \tag{i}$$

I now show that, as soon as equation (i) holds generally for every magnitude c, equations (b) and (c) are also satisfied, and thus the required problem is solved, provided λ is different from zero.[116]

Now $\delta' x$ is the same species as x, and therefore, if equation (i) holds generally, it must be possible to substitute it for c in this equation, whereby one obtains equation (h); thus equation (g) also holds, since (g) is equivalent to (h). Further,

δx is also the same species as x, and thus can also replace c in equation (i). But then (according to 505) its right side is zero, whence one obtains $\lambda X dx = 0$, and thus, since $\lambda \neq 0$ (by assumption), there follows $X dx = 0$, that is, equation (b) holds. But then so also does equation (f), derived from (b). Upon the addition of equations (f) and (g) there results equation (e). Now if one sets $\log \mathsf{N} = d^{-1}\lambda dt$, equation (d) will also be satisfied, and when one substitutes the value for λ flowing from it into (e), so also will (c), and thus

$$\frac{X\delta' x}{\mathsf{N}} = 0$$

is the equation transformed out of $X dx = 0$ that now contains only a, and thus a magnitude derivable from $n-1$ units, as a variable, and the problem is solved.

Represented as a theorem:

If the numerical equation

$$(a) \qquad X dx = 0,$$

in which X is assumed to be a function of x, and x is derivable from m units, is assumed, and one defines x as a function of a variable a, derivable from $m-1$ units, and a variable number t, so that, if δ' denotes the differential with respect to a with t set to a constant, and δ the differential quotient with respect to t with a set to a constant, and c is an arbitrary magnitude of the same species as x, λ an unknown but nonzero numerical magnitude, then if the equation

$$(i) \qquad \lambda X c = 2\left[\frac{d}{dx}X \cdot c\delta x\right]$$

is satisfied {for every c}, equation (a) is replaced by the equation

$$(k) \qquad \frac{X\delta' x}{\mathsf{N}} = 0,$$

in which $\frac{X\delta' x}{\mathsf{N}}$ no longer depends on t, and

$$(l) \qquad \log \mathsf{N} = d^{-1}\lambda dt.$$

515. Continuation. First of all it is a matter of using equation 514(i) to make δx stand alone on one side. We carry this out at first *under the assumption that $m = 2n$.*

If one replaces c by the units $e_1, \ldots, e_{2n}$ one by one, that equation contains $2n$ numerical equations, by which the magnitudes $\delta x_1, \ldots, \delta x_{2n}$, which are contained in $\delta x = e_1 dx_1 + \cdots + e_{2n} dx_{2n}$, can be expressed in general.[117] This is possible in an extremely simple way, provided it is assumed that

$$(a) \qquad \left[X\left(\frac{d}{dx}X\right)^{n-1}\right] \neq 0,$$

$$(b) \qquad \left[\left(\frac{d}{dx}X\right)^{n}\right] \neq 0.$$

In fact, in order to find δx_r it is only necessary to replace c in equation (i) with the value

$$\left[\left(\frac{d}{dx}X\right)^{n-1} E_r\right],$$

where $[e_r E_r] = 1$ and E_r contains the $2n-1$ units different from E_r as factors. Thus since

$$\left(\frac{d}{dx}X\right)^{n-1}$$

contains $2n-2$ openings in all, and E_r is a product of $2n-1$ units, it follows (according to 504) that

$$\left[\left(\frac{d}{dx}X\right)^{n-1} E_r\right]$$

is a multiple sum of the units from which x is derived, and thus is of the same family as x and can therefore be substituted for c in equation 514(i). This is then transformed into

$$\lambda X\left[\left(\frac{d}{dx}X\right)^{n-1} E_r\right] = 2\left[\frac{d}{dx}X \cdot \left[\left(\frac{d}{dx}X\right)^{n-1} E_r\right]\delta x\right].$$

If one transforms the left side of this equation (according to 509) and the right (according to 510), where one recognizes that $\frac{d}{dx}X$ is an expression with $2n$ openings, and both X and $\frac{d}{dx}X$ are numerical magnitudes after the filling of their openings, then that equation is transformed into

$$\lambda\left[X\left(\frac{d}{dx}X\right)^{n-1} E_r\right] = 2\left[\left(\frac{d}{dx}X\right)^{n} E_r\delta x\right].$$

If in this we replace δx by its value $e_1 dx_1 + \cdots + e_{2n}\delta x_{2n}$, then, since $[E_r e_s]$ contains equal factors if r is different from s, the product

$$\left[\left(\frac{d}{dx}X\right)^{n} E_r e_s\right]$$

will remain zero {according to 505}, and (according to 58) $[E_r e_r] = -[e_r E_r] = -1$,

$$(c) \qquad \lambda\left[X\left(\frac{d}{dx}X\right)^{n-1} E_r\right] = -2\left[\left(\frac{d}{dx}X\right)^{n}\right]\delta x_r.$$

Now if equations (a) and (b) are satisfied, and one sets

$$(d) \qquad -n\lambda : \left[\left(\frac{d}{dx}X\right)^{n}\right] = \mu,$$

one obtains

$$(d^*) \qquad \delta x_r = \frac{\mu}{2n}\left[X\left(\frac{d}{dx}X\right)^{n-1} E_r\right];$$

thus

$$\delta x = \sum e_{\mathfrak{a}}\delta x_{\mathfrak{a}} = \frac{\mu}{2n}\sum e_{\mathfrak{a}}\left[X\left(\frac{d}{dx}X\right)^{n-1} E_{\mathfrak{a}}\right],$$

that is (according to 504 and 506),

$$(e) \qquad \delta x = \mu\left[X\left(\frac{d}{dx}X\right)^{n-1}\right],$$

or (according to 2)

$$(f) \qquad \delta x \equiv \left[X\left(\frac{d}{dx}X\right)^{n-1}\right].$$

It remains to show that the value of δx in equation (e), in which μ has the meaning expressed by (d), when substituted into equation 514(i), makes this an identity *for any arbitrary value of c.*

First, if one replaces c by one of the units, for example c_r, then the right side of equation 514(i) is

$$2\left[\frac{d}{dx}X \cdot e_r \delta x\right] = 2\mu\left[\frac{d}{dx}X \cdot e_r\left[X\left(\frac{d}{dx}X\right)^{n-1}\right]\right].$$

Since {the expression}

$$X\left(\frac{d}{dx}X\right)^{n-1}$$

is a numerical magnitude after fillling its $2n-1$ openings, {we can, without changing its value, append to it another opening (according to 504). If this opening is the same family as the others, then (according to 504)}[118]

$$\left[X\left(\frac{d}{dx}X\right)^{n-1}\right] = \left[lX\left(\frac{d}{dx}X\right)^{n-1}\right] = -\left[Xl\left(\frac{d}{dx}X\right)^{n-1}\right]$$

(according to 508), and again (according to 509), this

$$= -\frac{1}{2n}\sum Xe_{\mathfrak{a}}\left[l\left(\frac{d}{dx}X\right)^{n-1}E_{\mathfrak{a}}\right],$$

where $E_{\mathfrak{a}}$ has the units different from $e_{\mathfrak{a}}$ as factors, and $[e_{\mathfrak{a}}E_{\mathfrak{a}}] = 1$; thus, since $Xe_{\mathfrak{a}}$ is a numerical magnitude, and therefore can assume any position in the product, one obtains

$$2\left[\frac{d}{dx}X \cdot e_r\delta x\right] = -\frac{\mu}{n}\sum Xe_{\mathfrak{a}}\left[\frac{d}{dx}X \cdot e_r\left[l\left(\frac{d}{dx}X\right)^{n-1}E_{\mathfrak{a}}\right]\right]$$

{or, if one now drops the l once more and then applies No. 510},

$$= -\frac{\mu}{n}\sum Xe_{\mathfrak{a}}\left[\left(\frac{d}{dx}X\right)^{n}e_{\mathfrak{a}}E_{\mathfrak{a}}\right]. \qquad [510]$$

Now since $E_{\mathfrak{a}}$ contains all the units different from $e_{\mathfrak{a}}$ as factors, it also contains e_r if $\mathfrak{a}$ is different from r; but then (according to 60) $[e_rE_{\mathfrak{a}}]$ is zero, and thus (according to 505) so is the whole expression in which $e_rE_{\mathfrak{a}}$ appears, and consequently the above sum reduces to the term for which $\mathfrak{a} = r$; but since $[e_rE_r] = 1$, one finds that the above expression {according to 506}

$$= -\frac{\mu}{n}Xe_r\left[\left(\frac{d}{dx}X\right)^{n}\right].$$

If in this one replaces μ by its value from (d), the last expression

$$= \lambda Xe_r.$$

Consequently equation 514(i) holds for whatever unit e_r may be substituted for c, and thus for every multiple sum of these units, that is, for every magnitude

c of the same family as x. Thus it is proved that expression (e) solves equation 514(i) generally under the assumptions made for δx.

REMARK. In the article mentioned (*Crelle* **2**, p. 354 {*Werke*, Bd. 4, p. 25}), JACOBI likewise tacitly assumes the assumptions made here, expressed by equations (a) and (b), and does not treat the remaining cases, where these assumptions are not made, at all.

The resulting formulas (e) or (f) yield, expressed in terms of ordinary analysis, the same equations that JACOBI formulates there (*Crelle* **2**, p. 354ff. {*Werke*, Bd. 4, p. 25ff}), without however providing a proof. The determinant, which is derived directly from the $2n$ numerical equations contained in 514(i), includes twice as many factors as the expression serving for the solution of the equations requires; but this determinant can be represented as the square of an expression which at the same time enters as a factor in all the numerical determinants for the fractions $\delta x_1, \ldots, \delta x_{2n}$. This factor is therefore removed from the fractions in question, and there remain in their numerators expressions which contain in each term only half as many factors as in the original numerators, and which correspond precisely to the expressions

$$\left[X\left(\frac{d}{dx}X\right)^{n-1}E_r\right] \qquad (r = 1, \ldots, 2n)$$

given above in (d^*). All these transformations are eliminated by the method applied above, as it makes clear itself.[119]

The expression for δx stands in a remarkable relation to the equations of condition of No. 511, for which in addition the grounds are provided. It still remains to complete the method for the case that the assumptions represented by the above inequalities (a) and (b) are not satisfied.

516. CONTINUATION.[120] Thus if, as in the preceding Number,

$$(a) \qquad \left[X\left(\frac{d}{dx}X\right)^{n-1}\right] \neq 0,$$

but

$$(b) \qquad \left[\left(\frac{d}{dx}X\right)^{n}\right] = 0,$$

then equation 515(c), which is independent of assumptions (a) and (b), yields the value zero for λ. Consequently we no longer have to refer back to equation 514(i), since this only leads to a solution of the problem for the case that $\lambda \neq 0$. But it can be shown that equation {515(e), that is the equation}

$$(c) \qquad \delta x = \mu\left[X\left(\frac{d}{dx}X\right)^{n-1}\right],$$

which one can also replace by the congruence

$$\delta x \equiv \left[X\left(\frac{d}{dx}X\right)^{n-1}\right],$$

{still} yields the solution to problem 514, that is it makes equations 514(b) and (c) identities.

For then

$$X\delta x = \mu X\left[X\left(\frac{d}{dx}X\right)^{n-1}\right],$$

$$= \mu\left[X^2\left(\frac{d}{dx}X\right)^{n-1}\right], \qquad [509]$$

(*d*) $$X\delta x = 0. \qquad [508]$$

Further, on the same grounds as in 515,

$$2\left[\frac{d}{dx}X \cdot e_r\delta x\right] = -\mu X e_r\left[\left(\frac{d}{dx}X\right)^n\right]$$
$$= 0,$$

according to the second assumption (*b*). This equation holds for every unit $e_r = e_1$, $\ldots, e_{2n}$, and thus also for an arbitrary multiple sum of these units, whence also for $\delta' x$, since this is the same family as x, and is therefore also numerically derivable from the units $e_1, \ldots, e_{2n}$. Thus

$$\left[\frac{d}{dx}X \cdot \delta' x \cdot \delta x\right] = 0,$$

that is

$$\frac{d}{dx}X \cdot \delta' x \cdot \delta x - \frac{d}{dx}X \cdot \delta x \cdot \delta' x = 0.$$

But $\frac{d}{dx}X \cdot \frac{\delta x}{l_2} = \delta X$ and $\frac{d}{dx}X \cdot \frac{\delta' x}{l_2} = \delta' X$, {in which the appended denominator l_2 is to indicate that the filling magnitudes δx and $\delta' x$ are to enter in the second opening, that is, into the one resulting by differentiation}; thus one has

$$\delta X \cdot \delta' x - \delta' X \cdot \delta x = 0.$$

If one differentiates equation (*d*) with respect to a, while t is set to a constant, then since δ' was the symbol for this differentiation, one obtains

$$\delta' X \cdot \delta x + X\delta\delta' x = 0.$$

If one adds this equation to the previous one, one has

$$\delta X \cdot \delta' x + X\delta\delta' x = 0,$$

that is

(*e*) $$\delta(X\delta' x) = 0,$$

which means that $X\delta' x$ is independent of t, and thus, since (according to 514) $X dx = X\delta' x + X\delta x \cdot dt$, and $X\delta x = 0$,

$$X dx = X\delta' x,$$

where the last expression is independent of t, and thus only depends on $2n - 1$ variables.

REMARK. Equation (*b*) is therefore (under assumption (*a*)) the equation of constraint in order that the expression $X dx$ can *immediately (without the introduction of factors)* be transformed into an expression that only contains $2n-1$ variable numerical magnitudes; if on the other hand equation (*b*) is not satisfied, then this transformation is only achieved by means of a variable factor whose reciprocal value was denoted above by N. If in particular $n = 1$, that is, if Xdx has the form $X_1dx_1 + X_2dx_2$, then there results the equation

$$\left[\frac{d}{dx}X\right] = 0,$$

as the equation of condition, since $X_1dx_1 + X_2dx_2$ is transformed into a differential expression Udu with *only* one variable numerical magnitude (u), and thus is completely integrable. But the equation

$$\left[\frac{d}{dx}X\right] = 0$$

expresses that $\frac{d}{dx}X$ contains two interchangeable {bound} openings, which agrees with No. 486.

Before I show how the problem is to be solved when equation (a) also drops out, I will by integration of equation 515(e) actually complete the transformation indicated, wherein I make use of JACOBI's method (in *Crelle* **17**, p. 138 {*Werke*, Bd. 4, p. 101}).

517. CONTINUATION. The congruence or equation

$$(a) \qquad \delta x \equiv \left[X\left(\frac{d}{dx}X\right)^{n-1}\right] \quad \text{or} \quad \delta x = \mu\left[X\left(\frac{d}{dx}X\right)^{n-1}\right]$$

defines only the ratios of the differential quotients $\delta x_1, \ldots, \delta x_{2n}$; consequently we can assume one of them arbitrarily, that is, we can choose t arbitrarily.

If we set $t = x_{2n}$, then $\delta x_{2n} = 1$. If we then set $x_1e_1 + \cdots + x_{2n-1}e_{2n-1} = y$ for the moment, then $x = y + te_{2n}$, and $\delta x = \delta y + e_{2n}$, and thus one obtains

$$\delta y = \mu\left[X\left(\frac{d}{dx}X\right)^{n-1}\right] - e_{2n}.$$

Here μ is defined by the assumption $\delta x_{2n} = 1$; if in this equation one replaces δx_{2n} by its value from 515(d^*), one obtains for the definition of μ the equation

$$1 = \frac{\mu}{2n}\left[X\left(\frac{d}{dx}X\right)^{n-1} E_{2n}\right],$$

where

$$E_{2n} = -[e_1e_2\cdots e_{2n-1}];$$

consequently we obtain

$$\delta y = 2n\left[X\left(\frac{d}{dx}X\right)^{n-1}\right] : \left[X\left(\frac{d}{dx}X\right)^{n-1} E_{2n}\right] - e_{2n}.$$

Here the right side is a function of x, and thus of y and t; and therefore this equation can be integrated by the method of 494. Thus y is (according to 494(c)) given in the form

$$y = a + t\phi,$$

where ϕ is yet again a function of a and t, that is of x, and where a is the value that y, and thus also x, assumes for $t = 0$.

At the same time, formula 494(d) informs us how to find a as a function of y and t, that is here as a function of x. For, since δ' denotes differentiation with respect to a in which t is constant, $\delta'x$ is equal to $\delta'y = da + t\delta'\phi$. Therefore $X\delta'x : \mathsf{N}$ which, as shown, is independent of t, is equal to $X(da + t\delta'\phi) : \mathsf{N}$. Thus if we write $X(x)$, $\mathsf{N}(x)$ for X and N in order to denote them as functions of x, we obtain

$$\frac{X\delta' x}{\mathsf{N}} = \frac{X(x)(da + t\delta'\phi)}{\mathsf{N}(x)}.$$

But since the expression on the left is independent of t, so also must be the expression on the right, and thus it must maintain the same value that it has for $t = 0$. Since in this case $y = a$, and thus also x $(= y + te_{2n})$ is equal to a, one obtains

$$\frac{X\delta' x}{\mathsf{N}} = \frac{X(a)da}{\mathsf{N}(a)}.$$

Now since $X\delta x = 0$, $Xdx = X\delta' x = 0$, whence one obtains

$$X(a)da = 0$$

as the transformation of $Xdx = 0$; and indeed in it a is derivable from the units e_1, $\ldots$, e_{2n-1}, and thus is only dependent on $2n - 1$ variable numerical magnitudes, as was required.

REMARK. For the present purpose we could also have given the theorem in 494 that we make use of here the form: If x is derivable from several (m) units, and $dx \equiv f(x)$ is given, then this congruence is integrated by a function $F(x)$, which is set equal to a constant a derivable from $m - 1$ units; and in particular one can give this integrating equation $F(x) = a$ (which contains $m - 1$ numerical equations) the form that a is the value that x assumes for the case that one of the derivation numbers of x, for example x_m, is zero.

I will utilize the theorem in this form in the following, since the proof of the theorem in this altered form is completely contained in what has been said above. Now in order to be able to solve the present problem for the previously excluded case I will formulate an auxiliary theorem, which is also of interest in itself.

518.[121] *If*

$$\left[X\left(\frac{d}{dx}X\right)^{n}\right] = 0,$$

so also is

$$\left[\left(\frac{d}{dx}X\right)^{n+1}\right] = 0.$$

PROOF. 1. The openings of the first of these expressions are filled by $2n + 1$ factors, those of the second by $2n+2$. Now I show first, that the second expression, after filling its openings by any $2n+2$ factors a_1, $\ldots$, a_{2n+2}, can be represented as a multiple sum of expressions of the first type.

In order to mediate both types of expression I proceed from the expression

$$Xa_1\left[\left(\frac{d}{dx}X\right)^{n+1} a_1 a_2 \cdots a_{2n+2}\right].$$

To this end I will denote by F_r the product of the magnitudes a_1, $\ldots$, a_{2n+2} different from a_r, and in fact take this product so that

$$[a_r F_r] = [a_1 a_2 \cdots a_{2n+2}],$$

and likewise by F_{rs} the product of those magnitudes different from a_r and a_s, and in fact so that

$$[a_r a_s F_{rs}] = [a_1 a_2 \cdots a_{2n+2}].$$

Then the above expression

$$= Xa_1 \left[\left(\frac{d}{dx} X \right)^{n+1} a_1 F_1 \right].$$

Moreover since, if $\mathfrak{b} \neq 1$, $[a_1 F_\mathfrak{b}]$ necessarily contains the factor a_1 twice, and therefore (according to 505) the above expression is zero in this case, if we write $a_1 F_\mathfrak{b}$ instead $a_1 F_1$ and simultaneously $Xa_\mathfrak{b}$ instead of Xa_1, we can add arbitrarily many expressions of this type; and for the above expression we obtain

$$= \sum Xa_\mathfrak{b} \left[\left(\frac{d}{dx} X \right)^{n+1} a_1 F_\mathfrak{b} \right],$$

where the sum refers to all values $\mathfrak{b} = 1, \dots, 2n+2$. But {(according to 510) this expression

$$= \sum Xa_\mathfrak{b} \left[\frac{d}{dx} X \cdot a_1 \left[\left(\frac{d}{dx} X \right)^{n} F_\mathfrak{b} \right] \right]$$

and (according to 504) this is again}

$$= \frac{1}{2n+1} \sum Xa_\mathfrak{b} \sum \left[\frac{d}{dx} X \cdot a_1 a_\mathfrak{a} \right] \left[\left(\frac{d}{dx} X \right)^{n} F_{\mathfrak{b}\mathfrak{a}} \right],$$

if $F_{\mathfrak{b}\mathfrak{a}}$ has the meaning given above, that is

$$[a_\mathfrak{b} a_\mathfrak{a} F_{\mathfrak{b}\mathfrak{a}}] = [a_1 a_2 \cdots a_{2n+2}],$$

whence $[a_\mathfrak{a} F_{\mathfrak{b}\mathfrak{a}}] = F_\mathfrak{b}$, whereby it follows that $\mathfrak{a}$ must be different from $\mathfrak{b}$ and therefore {for each of the $2n+2$ values of $\mathfrak{b}$} can only assume $2n+1$ different values.

Now if (according to 509) we combine the first and third of the factors under the summation sign, {that is, if we carry out the summation with respect to $\mathfrak{b}$ with fixed $\mathfrak{a}$,} then (according to 509) we obtain that the equation found

$$= - \sum \left[\frac{d}{dx} X \cdot a_1 a_\mathfrak{a} \right] \left[X \left(\frac{d}{dx} X \right)^{n} F_\mathfrak{a} \right],$$

where {$\mathfrak{a}$ now has to run through all values $1, \dots, 2n+2$, and where} the minus sign is to be used, since from

$$[a_\mathfrak{b} a_\mathfrak{a} F_{\mathfrak{b}\mathfrak{a}}] = [a_1 a_2 \cdots a_{2n+2}]$$

it follows that

$$[a_\mathfrak{a} a_\mathfrak{b} F_{\mathfrak{b}\mathfrak{a}}] = -[a_1 a_2 \cdots a_{2n+2}]$$

and thus according to the principle of the above notation it is necessary that $F_\mathfrak{a} = -[a_\mathfrak{b} F_{\mathfrak{b}\mathfrak{a}}]$. The formula obtained is therefore

$$(a) \qquad Xa_1 \left[\left(\frac{d}{dx} X \right)^{n+1} a_1 F_1 \right] = - \sum \left[\frac{d}{dx} X \cdot a_1 a_\mathfrak{a} \right] \left[X \left(\frac{d}{dx} X \right)^{n} F_\mathfrak{a} \right],$$

where according to the above $F_\mathfrak{a}$ is a product whose factors are taken from an arbitrarily chosen series of factors $a_1, a_2, \dots, a_{2n+2}$ in such a way that except for $a_\mathfrak{a}$ all the factors of this series appear in it, and $[a_\mathfrak{a} F_\mathfrak{a}] = [a_1 a_2 \cdots a_{2n+2}]$.

2. Now if

$$\left[X \left(\frac{d}{dx} X \right)^{n} \right] = 0,$$

then the right side of equation (a) is also zero, whence so too is the left. Consequently either Xa_1 or

$$\left[\left(\frac{d}{dx}X\right)^{n+1} a_1 F_1\right],$$

that is

$$\left[\left(\frac{d}{dx}X\right)^{n+1} a_1 a_2 \cdots a_{2n+2}\right],$$

must be zero. Should the former be the case, one could replace a_1 by any other of the magnitudes $a_1, a_2, \ldots, a_{2n+2}$; and even if the product $Xa_r \neq 0$ for only one of them, it follows that the second factor

$$\left[\left(\frac{d}{dx}X\right)^{n+1} a_1 \cdots a_{2n+2}\right]$$

is equal to zero. But if $Xa_1, Xa_2, \ldots, Xa_{2n+2}$ should all be zero, $\frac{d}{dx}(Xa_r)$ would be zero for every index r, and therefore

$$\frac{d}{dx}X \cdot a_r a_s$$

{would be zero for every r and s,} also

$$\left[\left(\frac{d}{dx}X\right)^{n+1} a_1 \cdots a_{2n+2}\right]$$

would equal zero. This expression is therefore zero for every series of factors a_1, $\ldots, a_{2n+2}$, and thus (according to 506)

$$\left[X\left(\frac{d}{dx}X\right)^{n+1}\right]$$

is itself zero.

519. Continuation of the Problem 514–517. As in 514, consider the numerical equation

$$(a) \qquad X dx = 0,$$

in which, as there, x is derivable from a system of m units. There is always a given value of n such that

$$(b) \qquad \left[X\left(\frac{d}{dx}X\right)^{n}\right] = 0,$$

$$(c) \qquad \left[X\left(\frac{d}{dx}X\right)^{n-1}\right] \neq 0.$$

For inequality (c) is always satisfied if $n = 1$, where it reduces to $[X] \neq 0$; and equation (b) is always satisfied if $2n + 1 > m$; for then there is a numerical relation between every $2n + 1$ magnitudes that might fill the openings of the expression

$$X\left(\frac{d}{dx}X\right)^{n}$$

since it was numerically derivable from less than $2n + 1$, namely m units, whence (according to 505) that expression multiplied by any $2n + 1$ magnitudes that fill its openings yields zero; thus it is itself zero (according to 506). This therefore always occurs if $2n+1 > m$, that is $n > \frac{1}{2}(m-1)$, consequently between 1 and $\frac{1}{2}(m+1)$ {for even m between 1 and $\frac{1}{2}m$, the limits always included} there must be a value n for which the above expressions (b) and (c) are satisfied. {Finally, since the existence of equation (b) manifestly follows from the vanishing of the expression

$$\left[X\left(\frac{d}{dx}X\right)^{n-1}\right],$$

there is also *only* one such value n.} Let this value be assumed for n.

Now (according to 514) it is a matter of satisfying the equation

$$2\left[\frac{d}{dx}X \cdot e_s\delta x\right] - \lambda X e_s = 0$$

for every s from 1 to m; since, as soon as this is satisfied *and λ is not zero*, equation (a) is replaced by the equation

$$\frac{X\delta' x}{\mathsf{N}} = 0,$$

in which the left side no longer depends on t and N is defined by formula 514(l).

If for short we denote by G_s the expression

$$G_s = 2\left[\frac{d}{dx}X \cdot e_s\delta x\right] - \lambda X e_s,$$

the above equation can be written

$$0 = G_1 = G_2 = \cdots = G_m.$$

Now I show that, if $m > 2n$, *there is a numerical relation between every* $2n+1$ *of the magnitudes* $G_1, \ldots, G_m$. Assuming there is not yet any numerical relation between the $2n$ magnitudes $G_1, \ldots, G_{2n}$, I show that each of the remaining magnitudes, for example G_m, can be represented as multiple sums of $G_1, \ldots, G_m$.

If we denote by E the product $[e_1e_2\cdots e_{2n}e_{2m}]$ and by $F_\mathfrak{a}$ the product of all the factors of the product E different from $e_\mathfrak{a}$, and in the sense that $[e_\mathfrak{a}F_\mathfrak{a}] = E$, and finally if we denote the expression

$$\left[\left(\frac{d}{dx}X\right)^n F_\mathfrak{a}\right]$$

by $\alpha_\mathfrak{a}$, then, if one refers the following sums to the values 1, 2, ..., $2n$ and m, which $\mathfrak{a}$ is to assume one by one,

$$\sum \alpha_\mathfrak{a} G_\mathfrak{a} = -2\sum\left[\frac{d}{dx}X \cdot \delta x \cdot e_\mathfrak{a}\right]\left[\left(\frac{d}{dx}X\right)^n F_\mathfrak{a}\right] - \lambda\sum X e_\mathfrak{a}\left[\left(\frac{d}{dx}X\right)^n F_\mathfrak{a}\right],$$

which, according to the concept of the product denoted by brackets,

$$= -2(2n+1)\left[\left(\frac{d}{dx}X\right)^{n+1}\delta x \cdot E\right] - \lambda(2n+1)\left[X\left(\frac{d}{dx}X\right)^n E\right].$$

Now (according to (b))

$$\left[X\left(\frac{d}{dx}X\right)^n\right] = 0,$$

and thus (according to 518) also

$$\left[\left(\frac{d}{dx}X\right)^{n+1}\right] = 0,$$

whence the whole right side is equal to zero, and thus also

$$\sum \alpha_{\mathfrak{a}} G_{\mathfrak{a}} = 0,$$

that is, there is a numerical relation between the magnitudes $G_1, \ldots, G_{2n}$ and G_m and generally between any $2n+1$ of the magnitudes $G_1, \ldots, G_m$. Thus among them $2n$ can be taken, say $G_1, \ldots, G_{2n}$, from which the rest are numerically derivable.

Thus if the equations $G_1 = 0, \ldots, G_{2n} = 0$ are satisfied, so also are the rest out to $G_m = 0$.[122] Therefore, from the magnitudes $\delta x_1, \ldots, \delta x_m$, whose ratios are to be defined from the equations $G_1 = 0, \ldots, G_m = 0$, we can choose the magnitudes $\delta x_{2n+1}, \ldots, \delta x_m$ at will, for example all equal to zero, that is, we can regard $x_{2n+1}, \ldots, x_m$ as constant with respect to the differentiation expressed by δ. Therefore, *always under the assumption that* $\lambda \neq 0$, we have the equations $G_1 = 0, \ldots, G_{2n} = 0$ with only $2n$ variables (since we still set $x_{1n+1}, \ldots, x_m$ as constants) and with the condition

$$\left[X\left(\frac{d}{dx}X\right)^{n-1}\right] \neq 0.^{123}$$

Thus (according to 515) we obtain

$$(d) \qquad \delta x \equiv \left[X\left(\frac{d}{dx}X\right)^{n-1} e_1 \cdots e_{2n}\right]$$

or

$$\delta x = \mu \left[X\left(\frac{d}{dx}X\right)^{n-1} e_1 \cdots e_{2n}\right]$$

as a magnitude that satisfies the equations $G_1 = 0, \ldots, G_{2n} = 0$, and thus also (as just shown) the equations $G_1 = 0, \ldots, G_m = 0$, whence also the expression $X\delta' x : \mathsf{N}$ is independent of t, and lets $X dx$ be zero.

If one integrates this congruence according to the method of No. 517 (cf. Remark) by an equation of the form $F(x) = b$, where b denotes the value $x_1 e_1 + \cdots + x_{2n-1} e_{2n-1}$ assumes for $t = x_{2n} = 0$, and sets $a = b + x_{2n+1} e_{2n+1} + \cdots + x_m e_m$, then since $x_{2n+1}, \ldots, x_m$ are independent of t, a is the value that x assumes for $t = 0$; and since then, by virtue of the equations $G_1 = \cdots = G_m = 0$, the magnitude $X\delta' x : \mathsf{N}$ is independent of t, one obtains, on the same grounds as in 517, $X(a)da = 0$ as the equation that replaces the equation $X dx = 0$ or $X(x)dx = 0$, and which depends on only $m - 1$ variable numerical magnitudes, namely on the numerical magnitudes by which a is derivable from the units $e_1, \ldots, e_{2n-1}, e_{2n+1}, \ldots, e_m$.

It was still assumed here that $\lambda \neq 0$. We can replace this assumption by the assumption that

$$\left[\left(\frac{d}{dx}X\right)^{n}\right] \neq 0.$$

That is, just as in 515, one derives the equation (which was denoted by (c) there, namely)

$$\lambda\left[X\left(\frac{d}{dx}X\right)^{n-1}E_r\right] = -2\left[\left(\frac{d}{dx}X\right)^n\right]\delta x_r$$

from the equations $G_1 = G_2 = \cdots = G_{2n} = 0$;[124] in particular, if one sets $r = 2n$, then $\delta x_r = \delta x_{2n} = 1$, and one obtains

$$\lambda\left[X\left(\frac{d}{dx}X\right)^{n-1}E_{2n}\right] = -2\left[\left(\frac{d}{dx}X\right)^n\right].$$

Thus if

$$\left[\left(\frac{d}{dx}X\right)^n\right]$$

is different from zero, then λ must also be different from zero.

Thus *there remains only to investigate the case where*

$$\left[\left(\frac{d}{dx}X\right)^n\right] = 0.$$

But if this equation is satisfied, it follows easily that equation (d) likewise satisfies the problem. For it then follows, as in 516(d), that $X\delta x = 0$, likewise it follows that

$$2\left[\frac{d}{dx}X\cdot e_s\delta x\right] = 2\mu\left[\frac{d}{dx}X\cdot e_s\left[X\left(\frac{d}{dx}X\right)^{n-1}e_1\cdots e_{2n}\right]\right],$$

which, exactly as the corresponding expression in 515, is transformed into

$$= -\frac{\mu}{n}\sum Xe_{\mathfrak{a}}\left[\left(\frac{d}{dx}X\right)^n e_s E_{\mathfrak{a}}\right],$$

where $[e_{\mathfrak{a}}E_{\mathfrak{a}}] = [e_1\cdots e_{2n}]$ and s can be any of the numbers $1, \ldots, m$. The right side is zero under the assumptions made. Consequently

$$\left[\frac{d}{dx}X\cdot e_s\delta x\right] = 0,$$

where e_s can be replaced by each of the units $e_1, \ldots, e_m$, and thus also any multiple sum of them, and thus also $\delta' x$, whence one obtains

$$\left[\frac{d}{dx}X\cdot\delta' x\cdot\delta x\right] = 0,$$

from which, as in 516, there follows

$$\delta(X\delta' x) = 0 \quad \text{and} \quad Xdx = X\delta' x.$$ [125]

Thus the following theorem has been produced:

If in the equation

(a) $$Xdx = 0,$$

in which X is a function of x and x is derivable from the units $e_1, \ldots, e_m$ by the numbers $x_1, \ldots, x_m$, the magnitude X has the property that, for some value n smaller than $\frac{m}{2}$,

(b) $$\left[X\left(\frac{d}{dx}X\right)^n\right] = 0$$

and

(c) $$\left[X\left(\frac{d}{dx}X\right)^{n-1}\right] \neq 0,$$

then equation (*a*) *can be replaced by an equation*

(e) $$Ada = 0,$$

in which a is derivable from $m-1$ *units, that is, it includes only* $m-1$ *variable numbers, and A is the same function of a, as X is of x. And in fact one finds the magnitude a by the integration of the equation*

(d) $$\delta x = \mu\left[X\left(\frac{d}{dx}X\right)^{n-1} e_1\cdots e_{2n}\right],$$

in which δ *denotes the differential quotient with respect to one of the variables* x_1, …, x_{2n}, *for example* x_{2n}, *and thus* $\delta x_{2n} = 1$, *whereby* μ *is defined, and in which* $x_{2n+1}, x_{2n+2}, \ldots, x_m$ *are treated as constant magnitudes.* {*Thereby it is assumed that the units* $e_1, \ldots, e_{2n}$ *are chosen so that the factor of* μ *on the right side of* (*d*) *does not vanish.*}

Now if in this sense equation (d) is integrated by an equation of the form

(f) $$F(x) = b,$$

where b denotes the value that $x_1e_1 + \cdots + x_{2n-1}e_{2n-1}$ *assumes for* $x_{2n} = 0$, *then*

(g) $$a = F(x) + x_{2n+1}e_{2n+1} + \cdots + x_me_m,$$

whence a is derivable from the units $e_1, \ldots, e_{2n-1}, e_{2n+1}, \ldots, e_m$.

520. Corollary 1. *If in the previous number only equations* (*a*) *and* (*b*) *are satisfied, but not inequality* (*c*), *then, whether* $m = 2n$ *or* $> 2n$, *equation* (*a*) *can likewise be reduced to* $m-1$ *variable numerical magnitudes and in fact to an equation of the form* $Ada = 0$, *where A is the same function of a as X is of x, and a is derivable from* $m-1$ *units.*

Proof. For if

$$\left[X\left(\frac{d}{dx}X\right)^{n-1}\right] = 0$$

as well, then since ultimately $[X] \neq 0$, it must be possible to find a numerical value $n' < n$ such that expressions (*b*) and (*c*) hold when n' replaces n. Now since $m >$ or $= 2n$, m is always $> 2n'$, whence according to the preceding theorem one can likewise reduce the equation $Xdx = 0$ to $m-1$ variables, and so forth.

{Remark. According to 519, for odd m one can have $n = \frac{1}{2}(m+1)$, whence $m = 2n-1$. *Only for this case* the reduction of the equation $Xdx = 0$ to $m-1$ variable numerical magnitudes *is not carried out*; in fact, as is shown below (Remark following 525), it is then generally not possible.}[126]

521. Corollary 2. *In particular, one can, if* $m = 2n$, *always reduce the equation* $Xdx = 0$ *to* $m-1$ *variable numerical magnitudes.*

Proof. For then (according to 519) equation (*b*) always holds; and if inequality (*c*) also holds, then (according to 517) the reduction is found to obtain; but if that inequality does not hold, it proceeds according to 520.

522. *If equations* (*a*) *and* (*b*) *hold in the same sense as in* 519, *one can always reduce the equation*

$$Xdx = 0$$

to $2n-1$ *variable numerical magnitudes, and in fact to the form*

$$Ada = 0,$$

where A *is the same function of* a *as* X *is of* x, *and* a *is derivable from* $2n-1$ *units, that is, it includes only* $2n-1$ *variable numerical magnitudes.*

PROOF. Not only will the theorem be proved here, but also it will be shown how the new variable a can be found as a function of x.

According to 520 one can {*if* $m > 2n$} replace $Xdx = 0$ by an equation of the form $A_1da_1 = 0$, where a_1, which is derivable from $m-1$ units, is a known function of x, and A_1 is the same function of a as X is of x. Now since equation (b) holds for every value of x, it also holds if one replaces x by a_1, whence one obtains

$$\left[A_1\left(\frac{d}{da_1}A_1\right)^n\right] = 0.$$

Thus if $m-1$ (the number of units from which a_1 is derivable) is still greater than $2n$, one can apply the method in 519 or 520 once more, and thus obtain an equation of the form $A_2da_2 = 0$, where a_2, which is derivable from $m-2$ units, is a known function of a_1, and thus also of x, and A_2 is the same function of a_2 as A_1 is of a_1, and thus as X is of x. One can proceed in this way so long as the number of remaining variable numerical magnitudes is greater than $2n$; *in fact* (according to 521) *until this number* $= 2n$. By this procedure the number of variable numerical magnitudes is reduced to $2n-1$. If the resulting equation is the equation $Ada = 0$, then a is derivable from $2n-1$ units, and a known function of x, and A is the same function of a as X is of x.

{If $m = 2n-1$, then the original equation $Xdx = 0$ is already of the required form from the beginning.}

REMARK. This theorem includes the most general solution of the problem, begun in 514, of the reduction of the equation to the smallest possible number of variable numerical magnitudes. That, if

$$\left[X\left(\frac{d}{dx}X\right)^{n-1}\right]$$

is not equal to zero (519(c)), Xdx cannot be reduced to less than $2n-1$ variable numerical magnitudes, I will prove incidentally below {cf. the Remark to No. 525}.

But I remark that instead of setting $t = x_{2n}$, one could have set it equal to any given function of the variable numerical magnitudes. For one has only to express one of the latter, for example x_{2n}, by t and the remaining variables and to introduce this expression into the given equation, and then to follow exactly the method given previously. In particular, if the integration of a function should appear which is not continuous for $x_{2n} = 0$, one can set $t = x_{2n} - c$, where c denotes a constant; since then for $t = 0$, $x_{2n} = c$, and c can be chosen so that that function is continuous at $t = 0$, that is at $x_{2n} = c$.

523. PROBLEM. *To reduce the numerical expression* Xdx, *in which* X *is a function of the extensive magnitude* x, *to the form*

$$Xdx = U_1du_1 + \cdots + U_ndu_n,$$

where $U_1, \ldots, U_n, u_1, \ldots, u_n$ are numerical magnitudes, under the assumption that

$$(a) \qquad \left[X\left(\frac{d}{dx}X\right)^n\right] = 0.$$

SOLUTION. One can (according to 522) reduce the equation $Xdx = 0$ to $2n-1$ variable numerical magnitudes that are known functions of x. Denoting an arbitrary member of these variable numerical magnitudes by u_1, u_1 is likewise a known function of x. Now let u_1 be set to a constant; then there remain only $2n-2$ variable numerical magnitudes left. Consequently (according to 521) we can reduce the resulting equation (which according to the above theorem always has the form $Ada = 0$) to $2n-3$ variable numerical functions, which are known functions of the above $2n-1$ variables, and thus also are known functions of x. Let one of them be denoted by u_2, and let u_2 be set to a constant; then one has left only $2n-4$ variable numerical magnitudes, which can be reduced to $2n-5$ such, and so on.

If one has applied this method r times, and therefore so that one has one by one set the magnitudes $u_1, u_2, \ldots, u_r$, all of which are known functions of x, as constants, there still remain only $2(n-r)-1$ variable numerical magnitudes left over. Thus if one sets $r = n-1$, there remains only one variable numerical magnitude left over, and the resulting equation has the form $U_n du_n = 0$, where U_n depends only on the variable numerical magnitude u_n. Consequently if one also sets u_n to a constant, then all the differential equations, and thus in particular the first, will be satisfied, if the functions $u_1, u_2, \ldots, u_n$ are set equal to constants, whence according to the method {of No.} 502 {the expression} Xdx can be represented in the form

$$Xdx = U_1 du_1 + \cdots + U_n du_n,$$

and the problem is solved.

524. *The numerical expression Xdx, in which X is a function of the extensive magnitude x, is derivable from a sum of n terms of the form Udu, where U and u represent numerical magnitudes, if and only if*

$$(a) \qquad \left[X\left(\frac{d}{dx}X\right)^n\right] = 0.$$

PROOF. If equation (a) is satisfied, then the required reduction of Xdx to n terms of the form Udu is feasible (according to 523), and conversely if this reduction is possible, then (according to 511) equation (a) is satisfied.

525. COROLLARY *If x is derivable from $2n$ units, then Xdx can always be reduced to n terms of the form Udu, where U and u are numbers.*

PROOF. For, if x is derivable from $2n$ units, the equation

$$\left[X\left(\frac{d}{dx}X\right)^n\right] = 0$$

is always satisfied (according to 519), and thus (according to 524) Xdx is reducible to n terms of the form Udu.

REMARK. It follows immediately from these theorems that, if

$$\left[X\left(\frac{d}{dx}X\right)^{n-1}\right]$$

is different from zero, the equation $Xdx = 0$ cannot be reduced to less than $2n-1$ variable numerical magnitudes. For if this could be reduced to only $2n-2$ such, and were $Ada = 0$ the equation thus obtained, then (according to 525) Ada could be reduced to $n-1$ terms of the form Udu. But since the equations $Xdx = 0$ and $Ada = 0$ are mutually substitutable, they can only be distinguished by a numerical factor. Let $Xdx = \mathsf{N}Ada$; then Xdx would be reduced to $n-1$ terms of the form Udu; whence (according to 511)

$$\left[X\left(\frac{d}{dx}X\right)^{n-1}\right] = 0,$$

which conflicts with the assumption. Consequently, under the given assumptions the equation $Xdx = 0$ cannot be reduced to less than $2n-1$ variable numerical magnitudes.

This Remark forms a completion, already mentioned above {Remark to No. 522}, to the theorem 522.

526. PROBLEM. *To integrate completely the numerical equation $Xdx = 0$, in which X is a function of the extensive magnitude x.*

SOLUTION. There can always (according to 519) be given a value of n for which

$$\left[X\left(\frac{d}{dx}X\right)^{n}\right] = 0, \quad \left[X\left(\frac{d}{dx}X\right)^{n-1}\right] \neq 0.$$

Then (according to 523, 524) Xdx can always be reduced to n (but not less than n) terms of the form Udu (where U and u are numbers). Let

$$Xdx = U_1du_1 + \cdots + U_ndu_n = 0;$$

then if one seeks all the integrating assemblies, each of n equations, to the equation $U_1du_1 + \cdots + U_ndu_n = 0$, (according to 503) these assemblies also integrate the equation $Xdx = 0$ identical to it.

527. COROLLARY. *The equation*

$$\left[X\left(\frac{d}{dx}X\right)^{n}\right] = 0$$

is the necessary and also the sufficient equation of condition for the equation $Xdx = 0$ to be integrated by assemblies of n equations.

REMARK. According to 500, with the complete integration of the numerical equation $Xdx = 0$, that of the partial differential equation of first order is simultaneously completed; whereas (according to 501) the integration of the partial differential equation of higher order reduces to the integration of the extensive equation $Xdx = 0$, which we have excluded here.

Index of Technical Terms

(with reference to Numbers in which they are defined)

Editorial Notes

Numbers enclosed [in brackets] identify the text page where the Editorial Note is cited.

The *Ausdehnungslehre* of 1862, which, following GRASSMANN's precedent, we denote by A_2, bears the date 29 August 1861 under the Foreword, and was also published in 1861. GRASSMANN had had it printed at his own expense in a run of 300 copies, giving the commission for the printing to ENSLIN in Berlin; in any case the year 1862 appearing on the title page was chosen for the sake of the book dealers.

[xiii] 1. The articles in *Crelles Journal* relate to the generation of algebraic curves and surfaces, the so-called higher projectivity in the plane and in space, and to the various types of multiplication.

[xiii] 2. This article was later reprinted as Appendix III to the Second Edition of the A_1.

[xiv] 3. The *Lehrbuch der Arithmetik* was originally published in 1860 by R. GRASSMANN in Stettin; it was then reprinted on commission to ENSLIN, and provided a new title with the date 1861.

[xvi] 4. The planned work on the important branches of physics was never published. Because the *Ausdehnungslehre* at first received no attention from professional mathematicians, for a number of years GRASSMANN turned completely to the philological studies he had begun earlier. Later works related to physical problems are two articles on mechanics and one each on electrodynamics and acoustics.

[xvi] 5. GAUSS's letter is dated 14 December 1844. An extract follows:

" ... leafing through your book amongst a host of other heterogeneous works, I think I perceive that its tendency has *in part* been encountered on that path which I myself traveled almost half a century ago, of which admittedly only a small part is mentioned in the 1831 *Comment. der Göttingischen Societat* and again in the *Göttingischen Gelehrten Anzeigen* (1831, p. 64), that is the concentrated *metaphysics* of complex magnitudes; while the infinite fruitfulness of this principle for the investigation of spatial relations was abundantly dealt with in my lectures, proofs of it were communicated, with other motivations, only here and there, and discernible *as such* only to the observant eye. In them there is only a partial and distant similarity with your work; and indeed I perceive that, in order to discern the essential core of your work it is first necessary to become familiar with your special terminology. Since however that will require of me a time free of other duties, I must no longer delay expressing my sincere thanks to you for so kindly sending me your work ... "

The point in the *Göttingischen Anzeigen* to which GAUSS refers appears in his *Works* in Vol. II, p. 175ff.

[xvii] 6. The full title of this journal is *Annali delle scienze del Regno Lombardo-Veneto.* Opera periodica di alcuni collaboratori. Padova (from 1831). In it appear numerous articles by BELLAVITIS, of which the following are relevant:

"Sulla Geometria derivata", **2**, 250 (1831). This includes the presentation of the sum of two straight lines by the diagonal and its known significance for the imaginary magnitudes; this means one is led from that theorem on points in a *straight line* to such a theorem on points in a *plane*, whence the *derivata.*

"Saggio di applicazioni di un nuovo metodo di Geometria analitica (Calcolo delle Equipollenze); Memoria di Giusto Bellavitis di Bassano", **5**, 244–259 (1835). The author had already presented the method in the "Ateneo veneto" in September 1832, and used it for the derivation of the characteristics of conic sections, publishing in the "Poligrafo di Verona", Gennaio 1833.

"Teoria delle figure inverse e loco uso nella Geometria elementare", **6**, 126 (1836). The theory of reciprocal radius-vectors and of equipollences are interpreted as special cases of a general principle of transference (Geometria derivata).

"Memoria sul Metodo delle equipollenze", **7**, 243–261 (1837), **8**, 17–37, 85–121 (1838). Complete presentation of the theory sketched in Vol. **5**.

[xvii] 7. The title of this article is "Mémoire sur les sommes et les différences géometriques, et sur leur usage pour simplifier la Mécanique".

[xvii] 8. The first two articles are entitled "Sur les clefs algébriques", the third "Sur les advantages que presente, dans un grand nombre de questions, l'emploi des clefs algébriques". On the same subject there is another article in Vol. **36**, 38–45, "Sur la théorie des moments lineares des divers ordres", and another in Vol. **37**, 57–68 (1853), "Mémoire sur les differentielles et les variations employées comme clefs algébriques".

[xvii] 9. The session of the Academy referred to was on 17 April 1854. One reads the following in the *Comptes Rendus* reference cited:

"*Mathematical analysis.* Extract of a memoir of M. GRASSMANN.

"'I request that the Academy of Science take notice of the priority claim I find it necessary to make on the occasion of two articles: "Sur les clefs algébriques", by M. CAUCHY, and "De l'interprétation {géométrique} des clefs algebriques et des déterminants", by M. de SAINT-VENANT,* inserted in the *Comptes Rendus* **36**, pages 70, 129, 582. I have, in the year 1844, published the principles established in these articles, and the results deduced from them by the two geometers I have just named. I have the honor to present to the Academy the work in which these ideas are contained (1) and certain articles published subsequently on the same subject (2), and I would be pleased if the Academy of Sciences would accept these works. To support my claim I take the liberty of communicating to you an extract of my researches reporting on that subject and contained in the works cited, noting where in these works each question is to be found.

"'All these researches are based on quantities that I call *extensive magnitudes*, which are, in essence, no different from the *facteurs symboliques* and *clefs algébriques*

*In this article SAINT-VENANT does in fact mention GRASSMANN's name in a footnote on p. 584, but only with respect to the concept of the inner product, which GRASSMANN had developed in the *Geometrische Analyse.*

of M. CAUCHY. But as the point of view of these quantities that I envisage is completely different from that of M. CAUCHY it is necessary to go into certain details. That is the object of the Note that I have the honor to submit today to the judgment of the Academy.'

"That note, in its nature scarcely susceptible to analysis, cannot be reproduced *in extenso* on account of its length. It was referred to the examination of a Commission composed of MM. CAUCHY{!}, LAMÉ, and BINET."

MÖBIUS brought the articles by CAUCHY and SAINT-VENANT to GRASSMANN's attention in a letter of 2 September 1853, and urged him to pursue his priority claim.

[3] 10. *Barycentrische Kalkül*, Chapter 2, §15; *Werke*, Vol. I, p. 39.

[27] 11. By the "principle" that GRASSMANN posits here he excludes from the outset any consideration of numerical relations between the unit products and the original units, and limits himself to numerical relations between unit products of equally many factors. In this way he is barred from proceeding to the system of higher complex numbers, whose concept HAMILTON had already presented in full generality in his *Lectures on Quaternions* of 1853, and whose theory has been further developed more recently by various mathematicians.

In addition it is remarkable that in this general investigation of the various types of product structure GRASSMANN considered only products of *two* factors, and not also those of *three* factors. Not once does he mention that the product structure in which every set of three units satisfies the equation

$$(e_i e_k)e_j = e_i(e_k e_j),$$

whence the so-called *associative* law holds, likewise belong to the *linear* product structures, and only later, in No. 78, does he introduce the associative law, but without mentioning its significance. This is the more surprising, since in the A_1 he develops the associative law, or, as he says, the law of the combinability of the terms of a conjunction, right at the beginning (in §3) with the greatest acuteness and clarity.

Had GRASSMANN sought to define all possible species of linear product structures for products of *three* factors as well, he would perhaps have arrived at a different conception of the importance of his "principle", for he would then have hit upon completely new product structures, which he could not have provided for in his system. Naturally we ought not to engage upon treating the problem suggested here, so we will mention only two types of linear product structure that enter with products of three factors.

The defining equations of the first type have the form

$$(e_i e_k)e_j = \mathfrak{a} e_i(e_k e_j) \quad (i, k, j = 1, \ldots, n), \tag{1}$$

and those of the second read as follows:

$$\begin{cases} (e_i e_k)e_j + (e_k e_j)e_i + (e_j e_i)e_k = 0, \\ (e_i e_k)e_j = \mathfrak{b} e_j(e_i e_k), \end{cases} \tag{2}$$

where $\mathfrak{a}$ and $\mathfrak{b}$ denote numbers. In the first case one obtains the associative product structure for $\mathfrak{a} = 1$. In the second case, if one sets $\mathfrak{b} = -1$ and in addition adopts the defining equation

$$e_i e_k + e_k e_i = 0 \quad (i, k = 1, \ldots, n), \tag{3}$$

one obtains a linear product structure of a completely different type, which plays an important rôle in LIE's theory of transformation groups. In this product structure the units $e_1, \ldots, e_n$ are independent infinitesimal transformations of an arbitrary space and the product $e_i e_k$ is the POISSON bracket formed from two infinitesimal transformations; the first of equations (2) is then nothing but the famous JACOBI identity. In this connection, see Sophus LIE, *Theorie der Transformationsgruppen*, Vol. III, p. 747ff.

[33] 12. The sign convention chosen by GRASSMANN corresponds to the sign rule of CRAMER, cf. his *Introduction á l'analyse des lignes courbes*, Genf 1750, Appendix, p. 657f. CAUCHY's sign convention is found e.g. in BALTZER's *Determinanten*.

[33] 13. The multiplication theorem for determinants follows at once from this theorem. Thus if one sets

$$a_k = \beta_1^{(k)} b_1 + \cdots + \beta_n^{(k)} b_n,$$

then according to No. 63 one can express both sides of the final equation of this number by the combinatorial product $[b_1 \cdots b_n]$, and if this does not vanish, one obtains, with regard to No. 32, an equation that is nothing but the multiplication theorem for determinants. In his *Theorie der complexen Zahlensysteme*, Leipzig 1867, p. 122f, HANKEL published this proof of the multiplication theorem, probably as a consequence of a letter from GRASSMANN.

[38] 14. The concept of the elementary linear evolution is in the following always applied to the case that the linearly evolved series of magnitudes is formed from n extensive magnitudes that are derivable from the n original units and stand in no numerical relation to one another. Under this assumption the elementary linear evolution is equivalent to a linear homogeneous transformation of a special form.

In fact the aggregate of all magnitudes derivable from $a_1, \ldots, a_n$ form a domain of nth order, which (according to 21 and 24) coincides with the domain defined by $e_1, \ldots, e_n$; the general form of a magnitude of this domain is $x_1 a_1 + \cdots + x_n a_n$. Now if one subjects this series of magnitudes $a_1, \ldots, a_n$ to an elementary linear evolution, perhaps by replacing a_m with $a_m + \alpha a_{m+1}$ and leaving the remaining a_k unchanged, then each magnitude $\sum x_i a_i$ of our domain of nth order is changed into a magnitude $\sum x_i a_i + \alpha x_m a_{m+1}$, which likewise belongs to this domain. If we write these new magnitudes in the form $\sum x_i' a_i$, we obtain between the x_i and the x_i' the relations

$$(*) \; x_1' = x_1, \; \ldots, \; x_m' = x_m, \; x_{m+1}' = x_{m+1} + \alpha x_m, \; x_{m+2}' = x_{m+2}, \; \ldots, \; x_n' = x_n.$$

This however is evidently a linear homogeneous transformation of unit determinant, which indicates how the magnitudes of our domain of nth order interchange among themselves under that elementary linear evolution. If one applies several elementary linear evolutions, one after the other, this is equivalent to the execution of several of that type of linear homogeneous transformations, one after the other, and the linear evolution by which the first series is transformed into the last thus likewise amounts to a linear homogeneous transformation of unit determinant.

It should be explicitly observed that the ∞^1 linear homogeneous transformations of the special form $(*)$ taken together form a *one-parameter group* in the sense of LIE.

[38] 15. That linear evolutions in geometry can be carried out by means of straightedges is demonstrated in No. 254. On circular evolution cf. Nos. 154 and 331.

[44] 16. If $[a_1 \cdots a_m] = [b_1 \cdots b_m] \neq 0$, then according to 70 $b_1, \ldots, b_m$ can be derived from $a_1, \ldots, a_m$, whence

$$b_k = \alpha_{k1}a_1 + \cdots + \alpha_{km}a_m \quad (k = 1, \ldots, m)$$

where the determinant of the α_{kv} is certainly not zero and in addition by virtue of the equation $[a_1 \cdots a_m] = [b_1 \cdots b_m]$ has the value one according to Nos. 63 and 32. Simultaneously the domain of mth order defined by $a_1, \ldots, a_m$ is identical to the domain defined by $b_1, \ldots, b_m$. Thus if in an arbitrary magnitude $\sum x_v a_v$ of this domain one replaces the a_v by the b_v, then the magnitudes of the domain are interchanged among themselves and in fact, as one easily perceives, by a linear homogeneous substitution of unit determinant; for we can set $\sum x_v b_v = \sum x'_v a_v$ and obtain thereby between the x and x' the relation

$$x'_v = \alpha_{1v}x_1 + \cdots + \alpha_{mv}x_m \quad (v = 1, \ldots, m),$$

that is a genuine linear homogeneous transformation of this type.

If one recalls Editorial Note 14, then one recognizes that Theorem 76 is equivalent to the following: Every linear homogeneous substitution of unit determinant can be obtained by successively applying a number of substitutions of a special form, that is substitutions of the form (*), which according to Editorial Note 14 express an elementary linear evolution.

[44] 17. This corollary is necessary since the numbers in Nos. 84 and 95 are treated as magnitudes of zeroth order without comment, and especially since No. 95 refers to No. 77. But we observe that the product of an elementary magnitude and a number is again an elementary magnitude. For according to 46 one can move the number to any given factor of first order of the product, but this factor remains a factor of first order.

[45] 18. Here it is assumed that $[(pq)(pq)] = [pqpq]$, although as yet the multiplication of two magnitudes of higher order is not sufficiently defined that one knows whether the parentheses can be dropped or not; only in No. 87 will a stipulation on this point be found. GRASSMANN probably noticed this oversight later and consequently repeated the whole treatment in the Remark following No. 88.

[50] 19. From this theorem one easily obtains the following:

THEOREM. *In a principal domain of nth order every magnitude of $(n-1)$th order is elementary.*

Thus according to No. 88 it is clear, first of all, that every sum of arbitrarily many elementary magnitudes of $(n-1)$th order in a principal domain of nth order is again an elementary magnitude. Now since every magnitude of $(n-1)$th order can be represented as a sum of elementary magnitudes of $(n-1)$th order, that is as a sum of products of each unit of $(n-1)$th order and a numerical magnitude, and since (according to 77) these units and (according to the Remark following No. 77) these products are also elementary magnitudes, in general every magnitude of $(n-1)$th order in a principal domain of nth order is elementary.

In the Remark following No. 88 GRASSMANN mentions that a magnitude of mth order in a principal domain of nth order is not in general elementary if $1 < m < n-1$, but he demonstrates this only for the simplest case: $n = 4$, $m = 2$, and passes over the case $n > 4$ in silence. Only at a *single* later point in this work does GRASSMANN touch on the question, namely in No. 286, where he shows how one can determine whether a given sum of line elements in space is again a line element or not, or in other words, whether a given magnitude of second order in a domain of fourth order is elementary or not.

Moreover, precisely the same lacuna is already found in the A_1 (cf. there §51); there also only the case $n = 4$, $m = 2$ is considered (§124).

Later, in his last mathematical article,* which was inspired by the work of REYE, GRASSMANN returned once more to the question and settled it by specifying the conditions which a magnitude of qth order, expressed in terms of the units of qth order, in a principal domain of $(q + s)$th order, must satisfy if it is to be elementary. The conditions at which he arrived are certain equations of second, third, ... , qth degree between the coefficients of the magnitude of qth order under consideration.

[50] 20. The supplement symbol | was read as "in" by GRASSMANN's sons, a reading that probably originated with GRASSMANN himself.

[50] 21. The transition from the magnitude of a principal domain to its supplement is an operation which from the standpoint of projective geometry is considered as belonging to the reciprocities (dualistic transformations), and indeed to the special reciprocities that one calls *polar systems.*

If $a = \sum x_k e_k$ is an arbitrary magnitude of first order, then its supplement $|a = \sum\{x_k|e_k\}$ is a magnitude of $(n-1)$th order, which according to No. 88 and Editorial Note 19 is certainly elementary. The domain of these magnitudes consists of all magnitudes of first order $b = \sum y_k e_k$ subordinate to them, and which therefore satisfy the equation $[ba] = 0$. But now according to No. 143,

$$\left[\sum y_k e_k \middle| \sum x_j e_j\right] = \sum x_k y_k,$$

whence the domain of $|\sum x_j e_j$ consists of all magnitudes of first order $\sum y_j e_j$ for which $\sum x_k y_k = 0$. Thus if we interpret the units $e_1, \dots, e_n$ as n points of an $(n-1)$-fold extended plane that do not lie in an $(n-2)$-fold extended plane, and thereby the $x_1, \dots, x_n$ as homogeneous coordinates of the points of this space (cf. Nos. 232 and 238), the domain of the supplement of $\sum x_k e_k$ is nothing but the $(n-2)$-fold extended polar plane of the points $x_1, \dots, x_n$ with respect to the manifold of second degree $\sum x_k^2 = 0$.

We obtain another interpretation of the supplement if we take $e_1, \dots, e_n$ to be such displacements (infinitely distant points) of an n-fold extended space as are not parallel to an $(n-1)$-fold extended plane of this space (cf. No. 229), and if at the same time we interpret the numbers $x_1, \dots, x_n$ as right- or oblique-angle parallel coordinates. Then the supplement of $\sum x_k e_k$ is the product of $n-1$ displacements that are parallel to the $(n-1)$-fold extended polar plane of the infinitely distant points of the displacement $\sum x_k e_k$ with respect to an arbitrary member of the ∞^1 manifolds of second degree

$$\sum x_k^2 = \text{const.}$$

* "Verwendung der Ausdehnungslehre fur die allgemeine Theorie der Polaren und den Zusammenhang algebraischer Gebilde", *Crelles Journal* **84**, 273–283 (1878), esp. p. 281f.

Of these two, to be sure not essentially different, interpretations, GRASSMANN only applied the second in the A_2, and even this only for right-angle parallel coordinates. Cf. the Remark following No. 93 and Nos. 330, 331, 335. The first interpretation is found in GRASSMANN's article in *Crelles Journal* **84**, 273–283 (1878) cited in Editorial Note 19. Cf. also R. STURM's remarks in GRASSMANN's obituary notice, *Math. Ann.* **14**, 16 (1878).

[53] 22. By this definition all outer products of two factors, the sum of whose orders exceeds that of the principal domain, are excluded from consideration. This seems appropriate, since all such products would have the value zero.

[53] 23. This may be claiming too much; cf. Editorial Note 11.

[54] 24. In the appeal to No. 79 there is a fundamental idea that deserves to be expressed as a special theorem between Nos. 94 and 95, perhaps thus:

94A. *If the orders of the two elementary magnitudes A and B together are not greater than the order n of the principal domain, then one obtains the progressive product of A and B by combinatorially multiplying the series of the elementary factors of A with the elementary factors of B, thus for $q + r \leq n$:*

$$[(a_1 a_2 \cdots a_q)(b_1 b_2 \cdots b_r)] = [a_1 a_2 \cdots a_q b_1 b_2 \cdots b_r].$$

PROOF. According to No. 94 the progressive product of A and B is an outer product of these magnitudes; the rest follows from No. 79.

[56] 25. In the original edition of the A_2 GRASSMANN did not explicitly consider the case of even n; the material enclosed by braces was inserted by the editors of the Teubner Edition to fill the gap.

[57] 26. The converse in braces is necessary because later it is used several times, for example in Nos. 119, 119b, 120, and 124.

[58] 27. Cf. Nos. 112, 116b, 119b, 127–131, 306–322, 377, and the Remark following No. 383.

[59] 28. The product $[QR]$ is certainly progressive. Thus if one denotes the orders of E, F, G, Q, R by the corresponding lower case letters, then

$$q = n - (e + f), \quad r = n - (e + g),$$

whence

$$q + r = 2n - 2e - f - g,$$

or, since $e + f + g = n$, $q + r = n - e$, that is

$$q + r < n,$$

from which it follows that the product $[QR]$ is in fact progressive.

[61] 29. The beginning of the proof of No. 103 does not exactly harmonize with the rest, since later it is obviously assumed that each of the n factors of first order $a_1, a_2, \ldots$ is only included in *one* of the three magnitudes A, B, C, which cannot easily be inferred from the introductory words. This difficulty is eliminated if one gives the beginning of the proof the following wording: "Suppose $a_1, \ldots, a_n$ is any given series of n magnitudes of first order, and assume that formula 103 always holds if the n elementary factors included in A, B, C, taken in any given order, correspond to the series of magnitudes $a_1, \ldots, a_n$; then I will show" etc. One must then observe that the $a_1, \ldots, a_n$ are completely arbitrary, and thus may stand in

a numerical relation to one another, indeed they need not even be different from one another.

The proof of No. 103 is therefore the more remarkable, as it permits the invariant character of the equation

$$[AB \cdot AC] = [ABC]A$$

to appear most clearly. It is based on the fact that this equation continues to hold if one replaces the elementary factors of A, B, C with arbitrary magnitudes numerically derivable from them, and since the equation is proved in No. 102 for products of units, it is thereby proved in general. If we recall Editorial Note 16, we can also obviously say: *The above equation remains* ***invariant*** *if one subjects the magnitudes* $\sum x_k e_k$ *of the principal domain* $e_1, \ldots, e_n$ *to an arbitrary linear homogeneous transformation* $x_i' = \sum \alpha_{ik} x_k$.

We must leave it to the reader to make clear this invariant character of the equations of extension theory in each individual case. We will only indicate an especially striking example.

It should not go unremarked that the equation of No. 103 is introduced in the A_1 in a completely different way than it is in the A_2. In the A_1 this equation serves to define the regressive product (A_1, §132 and 133). In the A_2 on the other hand the regressive product is defined in the simplest conceivable way on the basis of a new concept, the concept of the *supplement*, in No. 94, and then the equation of No. 103 is *proved*, whereby the concept of the regressive product is again made independent of that of the supplement.

One can use the formula of No. 103,

$$[AB \cdot AC] = [ABC]A,$$

where the sum of the orders of A, B, C equals the order of the principal domain, for the derivation of a series of theorems about which indeed GRASSMANN said nothing explicitly, but which properly belong in his system, and of which he has implicitly made use, in part. We will formulate and prove these theorems here.

THEOREM 1. *A product of arbitrarily many elementary magnitudes is always again an elementary magnitude.*

We obviously need only prove this theorem for a product of two factors. Thus suppose L and M are two nonvanishing elementary magnitudes with orders λ and μ. If n is the order of the principal domain and $\lambda + \mu \leq n$, then the product $[LM]$ is progressive and is therefore certainly an elementary magnitude.* On the other hand, if $\lambda + \mu > n$, then according to No. 87 L and M can be represented in the form

$$L = [AL_1], \quad M = [AM_1],$$

where A, L_1, M_1 are elementary magnitudes and in particular A is of $(\lambda+\mu-n)$th order. Moreover,

$$[LM] = [AL_1 \cdot AM_1] = [AL_1M_1]A,$$

since the assumptions of No. 103 are satisfied, and here $[AL_1M_1]$ is a progressive product of nth order, and therefore a number. Consequently the product $[LM]$ is also an elementary magnitude in the case $\lambda + \mu > n$. Our theorem is therefore proved.

---

*For $\lambda + \mu = n$ the product is a number; but the numbers must be reckoned among the elementary magnitudes.

THEOREM 2. *The supplement of an elementary magnitude is always again an elementary magnitude.*

PROOF. That the supplement of a magnitude of *first order* is again an elementary magnitude is obvious; for this supplement is of $(n-1)$th order, and according to No. 88 and Editorial Note 19 every magnitude of $(n-1)$th order is elementary.

On the other hand, if A is an elementary magnitude of *mth order*, then one can set $A = [a_1 \cdots a_m]$, where $a_1, \ldots, a_m$ are magnitudes of first order. Then according to No. 98

$$|A = [|a_1|a_2 \cdots |a_m].$$

Here, as shown above, $|a_1$, $|a_2$, ... are each elementary magnitudes of $(n-1)$th order, whence so also is their product according to Theorem 1 above, that is the supplement of A is an elementary magnitude.

The third theorem we will prove here is the converse of No. 79B, and can be expressed in the following way:*

THEOREM 3. *If A and B are two nonzero elementary magnitudes and if A is subordinate to B, then A can be represented in the form*

$$A = [BC],$$

where C is an elementary magnitude and the product $[BC]$ is regressive.

PROOF. Suppose α and β are the orders of A and B, and according to 79B let B be represented as a progressive product in the form $B = [AL]$, where L is an elementary magnitude of $(\beta - \alpha)$th order. Further, suppose D is an elementary magnitude of $(n-\beta)$th order that has no domain of first order in common with B, nor therefore with A, so that $[BD]$ is a number different from zero and $[AD]$ is a progressive product different from zero. In particular we imagine D so chosen that $[BD] = 1$. The sum of the orders of A, L, D is then equal to n and therefore according to No. 103

$$[B \cdot AD] = [AL \cdot AD] = [ALD]A = [BD]A = A,$$

or, setting the progressive product $[AD] = C$,

$$A = [BC].$$

The representation claimed for A is thus constructed, for the product $[BC]$ is manifestly regressive, since the order of C is $\alpha + n - \beta$.

From the last equation it follows according to 97 that

$$|A = [|B|C],$$

where $|A$, $|B$, $|C$ are again elementary magnitudes according to Theorem 2, and where (according to 115) the product $[|B|C]$ is now regressive. From this there follows

THEOREM 4. *If A and B are elementary and A is subordinate to B, then the supplement $|A$ is superordinate to the supplement $|B$.*

Finally, from Theorems 2 and 4 there follows

THEOREM 5. *The supplements of two incident elementary magnitudes are again incident elementary magnitudes.*

---

*In the A_1 this is found completely incidentally as a footnote to §153.

[63] 30. It is assumed here that

$$[ACD] = [A \cdot CD] \quad \text{and} \quad [A \cdot BC] = [ABC],$$

which is not permitted without further details. One can remove this objection if, as with the proof of formula 105, one distinguishes the two cases in which the sum of the orders of A, B, C equals n, and when it equals $2n$.

In the *first* case the proof follows exactly as given in the text; for if one sets $[BC] = [CD]$, then D is of the same order as B, whence

$$[AC \cdot BC] = [AC \cdot CD] = [ACD]C.$$

Now here the products $[AC]$ and $[ACD]$ are progressive, since for both, by virtue of the assumption, the sum of the orders of their factors is not greater than n. Thus (according to 80)

$$[ACD] = [A \cdot CD],$$

or, since $[CD] = [BC]$,

$$[ACD] = [A \cdot BC] = [ABC],$$

and thus as a matter of fact

$$[AC \cdot BC] = [ABC]C.$$

In the *second* case the sum of the orders of A, B, C is equal to $2n$, whence one proceeds exactly as in the proof of No. 105.

[63] 31. As before let α, β, γ be the orders of the elementary factors A, B, C. Since the two cases $\alpha + \beta + \gamma = n$ and $= 2n$ were already disposed of in Nos. 105–107, we need only consider the cases in which $\alpha+\beta+\gamma < n$, $n < \alpha+\beta+\gamma < 2n$, and $\alpha + \beta + \gamma > 2n$. However, by virtue of No. 101 the last immediately reduces to the first if one takes the supplements on both sides of the equations to be proved. For the supplements $|A$, $|B$, $|C$ are according to Theorem 2 of Editorial Note 29 again elementary magnitudes, and of orders $n - \alpha$, $n - \beta$, $n - \gamma$, and from $\alpha + \beta + \gamma > 2n$ it follows that

$$n - \alpha + n - \beta + n - \gamma < n.$$

Thus only the first two cases remain to be considered.

We begin with the proof of the *first* formula,

$$[AB \cdot AC] = [A \cdot ABC].$$

Case I. $\alpha + \beta + \gamma < n$. If $2\alpha + \beta + \gamma \leq n$, then both products $[AB \cdot AC]$ and $[A \cdot ABC]$ are progressive, and according to No. 109 both are zero, since the domains of the two factors always have a domain of αth order in common. If on the other hand $2\alpha + \beta + \gamma > n$, then both products are regressive and again according to 109 are zero, since the elementary magnitudes A, $[AB]$, $[AC]$, $[ABC]$ can all be contained in a domain of $(\alpha + \beta + \gamma)$th order at most.

Case II. $n < \alpha + \beta + \gamma < 2n$. If $[AB]$ vanishes, then both sides of the equation to be proved are zero; on the other hand if $\alpha + \beta = n$, then $[AB]$ is a number that can be placed anywhere in the product, and since

$$[ABC] = [(AB)C],$$

the equation to be proved is correct as well. *We can therefore assume* that $[AB] \neq 0$ and $\alpha + \beta \neq n$.

We must now consider separately four different subcases, when $\alpha + \beta < n$ or $> n$, and $\alpha + \gamma \leq n$ or $\geq n$.

Subcase 1. $\alpha+\beta<n$, $\alpha+\gamma\leq n$, whence $2\alpha+\beta+\gamma<2n$. If A, B, and C are included in a domain of lower than nth order, then this is also true of the magnitudes C, $[AB]$, and $[AC]$; the regressive products $[AB\cdot AC]$ and $[(AB)C]=[ABC]$ are therefore according to 109 both zero. We therefore assume that A, B, C *can be contained in no domain of less than nth order.*

Under this assumption $[AB]$ and C have a domain of $(\alpha+\beta+\gamma-n)$th, but no higher, order in common, and therefore according to No. 87 can be represented in the form

$$[AB]=[DL],\ C=[DC_1],$$

where the sum of the orders of D, L, and C_1 is equal to n. Thus according to No. 108

$$[ABC]=[DL\cdot DC_1]=[DLC_1]D,$$

where $[DLC_1]=[ABC_1]$ is a number.

Now it is a question of whether A and C have a domain in common or not.

If A and C have a domain of first or higher order in common, then the same is true of A and D, and thus according to No. 109 the progressive products $[AC]$ and $[AD]$ are both zero. Consequently

$$[A\cdot ABC]=[DLC_1][AD]=0=[AB\cdot AC],$$

that is, the equation to be proved is satisfied.

If however A and C have no domain of first or higher order in common, then the same is true of A and D, so that $[AD]$ and $[AC]$ are both $\neq 0$, and in fact

$$[AC]=[A(DC_1)]=[(AD)C_1],$$

since the product $[A(DC_1)]$ is purely regressive (cf. Nos. 116 and 119). On the other hand, according to our assumption $[AB]$ and $[AC]$ are included in no domain of lower than nth order and therefore have a domain of $(2\alpha+\beta+\gamma-n)$th, but no higher, order in common; at the same time it is clear that $[AD]$ is a magnitude of $(2\alpha+\beta+\gamma-n)$th order in this domain. Consequently $[AB]$ can be represented in the form

$$[AB]=[(AD)B_1],$$

and it follows from 103 that

$$[AB\cdot AC]=[(AD)B_1\cdot(AD)C_1]=[(AD)B_1C_1][AD]=[ABC_1][AD],$$

while the value previously found for $[ABC]$ yields

$$[A\cdot ABC]=[DLC_1][AD]=[ABC_1][AD].$$

Thus the two expressions are equal.

The first subcase is therefore completed.

Subcase 2. $\alpha+\beta>n$, $\alpha+\gamma\leq n$. Since $[AB]\neq 0$, the domains of A and B have a domain of $(\alpha+\beta-n)$th, but no higher, order in common. If M is a magnitude in this domain, then we can set

$$A=[MA_1],\ B=[MB_1],$$

and according to No. 103

$$[AB]=[MA_1\cdot MB_1]=[MA_1B_1]M,$$

where $[MA_1B_1]$ is a number, whence

$$[AB\cdot AC]=[MA_1B_1][M\cdot AC].$$

On the other hand one has

$$[A \cdot ABC] = [MA_1B_1][A \cdot MC].$$

But $[A \cdot MC] = [MA_1 \cdot MC]$, and here the sum of the orders of M, A_1, and C equals $\alpha + \gamma$, and thus $\leq n$, whence we come back either to Case I or to No. 103 and find that

$$[A \cdot MC] = [MA_1 \cdot MC] = [M \cdot MA_1C] = [M \cdot AC],$$

and thus again it follows that

$$[AB \cdot AC] = [A \cdot ABC].$$

Subcases 3 and 4. The two remaining subcases, $\alpha + \beta > n$, $\alpha + \gamma \geq n$ and $\alpha + \beta < n$, $\alpha + \gamma \geq n$, reduce to 1 and 2 if one takes the supplement on both sides of the equation to be proved.

The *second* formula is an immediate consequence of the first. Thus one has $[ABC] = [(AB)C]$ and $[AB] = \pm[BA]$, whence from the first formula it follows that

$$[BA \cdot AC] = [A \cdot (BA)C] = [A \cdot BAC],$$

which is precisely the second formula.

The *third* formula,

$$[AC \cdot BC] = [C \cdot ABC],$$

can, for the cases $\alpha + \beta + \gamma \leq n$ and $\geq 2n$ be proved precisely as was the first. However, *in the case $n < \alpha + \beta + \gamma < 2n$ it is not always valid*, as one can best see by considering an example.

Let $n = 5$ and

$$A = [e_1e_2], \; B = [e_1e_3], \; C = [e_4e_5];$$

then $[AB]$ and $[ABC]$ are both zero, but on the other hand

$$[AC] = [e_1e_2e_4e_5] = [e_1e_4e_5e_2], \; [BC] = [e_1e_3e_4e_5] = [e_1e_4e_5e_3],$$

and therefore

$$\begin{aligned}[AC \cdot BC] &= [e_1e_4e_5e_2 \cdot e_1e_4e_5e_3] \\ &= [e_1e_4e_5e_2e_3][e_1e_4e_5] \\ &= [e_1e_4e_5],\end{aligned}$$

and is thus not zero, whence the third formula is proved incorrect for this case.

[63] 32. Both formulas still remain valid if B is superordinate to A, *and thus generally if A and B are incident.* Since the first of the two formulas in No. 129 applies in this more general case, we will at once produce the proof of the two formulas for the case not treated in the text.

Thus suppose A, B, C are elementary magnitudes with orders α, β, γ, the sum of the orders of A and C, that is $\alpha + \gamma$, equals n, and A is subordinate to B. Thus according to Editorial Note 29, $|A$, $|B$, $|C$ are elementary magnitudes and $|B$ is subordinate to $|A$, and in addition the sum of the orders of $|A$ and $|C$ equals $n - \alpha + n - \gamma = 2n - n = n$. Consequently, according to No. 108,

$$[|A \cdot (|B|C)] = [|A|C]|B,$$
$$[|C|B \cdot |A] = [|C|A]|B,$$

whence according to No. 101 the equations of No. 108 also remain valid in the case just considered.

[66] 33. The editors of the Teubner Edition replaced the material in braces with the following:

First, if $[AB] = 0$, then since $\alpha + \beta > n$, the domains of A and B must (according to 109) be included in a domain of lower than nth order. Therefore not all n elementary factors $a_1, \dots, a_n$ appear in A and B, but rather only a certain number δ of them, where $\delta < n$, while the other $n - \delta$ are missing in both A and B. Now since IA includes only the factors not appearing in A, and correspondingly for IB, IA and IB have these $n - \delta$ factors missing from A and B. Since in addition the sum of the orders of IA and IB has the value $n - \alpha + n - \beta$ and thus is less than n, it follows (according to 109) that

$$[IA \cdot IB] = 0.$$

But since $[AB] = 0$ as well, and (according to b) the supplement of a number in the same sense is also equal to this number, so that that of zero is again zero, it follows that

$$I[AB] = [IA \cdot IB].$$

Second, if $[AB]$ differs from zero, then (according to 109) the covering domain of the domains of A and B has the order n, their common domain the order $\alpha+\beta-n$. Now since the magnitudes A and B are both products of factors from the series of n magnitudes $a_1, \dots, a_n$, it is clear that each of the n factors $a_1, \dots, a_n$ appears in at least one of the two magnitudes A and B and that in addition A and B have precisely $\alpha + \beta - n$ of these elementary factors in common. If C is the product of these $\alpha + \beta - n$ factors, then A and B can be represented in the forms

$$A = [CA_1], \; B = [CB_1],$$

where $[A_1CB_1]$ includes all elementary factors and is therefore a number. But (according to 105)

$$[AB] = [CA_1 \cdot CB_1] = [CA_1B_1]C,$$

and further (according to c)

$$I[AB] = [CA_1B_2]IC,$$

or, since (according to a)

$$IC = [CA_1B_1][A_1B_1],$$

one has

$$I[AB] = [CA_1B_1][CA_1B_1][A_1B_1].$$

On the other hand

$$IA = I[CA_1] = [CA_1B_1]B_1, \; IB = I[CB_1] = [CB_1A_1]A_1,$$

whence

$$[IA \cdot IB] = [CA_1B_1][CB_1A_1][B_1A_1],$$

and since this expression (according to 58) does not change its value if one interchanges A_1 and B_1 twice, it follows again that

$$I[AB] = [IA \cdot IB].$$

[67] 34. By virtue of this theorem one can represent every elementary magnitude in a principal domain of nth order as a pure regressive product of elementary magnitudes of $(n-1)$th order. Thus, given an elementary magnitude of mth order $C = [a_1, \dots, a_n]$, one need only define $n - m$ magnitudes of first order $a_{m+1}, \dots, a_n$ such that they stand in no numerical relation to one another or to $a_1, \dots, a_m$, and

that $[a_1a_2\cdots a_m]=1$. If one then defines the magnitudes $A_1,\ldots,A_n$ as in No. 112, one has $C=[A_nA_{n-1}\cdots A_{m+1}]$. At the same time the possibility is hereby given of representing the supplement of C in a simple way as a combinatorial product of magnitudes of first order. Thus according to No. 98

$$|C=[|A_n|A_{n-1}\cdots|A_{m+1}],$$

where the $|A_k$ are magnitudes of first order that can be written down immediately, once one has ascertained in what way $A_1,\ldots,A_n$ are derived from the n units of $(n-1)$th order.

It is clear that one is now in a position to take either magnitudes of first order, or of $(n-1)$th, as fundamental, at pleasure. In the language of modern projective geometry, it depends on whether one calculates with point or with plane coordinates.

Finally, two special cases of the general formula proved in theorem 112 should be mentioned. They are the two formulas

$$[A_nA_{n-1}\cdots A_2]=a_1$$

and

$$[A_nA_{n-1}\cdots A_1]=1,$$

which will be much used later on (cf. Nos. 292, 299, and 300).

Theorem 112 was later employed by CLEBSCH in the *Göttinger Abh.* **17** (1872) and by F. MEYER in his book *Apolarität und rationale Curven*, Tubingen 1883, and in fact in the form of a theorem on matrices.

[68] 35. Since A is of order α, $n-\alpha$ is the order of D_r and $m-(n-\alpha)=m+\alpha-n$ that of C_r; therefore $m+\alpha$ is necessarily $>n$ and consequently the product $[AB]$ is regressive. Now since the $[AD_r]$ are numbers and the C_r belong to the domain of the elementary magnitude B, the whole theorem amounts to the following: *If an arbitrary magnitude A is multiplied by an elementary magnitude B and the product $[AB]$ is **regressive**, then $[AB]$ is a magnitude belonging to the domain of B, for which an explicit representation is given in the theorem.*

To grasp the theorem one notes that, in the case where the order of C_r equals $m-1$, each D_r includes only one of the elementary factors of B, and that one can satisfy the equation $[C_rD_r]=B$ under these circumstances just by a change in the algebraic sign of the elementary factors in question.

There is a gap in the proof of No. 113. Thus if, following the prescription on p. 67, lines 7–6 from the bottom, one forms all combinations A_1, A_2, ... on the condition that A_r always consists of the magnitudes of $b_1,\ldots,b_n$ that are missing in D_r, then one by no means obtains *all* combinations of $b_1,\ldots,b_n$ of αth class, although A is in general only numerically derivable from *all* these combinations. This gap is however easily filled.

We can always write A in the form

$$A=\sum\alpha_rA_r+\sum\alpha'_{\mathfrak{a}}A'_{\mathfrak{a}},$$

where each A_r includes precisely $m-(n-\alpha)$ and no more of the factors $b_1,\ldots,b_n$, while each $A'_{\mathfrak{a}}$ contains at least $m-(n-\alpha)+1$ of these factors. Then each $A'_{\mathfrak{a}}$ has at least one factor in common with each D_r and the progressive products $[A'_{\mathfrak{a}}D_r]$ are therefore all zero; furthermore, B has in common with each $A'_{\mathfrak{a}}$ a domain of higher than $(m+\alpha-n)$th order, so that the regressive products $[A'_{\mathfrak{a}}B]$ are also all zero. Now if we form the product $[AB]$, then, as in the text,

$$[AB] = \sum \alpha_r [A_r B] = \sum \alpha_r [A_r D_r] C_r = \sum_{rs} \alpha_s [A_s D_r] C_r,$$

which, since $[A'_{\mathfrak{a}}D] = 0$, one can also write as

$$[AB] = \sum_r \left[\left(\sum \alpha_s A_s + \sum \alpha'_{\mathfrak{a}} A'_{\mathfrak{a}}\right) D_r\right] C_r = \sum [AD_r] C_r.$$

[73] 36. From this Number on, the concept of congruence is used with a somewhat broader meaning than was expressed in the definition of the congruence concept stated in No. 2. Thus while there only two *nonzero* magnitudes are called congruent if they stand in a numerical relation, from now on *two vanishing* magnitudes may be said to be congruent (see for example the second part of the proof of No. 121, and the Remark following No. 313).

[78] 37. If in the case e) $q+r$ and $q+s$ are both $< n$, then

$$\begin{aligned} [BAC] &= (-1)^{qr}[ABC] && [58] \\ &= (-1)^{qr+(r-t)(s-t)}[ACB], && [124e] \end{aligned}$$

and further

$$[ACB] = (-1)^{(n-q-s)(n-r)}[B \cdot AC] \qquad [120, \text{Proof } 1]$$

and thus, since $n-q-s = r-t$ and $s-t+n-r = 2n-2r-q$, one has in fact

$$[BAC] = (-1)^{qt}[B \cdot AC].$$

The case where $q+r$ and $q+s$ are both $> n$ reduces to the one just settled if one takes the supplements on both sides.

If in the case f) $q+r < n$ and $q+s < n$, then $q+s-n+r < n$, whence

$$\begin{aligned} [BAC] &= (-1)^{qr}[ABC] = (-1)^{qr}[ACB] && [124f] \\ &= (-1)^{qr+(q+s-n)r}[B \cdot AC] && [95, 58] \\ &= (-1)^{r(n-s)}[B \cdot AC]. \end{aligned}$$

The case where $q+r > n$ and $q+s < n$ reduces to the one just settled if one takes the supplements on both sides.

If one of the two sums $q+r$ or $q+s$ equals n, then B and C are also incident in the case e) and there is no longer a distinction between the cases e) and f). Thus for $q+r = n$, where $s = t$, the first formula of e) coincides with the second of f), and the second formula of e) with the first of f). If $q+s = n$, the corresponding follows.

[79] 38. Refer to No. 123, Proof, Parts 2 and 3.

[81] 39. It is interesting to compare this expression for the shadow (Zurückleitung) with that (Abschattung) given by GRASSMANN in the A_1.

The magnitudes denoted by A, A', B, C in No. 129 are in the A_1 denoted by A, A', G, L (A_1, §149), and upon using these characters the expression given in No. 129 for A' reads

$$A' = \frac{[G \cdot AL]}{[GL]}.$$

Now let p and m be the orders of A and G and thus $n-m$ be the order of L. Then (according to No. 128) for the case of the *progressive* shadow $m \geq p$, whence $p+(n-m) \leq n$, and thus (according to 58 and Part 1 of Proof, 120)

$$[GL] = (-1)^{m(n-m)}[LG], \quad [AL] = (-1)^{p(n-m)}[LA],$$
$$[G \cdot LA] = (-1)^{(n-m)(m-p)}[LA \cdot G].$$

On the other hand, for the case of the *regressive* shadow $m \leq p$, whence $p+(n-m) \leq n$, and thus

$$[GL] = (-1)^{m(n-m)}[LG], \quad [AL] = (-1)^{(n-p)m}[LA],$$
$$[G \cdot LA] = (-1)^{m(p-m)}[LA \cdot G],$$

since according to No. 95 [LA] is a magnitude of $(p-m)$th order and $m+(p-m) \geq n$, so that the product $[G \cdot LA]$ is progressive. Thus in both cases

$$A' = \frac{[LA \cdot G]}{[LG]},$$

which is precisely the same expression as appears in the A_1.

[81] 40. "The products $[A_{u+1}C], \ldots, [A_vC]$ are however pure with respect to the factors $a_1, \ldots, a_n$ (according to 114, 127, 128, and 119b)." Thus if one denotes the orders of A, B, C by α, β, γ and *first, if* $a_1, \ldots, a_n$ *are magnitudes of first order*, then (according to 127) the shadow is progressive and thus (according to 128)

$$\alpha \leq \beta,$$

or, since the order γ of the excluded domain C supplements the order of B up to that of the principal domain, that is $\beta = n - \gamma$, it follows that

$$\alpha \leq n-\gamma \quad \text{or} \quad \alpha + \gamma \leq n.$$

The products $[A_{u+1}C], \ldots$ are therefore progressive with respect to the factors A_{u+1} and $C, \ldots$, and thus (according to 119b) also remain purely progressive if one decomposes their factors into pure factors of first order.

Second, if the factors $a, \ldots, a$ are *magnitudes of* $(n-1)$*th order*, then (according to 127) the shadow is regressive and therefore (according to 128)

$$\alpha \geq \beta \text{ or } \alpha \geq n-\gamma, \text{ that is } \alpha+\gamma \geq n.$$

The products $[A_{u+1}C], \ldots$ are therefore regressive with respect to the factors A_{u+1} and $C, \ldots$, and thus (according to 119b) also remain regressive if one decomposes their factors A_{u+1} and C into products of $(n-1)$th order, that is, the products $[AC], \ldots$ are also purely regressive with respect to the factors $a_1, \ldots, a_n$.

[82] 41. The formula utilized here is in No. 108 only proved for the case that B is *subordinate* to A, but as is shown in Editorial Note 33 it is also valid if B is *superordinate* to A.

[82] 42. {In this Note GRASSMANN's son, Hermann Junior, illustrated the concept of the shadow in ordinary three-dimensional geometry with examples, since contemporary readers found the presentation in Numbers 127–129 too abstract. These are omitted here, as they are only of didactic interest. *Tr.*}

[84] 43. If $m = n-1$, each F_k includes only one of the units $e_1, \ldots, e_n$ as a factor; one may therefore not always be able to satisfy the equations $[E_kF_k] = 1$ by

a suitable ordering of the factors of F_k; instead one must in certain circumstances take F_k to be a negative unit. Cf. Editorial Note 35.

[87] 44. It still remains to be shown that the given values of $x_1, x_2, \dots, x_r$ together with the completely arbitrary values of $x_{r+1}, \dots, x_n$ actually satisfy equation (c).

It was already shown that, if equations (a) do not contain a contradiction, it must be possible to numerically derive the magnitude b from the $a_1, \dots, a_r$. Let

$$b = y_1 a_1 + y_2 a_2 + \cdots + y_r a_r;$$

and further, let

$$(*) \qquad x_{r+1} a_{r+1} + x_{r+2} a_{r+2} + \cdots + x_n a_n = d.$$

Then, since by assumption the magnitudes $a_{r+1}, a_{r+2}, \dots, a_n$ are numerically derivable from the $a_1, \dots, a_r$, the magnitude d is likewise derivable from these magnitudes, and thus can be represented in the form

$$d = z_1 a_1 + z_2 a_2 + \cdots + z_r a_r.$$

Now by virtue of $(*)$, $c = b - d$, and thus

$$c = (y_1 - z_1) a_1 + (y_2 - z_2) a_2 + \cdots + (y_r - z_r) a_r.$$

But if one substitutes this value for c into the expressions for $x_1, x_2, \dots, x_r$ given in (f), one obtains

$$x_1 = y_1 - z_1, \; x_2 = y_2 - z_2, \; \dots, \; x_r = y_r - z_r;$$

and if one multiplies these equations respectively by $a_1, a_2, \dots, a_r$ and adds them, there results

$$x_1 a_1 + x_2 a_2 + \cdots + x_r a_r = b - d,$$

that is, referring back to $(*)$, the equation

$$(c) \qquad x_1 a_1 + x_2 a_2 + \cdots + x_r a_r + x_{r+1} a_{r+1} + x_{r+2} a_{r+2} + \cdots + x_n a_n = b.$$

Thus equation (c) is actually satisfied by the values (f).

[89] 45. The two methods of solution of a system of linear equations admit a *geometric interpretation* in case $n < 4$, which will be developed here for the case $n = 3$.

In order to interpret the *first solution* geometrically, one regards the magnitudes $e^{(1)}, e^{(2)}, e^{(3)}$ as three noncoplanar displacements. The three magnitudes a_1, a_2, a_3, which according to equations (b) are numerically derived from the displacements $e^{(1)}, e^{(2)}, e^{(3)}$ by the coefficients standing in the *columns* of equations (a), are therefore likewise displacements, and, provided at first it is assumed that $[a_1 a_2 a_3] \neq 0$, they are noncoplanar. In addition, equation (c) implies that the displacement b is numerically derivable from the three displacements a_1, a_2, a_3, and the derivation numbers x_1, x_2, x_3 are the unknowns sought. Thus if one decomposes the displacement b into three summands b_1, b_2, b_3, which run parallel to the displacements a_1, a_2, a_3 respectively (see Fig. 22), which is only possible in one way so long as $[a_1 a_2 a_3] \neq 0$, then the three ratios of the three pairs of correlated displacements b_i and a_i are the unknowns x_i sought, that is,

$$x_1 = \frac{b_1}{a_1}, \; x_2 = \frac{b_2}{a_2}, \; x_3 = \frac{b_3}{a_3}.$$

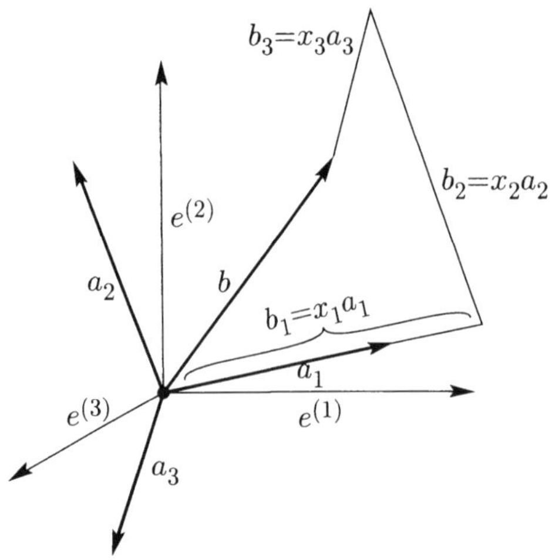

FIGURE 22

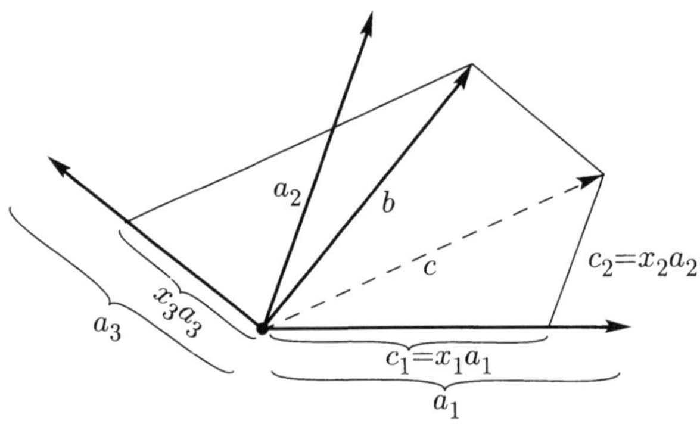

FIGURE 23

In case the combinatorial product $[a_1a_2a_3] = 0$, but two of the magnitudes a_1, a_2, a_3, say a_1 and a_2, still stand in no numerical relation to one another, one again constructs the four displacements a_1, a_2, a_3 and b, corresponding to equation (b). Thereby, if equations (a) do not contain a contradiction, one obtains four displacements in the same plane (see Fig. 23). One then assigns the magnitude x_3 a value completely arbitrarily, reduces b by x_3a_3 and sets $b - x_3a_3 = c$. Finally, if one represents c as a multiple sum of a_1 and a_2, that is, as a sum of two displacements c_1 and c_2 that run parallel to a_1 and a_2 respectively, then $x_1 = \frac{c_1}{a_1}$ and $x_2 = \frac{c_2}{a_2}$ are the values of x_1 and x_2 corresponding to the arbitrary value of x_3 chosen. If one chooses a common initial point for the displacements a_1, a_2, a_3, b, and c (as in Fig. 23), then one finds all values of x_1, x_2, x_3 that satisfy the given equation (a) if one displaces the final point of the displacement c parallel to a_3 and for each position of c decomposes the displacement b into the sum $b = c + x_3a_3$ and decomposes the displacement c itself into the sum $c = x_1a_1 + x_2a_2$. One proceeds similarly if each pair of the magnitudes a_1, a_2, a_3 stand in a numerical relation.

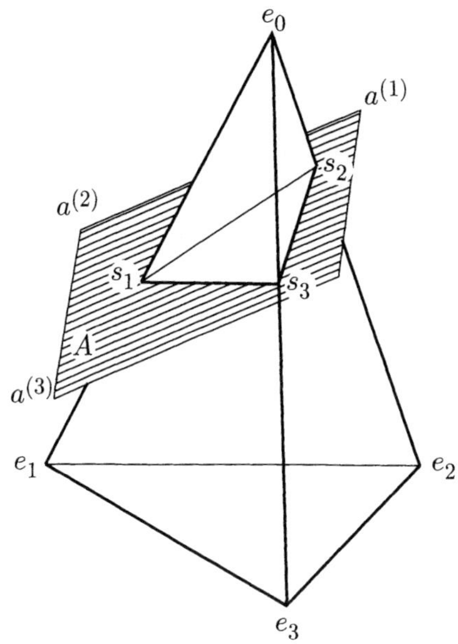

FIGURE 24

In order to find a geometric interpretation for the *second solution* of the system of linear equations as well, one regards the four units e_0, e_1, e_2, e_3, whose combinatorial product was set to unity, as elementary points that form the vertices of a tetrahedron (see Fig. 24). Then the equations

$$a^{(1)} = \alpha_0^{(1)} e_0 + \alpha_1^{(1)} e_1 + \alpha_2^{(1)} e_2 + \alpha_3^{(1)} e_3,$$
$$a^{(2)} = \alpha_0^{(2)} e_0 + \alpha_1^{(2)} e_1 + \alpha_2^{(2)} e_2 + \alpha_3^{(2)} e_3,$$
$$a^{(3)} = \alpha_0^{(3)} e_0 + \alpha_1^{(3)} e_1 + \alpha_2^{(3)} e_2 + \alpha_3^{(3)} e_3,$$

which are derived from the units e_0, e_1, e_2, e_3 by the coefficients appearing in the *rows* of equations (α), are again points, which, provided the product $[a^{(1)}a^{(2)}a^{(3)}] \neq 0$, define a surface element $A = [a^{(1)}a^{(2)}a^{(3)}]$. If in addition one again sets

$$X = x_0|e_0 + x_1|e_1 + x_2|e_2 + x_3|e_3,$$

whence X represents a surface element, and where $x_0 = 1$, then the given equations (α) are identical to the equations

$$(\delta) \qquad [a^{(1)}X] = 0, \ [a^{(2)}X] = 0, \ [a^{(3)}X] = 0,$$

which express that the plane of the surface element X goes through the three points $a^{(1)}$, $a^{(2)}$, $a^{(3)}$. This surface element can therefore be expressed in the form

$$(\varepsilon) \qquad X = \lambda \left[a^{(1)}a^{(2)}a^{(3)}\right] = \lambda A,$$

where λ denotes a number whose value can be ascertained with the help of the equation $[e_0 X] = x_0 = 1$. The expressions

$$(\theta) \qquad x_1 = \frac{e_1 A}{e_0 A}, \ x_2 = \frac{e_2 A}{e_0 A}, \ x_3 = \frac{e_3 A}{e_0 A}$$

for the unknowns x_1, x_2, x_3 resulting from equation (ε) then represent the unknowns as ratios of those four parallelepipeds defined by the surface element A and each vertex of the basis tetrahedron. But two such parallelepipeds, for example $[e_1 A]$ and $[e_0 A]$, are to one another as are the two segments $e_1 s_1$ and $e_0 s_1$ into which the edge $e_1 e_0$ of the tetrahedron is divided by its intersection s_1 with the plane A. The unknowns x_1, x_2, x_3 are therefore nothing but the ratios of the segments which the plane of the surface element A produces on the edges of the tetrahedron $[e_1 e_0]$, $[e_2 e_0]$, $[e_3 e_0]$, assuming that the segments are always reckoned from the vertices of the tetrahedron out to the point of the division.

H. Grassmann, Junior

[92] 46. CAUCHY encloses multiple products, such as would be denoted *outer* in GRASSMANN's sense, between two vertical lines; he also uses brackets in a similar but not identical way.

[93] 47. With the help of the concept of the supplement GRASSMANN can now define the inner products of two magnitudes directly, whereas in the *Geometrische Analyse* he had always to provide outer products to which the inner products considered were proportional.

[96] 48. The second part of the proof of this theorem would be considerably shorter if one referred to Theorem 5 of Editorial Note 29, and made use of the Corollary to No. 90.

[97] 49. The editors of the Teubner Edition replaced the material in braces with the following:

149. *If E, F, G are units, and if F is of* ***higher order*** *than G and the product $[EF]$ is progressive and not zero, then*

$$[EF|EG] = [F|G].$$

Further, if F is of ***lower order*** *than G and the product $[GE]$ is progressive and not zero, then*

$$[FE|GE] = [F|G].$$

Finally, if F and G are of ***the same*** *order, and both products $[EF]$ and $[GE]$ are progressive and not zero, then both formulas apply.*

PROOF. 1. If F and G are *not incident on one another*, then $[EF]$ and $[EG]$ are also not incident on one another, and thus (according to 147) both sides of the equations to be proved are zero.

2. If on the other hand F is *incident* on G, then the three cases distinguished in the theorem must be handled separately.

a. We assume, *first*, that *the product $[EF]$ is progressive and not zero, and that F is of higher order than G*; then G, which is at the same time incident on F, is subordinate to F, and one can therefore represent F in the form

$$F = [GH],$$

where the product $[GH]$ is progressive and H is again a unit, or perhaps a negative unit. Consequently

$$[EF|EG] = [E(GH)|EG],$$

or, since the product $[E(GH)]$ is purely progressive, according to 119

$$\begin{aligned}[EF|EG] &= [EGH|EG] \\ &= H\end{aligned}$$

(according to 148), for $[EGH] = [EF]$ is not zero by assumption, and $[EG]$ is itself a unit (according to 77) as a nonvanishing progressive product of two units. But the right side of our equation is (again according to 148)

$$= [GH|G] = [F|G].$$

b. Assuming, *second*, that *the product* $[GE]$ *is progressive and not zero, and that* F *is of lower order than* G, then F, which is at the same time incident on G, is subordinate to G, and one can therefore represent G in the form

$$G = [HF],$$

where the product $[HF]$ is progressive and H is a unit, or perhaps a negative unit. Consequently

$$[FE|GE] = [FE|HFE],$$

or, since the product $[HFE]$ is purely progressive, (according to 119)

$$\begin{aligned}[FE|GE] &= [FE|H(FE)] \\ &= |H\end{aligned}$$

(according to 148), for $[H(FE)] = [HFE] = [GE]$ is not zero. But (again according to 148) the right side is again

$$= [F|HF] = [F|G].$$

c. *Finally,* if we assume that *the products* $[EF]$ *and* $[GE]$ *are both progressive and not zero, and that* F *and* G *are of the same order,* then, since they are at the same time incident on one another, the domains of the two magnitudes must coincide; consequently G is subordinate to F and F to G, and thus according to Parts 2a and b of the Proof both formulas apply.

REMARK. A corresponding theorem also applies if $[EF]$ and $[EG]$ are *regressive* products, except that in this case interchange the expressions "higher" and "lower". The proof is precisely as in No. 148.

[97] 50. The formula also applies if $q = r$; for then $q(r-1)$ is certainly even, and thus the formula of No. 150 takes the form $[A|B] = |[B|A]$, which corresponds with Nos. 141 and 144.

[98] 51. That is, in the Proofs of Nos. 147 and 148.

[98] 52. The expression "elements of a domain" has not previously been used in a precise sense; what is intended one can however infer from the names "line element" and "surface element" introduced later (cf. Nos. 249 and 257). Thus by an element of a domain of qth order GRASSMANN obviously means an elementary magnitude of qth order belonging to the domain; thus he ought properly to say:

A domain of mth and a domain of qth order in a principal domain of nth order are called normal to one another if an elementary magnitude of the first domain is normal to an elementary magnitude of qth order of the second.

Thus one needs only a single "element" of the one domain to compare with a single "element" of the other, since according to No. 70 all elements of a domain are congruent in the sense of No. 2.

If one has two elementary magnitudes A and B, of mth and qth order respectively, and if $m \leq q$, then the product $[A|B]$ is progressive;* if however $m \geq q$, the product $[B|A]$ is progressive. *Thus according to No. 109 the two magnitudes A and B are normal to one another if and only if that one of them whose order is* ***no greater*** *than that of the other has a domain of first or higher order in common with the supplement of the other.*

[98] 53. This definition is based on an assumption that ought properly be proved, namely that two completely normal domains or magnitudes are also normal in the sense previously defined. Since one can only justify this assumption with the help of a part of the development in Nos. 153 to 167, it would have been better to introduce the definition of completely normal domains and magnitudes only after No. 167, and then to formulate a theorem, somewhat as follows:

THEOREM 1. *Two completely normal domains or magnitudes of a principal domain of nth order are also normal to one another in the sense of No. 152.*

This would have been the more appropriate as the concept "completely normal" does not appear at all in Nos. 153–167, and only plays a role in Nos. 171 and 172.

In order to prove Theorem 1 we will consider two completely normal domains of mth and qth orders respectively. Then according to No. 163 we can in each of the two domains find a normal system of the appropriate order and of numerical value one, and in fact let $u_1, \ldots, u_m$ and $v_1, \ldots, v_q$ be the two normal systems. According to the concept of complete normality every magnitude $\sum \alpha_k u_k$ is normal to every magnitude $\sum \beta_k v_k$, and thus in particular every u_k to every v_k. It therefore follows that in addition $u_1, \ldots, u_m, v_1, \ldots, v_q$ all together form a normal system of numerical value one, and (according to 157) at the same time there is no numerical relation between $u_1, \ldots, u_m, v_1, \ldots, v_q$. Consequently, on the basis of No. 161, we can append to $u_1, \ldots, u_m, v_1, \ldots, v_q$ other $n-m-q$ magnitudes $w_1, \ldots, w_{n-m-q}$ in such a way that a complete normal system of numerical value one results.

Furthermore, according to No. 167

$$(*) \qquad |[v_1 \cdots v_q] = \pm[u_1 \cdots u_m \cdot w_1 \cdots w_{n-m-q}]$$

and thus

$$[(u_1 \cdots u_m)|(v_1 \cdots v_q)] = 0.$$

For, if $m \leq q$, the product $[(u_1 \cdots u_m)|(v_1 \cdots v_q)]$ is progressive (cf. the previous Editorial Note) and therefore vanishes on account of $(*)$, according to No. 60; but if $m \geq q$, the product is regressive and vanishes according to 109, since on account of $(*)$ the covering domain of its factors is of $(n-q)$th order, and thus is smaller than n. Consequently the two completely normal magnitudes $[u_1 \cdots u_m]$ and $[v_1 \cdots v_q]$ and likewise their domains are really normal to one another in the sense of No. 152.

From the above considerations it follows that one of two completely normal magnitudes is always subordinate to the supplement of the other. Conversely it is clear that every elementary magnitude that is subordinate to the supplement of a magnitude $[v_1 \cdots v_q]$, is completely normal to that magnitude. Consequently we can also say:

THEOREM 2. *Two magnitudes are completely normal to one another if and only if one is subordinate to the supplement of the other.*

Finally we can express the following, no longer independent, theorem:

*Indeed, the supplement $|B$ is an elementary magnitude of $(n-q)$th order; cf. the Corollary to No. 90 and Editorial Note 29.

THEOREM 3. *If two domains are completely normal to one another, every magnitude of one of the domains is normal to every magnitude of the other.*

Only with this theorem does the nomenclature "completely normal" appear in its proper light.

[98] 54. These words are somewhat astonishing, for the "must" makes no sense; one could equally well choose three arbitrary noncoplanar displacements as units.

[99] 55. The circular evolution is, just like the linear (cf. No. 71), equivalent to a linear homogeneous substitution of a particular form. Thus if one subjects the series of n magnitudes $a_1, \dots, a_n$, which stand in no numerical relation to one another, to a *positive* circular evolution, say by replacing a_1 and a_2 with $\cos\alpha \cdot a_1 + \sin\alpha \cdot a_2$ and $\cos\alpha \cdot a_2 - \sin\alpha \cdot a_1$, then the magnitudes $\sum x_k a_k$ of the domain of nth order derivable from $a_1, \dots, a_n$ will be interchanged among themselves by the linear homogeneous transformation

$$(1) \qquad \begin{cases} x_1' = x_1 \cos\alpha - x_2 \sin\alpha, \\ x_2' = x_1 \sin\alpha + x_2 \cos\alpha, \\ x_3' = x_3, \\ \dots \\ x_n' = x_n. \end{cases}$$

This transformation has determinant 1 and obviously leaves the quadratic form $\sum x_k^2$ invariant, that is, it is an *orthogonal* substitution. The totality of all ∞^1 transformations of the form (1) constitute a one-parameter group in the sense of LIE.

Under a *negative* circular evolution the magnitudes a_1 and a_2 are replaced by the magnitudes $\cos\alpha \cdot a_1 + \sin\alpha \cdot a_2$ and $-(\cos\alpha \cdot a_2 - \sin\alpha \cdot a_1)$. It corresponds to a linear homogeneous transformation of the form

$$(2) \qquad \begin{cases} x_1' = x_1 \cos\alpha + x_2 \sin\alpha, \\ x_2' = x_1 \sin\alpha - x_2 \cos\alpha, \\ x_3' = x_3, \\ \dots \\ x_n' = x_n. \end{cases}$$

This is also orthogonal, but has determinant -1. It therefore follows that the totality of all ∞^1 transformations (2) do not form a group, but that the transformations (1) and (2) together form a non-continuous group.

[99] 56. Cf. No. 220.

[100] 57. If in a domain of mth order one has two normal systems of mth order, $a_1, \dots, a_m$ and $b_1, \dots, b_m$ of the same numerical value, then according to No. 157* neither the $a_1, \dots, a_m$ nor the $b_1, \dots, b_m$ stand in a numerical relation, and thus the $b_1, \dots, b_m$ can be numerically derived from the $a_1, \dots, a_m$:

$$b_\mu = \alpha_{\mu 1} a_1 + \cdots + \alpha_{\mu m} a_m \quad (\mu = 1, \dots, m),$$

where according to No. 63 the determinant of the $\alpha_{\mu v}$ is $\neq 0$. Now if in an arbitrary magnitude $\sum x_\mu a_\mu$ of the domain under consideration one replaces the magnitudes $a_1, \dots, a_m$ with the $b_1, \dots, b_m$, one obtains a new magnitude $\sum x_\mu b_\mu = \sum x_\mu' a_\mu$

*Numbers 157–159 would have been better placed before No. 154.

of the domain. Consequently the transition from one normal system to the other is equivalent to the linear homogeneous transformation

$$(1) \qquad x'_\mu = \alpha_{1\mu}x_1 + \cdots + \alpha_{m\mu}x_m \quad (\mu = 1, \ldots, m),$$

by virtue of which the magnitudes of first order of the domain under consideration will be interchanged among themselves. Since in addition

$$\left[\sum x_\mu b_\mu\right]^2 = \left[\sum x'_\mu a_\mu\right]^2$$

and under our assumptions the equations

$$b_1^2 = \cdots = b_m^2 = a_1^2 = \cdots = a_m^2,$$
$$[b_\mu | b_\nu] = 0, \ [a_\mu | a_\nu] = 0 \ (\mu \neq \nu)$$

are valid, it follows that $\sum {x'_v}^2 = \sum x_v^2$. Thus the linear homogeneous transformation (1) leaves the quadratic form $\sum x_v^2$ invariant. In other words, the substitution (1) is orthogonal.

Conversely, every real orthogonal substitution (1) corresponds to the transition from one normal system of mth order to another, numerically equal to it. In particular therefore, according to Editorial Note 55, a circular evolution must transform any given normal system into a numerically equal normal system; and this is demonstrated in 155.

[102] 58. If the two normal systems are of the same order, say the mth, then with this theorem the following is also proven:

Every real linear homogeneous substitution

$$x'_\mu = \alpha_{\mu 1}x_1 + \cdots + \alpha_{\mu m}x_m \ (\mu = 1, \ldots, m),$$

for which the quadratic form $\sum x_\mu^2$ *remains invariant, can thus be obtained if one carries out, one after the other, a series of linear homogeneous substitutions of the special forms* (1) *and* (2) *displayed in Editorial Note* 55. *It therefore follows in addition that the determinant of the substitution is* ± 1.

[102] 59. That is, replace a positive circular evolution with the corresponding negative one and conversely.

[103] 60. This only holds if the units are three equally long and mutually orthogonal displacements. Cf. Editorial Note 53.

[103] 61. One observes that the shadow is progressive or regressive according as the orders α and β of A and B stand in the relation $\alpha \leq \beta$ or $\alpha \geq \beta$. If on both sides of the equation for A' one takes the supplement, then one recognizes that $|A'$ is the normal shadow of $|A$ on the domain of $|B$. If A' was a progressive shadow, the shadow $|A'$ would naturally be regressive and conversely.

[104] 62. It is indeed shown by this theorem that the generalization of the concept of the supplement introduced in the proof of No. 110 coincides with the supplement defined in Nos. 89 and 90, provided $a_1, \ldots, a_n$ form a complete normal system of numerical value one, but a demonstration is still lacking that the normal system of numerical value one is the only one for which this happens (cf. p. 66, lines 3–4).

This demonstration is easily supplied. Thus if $a_1, \ldots, a_n$ are n magnitudes of the principal domain $e_1, \ldots, e_n$, whose combinatorial product has the value one, then according to the definition on p. 64,

$$Ia_k = [a_k a_1 \cdots a_{k-1} a_{k+1} \cdots a_n][a_1 \cdots a_{k-1} a_{k+1} \cdots a_n],$$

where the first factor has one of the two values ± 1; consequently

$$[a_k I a_k] = 1, \quad [a_k I a_j] = 0 \ (j \neq k).$$

Now for any arbitrary magnitude A it is supposed that always $IA = |A$, and so it is certainly necessary that always $Ia_k = |a_k$, and therefore that the equations

$$[a_k | a_k] = 1, \quad [a_k | a_j] = 0 \ (j \neq k)$$

obtain. It is therefore necessary that $a_1, \ldots, a_n$ form a normal system of numerical value one. That this is also sufficient is shown in No. 167.

[105] 63. Naturally the theorem also holds if B' is the regressive normal shadow of B on the domain of A. For then (according to Editorial Note 61) $|B'$ is the progressive normal shadow of $|B$ on the domain of $|A$, and thus according to No. 169

$$[|A||B] = [|A||B'], \ [|B||A] = [|B'||A],$$

and thus according to No. 101 the equations of No. 169 also apply in this case.

[106] 64. How this can be done is more precisely specified in the Proof of Theorem 1 in Editorial Note 52.

[107] 65. The editors of the Teubner Edition replaced the material in braces with the following:

One can now show that the product $[A_1 B_1 | A_1 C_r]$ satisfies all the conditions of the first formula of theorem 149. Let n be the order of the principal domain and A_1 the product of m magnitudes $a_1, \ldots, a_m$ of the normal system. Then (according to 159) all the magnitudes of first order normal to $a_1, \ldots, a_m$ belong to the domain of the $n - m$ remaining magnitudes $a_{m+1}, \ldots, a_n$ of this normal system. The sum of the orders of A_1 and B_1 can therefore be at most $m+n-m$, that is $= n$, whence the product $[A_1 B_1]$ is necessarily *progressive*. But (according to 109) it is also *not zero*, since the domains of the magnitudes A_1 and B_1 have no magnitude of first order in common. Thus since the domains of A_1 and B_1 are perfectly normal to one another, it follows (according to 152) that every magnitude of first order of the domain of A_1 is normal to every magnitude of first order of the domain of B_1. Consequently, if the two domains had a nonzero magnitude in common, this magnitude would be normal to itself, whence its inner square would be zero, which (according to 146) is impossible for a nonzero magnitude. In addition, by assumption B is of order greater than or equal to that of C, *whence* B_1 *is also of order greater than or equal to that of* C_r. Finally (according to 168) theorem 149, proved initially *for units of higher order*, that is for combinatorial products of the original units, still holds if, in the place of products of units, there appear *combinatorial products of the magnitudes of an elementary normal system*, and thus in our case; that is

$$[A_1 B_1 | A_1 C_r] = [B_1 | C_r],$$

for $r = 1, 2, \ldots$.

[108] 66. The editors of the Teubner Edition replaced the material in braces with the following:

First, let the elementary factors of $[AB]$ all be normal to one another and let the system of these factors $a_1, \ldots, a_m$ be extended to a complete normal system by the addition of the magnitudes $a_{m+1}, \ldots, a_n$; further, let $A, A_1, \ldots, A_q, A_{q+1}, \ldots, A_t$ be those multiplicative combinations of $a_1, a_2, \ldots, a_n$ whose orders are equal to that of A, and in particular let $A, A_1, \ldots, A_q$ be the combinations of the first m magnitudes $a_1, \ldots, a_m$. Then A, which is of the same order as A, is numerically derivable from the multiplicative combinations $A, A_1, \ldots, A_t$. Let

$$\mathrm{A} = \alpha A + \alpha_1 A_1 + \cdots + \alpha_q A_q + \alpha_{q+1} A_{q+1} + \cdots + \alpha_t A_t;$$

then

$$\begin{aligned}[AB|\mathrm{AB}] &= [AB|(\alpha A + \alpha_1 A_1 + \cdots + \alpha_q A_q + \alpha_{q+1} A_{q+1} + \cdots + \alpha_t A_t)\mathrm{B}] \\ &= \alpha[AB|A\mathrm{B}] + \alpha_1[AB|A_1\mathrm{B}] + \cdots + \alpha_q[AB|A_q\mathrm{B}] \\ &\quad + \alpha_{q+1}[AB|A_{q+1}\mathrm{B}] + \cdots + \alpha_t[AB|A_t\mathrm{B}].\end{aligned}$$

This development can however be simplified even further; for it can be shown that all the terms on the last line vanish.

Thus if one imagines the magnitudes $a_1, \ldots, a_n$ as forming the magnitudes B_i, whose order agrees with that of B, and B as represented by a multiple sum of the B_i, then from the terms of the last line of our development there result pure terms of the form

$$\alpha_{q+k}\beta_i[AB|A_{q+k}\mathrm{B}_i].$$

But here all the A_{q+k} include as a factor one of the magnitudes $a_{m+1}, \ldots, a_n$ that do *not* appear in the product $[AB]$, and these therefore also appear in the products $[A_{q+k}\mathrm{B}_i]$, which by assumption are progressive. These products are therefore *not subordinate* to the product $[AB]$. But they are also *not superordinate* to them; for, since by assumption B is of order less than or equal to that of B, every product $[A_{q+k}\mathrm{B}_i]$ must lack at least one factor of $[AB]$. The products $[A_{q+k}\mathrm{B}_i]$ are therefore *not incident* on $[AB]$, and thus (according to 147 and 168) generally

$$[AB|A_{q+k}\mathrm{B}_i] = 0.$$

The above development therefore simplifies to the following:

$$[AB|\mathrm{AB}] = \alpha[AB|A\mathrm{B}] + [AB|A_1\mathrm{B}] + \cdots + \alpha_q[AB|A_q\mathrm{B}].$$

Now since (by assumption) $[AB] = [A_1B_1] = \cdots$, we obtain the expression

$$= \alpha[AB|A\mathrm{B}] + \alpha_1[A_1B_1|A_1\mathrm{B}] + \cdots + \alpha_q[A_qB_q|A_q\mathrm{B}].$$

Because the elementary factors of $[AB]$ are all normal to one another, and identical to those of $\pm[A_1B_1], \ldots$ (by assumption), A is perfectly normal to B, and likewise A_1 to $B_1, \ldots$. Consequently (according to 171) the previous expression

$$= \alpha A^2[B|\mathrm{B}] + \alpha_1 A_1^2[B_1|\mathrm{B}] + \cdots + \alpha_q A_q^2[B_q|\mathrm{B}].$$

But now

$$[A_r|\mathrm{A}] = [A_r|(\alpha A + \alpha_1 A_1 + \cdots + \alpha_t A_t)] = \alpha_r A_r^2,$$

since (according to 147, 168) A_r inner multiplied with the magnitudes $A, A_1, \ldots$ gives zero, except for A_r. Thus in the equation previously established one can replace $\alpha_r A_r^2$ by $[A_r|\mathrm{A}]$, and that expression becomes

$$= [A|\mathrm{A}][B|\mathrm{B}] + [A_1|\mathrm{A}][B_1|\mathrm{B}] + \cdots + [A_q|\mathrm{A}][B_q|\mathrm{B}],$$

that is, formula (a) holds under our assumption.

[116] 67. According to Nos. 148 and 150 (see also Editorial Note 51),

$$[E|EG] = (-1)^{p(q-1)}|[EG|E] = (-1)^{p(q-1)}|G = [E'|E'G],$$

where p and q are understood to be the orders of E and EG.

[117] 68. It is easy to show that the value of the expression for $\cos\angle AB$ lies between the limits -1 and $+1$, and thus that $\angle AB$ is real. Thus if $A = \sum \alpha_k E_k$ and $B = \sum \beta_j E_j$, then according to Nos. 143, 151, and 146:

$$[A|B] = \sum \alpha_k\beta_k,\ \alpha^2 = \sum \alpha_k^2,\ \beta^2 = \sum \beta_k^2$$

and thence follows, as is known,

$$[A|B]^2 \leq \alpha^2\beta^2.$$

One misses here and in the following a definition of the angle between two magnitudes of differing order. Perhaps GRASSMANN, in the puzzling Remark following No. 213, had something of the sort in mind; but nothing certain on this point can be established.

If one wishes to define the angle between two elementary magnitudes A and B of orders α and β, two procedures offer themselves.* One can either define this angle as the angle between A and the normal shadow A' of A on the domain of B, or as the angle between B and the normal shadow B' of B on the domain of A. It can be shown that these two definitions express the same thing and at the same time include GRASSMANN's definition of the angle between two magnitudes of the same order.

Thus according to 165,

$$A' = \frac{[B(A|B)]}{B^{\underline{2}}},\ B' = \frac{[A(B|A)]}{A^{\underline{2}}},$$

and thus according to Nos. 195 and 98,

$$\cos\angle AA' = \frac{[A|A']}{\sqrt{A^{\underline{2}}A'^{\underline{2}}}} = \frac{[A(|B(|A||B))]}{B^{\underline{2}}\sqrt{A^{\underline{2}}A'^{\underline{2}}}},$$

$$\cos\angle B'B = \frac{[B'|B]}{\sqrt{B'^{\underline{2}}B^{\underline{2}}}} = \frac{[(A(B|A))|B]}{A^{\underline{2}}\sqrt{B'^{\underline{2}}B^{\underline{2}}}}.$$

If we assume for simplicity that $\alpha \leq \beta$, then the three magnitudes A, $|B$, $[|A||B]$ have the orders α, $n-\beta$, and $\beta-\alpha$, respectively,whence their product is according to No. 116 purely progressive, and thus, according to Nos. 119 and 97

$$[A(|B(|A||B))] = [(A|B)|(A|B)] = [A|B]^{\underline{2}}.$$

Likewise it follows, considering No. 124, Case a, that

$$[(A(B|A))|B] = (-1)^{(\beta-\alpha)(n-\beta)}[(A|B)(B|A)],$$

and on the other hand, according to 150 and 92,

$$\begin{aligned}|[A|B] &= (-1)^{\alpha(\beta-1)}||[B|A]\\ &= (-1)^{\alpha(\beta-1)+(\beta-\alpha)(n-\beta+\alpha)}[B|A],\end{aligned}$$

*In ordinary space one defines the angle between a straight line and a plane as the angle between the straight line and its orthogonal projection to the plane. The following treatment is just the natural generalization of this definition to a domain of nth order.

and consequently also

$$[(A(B|A))|B] = [(A|B)|(A|B)] = [A|B]^{\underline{2}},$$

so that the numerators in the two expressions $\cos\angle AA'$ and $\cos\angle B'B$ correspond.

In order to prove the correspondence of the denominators as well, we must calculate $A'^{\underline{2}}$ and $B'^{\underline{2}}$.

One has

$$A'^{\underline{2}} = [A'|A'] = \frac{[A'(|B(|A||B))]}{B^{\underline{2}}}.$$

Here the product of the three magnitudes A', $|B$, and $[|A||B]$ is purely regressive, whence

$$A'^{\underline{2}} = \frac{[(A'|B)|(A|B)]}{B^{\underline{2}}} = \frac{[A|B]^{\underline{2}}}{B^{\underline{2}}},$$

since according to No. 169 and Editorial Note 63, $[A'|B] = [A|B]$. Likewise

$$B^{\underline{2}} = \frac{[B|A]^{\underline{2}}}{A^{\underline{2}}} = \frac{[A|B]^{\underline{2}}}{A^{\underline{2}}}.$$

On account of these formulas one obtains at once

$$\cos\angle AA' = \cos\angle B'B = \sqrt{\frac{[A|B]^{\underline{2}}}{A^{\underline{2}}B^{\underline{2}}}},$$

where the positive value of the root is to be taken.

Thereby one may, if A and B are of arbitrary order, define the angle AB by the equation

$$\cos\angle AB = \sqrt{\frac{[A|B]^{\underline{2}}}{A^{\underline{2}}B^{\underline{2}}}}.$$

Here one can, so long as A and B are of different orders, leave the sign of the root unspecified; but if A and B are of the same order and thus $[A|B]$ is a number, one must set the square root of the numerator $= [A|B]$ and choose the positive square root in the denominator.

[117] 69. Here too it must be shown that $\sin(abc\cdots)$ belongs to a real angle, thus that the numerical value of $[abc\cdots]$ is not greater than the product $\alpha\beta\gamma\cdots$ of the numerical values of the individual factors a, b, c, ... In GRASSMANN's style the proof can be given as follows:

We imagine each of the magnitudes a, b, c, ... divided by its numerical value, so that a product $[a_1\cdots a_n]$ $(m \leq n)$ results, whose factors of first order, $a_1, \ldots, a_m$ are all numerically equal to one; then we need only show that the numerical value of $[a_1\cdots a_m]$ can be no greater than one. For this we may assume that the product $[a_1\cdots a_m]$ does not vanish, for if it were its numerical value would be zero.

If $a_1, \ldots, a_m$ are normal to one another, then according to No. 175

$$[(a_1\cdots a_m)|(a_1\cdots a_m)] = [a_1\cdots a_m]^{\underline{2}} = 1.$$

If on the other hand they are not normal to one another, we may assume without loss of generality that a_1, say, is not normal to all the $m-1$ magnitudes $a_2, \ldots, a_m$, and in addition, according to Nos. 160 and 163, we can find a normal system of mth order $a_1, u_2, \ldots, u_m$ of unit numerical value, whose domain coincides with the domain of the magnitudes $a_1, \ldots, a_m$. Then

$$a_k = \lambda_k a_1 + \sum_{v=2}^{m} \lambda_{kv} u_v = \lambda_k a_1 + a_k' \quad (k = 2, \ldots, m),$$

where the $\lambda_2, \ldots, \lambda_n$ are certainly not all zero. From the equations

$$a_k^2 = 1 = \lambda_k^2 + \sum_v \lambda_{kv}^2, \quad a_k'^2 = \sum_v \lambda_{kv}^2$$

it therefore follows that the product $\rho_2 \cdots \rho_n$ of the numerical values $\rho_2, \ldots, \rho_n$ of $a_2', \ldots, a_n'$ is smaller than one. If we now set $a_k' = \rho_k a_k''$, then a_k'' is numerically equal to one, and according to No. 67 we get

$$\begin{aligned}[a_1 \cdots a_m] &= [a_1 a_2' \cdots a_m'] \\ &= \rho_2 \cdots \rho_m [a_1 a_2'' \cdots a_m''].\end{aligned}$$

Since here the numerical value of the product $[a_1 a_2'' \cdots a_m'']$ is obviously greater than that of $[a_1 \cdots a_m]$, we can say: "If one has a nonvanishing product of m magnitudes of first order ($m \leq n$), all of which have numerical value one, but which are not normal to one another, then in the domain of these magnitudes one can always find m magnitudes of first order of numerical value one, whose product has a greater numerical value than that of the given product." One can now continue this increase in the numerical value until one is led to a product whose factors are normal to one another; since on the other hand a product of the nature considered, whose factors are normal to one another, possesses the numerical value one, the following theorem holds:

THEOREM. *If $a_1, \ldots, a_m$ are nonzero magnitudes of first order in a principal domain of nth order ($m \leq n$), then the numerical value of the product $[a_1 \cdots a_m]$ is no greater than the product of the numerical values of $a_1, \ldots, a_m$, and in fact it is equal to this product if the $a_1, \ldots, a_m$ are normal to one another.*

[118] 70. That here and in No. 199 a, b, c, d are supposed to be magnitudes of *first* order follows both from the notation as lower-case Latin characters and from the application of the theorem 127.

Otherwise it is not without interest that the theorems 175–182, 185–194, 196, 198–199, 201–205, 208–210, 213–215 remain applicable if one replaces the magnitudes of first order and their numerical values appearing therein with magnitudes of $(n-1)$th order and their numerical values. Thus in the first place every magnitude of $(n-1)$th order is the supplement of a perfectly unique magnitude of first order. Second, according to No. 98 the supplement of a product equals the product of the supplements of its factors, and thus in particular the supplement of an inner product is equal to the inner product of the supplements of its two factors, whence it follows at once that the supplement of a magnitude always has the same numerical value as the magnitude itself. Finally it is also obvious that $\cos\angle ab = \cos\angle|a|b$, since $[a|b]$ is a number and thus $|[a|b] = [a|b]$. If one bears in mind that every equation in the Numbers referred to continue to hold if one takes the supplement on both sides, one perceives immediately that precisely the same equations also apply for magnitudes of $(n-1)$th order.

[118] 71. Here the *positive* value of the square root is to be understood.

[121] 72. If in ordinary space one chooses for units three mutually orthogonal displacements of length one, then a, b, c are three arbitrary displacements in space which one may produce from a point O, and $\sin(abc)$ is precisely the expression that v. STAUDT has denoted the sine of the vertex a, b, c (*Crelle's Journal* **24**, 255 (1842)).

[121] 73. The editors of the Teubner Edition dropped this Remark, since it is not clear what GRASSMANN really had in mind in this regard.

[121] 74. It seems worth while to say a few words in conclusion about the significance of the concepts introduced in Nos. 151–215.

If one thinks of the units $e_1, \ldots, e_n$ in an n-fold extended Euclidean space as n mutually orthogonal displacements, all of whose lengths are equal to the unit of length, then every magnitude $\sum x_v e_v$ is likewise represented by a displacement whose length is equal to the numerical value of $\sum x_v e_v$. Mutually orthogonal magnitudes of first order are described by mutually orthogonal displacements, and every elementary normal system by n mutually orthogonal displacements; the angle between two magnitudes of first order is equal to the angle between the corresponding displacements, and so forth. For the case $n = 3$ GRASSMANN himself has carried this out in detail.

Now, as LIE has emphasized,* all these considerations are grounds for saying that GRASSMANN worked on non-Euclidean geometry. Thus if Euclidean geometry applies in a space of n dimensions, then among the manifold of all the displacements of this space, or, what is the same thing, in the infinitely distant $(n-1)$-fold extended plane, the non-Euclidean geometry discovered by RIEMANN applies, and GRASSMANN has in fact developed this geometry here.

If we recall the connection between the normal system and the orthogonal substitutions (cf. Editorial Note 58) and consider that in an n-fold extended Euclidean space the orthogonal substitutions of unit determinant are nothing but the rotations about the origin of coordinates, then with STUDY (Editorial Note 4 to *Linear Extension Theory*, translation of the A_1) we can express the state of affairs thus: The investigations of Nos. 151–215 are contributions to the theory of the invariants of the group of all rotations about a point.

One is led to an interpretation somewhat more general but not essentially different from the above, if one interprets the coefficients $x_1, \ldots, x_n$ of the magnitude of first order $\sum x_v e_v$ as arbitrary homogeneous coordinates in an $(n-1)$-fold extended plane space. The concept of numerical value no longer has a geometric significance, since its analytical expression is not homogeneous of zeroth order in the coordinates. On the other hand, for example, the angle between two magnitudes of first order $\sum x_v e_v$ and $\sum y_v e_v$ coincides with the distance between the two points $x_1, \ldots, x_n$ and $y_1, \ldots, y_n$ if one takes as fundamental for the measure of this distance CAYLEY's definition of measure with respect to the fundamental manifold $\sum x_v^2 = 0$. To be sure, CAYLEY's definition of measure is older than the A_2, for it was originated in 1839.

[126] 75. This Number would be more easily understood if the special theorem denoted a Corollary had been placed at the head of the Number. Thus the thrust of the main theorem of No. 222 is only comprehensible from the Corollary; for it assumes one already knows that $A - R$, $B - R$, ... are displacements, which only emerges from the Corollary. The transposition could be done without hesitation, since the proof of the Corollary is completely independent of the main theorem.

[131] 76. Regarding BELLAVITIS, cf. Editorial Note 6.

The demonstration "that there is no other addition of points and lines than that given here" cannot be regarded as adequate, for the entire subsequent proof is

---

* *Theorie der Transformationsgruppen*, Bd. III, P. 534f.

only justified if one, first, makes the various special assumptions that GRASSMANN introduces, which can have no basis in the nature of the subject; second, as is done in the text, inserts the word "simple" throughout; and finally, assumes Euclidean geometry from the outset. Other assumptions produce yet other types of addition of points, cf. STUDY, *Wiener Berichten* **91**, 111 (1885).

[143] 77. After Nos. 254 and 262 one anticipates the introduction of a name for the products of two and three displacements corresponding to the names already introduced in Definitions 249, 257, and 265 for the products of two, three, and four points, say in the form:

We call the product [ab] of two displacements a ***surface area****, the area of the parallelogram ab its content and the position of this parallelogram its position.*

Further: *We call the product [abc] of three displacements a, b, c a* ***volume*** *and the content of the parallelepiped its content.*

In fact, in subsequent Numbers (cf. 330, 346, and 347) the expressions surface area and volume are often used for these products, but without these names being specifically introduced as technical terms. To be sure, on account of their similarity to the expressions surface element and volume element they are not especially suitable. On this account R. MEHMKE has introduced (first in a lecture at the Stettin Polytechnic, Summer 1881) a new name, "the field" {das Feld}, for the product of two displacements, and he has been joined by G. MAHLER and F. KRAFT. From the other side the expression "the cell" {das Fach} has been proposed for the product of three displacements.

H. Grassmann, Junior

[144] 78. "One could have defined ... ": This GRASSMANN had originally done. In the language of the A_1 the content of the triangle ABC is designated the "extension" of the outer product $[ABC]$. On the other hand, in his Notice of the A_1 published in *Grunerts Archiv*, GRASSMANN said explicitly: "The product ABC denotes the triangle with vertices A, B, C, understood as (etc.)."

[146] 79. The Definition given in No. 71 of the concept of the *elementary* linear evolution is not rigorously followed here and later (cf. No. 505); for the requirement that the magnitude whose multiple is to be added be a *neighbor* of the incremented magnitude, is dropped.

[150] 80. By "the infinitely distant line element" is of course meant a line element resulting from the multiplication of two infinitely distant points, and thus the product of two displacements.

[160] 81. Regarding the Proof, refer to Editorial Note 34.

[163] 82. In the Proof it should have been mentioned that the homogeneous equation of nth degree between x_1, x_2, x_3 includes a power of x_1, say x_1^m, as a factor; thus in the "ordinary coordinates" one gets an algebraic curve of $(n-m)$th order, and only if one appends to this curve the infinitely distant straight line counted m times, does one obtain the structure of nth order represented by the equation $\mathfrak{P}_{n,x} = 0$. The Proof of No. 311 provides an occasion for a corresponding remark. Cf. also the Remark to No. 329.

[166] 83. It should have been noted that according to No. 320 the product $[paBc_1D] = q$ certainly does not vanish, and that on the same grounds $[qe]$ also does not vanish.

[175] 84. "Conversely, etc." Thus the last equation says that $[(g)Lk] = 0$, and since $[(g)C]$ manifestly vanishes, $(g) \equiv [L_k C]$. On the other hand $(h) \equiv [L_k C]$, whence $(g) \equiv (h)$, or, since (g) and (h) are simple points, $(g) = (h)$, whence $g = h$ follows immediately.

[181] 85. This Remark is already unclear in its second paragraph, and the fourth can only be preserved by main force despite the interpolations. Perhaps it would have been better to suppress the Remark completely except for its first and last paragraphs, for the type of normal shadow on points, lines and planes described in the Remark can only be realized if one replaces the concept of the supplement previously introduced by a completely different one.

Thus it would be necessary, besides the system of original units a, b, c, d consisting of a simple point a and three mutually orthogonal displacements b, c, d, *to adopt, for each individual finitely distant point x in space, yet another system of units that specify the supplements of the point x itself and the line and surface elements going through it*, and indeed the special system of units belonging to the point x must, besides the unit displacements b, c, d established for all points of space, include as first unit *the simple point congruent with the point x*. The supplements referred to this variable system of units — for the moment let them be denoted by a $\mathfrak{J}$ — then to be sure yield, when applied to line and surface elements, the fields and displacements orthogonal to these structures, and the normal shadow belonging to this type of supplement actually does acquire the meaning given in the text. On the other hand, for the new symbol $\mathfrak{J}$ even the fundamental equation of the supplement,

$$\mathfrak{J}[AB] = [\mathfrak{J}A \cdot \mathfrak{J}B],$$

is no longer valid, so that one would also be required first to develop the calculus for the symbol $\mathfrak{J}$ completely anew.

The difficulty that arises if one wants to apply the concepts of the supplement and normal to the points of Euclidean space rests on the following circumstance: GRASSMANN's calculus, insofar as it makes use of these concepts, is tailored to the group consisting of all rotations of a Euclidean space about a point (cf. Editorial Note 74). However in the geometry of Euclidean space there is treated not just this group, but rather the more extensive group of *all* motions, which include, besides rotations, parallel displacement and twists. Now since both groups transform infinitely distant points, that is displacements in Euclidean space, in the same way, GRASSMANN's calculus applies in its full extent to the displacements in Euclidean space. If however one wants to apply it as well to the finitely distant points of Euclidean space, one must remodel it, and that would be achieved in the text if the concepts "supplement" and "normal" were always applied only to displacements and products of displacements.

[183] 86. The idea introduced here is carried out in GRASSMANN's article on quaternions, *Math. Ann.* **12**, 384–386 (1877). Regarding the use of the exterior angles of the spherical triangle instead of the interior angles, see also GRASSMANN's presentation in his *Lehrbuch der Trigonometrie für höhere Schulen*, Enslin, Berlin (1865), pp. 100–115. However MÖBIUS was the first to make serious use of the exterior angles of the triangle instead of the interior in spherical trigonometry (in the *Analytische Sphärik*, 1846, and in the *Grundformeln der sphärischen Trigonometrie*, 1860).

[186] 87. For if one writes the first equation in the form $-\alpha a = \beta b + \gamma c + \cdots$, then the point a endowed with the coefficient $-\alpha$ appears as the summation point of the multiple points βb, γc, ... But according to 222 the numerical coefficient of the summation point is equal to the sum of the coefficients of the summand points, that is

$$-\alpha = \beta + \gamma + \cdots, \text{ whence } \alpha + \beta + \gamma + \cdots = 0.$$

[188] 88. In fact the equation

$$2[(s - s')|r] + \mu = 0 \tag{1}$$

with constant s and r and variable s' represents a plane orthogonal to r. This plane goes through the point

$$s + \frac{\mu r}{2\rho^2} = p; \tag{2}$$

thus from equation (2) it follows that

$$2(s - p) + \mu\frac{r}{\rho^2} = 0$$

and from this upon inner multiplication with r, since $r^{\underline{2}} = \rho^2$, there results the equation

$$2[(s - p)|r] + \mu = 0,$$

which expresses precisely that the plane (1) goes through the point p.

[188] 89. (Cf. also No. 286.) A sum of line elements that cannot be reduced to a single line element HYDE, in agreement with the usage of BALL, calls a *screw*, and the representation of such a sum as the sum of a line element and a field orthogonal to it the *normal form of a screw.**

If one combines HYDE's technical word *screw* with the above chosen (Editorial Note 77) abbreviated names *field* and *cell* and adds the expressions *rod* {Stab} and *sheet* {Blatt} proposed by the undersigned** for the concepts "line element" and "surface element", one obtains the following scheme of abbreviated names for the application of the fundamental concepts of extension theory to geometry:

Displacement {Strecke};
Field {Feld} (product of two displacements, surface area);
Cell {Fach} (product of three displacements, volume);
Point;
Rod {Stab} (product of two points, line element);
Screw {Schraube} (sum of rods);
Sheet {Blatt} (product of three points, surface element), to which one could perhaps add
Block (product of four points, volume element).

H. Grassmann, Junior

*Cf. E. W. HYDE, The directional theory of screws, *Annals of Math.* **4**, 137 (1888), and the independent Work by the same author, *The Directional Calculus, based upon the methods of Hermann Grassmann*, Boston, 1890, Arts. 61 and 67.

**Cf. H. GRASSMANN, *Punktrechnung und projektive Geometrie.* Erster Teil: Punktrechnung, Halle a/S (1894). Contribution for the Festschrift for the Lateinischen Hauptschule for the two hundredth anniversary of the University at Halle, pp. 81 and 92, and separate part, pp. 7 and 18.

[196] 90. For the expression "interchangeable openings", cf. Note 105 (to No. 485).

[221] 91. In the original edition this special type of affinity was denoted "equality" {Gleichheit} according to the prior usage of MÖBIUS (*Barycentrische Kalkül* (*Werke*, Bd. I, p. 194ff)).

[221] 92. {In this Note Hermann GRASSMANN Junior applies the general theory of geometric relations developed in Nos. 377–390 to *collineations in space*. This lengthy note — occupying over 24 pages of the Teubner Edition — is omitted here as being only of technical interest. *Tr.*}

[223] 93. The editors of the Teubner Edition replaced the material in braces with the following:

It will now be shown that such an assembly continues to be such an assembly under a certain special type of circular evolution of the magnitudes appearing in it, and indeed such that the number of reals among the n magnitudes of the one assembly is the same as that in the other. This special type of circular evolution is to transform two magnitudes a_1 and a_2, when both are real or both simple imaginaries, into two other magnitudes b_1 and b_2, where

$$(h) \qquad b_1 = xa_1 + ya_2, \; b_2 = xa_2 - ya_1,$$

where x and y are both real, and the sum of their squares is one, thus

$$x^2 + y^2 = 1.$$

On the other hand if one of the two magnitudes a_1 and a_2 is real, the other imaginary, then the two magnitudes a_1 and a_2 are to transform into the magnitudes

$$b_1 = xa_1 + yia_2, \; b_2 = xa_2 - yia_1,$$

where $i = \sqrt{-1}$, x and y are real, and $x^2 - y^2 = 1$, that is, $x_2 + (yi)^2 = 1$. One can therefore also say: Equations (h) each represent a circular evolution of a_1 and a_2 if x is always real, but y is only imaginary, and indeed simple imaginary, if one of the magnitudes a_1 and a_2 is real and the other is simple imaginary.

[226] 94. Let $e_1, \ldots, e_n$ be the original units and

$$\boldsymbol{Q}e_\kappa = \sum_{\nu}^{1\ldots n} \alpha_{\kappa\nu} e_\nu \quad (\kappa = 1, \ldots, n),$$

where the $\alpha_{\kappa\nu}$ are to signify *real* numbers. If one then sets $a = \sum x_\kappa e_\kappa$ and $b = \sum y_j e_j$, the expression $[\boldsymbol{Q}a|b]$ takes the form

$$[\boldsymbol{Q}a|b] = \sum_{\kappa;\nu;j} \alpha_{k\nu} x_\kappa y_j [e_\nu|e_j] = \sum_{\kappa;j} \alpha_\kappa x_\kappa y_j,$$

and the requirement that, for arbitrary a and b, $[\boldsymbol{Q}_a|b] = [\boldsymbol{Q}b|a]$ always, can therefore be replaced by the equations $\alpha_{\kappa j} = \alpha_{j\kappa}$ $(\kappa, j = 1, \ldots, n)$. Again, these last say that $[\boldsymbol{Q}a|b]$ is the first polar of b with respect to the quadratic form $[\boldsymbol{Q}a|a] = \sum \alpha_{\kappa j} x_\kappa x_j$.

The other requirement, that $[\boldsymbol{Q}a|a]$ is to be different from zero for arbitrary a, amounts to the requirement that the quadratic form $\sum \alpha_{\kappa j} x_\kappa x_j$ can only vanish for real x if all the x are zero, and that it is therefore a positive definite or negative definite form. This assumption is utilized several times by GRASSMANN in the proof of No. 391, but would, as will be shown, have the consequence that the principal numbers of the quotient $\boldsymbol{Q}$ be all positive or all negative, and must therefore, if one would retain the theorem 391 in its present general form, be replaced by another, namely that *the determinant of the quadratic form* $\sum \alpha_{\kappa j} x_\kappa x_j$ *or*, what amounts to the same thing, *the power of the fraction* $\boldsymbol{Q}$ *is to be different from zero.*

We will show later how the proof of the theorem 391 proceeds if one replaces GRASSMANN's assumption by this last. First however we will clarify for ourselves what properly is the sense and aim of the theorem. Thus we will assume at once that the power of the fraction $\boldsymbol{Q}$, that is the determinant of $\alpha_{\kappa j}$, is different from zero.

Besides $\boldsymbol{Q}$, the quadratic form $\sum \alpha_{\kappa j} x_\kappa x_j$ stands in a relation with yet another quotient, that is with the quotient $\boldsymbol{P}$, which is defined by the equations

$$\boldsymbol{P} e_\kappa = \sum_{\nu}^{1\dots n} \alpha_{\kappa\nu} | e_\nu \quad (\kappa = 1, \dots, n),$$

and whose denominators are of $(n-1)$th order, while its numerators, as with $\boldsymbol{Q}$, are of first order.

This quotient $\boldsymbol{P}$ has a simple geometric meaning. Thus if we think of $e_1, \dots, e_n$ in an n-fold extended Euclidean space as n mutually orthogonal displacements of unit length through a point O and correspondingly interpret $x_1, \dots, x_n$ as rectangular coordinates in this space, with O as origin, then $\boldsymbol{P}$ coordinates with each displacement going through O the $(n-1)$-fold extended plane that includes all conjugate diameters of this displacement with respect to the ∞^1 manifolds of second degree $\sum \alpha_{\kappa j} x_\kappa x_j = \text{const.}$*

The relation between the fraction $\boldsymbol{P}$ and the quadratic form $\sum \alpha_{\kappa j} x_\kappa x_j$ depends on the choice of the units $e_1, \dots, e_n$. Thus let $c_1, \dots, c_n$ be n magnitudes of first order whose combinatorial product is equal to one; then

$$c_\kappa = \sum_{\nu}^{1\dots n} \lambda_{\kappa\nu} e_\nu \quad (\kappa = 1, \dots, n),$$

where the determinant of the $\lambda_{\kappa\nu}$ is equal to one, and conversely let

$$e_\nu = \sum_{j}^{1\dots n} \Lambda_{j\nu} c_j \quad (\nu = 1, \dots, n),$$

and finally let I be symbol for the supplement belonging to the system of units $c_1, \dots, c_n$ (see No. 110). Then necessarily

$$I c_\kappa = \sum_{\nu}^{1\dots n} \rho_{\kappa\nu} | e_\nu,$$

or, if one multiplies this equation by $\sum \Lambda_{j\mu} e_j = e_\mu$, $\Lambda_{\kappa\mu} = \rho_{\kappa\mu}$, whence

$$I c_\kappa = \sum_{\nu}^{1\dots n} \Lambda_{\kappa\nu} | e_\nu .$$

*If one interprets $x_1, \dots, x_n$ as homogeneous coordinates in an R_{n-1}, then $\boldsymbol{P}$ transforms each point $x_1, \dots, x_n$ into its $(n-2)$-fold extended polar plane with respect to the manifold $\sum \alpha_{\kappa j} x_\kappa x_j = 0$, whereas $\boldsymbol{Q}$ transforms the point $x_1, \dots, x_n$ into the pole of the same plane with respect to the manifold $\sum x_\nu^2 = 0$.

Thus upon solution it follows that

$$|e_\nu = \sum_j \lambda_{j\nu} I c_j.$$

Now one finds

$$\begin{aligned} \boldsymbol{P}c_\kappa &= \sum_\nu \lambda_{\kappa\nu} \boldsymbol{P}e_\nu = \sum_{\mu\nu} \lambda_{\kappa\nu}\alpha_{\nu\mu}|e_\mu \\ &= \sum_{\mu\nu\tau} \lambda_{\kappa\nu}\lambda_{\tau\mu}\alpha_{\nu\mu} I c_\tau \end{aligned}$$

On the other hand, $a = \sum x_\nu e_\nu = \sum x'_\kappa c_\kappa$, whence $x_\nu = \sum \lambda_{\kappa\nu} x'_\kappa$, and thus the quadratic form $\sum \alpha_{\kappa\nu} x_\kappa x_\nu$ is transformed, upon introduction of the new units c_1, ..., c_n, into a form $\sum \alpha'_{\kappa\nu} x'_\kappa x'_\nu$, where

$$\alpha'_{\kappa\tau} = \sum_{\nu\mu} \lambda_{\kappa\nu}\lambda_{\tau\mu}\alpha_{\nu\mu},$$

and if one compares this with the previously found expression for $\boldsymbol{P}c_\kappa$, one obtains

$$(*) \qquad \boldsymbol{P}c_\kappa = \sum_\nu \alpha'_{\kappa\nu} I c_\nu.$$

But these equations say that the relation between the quotients $\boldsymbol{P}$ and the quadratic form $\sum \alpha_{\kappa\nu} x_\kappa x_\nu$ continues to hold upon the introduction of new units.

It is otherwise with the relation between the quotient $\boldsymbol{Q}$ and our quadratic form. Thus according to the definitions of $\boldsymbol{Q}$ and $\boldsymbol{P}$, $|\boldsymbol{Q}c_\kappa = \boldsymbol{P}c_\kappa$. Now if the relation between the quotient $\boldsymbol{Q}$ and the form $\sum \alpha_{\kappa\nu} x_\kappa x_\nu$ is to continue to hold upon the transformation to the units $c_1, \ldots, c_n$, then necessarily

$$\boldsymbol{Q}c_\kappa = \sum_\nu \alpha'_{\kappa\nu} c_\nu,$$

whence

$$|\boldsymbol{Q}c_\kappa = \sum_\nu \alpha'_{\kappa\nu}|c_\nu = \sum_\nu \alpha'_{\kappa\nu} I c_\nu.$$

But since according to No. 384 the determinant of the $\alpha'_{\kappa\nu}$ is equal to the power of $\boldsymbol{Q}$ and thus certainly does not vanish, the last equation can only hold if $Ic_\nu = |c_\nu$, in other words (cf. No. 167 and Editorial Note 62) if $c_1, \ldots, c_n$ form a complete normal system of numerical value one, or what is the same thing, if the transformation $x_\nu = \sum \lambda_{\kappa\nu} x''_\kappa$ is orthogonal (cf. Editorial Note 57).*

Following these preparations we can now express the content of the theorem 391 in a form independent of the symbols of the *Ausdehnungslehre*.

First it is claimed in the theorem that one can always specify n magnitudes c_1, ..., c_n that stand in no numerical relation and for which $[\boldsymbol{Q}c_\kappa|c_j] = 0$, provided $\kappa \neq j$. If we now imagine, as can be done without loss of generality, that the $c_1, \ldots,$

---

*One can also express this as follows: The quotient $\boldsymbol{Q}$ is understood as a simultaneous covariant of the two quadratic forms $X = \sum \alpha_{\kappa j} x_\kappa x_j$ and $U = \sum u_\nu^2$, under the contragradient variables x and u. In fact, if one interprets the x as point and the u as plane coordinates, the covariant

$$\frac{1}{2}\sum \frac{\partial X}{\partial x_j}\frac{\partial U}{\partial u_j} = \sum \alpha_{j\kappa} x_\kappa u_j$$

set equal to zero represents precisely the linear substitution whose symbol is the quotient $\boldsymbol{Q}$. In this connection cf. GUNDELFINGER, *Vorlesungen aus der analytischen Geometrie der Kegelschnitte*, edited by DINGELDEY, Leipzig, Teubner, 1895, p. 70ff.

c_n are chosen so that their combinatorial product is equal to one, then equation $(*)$ holds for the c_κ, and since according to what went before, $|Qc_\kappa = Pc_\kappa$ and according to No. 144, $[Qc_\kappa|c_j] = [c_j|Qc_\kappa]$,

$$[Qc_\kappa|c_j] = [c_j \cdot Pc_\kappa] = \sum_\nu \alpha'_{\kappa\nu}[c_j I c_\nu] = \alpha'_{\kappa j}.$$

But since the expressions $[Qc_\kappa|c_j]$ all vanish for $\kappa \neq j$, there follows from $(*)$ that

$$Pc_\kappa = \alpha'_{\kappa\kappa} I c_\kappa,$$

and the quadratic form $\sum \alpha_{\kappa j} x_\kappa x_j$, upon the transformation to the units $c_1, \dots, c_n$, obtains the form $\sum \alpha'_{\kappa\kappa} {x'_\kappa}^2$. Here, finally, none of the magnitudes $\alpha'_{\kappa\kappa}$ can vanish, for according to No. 92,

$$|Pc_\kappa = ||Qc_\kappa = (-1)^{n-1} Qc_\kappa,$$

whence

$$\begin{aligned} [Qc_1 \cdot Qc_2 \cdots Qc_n] &= (-1)^{n(n-1)} |[Pc_1 \cdots Pc_n] && [98] \\ &= \alpha'_{11} \cdots \alpha'_{nn} |I[c_1 \cdots c_n] && [110] \\ &= \alpha'_{11} \cdots \alpha'_{nn}, \end{aligned}$$

but this expression cannot vanish, since it represents the power of Q.

Thus if one specifies $c_1, \dots, c_n$ as above, one basically solves the problem of *reducing the quadratic form* $\sum \alpha_{\kappa j} x_\kappa x_j$, *to a sum of n squares by a real homogeneous substitution of determinant one.** The number of positives among these n squares is therefore equal to the number of positives among the n nonzero real numbers $[Qc_\kappa|c_\kappa] = \alpha'_{\kappa\kappa}$.

In our theorem it is claimed in addition that there is always a normal system $e_1, \dots, e_n$ such that $Qe_r = \rho_r e_r$, where the ρ_r are real and among them as many are positive as there are among the magnitudes $[Qc_\kappa|c_\kappa] = \alpha'_{\kappa\kappa}$. If, as is always feasible, we choose the normal system $e_1, \dots, e_n$ so that its numerical value is equal to one, then from what has been said earlier it follows that under the transformation to the new units $e_1, \dots, e_n$ the quadratic form $\sum \alpha_{\kappa j} x_\kappa x_j$ takes the form $\sum \rho_r x_r^2$.** Thereby the specification of such a normal system $e_1, \dots, e_n$ basically solves the problem, *to reduce the quadratic form* $\sum \alpha_{\kappa j} x_\kappa x_j$, *to a sum of squares by a real substitution under which the form* $\sum x_\nu^2$ *remains invariant, that is, by an orthogonal substitution.* But one can also say that here the problem is solved, *to specify the principal axes of the* ∞^1 *manifolds of second degree* $\sum \alpha_{\kappa j} x_\kappa x_j = \text{const}$ *of* R_n, and at the same time is proved the well-known theorem that the equation of nth degree, on which the definition of these principal axes depends, has purely real roots (cf. p. 226, line 9, where after "second order" there ought properly to have been added "with center").

Finally, the fact that among the ρ_r the principal numbers of the quotient Q, there are precisely as many positive as there are among the magnitudes $[Qc_\kappa|c_\kappa] = \alpha'_{\kappa\kappa}$, is obviously equivalent to SYLVESTER's *law of inertia of quadratic forms* (cf.

---

*The transformation to the coordinates x' is naturally equivalent to relating the ∞^1 manifolds $\sum \alpha_{\kappa j} x_\kappa x_j = \text{const}$ to a system of conjugate diameters, whence the expression "conjugate assembly" on p. 223, line 2.

**If one sets $a = \sum x_r e_r$, then $[Qa|a] = \sum \rho_r x_r^2$, an expression that comes out positive for every arbitrary a only if the ρ_r are all positive or all negative. This confirms what was said in the second paragraph of this Note.

p. 226, line 7). But as is well known, one can utilize this law of inertia to ascertain how many real roots of an equation lie between two given limits (cf. HERMITE, *Comptes Rendus* **36**, 294 (1853)), and thereby the connection with STURM's theorem is cleared up (cf. p. 226, lines 7–8).

It still remains for us to show how the proof of No. 391 proceeds when one replaces the assumption that $[\boldsymbol{Q}a|a]$ is different from zero for every arbitrary a, with the other one, that the power of $\boldsymbol{Q}$, and thus the determinant of the $\alpha_{\kappa j}$, is not to vanish.

First we seek n magnitudes $c_1, \ldots, c_n$ that stand in no numerical relation, and for which $[\boldsymbol{Q}c_\kappa|c_j] = 0$ so long as $\kappa \neq j$. If there are n such magnitudes, then according to the above the n expressions $[\boldsymbol{Q}c_\kappa|c_\kappa]$ are all different from zero, and thus we must from the beginning make sure that this condition is satisfied with the choice of the c_κ.

We will assume that we have already found m magnitudes $c_1, \ldots, c_m$ that stand in no numerical relation and satisfy the equations $[\boldsymbol{Q}c_\mu|c_\nu] = 0$ for $\mu \neq \nu$, while all $[\boldsymbol{Q}c_\mu|c_\mu] \neq 0$. We claim that a magnitude c_{m+1} not derivable from $c_1, \ldots, c_m$ can then always be specified that satisfies the m equations $[\boldsymbol{Q}c_\mu|c_{m+1}] = 0$ $(\mu = 1, \ldots, m)$ and thus also the m equations $[\boldsymbol{Q}c_{m+1}|c_\mu] = 0$, while $[\boldsymbol{Q}c_{m+1}|c_{m+1}] \neq 0$.

In order to prove our claim, we set $\boldsymbol{Q}c_\mu = k_\mu$ $(\mu = 1, \ldots, m)$; then $k_1, \ldots, k_m$ also certainly stand in no numerical relation, since otherwise the power of $\boldsymbol{Q}$ could not be different from zero. In addition we specify $n - m$ magnitudes of first order $b_{m+1}, \ldots, b_n$ so that they stand in no numerical relation and satisfy the equations

$$[k_\mu|b_{m+j}] = 0 \ (\mu = 1, \ldots, m; \ j = 1, \ldots, n - m).$$

According to Nos. 163 and 159 this is always possible, for in the domain of $k_1, \ldots, k_m$ we need only take any normal system of mth order $k'_1, \ldots, k'_m$ and to supplement this to a complete normal system $k'_1, \ldots, k'_m, b_{m+1}, \ldots, b_n$, and then the $b_{m+1}, \ldots, b_n$ are magnitudes with the desired property and indeed are normal to one another. Between $c_1, \ldots, c_m$ and the magnitudes $b_{m+1}, \ldots, b_n$ just found there can exist no numerical relation

$$\gamma_1 c_1 + \cdots + \gamma_m c_m + \beta_{m+1} b_{m+1} + \cdots + \beta_n b_n = 0,$$

for otherwise one would have

$$\sum_\mu \gamma_\mu [k_\nu|c_\mu] + \sum_j \beta_{m+j} [k_\nu|b_{m+j}] = \gamma_\nu [k_\nu|b_\nu] = 0$$

for $\nu = 1, \ldots, m$, whence $\gamma_1, \ldots, \gamma_m$ would all have to vanish, and thus among the $b_{m+1}, \ldots, b_n$ alone there must be a numerical relation, which is not the case. Further, the equation $[\boldsymbol{Q}b|b] = 0$ cannot be satisfied for any magnitude, since otherwise

$$[\boldsymbol{Q}b|b] = \sum_{\kappa;j}^{1\ldots n-m} \lambda_{m+\kappa} \lambda_{m+j} [\boldsymbol{Q}b_{m+\kappa}|b_{m+j}] = 0$$

for all values of $\lambda_{m+\kappa}$, but then $[\boldsymbol{Q}b_{m+\kappa}|b_{m+j}] = 0$ for $\kappa, j = 1, \ldots, n - m$, whence each of the magnitudes $\boldsymbol{Q}b_{m+\kappa}$ would be normal to all the magnitudes $b_{m+1}, \ldots, b_n$, with the consequence that all the $\boldsymbol{Q}b_{m+j}$ would belong to the domain of the magnitudes $k'_1, \ldots, k'_m$ or, what is the same thing, to the domain of the magnitudes $k_1, \ldots, k_m$:

$$Qb_{m+j} = \sum_{\mu}^{1\ldots m} \delta_{\mu j} k_j,$$

from which it again follows that the power of Q would be zero. We can therefore always choose $c_{m+1} = \sum \lambda_{m+j} b_{m+j}$ so that $[Qc_{m+1}|c_{m+1}] \neq 0$, and then the c_{m+1} will stand in no numerical relation with $c_1, \ldots, c_m$ and in addition satisfy the equations $[Qc_\mu|c_{m+1}] = [Qc_{m+1}|c_\mu]$ for $\mu = 1, \ldots, m$.

Our claim is therefore proven. Now since the assumption made above is certainly permitted for $m = 1$, since one can always choose c_1 so that $[Qc_1|c_1] \neq 0$, it is obviously proven that it is always possible to specify n magnitudes $c_1, \ldots, c_n$ with the desired property. We imagine them specified and besides for the sake of simplicity that $c_1, \ldots, c_n$ are chosen so that their combinatorial product is one. If as before* we then set $[Qc_\nu|c_\nu] = \alpha'_{\nu\nu}$, the power of our fraction Q is equal to the product $\alpha'_{11} \cdots \alpha'_{nn}$.

This replaces the development on p. 222, line 3 from the top to line 11 from the bottom. The rest of GRASSMANN's proof requires no other changes, and requires comment only on two other points.

P. 224, lines 11–20. Still missing here is the proof that the equation $x^2 + y^2 = 1$ can always be satisfied, and thus that γ cannot equal i. In order to show this, one observes that for $\gamma = i$ either $(r + r')^{\underline{2}}$ or $(r - r')^{\underline{2}}$ must be zero, or, since r and r' are real, either $r + r'$ or $r - r'$ equals zero. This is however impossible, since each of the magnitudes r and r' results from one of the magnitudes $c_1, \ldots, c_n$ by multiplication by a real number, and there is certainly no numerical relation between the $c_1, \ldots, c_n$.

P. 225, lines 6–7. "Now since there must be a minimum … ". According to our notation $[Qc_\kappa|c_\kappa] = \alpha'_{\kappa\kappa}$ the magnitudes $a_1, \ldots, a_n$ have the values

$$a_\kappa = c_\kappa : \sqrt{\alpha'_{\kappa\kappa}} \quad (\kappa = 1, \ldots, n),$$

whence the product $\alpha_1 \cdots \alpha_n$ of their numerical values is equal to the product of the numerical values of $c_1, \ldots, c_n$ divided by $\sqrt{\pm\alpha'_{11} \cdots \alpha'_{nn}}$, where the sign under the root is chosen so that the root is real, and the root itself is taken to be positive. But now (cf. Editorial Note 69) the product of the numerical values of $c_1, \ldots, c_n$ is certainly not less than the numerical value of the product $[c_1 \cdots c_n]$, which is equal to one, and on the other hand the product $\alpha'_{11} \cdots \alpha'_{nn}$ is equal to the power of Q, and thus always has the same numerical value, whence it yields a lower limit for the product $\alpha_1 \cdots \alpha_n$ defined by the power of Q.

[231] 95. E. MÜLLER, in his article "Die Kugelgeometrie nach den Principien der GRASSMANNschen *Ausdehnungslehre*" (*Monatshefte für Mathematik und Physik*, Jahrgang 3 u. 4, 1892 u. 1893), observes that it is more appropriate to interpret the orthogonal circle of three circles as *the* ***supplementary*** *domain of the domain of those circles.*

[235] 96. Compare this with the presentation in the A_1, §154ff.

[236] 97. The concept of the syncyclic relation of circles is noteworthy in that it was formulated by GRASSMANN on the basis of a general principle of transformation.

*On p. 222, line 10 from the bottom, GRASSMANN sets $[Qc_r|c_r] = \alpha_r$, but later, on p. 224f, uses the same notation α_r in a completely different context.

On the other hand its practical significance is limited, since, under the transformation from an assembly of circles to a syncyclically related assembly, tangent circles do not in general transform into tangent circles.

[244] 98. This nomenclature is not very happily chosen; instead of "$f(q)$ vanishes with q" one ought to say: "$f(q)$ is infinitely small with q".

[263] 99. According to this Definition every true series is unconditionally convergent, but not conversely is every unconditionally convergent series true.

[265] 100. One misses here the proof that every power series, so long as it is true, is a continuous function of its argument,* and that the termwise differentiated series is really the differential quotient of the series. It cannot be our problem to fill this and similar gaps in GRASSMANN's establishment of differential calculus and the theory of functions, but it may be mentioned that GRASSMANN's proofs of these two theorems are found in the *Nachlaß*, and thus that GRASSMANN himself had perceived the necessity of proofs for them.

[268] 101. Here it must be noted that the left side also remains zero for infinite n, since

$$\lim_{n\to\infty} n(\Theta - 1) = 2\pi i.$$

[278] 102. Thus if one replaces y by $y+x_1\tau$, where x_1 is an arbitrary magnitude of first order and τ is an arbitrary number, then one recognizes at once that

$$f^{(n)}(0)x_1y^{n-1} = n!a_nx_1y^{n-1},$$

and by continuing this procedure one finally obtains

$$f^{(n)}(0)x_1\cdots x_n = n!a_nx_1\cdots x_n,$$

whereupon No. 357 can be applied directly.

[283] 103. If x is an extensive magnitude derivable from $n > 1$ units, the integral of $f(x)dx$ has no definite sense in and of itself; rather, the path of integration must be prescribed. The path of integration here is so chosen by GRASSMANN that, if one interprets $x_1, \ldots, x_n$ as rectangular coordinates of an R_n, it coincides with the straight line from the origin of coordinates to the point x. In this way $d^{-1}f(x)dx$ gets a definite value, which on account of $t = \sqrt{x^{2}}$ and $e = x : \sqrt{x^{2}}$ can again be represented as a function of x alone. — The requirement of complete integrability of the differential $f(x)dx$ is equivalent to the requirement that the value of the integral of $f(x)dx$ depends only on the initial and final points, but not on the shape of the path of integration.

[284] 104. Here in the original edition, besides No. 433, No. 431c was cited, a citation that does not fit at all, for the expression $\frac{a}{l}$ is not an open expression, into whose opening a magnitude is to enter, but conversely that a is to enter in the opening l. One must therefore instead reason thus: Since a is a constant, it is unimportant whether it is to enter the opening before or after the differentiation.

[284] 105. The editors of the Teubner Edition replaced the material in braces with the following:

Since it is necessary here to establish a definite coordination between the individual openings of the function $f(x)$ and its filling magnitudes a and dx, the open

*In the Remark following No. 461, GRASSMANN assumes this theorem as self-evident.

expression $f(x)$ — we will call it an expression with "*bound*" openings — appears in contrast to the previously considered open expressions with "free" openings, with which a linking of openings to individual filling magnitudes does not obtain.* The difference of the two types of open expression is immediately apparent, in that an expression with n bound openings under multiplication with the product of n filling magnitudes always yields only a single term, whereas an expression with n free openings under such a multiplication yields a group of $n!$ terms, whose arithmetic mean is to be taken.

If one sticks to the above chosen notation, in which the openings are distinguished from one another by indices, while their linkage with the filling magnitudes is marked by the denominators appended to the filling magnitudes, it remains completely arbitrary whether one places the filling magnitudes before or after the open expression, since for each of them the place it is to obtain upon its entry into the open expression is exactly prescribed. Still, in the following another, more convenient method of notation will find its application. Thus for fixing the coordination between the openings and their filling magnitudes it suffices if one bestows on the openings, perhaps again by appending indices, a certain *rank ordering*, and places the filling factors, but without appending denominators, after the open expression, with the condition that the first listed filling magnitude is to enter into the first opening, the second into the second, and so on.

Whether the openings of an open expression are to be interpreted as free or bound usually follows from the way the expression originates. An example may clarify this.

Let $f(x)$ be an expression with one or more openings, and

$$x = x_1e_1 + x_2e_2 + \cdots ,$$

where the magnitudes e_1, e_2, ... form an elementary normal system. Then (according to 438 and 382) the differential quotient

$$f'(x) = [l|e_1]\frac{d}{dx_1}f(x) + [l|e_2]\frac{d}{dx_2}f(x) + \cdots .$$

It therefore includes one more opening than the function $f(x)$ itself. Further (according to 435)

$$f'(x)dx = d_xf(x).$$

It is therefore self-evident that the opening l coming into the expression $f'(x)$ by differentiation must be assigned to the filling magnitude dx; and, since the magnitude dx is usually placed immediately after the differential quotient, the coordination between the opening l and the filling magnitude dx is formally established by the condition that l is to be interpreted as the *first* opening of $f'(x)$.

Of particular importance for the following is the case where an expression with bound openings has the property that the same resultant always follows, whether one introduces the filling magnitudes into the openings in the order described or in another; then its openings are called *interchangeable*. Thus for example in an expression A_{ll_1} with two bound openings l and l_1, the first of which is l, these openings are *interchangeable* if for every two arbitrary filling magnitudes a and b the equation

$$A_{ll_1}ab = A_{ll_1}ba$$

*In the text the open expressions of the latter type were until now less significantly called expressions with "interchangeable" openings.

holds, that is, if for arbitrary values of a and b

$$(*) \qquad A_{ab} = A_{ba}.$$

In this case, then, multiplication of the expression A_{ll_1} by the product of the filling magnitudes a, b yields precisely the same product as if the expression possessed two *free* openings. For, if one denotes the expression that results from the open expression A_{ll_1} upon suppression of the linkage of its openings by A_{ll}, then

$$A_{ll}ab = \frac{1}{2}\{A_{ab} + A_{ba}\},$$

that is (according to $*$)

$$\begin{aligned} &= A_{ab} \\ &= A_{ll_1}ab. \end{aligned}$$

According to the general concept of the open expression (No. 357), which is also to be retained in this form for expressions with bound openings, one can in this case set

$$A_{ll_1} = A_{ll};$$

that is: *Interchangeable bound* openings can always be replaced by *free* openings.* Thus if, in the formula of the above theorem (that is, in 485), the two openings of the function $f(x)$ are to be interchangeable, the distinction between its openings is superfluous.

For the rest, it is often advantageous in an expression with interchangeable bound openings to retain the linkages of its openings, since the multiplication of such an expression proceeds more simply, both conceptually and formally, than with an expression having free openings. The importance of the concept of interchangeable bound openings is demonstrated in the following theorem (that is, No. 486).

[291] 106. The proof of this theorem has been found in GRASSMANN's *Nachlaß*, but it is too long and the whole theorem is of too little practical importance to reproduce here.

[295] 107. The extremely fruitful idea of reducing the integration of a partial differential equation of first order with an unknown function to the more general problem of the integration of an equation of the form

$$(*) \qquad X_1(x_1, \ldots, x_n)dx_1 + \cdots + X_n(x_1, \ldots, x_n)dx_n = 0$$

is found first in the celebrated article of PFAFF: "Methodus generalis, aequationes differentiarum partialum ... primi ordinis, inter quotcumque variabiles integrandi," *Abhandlungen der Berliner Akademie* 1815–16, pp. 76–136. In general it is therefore now customary to call an equation such as $(*)$ a PFAFFian equation.—The two articles by JACOBI referred to in the text have the following titles: "Ueber die PFAFFsche Methode, eine gewohnliche lineare Differentialgleichung zwischen $2n$ Variabeln durch ein System von n Gleichungen zu integriren" and "Ueber die Reduction der Integration der partiellen Differentialgleichungen erster Ordnung ...

*On the other hand, between an expression A_{ll_1} with two *non-interchangeable bound* openings and the corresponding expression A_{ll} with free openings there is the relation

$$A_{ll} = \frac{1}{2}\{A_{ll_1} + A_{l_1l}\},$$

a relation that one can easily transpose to expressions with arbitrarily many openings.

auf die Integration eines einzigen Systemes gewohnlicher Differentialgleichungen." They were published in the years 1827 and 1837.

[296] 108. If one imagines the given partial differential equation of second order solved by $t : t = F(x, y, z, p, q, r, s)$, then one has to set:

$$\begin{aligned} u &= e_1 x + e_2 y + e_3 z + e_4 p + e_5 q + e_6 r + e_7 s, \\ U_1 &= [l|e_3] - p[l|e_1] - q[l|e_2], \\ U_2 &= [l|e_4] - r[l|e_1] - s[l|e_2], \\ U_3 &= [l|e_5] - s[l|e_1] - F[l|e_2], \\ U &= e_1 U_1 + e_2 U_2 + e_3 U_3, \end{aligned}$$

and in this way one obtains the equation

$$U du = 0,$$

where the product $U du$ is to be understood such that du is to enter in the openings of U.

[297] 109. In the original edition this read just "where $c_1, \ldots, c_n$ are constants," but there can be no doubt that GRASSMANN had thought of them as *arbitrary* constants. Thus in the Proof he obviously assumes that the equations $u_1 = c_1$, $\ldots$, $u_n = c_n$ represent an integrating assembly, whatever values the constants c_1, $\ldots$, c_n may have; otherwise he could not have concluded that the expression $X dx$ is *identically* zero as soon as one expressed n of the differentials dx_ν by the other $m - n$.

[298] 110. This theorem is not correct in the generality expressed here. In fact the very beginning of the Proof is open to criticism.

Thus from No. 502 it only follows that, under the assumptions of No. 503, $X dx = 0$ cannot be integrated by an assembly of $n' < n$ equations that contain n' arbitrary constants *and which are solvable for these constants.* On the other hand it is perfectly conceivable that an integrating assembly of $n' < n$ equations exists, from which an equation free of constants can be derived. This case occurs, for example, if in the expression $X dx = X_1 dx_1 + \cdots + X_m dx_m$, the functions X_1, $\ldots$, X_{m-1} all vanish for $x_m = 0$, while X_m is not identically zero for $x_m = 0$, for then $x_m = 0$ is obviously an integral equation of $X dx = 0$.

One cannot say that GRASSMANN had never thought of the possibility of the occurrence of such an integrating assembly, for in No. 503 he remarks explicitly that $U_1 = 0, \ldots, U_n = 0$ is an integrating assembly: It therefore by no means escaped him that in certain cases one obtains an integrating assembly by setting to zero all the coefficients of the differential expression under consideration. He had however not noticed that in this way, even under the assumptions made in No. 503, one can sometimes find integrating assemblies of less than n independent equations, and indeed that this is especially implicit in another circumstance that he had overlooked, namely in the circumstance that an integrating assembly can be lost when one replaces the original equation $X dx = 0$ by the equation

$$U_1 du_1 + \cdots + U_n du_n = 0.$$

This possibility however always exists when some of the magnitudes $u_1, \ldots, u_n$, $U_1, \ldots, U_n$ assume infinitely large values for all systems of values $x_1, \ldots, x_m$ that satisfy the integrating assembly under consideration.

The circumstance just mentioned, which GRASSMANN overlooked, has the consequence that the proof of No. 503 itself is not perfectly free of objections, if one completely excludes from consideration the possibly existing case of integrating assemblies that contain less than n mutually independent equations. Thus if $v_1 = 0$, $\dots$, $v_n = 0$ is an integrating assembly with n independent equations, one still cannot know whether the functions $u_1, \dots, u_n$ ultimately remain for the system of values $x_1, \dots, x_m$ that satisfy the equations $v_1 = 0, \dots, v_n = 0$; it is therefore by no means certain that the system of equations $v_1 = 0, \dots, v_n = 0$ can be replaced by an assembly of equations between $u_1, \dots, u_n$ and $m - n$ of the x, as is assumed on p. 298, lines 7–1 from the bottom.

Fundamentally, therefore, GRASSMANN's proof is only applicable if from the beginning one is restricted to integrating assemblies of the form $v_1 = c_1, \dots, v_n = c_n$, where $c_1, \dots, c_n$ are arbitrary constants. If one modifies the treatment of the proof in this way, it will be completely rigorous and will show that every integrating assembly of the given form can be cast into the form of the set (b) on p. 298, where r is one of the numbers $1, \dots, n$ and where the arbitrary constants $c_1, \dots, c_n$ only appear in the functions $\phi_1, \dots, \phi_r$. One can then still observe that equations (b) represent an integrating assembly, whatever functions one may choose for $\phi_1, \dots, \phi_r$, and that the case $r = 0$ also yields an integrating assembly.

Despite this exception we have had to take to the theorem 503 and its proof, the theorem and proof are still noteworthy. First, it certainly represents a certain advance over the what JACOBI had said in the articles mentioned on p. 299, although naturally GRASSMANN's claim "that other than the assembly of equations (b) there is no other assembly of integrating equations" requires a qualification. Second, however, it is noteworthy that with the proof of No. 503 there is implicitly solved a problem that later played a role as an "auxiliary problem" in the investigations of LIE on partial differential equations and contact transformations, namely the problem of defining all the assemblies of at most n equations between the $2n$ independent variables $u_1, \dots, u_n, U_1, \dots, U_n$, by virtue of which the equation $U_1 du_1 + \cdots + U_n du_n = 0$ is integrated. The considerations in No. 503 show in fact that such an integrating assembly contains at least n equations, and that, if it contains exactly n equations, it can maintain the form (b), where r is one of the numbers $0, 1, \dots, n$. On various occasions LIE had commented on this Number, cf. *Archiv for Mathematik og Naturvidenskab 2*, p. 341, Kristiania 1877, and *Abh. d. Ges. d. Wiss. zu Leipzig, math.-phys. Klasse XIV*, No. XII, p. 539 (1888). In this last LIE also mentioned "that GRASSMANN's proof is incorrect and his theorem does not hold generally."

[300] 111. If $\boldsymbol{L}$ has exactly n openings and upon filling these openings is a numerical magnitude, then one can write the expression $[\boldsymbol{L}a_1 \cdots a_n]$ as follows:

$$[\boldsymbol{L}a_1 \cdots a_n] = \frac{1}{n!} \sum \pm \phi(i_1 \cdots i_n).$$

Here for $i_1, \dots, i_n$ one has to set all interchanges of the numbers $1, 2, \dots, n$ one by one and to choose the plus or minus sign according as the particular interchange is even or odd; finally, $\phi(i_1, \dots, i_n)$ denotes the numerical magnitude that one obtains if one lets the magnitudes $a_{i_1}, a_{i_2}, \dots, a_{i_n}$ enter sequentially into the first, second, $\dots$, nth openings of $\boldsymbol{L}$, whence (cf. the Remark following 428)

$$\boldsymbol{L}a_{i_1} a_{i_2} \cdots a_{i_n} = \phi(i_1 \cdots i_n).$$

In particular, if $\boldsymbol{L}$ is a product of n expressions, each with one opening, which upon filling their openings are all numerical magnitudes, then the function ϕ has the form

$$\phi(i_1 \cdots i_n) = \phi_1(i_1) \cdots \phi_n(i_n),$$

and the expression $[\boldsymbol{L}a_1 \cdots a_n]$ is an ordinary determinant, which to be sure is divided by $n!$.

GRASSMANN's symbol $[\boldsymbol{L}a_1 \cdots a_n]$ is therefore a generalization of the concept of the determinant, and indeed is in essence precisely the generalization that CAYLEY was already aiming at in 1848, in his investigation of the JACOBIan symbol* (1, 2, ..., $2n$) (cf. the short article "Sur les déterminants gauches", *Crelles Journal* **38**, pp. 93–96, and *The Collected Mathematical Papers of Cayley*, Vol. I, pp. 410–413).

[304] 112. The proof of this theorem is not completely satisfactory, since it takes no regard of the numerical factor which, according to No. 504, must be appended to the development of every relative open product.

[305] 113. It should have been remarked here that the procedure in the text also yields, without further ado, to the constraint

$$\left[\left(\frac{d}{dx}X\right)^{n+1}\right] = 0,$$

but that this, as it turns out later, is a consequence of the equation

$$\left[X\left(\frac{d}{dx}X\right)^{n}\right] = 0.$$

[305] 114. If not only x, but also Xdx is an extensive magnitude, then the equation $Xdx = 0$ is equivalent to a system of PFAFFian equations:

$$\sum_{\mu=1}^{m} X_{\kappa\mu} dx_\mu = 0 \quad (\kappa = 1, \ldots, h). \tag{1}$$

Thus if we set $\sum x_\mu e_\mu = x$ and

$$X_\mu = \sum_{\mu=1}^{m} X_{\kappa\mu}[l|e_\mu] \quad (\kappa = 1, \ldots, h),$$

then we can write the system (1) in the form

$$X_1 dx = 0, \ \ldots, \ X_h dx = 0, \tag{2}$$

where $X_1dx, \ldots, X_hdx$ are numerical magnitudes; thus if we set

$$e_1 X_1 + \cdots + e_h X_h = X,$$

then the system (2) is equivalent to the equation $Xdx = 0$.

Now should the system (1) or (2) be reducible to a system in which $n < m$ differentials appear, then it must be possible to define the magnitudes u_ν and $U_{\kappa\nu}$ as functions of $x_1, \ldots, x_m$ or, what amounts to the same thing, as functions of x, in such a way that

$$X_\kappa dx = U_{\kappa 1} du_1 + \cdots + U_{\kappa n} du_n \quad (\kappa = 1, \ldots, h). \tag{3}$$

---

*Later CAYLEY introduced the name "PFAFFian" for this symbol; we have made no use of this name, since with LIE we reserve the name "PFAFFian expression" exclusively for expressions of the form $\sum X_\nu dx_\nu$.

But then it follows that

$$(4)\quad X = \sum_{\nu=1}^{n} U_{\kappa\nu} \frac{d}{dx} u_\nu, \quad \frac{d}{dx} X_\nu = \sum_{\nu=1}^{n} \left(\frac{d}{dx} U_{\kappa\nu} \frac{d}{dx} u_\nu + U_{\kappa\nu} \frac{d^2}{dx^2} u_\nu \right) \quad (\kappa = 1, \dots, h),$$

and if one handles these equations according to the pattern of No. 511, one obtains as the necessary condition for the possibility of that reduction the following:

$$(5)\qquad \left[X_1^{\varepsilon_1} X_2^{\varepsilon_2} \cdots X_h^{\varepsilon_h} \left(\frac{d}{dx} X_1 \right)^{r_1} \cdots \left(\frac{d}{dx} X_h \right)^{r_h} \right] = 0,$$

where each ε_i has one of the values 0 or 1, and each r_i one of the values 0, 1, ... , $n+1$, and one has to choose $\varepsilon_1, \dots, \varepsilon_h, r_1, \dots, r_h$ in all possible ways such that

$$\varepsilon_1 + \cdots + \varepsilon_h + r_1 + \cdots + r_h = n + 1.$$

These are obviously the conditions that GRASSMANN had in mind; possibly he had even reduced it to equations for the magnitude X alone:

$$(5')\qquad \left[X^\mu \left(\frac{d}{dx} X \right)^{n+1-\mu} \right] = 0 \quad (\mu = 0, 1, \dots, h),$$

where the expression in brackets is interpreted as an algebraic product in the sense of Nos. 364–371.

Equation of condition (5), whose appearance seems to have been completely unnoticed previously, deserves a closer investigation; in particular it would be of interest to know whether it is not only necessary, but also sufficient, in cases where $h > 1$.

[307] 115. With JACOBI the expression (2, 3, ... , $2n+1$) was at first defined in a somewhat different way (see the Remark following No. 510).

[309] 116. It should have been said right here that in the case $\lambda = 0$ the problem is solved, provided it is possible to define δx so that it satisfies the equation

$$\left[\frac{d}{dx} X \cdot c \cdot \delta x \right] = 0$$

for every value of the magnitude c and in addition the equation $X dx = 0$. Thereby the representation in No. 516 would essentially have been achieved.

[310] 117. These numerical equations read

$$\lambda \cdot X e_\nu = 2 \left[\frac{d}{dx} X \cdot e_\nu \cdot \delta x \right] = 2 \sum_{\kappa=1}^{2n} \left[\frac{d}{dx} X \cdot e_\nu \cdot e_\kappa \right] \delta x_\kappa .$$

Under the assumptions made here one can, according to the procedure leading up to Theorem 1 of Editorial Note 112, write down the solution of this equation directly, and at the same time survey how the solution comes to its form. In GRASSMANN's presentation, on the other hand, everything depends on an artifice that is suited only to impede one's understanding. The application of this artifice has in addition the drawback that GRASSMANN must still first prove in particular that the value of δx found actually satisfies the given equations.

[312] 118. The editors of the Teubner Edition replaced the material in braces with the following: we can, without changing the meaning of the product

$$\left[X\left(\frac{d}{dx}X\right)^{n-1}\right],$$

(according to 504) introduce within the brackets yet another opening l as the first factor of the same family as the other openings. Then

[313] 119. Compare with Editorial Note 112. That the method used in No. 515 "makes clear itself" one can scarcely grant; but one must admit in any case that GRASSMANN's symbols lead directly to the result of No. 515.

[313] 120. Compare this with Editorial Notes 112 and 116.

[316] 121. Compare this with Theorem 2 of Editorial Note 112.

[320] 122. If one inserts for δx its value $\delta x = \sum e_\mu \delta x_\mu$, the equations $G_s = 0$ take the form

$$G_s = 2\sum_{\mu=1}^{m}\left[\frac{d}{dx}X \cdot e_s \cdot e_\mu\right]\delta x_\mu - \lambda X e_s = 0 \quad (s = 1, \ldots, m),$$

and the theorem proved here simply expresses that, in the matrix belonging to these m linear homogeneous equations, all $(2n+1)$-row determinants vanish, provided equation (b) is satisfied. In the proof it is tacitly assumed that the expression

$$\left[\left(\frac{d}{dx}X\right)^{n} e_1 \cdots e_{2n}\right] \neq 0,$$

for only under this assumption does the equation $\sum \alpha_\mathfrak{a} G_\mathfrak{a} = 0$, derived on p. 320 represent a numerical equation in which G_m actually appears. Only later (on the same page) does GRASSMANN take note of this assumption.

[320] 123. It would have been better if this constraint had been written in the form

$$\left[X\left(\frac{d}{dx}X\right)^{n-1} e_1 e_2 \cdots e_{2n}\right] \neq 0,$$

[321] 124. It is obviously assumed that, according to p. 320, lines 15–16, one has already set $\delta x_{2n+1}, \ldots, \delta x_m$ equal to zero in the equations $G_1 = 0, \ldots$.

[321] 125. It is also assumed here that

$$\left[X\left(\frac{d}{dx}X\right)^{n-1} e_1 e_2 \cdots e_{2n}\right] \neq 0.$$

Obviously this assumption is no restriction of generality, since from the beginning it is assumed that

$$\left[\left(\frac{d}{dx}X\right)^{n-1}\right]$$

does not vanish.

[322] 126. This Corollary was written in order to explain why GRASSMANN did not mention the case where $m = 2n - 1$. It is completely wrong to conclude from the absence of any mention of this case that GRASSMANN had excluded it, and it

is not really comprehensible how FORSYTH* can claim: "It is assumed implicitly that, if the coefficients of an equation satisfy no characteristic condition, then the number of variables is even; so GRASSMANN practically considers only the even classes of unconditioned equations." Yet in No. 512 GRASSMANN says explicitly that for $m < 2n+1$, and thus in particular for $m = 2n-1$, no equation of condition at all enters, and thus it is from this completely clear that the equation $X_1dx_1 + \cdots + X_{2n-1}dx_{2n-1} = 0$ can always be integrated by an assembly of n equations, now whether

$$\left[X\left(\frac{d}{dx}X\right)^{n-1}\right]$$

is equal to zero or different from zero.

Grassmann's investigations into Pfaff's problem

We would like at least to make GRASSMANN's heretofore so little noticed investigations into PFAFF's problem reasonably accessible to those that are unwilling to assimilate GRASSMANN's calculus. To this end we will now attempt to present the contents of Nos. 502–527, insofar as it relates to PFAFF's problem, in the language of ordinary analysis. For this we will make use of JACOBI's symbol $(1, 2, \ldots, n)$, and in fact in the way that CAYLEY has presented it. With the help of this symbol we are in a position to replace all GRASSMANN's calculations and transformations with complete equivalents.

Otherwise we by no means intend to follow GRASSMANN step by step — this would lead us too far afield — rather, we will only reproduce his train of thought as completely as possible, in order to show what GRASSMANN had actually accomplished.

In No. 502 GRASSMANN first showed that the equation

$$X_1dx_1 + \cdots + X_mdx_m = 0 \tag{1}$$

can be integrated by an assembly of n equations

$$u_1(x_1, \ldots, x_m) = \text{const}, \quad \ldots, \quad u_n(x_1, \ldots, x_m) = \text{const} \tag{2}$$

if and only if the expression Xdx is reducible to the form

$$\sum_{\mu=1}^{m} X_\mu dx_\mu = \sum_{\nu=1}^{n} U_\nu(x_1, \ldots, x_n)du_\nu \tag{3}$$

with only n differentials.** The proof is perfectly comprehensible.

If $\sum X_\mu dx_\mu$ can be cast into the form (3), the equations

$$\begin{cases} X_\mu = \displaystyle\sum_{\nu=1}^{n} U_\nu \frac{\partial u}{\partial x_\mu}, \\ \dfrac{\partial X_\mu}{\partial x_\kappa} = \displaystyle\sum_{\nu=1}^{n} \frac{\partial U_\nu}{\partial x_\kappa}\frac{\partial u_\nu}{\partial x_\mu} + \sum_{\nu=1}^{n} U_\nu \frac{\partial^2 u_\nu}{\partial x_\mu \partial x_\nu} \end{cases} \tag{4}$$

---

* *Theory of Differential Equations*, Part I, Cambridge 1890, p. 83.

**On No. 503 see Editorial Note 110.

must exist. In order to eliminate the U and u from these, GRASSMANN forms the expression

$$\sum \pm X_{\mu_1} \frac{\partial X_{\mu_2}}{\partial x_{\mu_3}} \cdots \frac{\partial X_{\mu_{2n}}}{\partial x_{\mu_{2n+1}}} \tag{5}$$

where $\mu_1, \ldots, \mu_{n+1}$ are any $2n+1$ of the numbers $1, 2, \ldots, m$, and where the sum is to be formed in such a way that one permutes $\mu_1, \ldots, \mu_{n+1}$ in all possible ways, giving each even permutation the plus sign, each odd permutation the minus sign. With this method of formation of expression (5) all the second differential quotients of u_ν drop out, and there remains

$$\sum_{\nu_1 \ldots \nu_{n+1}}^{1 \ldots n} U_{\nu_1} \sum \pm \frac{\partial u_{\nu_1}}{\partial x_{\mu_1}} \frac{\partial u_{\nu_2}}{\partial x_{\mu_2}} \frac{\partial U_{\nu_2}}{\partial x_{\mu_3}} \cdots \frac{\partial u_{\nu_{n+1}}}{\partial x_{\mu_{2n}}} \frac{\partial U_{\nu_{n+1}}}{\partial x_{\mu_{2n+1}}},$$

where the inner sum is formed exactly as with (5). Now since among the $n+1$ indices $\nu_1, \ldots, \nu_{n+1}$ there are always at least two that are equal, the inner sum vanishes identically, and it therefore follows that all expressions of the form (5) must have the value zero if an equation of the form (3) is to exist. Naturally this condition is only in force if $m \geq 2n+1$.

This is the content of No. 511. The derivation of the necessary conditions for the existence of (3) is completely GRASSMANN's own, and is decidedly most noteworthy, especially as it can be applied to systems of PFAFFian equations (cf. Editorial Note 114).

In No. 512 GRASSMANN shows how one can write the equations of condition found in No. 511 with the help of JACOBI's symbols $(1, 2, \ldots, 2n)$. Thus since with JACOBI he sets

$$\frac{\partial X_\mu}{\partial x_\kappa} - \frac{\partial X_\kappa}{\partial x_\nu} = (\mu, \kappa) \tag{6}$$

and utilizes that symbol,* he brings the equation that results from setting expression (5) to zero to the form

$$\sum X_{\mu_1}(\mu_1, \mu_2, \ldots, \mu_{2n+1}) = 0, \tag{7}$$

where now the sum consists of all the expressions that one obtains from the one as written by once, twice, ..., $(2n+1)$ times cyclically interchanging the indices μ_1, $\mu_2, \ldots, \mu_{n+1}$.

We do not occupy ourselves with establishing this result, and we will therefore just note that, with the help of an artifice due to CAYLEY,** equation (7) can be written in a very elegant form, which will serve us well later on. Thus if with CAYLEY we set

$$(0, \mu) = X_\mu, \quad (\mu, 0) = -X_\mu, \tag{6'}$$

*This symbol is a complete function of nth degree of the expression (μ, κ) and is defined by the recursion formula

$$(1, 2, \ldots, 2n) = \sum (1, 2)(3, 4, \ldots, 2n),$$

where the sum on the right consists of all expressions that one obtains if, in the expression as written, one cyclically interchanges the indices $2, 3, \ldots, 2n$ once, twice, ..., $(2n-1)$ times (Cf. the development in Editorial Note 112.) It then easily follows that the expression $(\mu_1, \ldots, \mu_{2n})$ changes its sign under interchange of two indices and also under cyclic interchange of all $2n$ indices, and that it vanishes if two of the indices are equal. Refer to the article by CAYLEY cited in Editorial Note 111.

Cf. CAYLEY, *Crelle Journal* **57, 275 (1860), Mathematical Papers IV, p. 361.

then (7) assumes the form

$$(7') \qquad (0, \mu_1, \mu_2, \dots, \mu_{n+1}) = 0.$$

We now turn from this short detour back to No. 512.

Properly one has to form all equations (7) or (7′) that one obtains when one sets for $\mu_1, \dots, \mu_{2n+1}$, any $2n+1$ of the numbers 1, 2, ..., m. But GRASSMANN proves *that under the assumption $X_1 \neq 0$, the existence of those equations (7) or (7′) in which one of the numbers $\mu_1, \dots, \mu_{2n+1}$ has the value one entails that of all the other equations.*

If we translate GRASSMANN's proof into the notation chosen here, it just comes to the following: the equation

$$(0, 0, 1, \mu_1, \mu_2, \dots, \mu_{2n+1}) = 0$$

is an identity, since JACOBI's symbol on the left side contains two equal indices. But if one develops this equation according to the rules holding for JACOBI's symbols, one obtains (since the first term vanishes)

$$(0,1)(\mu_1, \mu_2, \dots, \mu_{2n+1}, 0) + \sum_{\kappa=1}^{2n+1} (0, \mu_\kappa)(\mu_{\kappa+1}, \dots, \mu_{2n+1} 0, 1, \mu_1, \dots, \mu_{\kappa-1}) = 0,$$

and this equation is equivalent to the following:

$$X_1(0, \mu_1, \dots, \mu_{2n+1}) = \sum_{\kappa=1}^{2n+1} (-1)^{\kappa-1} X_\kappa (0, 1, \mu_1, \dots, \mu_{\kappa-1}, \mu_{\kappa+1}, \dots, \mu_{2n+1}),$$

from which the theorem to be proved follows immediately.

No. 514 considers the problem of *reducing the equation*

$$X_1 dx_1 + \cdots + X_m dx_m = 0,$$

by the introduction of new variables, to an equation in which only $m-1$ variables appear. Analytically this problem amounts to *bringing the expression $\sum X_\mu dx_\mu$, by the introduction of the new variables $y_1, \dots, y_{m-1}, t$, to the form*

$$(8) \qquad \sum_{\mu=1}^{m} X_\mu dx_\mu = \mathsf{N} \sum_{\nu=1}^{m-1} Y_\nu(y_1, \dots, y_{m-1}) dy_\nu,$$

where t only appears in the factor N. The whole problem and the procedure used originated with PFAFF,* so the content of No. 514 is in essence just a translation of PFAFF's ideas into GRASSMANN's language.

The problem would be solved if one has satisfied the equations

$$(9) \qquad \sum_{\mu=1}^{m} X_\mu \frac{\partial x_\mu}{\partial t} = 0, \quad \frac{\partial}{\partial t}\left(\frac{1}{\mathsf{N}} \sum_{\mu=1}^{m} X_\mu \frac{\partial x_\mu}{\partial y_\nu}\right) = 0 \quad (\nu = 1, \dots, m-1).$$

Now if one sets

$$(10) \qquad \frac{1}{\mathsf{N}} \frac{\partial \mathsf{N}}{\partial t} = \lambda$$

one obtains, upon combining with equations (8), the following:

$$(11) \qquad \lambda \sum_{\mu=1}^{m} X_\mu \frac{\partial x_\mu}{\partial y_\nu} = \sum_{\mu\kappa}^{1\dots m} \left(\frac{\partial X_\mu}{\partial x_\kappa} - \frac{\partial X_\kappa}{\partial x_\mu}\right) \frac{\partial x_\kappa}{\partial t} \frac{\partial x_\mu}{\partial y_\nu} \quad (\nu = 1, \dots, m-1),$$

*Cf. the article mentioned in Editorial Note 107.

which is obviously satisfied identically, if one can satisfy the equations

$$\lambda(0,\mu)=\sum_{\kappa=1}^{m}(\mu,\kappa)\frac{\partial x_\kappa}{\partial t}\quad(\mu=1,\dots,m), \tag{12}$$

where (μ,κ) and $(0,\kappa)$ have the meaning given in (6) and (6′).

Now conversely, if it is possible to define the x_κ as functions of t and $y_1,\dots,y_{m-1}$ in such a way that (12) is satisfied and that λ does not vanish, then obviously

$$\sum_{\mu=1}^{m}(0,\mu)\frac{\partial x_\mu}{\partial t}=\sum_{\mu=1}^{m}X_\mu\frac{\partial x_\mu}{\partial t}=0, \tag{13}$$

and at the same time relations (11) obtain. But from (13) and (11) there follows, upon using (10), equations (9), whence the problem is solved.

On the other hand, if equations (12) are so constituted that the vanishing of λ follows from them, then one must define the x_κ so that besides the equations

$$0=\sum_{\kappa=1}^{m}(\mu,\kappa)\frac{\partial x_\kappa}{\partial t}\quad(\mu=1,\dots,m), \tag{12′}$$

they also satisfy the equation

$$\sum_{\mu=1}^{m}(0,\mu)\frac{\partial x_\mu}{\partial t}=0. \tag{13′}$$

Thus (11) would then also be satisfied if one set $\lambda=0$, and since in the case $\lambda=0$ equation (10) expresses that N is free of t, it again follows from (11) and (13′) or (13) that equations (9) obtain and the problem is solved.*

The case that m *is even* is in Nos. 515–517 completely disposed of under certain assumptions.

If $m=2n$, then it follows from (12) that

$$\lambda\sum^{(c)}(0,1)(2,3,\dots,\mu-1,\mu+1,\dots,2n)$$
$$=\sum_{\kappa=1}^{2n}\sum^{(c)}(1,\kappa)(2,3,\dots,\mu-1,\mu+1,\dots,2n)\frac{\partial x_\kappa}{\partial t},$$

where $\sum^{(c)}$ signifies that one is to form the sum of all expressions that result from once, twice, ..., $(2n-1)$ times cyclically interchanging the indices $1, 2, \dots, \mu-1, \mu+1, \mu+2, \dots, 2n$ in the expression as written. But the last equation can be

*To be sure the case $\lambda=0$ is only considered in No. 516, but it is better to mention it right here. Naturally it would be even more appropriate to say that the problem is solved if one has defined λ and $x_1,\dots,x_m$ as functions of $y_1,\dots,y_{m-1}$ and t such that the equations

$$\lambda(0,\mu)+\sum_{\kappa=1}^{m}(\kappa,\mu)\frac{\partial x_\kappa}{\partial t}=0\quad(\mu=1,\dots,m)$$

are identically satisfied; but this would be a real deviation from GRASSMANN's train of thought, and we only mention it here because one thereby recognizes that CAYLEY's notation $X_\mu=(0,\mu)$ is grounded in the nature of the subject.

written

$$\lambda(0,1,2,\dots,\mu-1,\mu+1,\dots,2n) = -\sum_{\kappa=1}^{2n}(\kappa,1,2,\dots,\mu-1,\mu+1,\dots,2n)\frac{\partial x_\kappa}{\partial t},$$

and since here on the right all those terms vanish in which $\kappa \neq \mu$, there results

$$\lambda(0,1,2,\dots,\mu-1,\mu+1,\dots,2n) = (-1)^\mu(1,2,\dots,2n)\frac{\partial x_\mu}{\partial t} \quad (\mu=1,2,\dots,2n). \tag{14}$$

On the other hand,

$$(\kappa,0,1,2,\dots,2n)=0,$$

whatever one may set κ to among the numbers 0, 1, 2, … , $2n$, but from this, if one develops the left side according to the recursion formula in the footnote on p. 379, it follows that

$$(\kappa,0)(1,2,\dots,2n)+\sum_{\mu=1}^{2n}(\kappa,\mu)(\mu+1,\dots,2n,0,1,\dots,\mu-1)=0,$$

or

$$\sum_{\mu=1}^{2n}(-1)^\mu(\kappa,\mu)(0,1,\dots,\mu-1,\mu+1,\dots,2n)=(0,\kappa)(1,2,\dots,2n) \quad (\kappa=0,1,2,\dots,2n). \tag{15}$$

If therefore not all $2n$ expressions

$$(0,1,\dots,\mu-1,\mu+1,\dots,2n) \quad (\mu=1,\dots,2n) \tag{16}$$

are equal to zero and if $(1,2,\dots,2n)\neq 0$ as well, then equations (12) are satisfied by the values of the differential quotients of $x_1, \dots, x_{2n}$ following from (14), provided λ does not vanish, and in fact one can, if say

$$(0,1,2,\dots,2n-1) \tag{17}$$

is different from zero, set $x_{2n}=t$, and thus obtain λ and the differential quotients of $x_1, \dots, x_{2n-1}$ with respect to t, represented as functions of $x_1, \dots, x_{2n-1}$ and $x_{2n}=t$.

If on the other hand not all of the expressions (16) vanish, but (1, 2, … , $2n$) is zero, and therefore on account of (14) so is λ, then equations (15) show that (12′) and (13′) are identically satisfied if one sets

$$\frac{\partial x_\mu}{\partial t}=\rho(-1)^\mu(0,1,\dots,\mu-1,\mu+1,\dots,2n), \tag{18}$$

meaning by ρ an arbitrary magnitude. Thus if, in particular, expression (17) is different from zero, one can again set $x_{2n}=t$ and obtain ρ and the differential quotients of $x_1, \dots, x_{2n-1}$ with respect to t expressed by $x_1, \dots, x_{2n-1}$ and $x_{2n}=t$.

Thus if $m=2n$ and not all the expressions (16) are zero, then one always obtains a system of equations

$$\frac{\partial x_\mu}{\partial t}=\xi_\mu(x_1,\dots,x_{m-1},t) \quad (\mu=1,\dots,m-1), \tag{19}$$

which together with $x_m = t$ satisfy either equations (12) with nonvanishing λ or equations (12′) and (13′); if the latter case occurs, one has to set $\lambda = 0$.

Differential equations (19) are now integrated on the basis of the initial conditions* $x_\nu = y_\nu$ for $t = 0$ $(\nu = 1, \ldots, m-1)$. If in addition one calculates N from equation (10), then $x_1, \ldots, x_m$, as functions of $y_1, \ldots, y_{m-1}$ and t, satisfy equations (9) and thus there exists an equation of the form (8). Finally, if in (8) one makes the substitution $t = x_{2n} = 0$, one recognizes at once that, upon the transition to the new variables $y_1, \ldots, y_{m-1}$, the equation $\sum X_\mu dx_\mu = 0$ takes the form

$$\sum_{\nu=1}^{m-1} X_\nu(y_1, \ldots, y_{m-1}, 0)dy_\nu = 0. \tag{20}$$

In addition GRASSMANN explicitly takes note that, in the case where $(1, 2, \ldots, 2n)$ and thus also λ vanish, the multiplier N is free of t, and therefore that, should not all the expressions (16) be zero, the equation $(1, 2, \ldots, 2n) = 0$ is necessary and sufficient for the *Pfaffian expression* $\sum X_\mu dx_\mu$ to be reduced, by the introduction of new variables $y_1, \ldots, y_{m-1}, t$, to an expression containing only $m-1$ variables.

The content of Nos. 515–517 is for the most part just a broadening and completion of JACOBI's ideas. For JACOBI had already taken notice of the special properties of equations of the form (12), and had also showed that by the introduction of the initial values one can find the new form (20) of the PFAFFian equation $\sum X_\mu dx_\mu = 0$ (cf. *Crelle* **17**, p. 157ff, *Werke*, Bd. 4, p. 121ff).

GRASSMANN prepares for the general solution of the problem posed in No. 514 (see p. 412f) in No. 518 by proving the theorem

If all expressions of the form

$$(0, \mu_1, \mu_2, \ldots, \mu_{2n+1}) \tag{21}$$

vanish, then so also do all expressions of the form

$$(\mu_1, \mu_2, \ldots, \mu_{2n+2}) \tag{22}$$

meaning by μ_1, μ_2, ... *any of the numbers* 1, 2, ... , m.

GRASSMANN's proof of this theorem** emerges from the identity (15), in which one has only to replace n by $n+1$. Thus if i is one of the numbers $1, \ldots, 2n+2$, then it is of course true that

$$(\mu_i, 0, \mu_1, \ldots, \mu_{2n+2}) = 0 = (\mu_i, 0)(\mu_1, \ldots, \mu_{2n+2}),$$

whence if not all the expressions $(\mu_1, 0), \ldots, (\mu_{2n+2}, 0)$ vanish, it follows at once that $(\mu_1, \ldots, \mu_{2n+2}) = 0$. But if these expressions are all zero, then one cannot come to this conclusion; rather, one must form the expression

$$(\mu_{2n+3}, 0, \mu_1, \ldots, \mu_{2n+2}),$$

which under the assumptions of the theorem reduces to

$$(\mu_{2n+3}, 0)(\mu_1, \ldots, \mu_{2n+2}),$$

*Here it is tacitly assumed that the functions $\xi_1, \ldots, \xi_{m-1}$ remain finite and continuous for $t = 0$, an assumption of which GRASSMANN himself took note later on (in the Remark following No. 522). It should also be noted that in No. 494 GRASSMANN had not actually proved the *existence* of the solutions that he makes use of here.

**For the method of proof used here this theorem appears as the immediate broadening of a theorem proved in No. 512 (cf. p. 380, equation (7′)ff).

which on the other hand can be written in the form

$$(\mu_1, 0, \mu_2, \ldots, \mu_{2n+3}),$$

and consequently is also equal to

$$(\mu_1, 0)(\mu_2, \ldots, \mu_{2n+3}).$$

Now if all the magnitudes $(\mu, 0)$, ... , $(\mu_{2n+2}, 0)$ vanish, but not $(\mu_{2n+3}, 0)$, it now also follows that $(\mu_1, \ldots, \mu_{2n+2}) = 0$.

In No. 519 GRASSMANN shows that *to each **Pfaffian** equation* $X_1 dx_1 + \cdots + X_m dx_m = 0$ *belongs a completely definite number* n *such that all expressions of the form*

$$(0, \mu_1, \mu_2, \ldots, \mu_{2n+1}) \tag{21}$$

vanish, but not all expressions of the form

$$(0, \mu_1, \mu_2, \ldots, \mu_{2n-1}). \tag{23}$$

Thus if all expressions of the form (23) are equal to zero, then as was just shown so also do all expressions $(\mu_1, \mu_2, \ldots, \mu_{2n+1})$ and thus also all expressions (21). On the other hand for $2n + 1 > m$ every expression of the form (21) certainly has the value zero, since it then contains two equal indices, whereas for $n = 1$ it is certain that not all expressions (23), that is, not all magnitudes X_1, ... , X_m, vanish.

Consequently between the limits $1 \leq n \leq \frac{1}{2}(m+1)$ there is given a completely definite value of n that possesses the required property. GRASSMANN supposes this value as chosen for n.

Now the problem posed in No. 514 is already settled for the case where m is even and $n = \frac{1}{2}m$ (cf. 515–517 and p. 381f). Thus there remains only the two cases $m = 2n - 1$ and $m > 2n$. Of these, GRASSMANN considers only the second in No. 519, and on very good grounds indeed, for in the case $m = 2n - 1$ the problem posed is generally not solvable. To be sure GRASSMANN did not mention this explicitly, but there is no doubt that he was completely clear in his mind about it.

Thus let $m > 2n$. According to what has gone before, everything comes down to satisfying the equations

$$\lambda(0, \mu) = \sum_{\kappa=1}^{m} (\mu, \kappa) \frac{\partial x_\kappa}{\partial t} \quad (\mu = 1, \ldots, m), \tag{12}$$

and if this should only be possible for $\lambda = 0$, in addition the equation

$$\sum_{\mu=1}^{m} (0, \mu) \frac{\partial x_\mu}{\partial t} = 0. \tag{13'}$$

But GRASSMANN shows that these requirements can be reduced to the simpler case already solved previously, where $m = 2n$.

Under the given assumptions all expressions of the form (21) equal zero, but not all of the form (23). Further we will at first assume as well that not all expressions of the form $(\mu_1, \mu_2, \ldots, \mu_{2n})$ vanish, and in fact that, say,

$$(1, 2, 3, \ldots, 2n) \tag{24}$$

may be different from zero. Then it can be shown, as GRASSMANN proves, that all the equations (12) are a consequence of the first $2n$ of them.

Thus if, with GRASSMANN, we set

$$G_\mu = \lambda(0,\mu) + \sum_{\kappa=1}^{m} (\kappa,\mu)\frac{\partial x_\kappa}{\partial t},$$

and, if by $\sum^{(c)}$ we mean the sum of all expressions that one obtains if one once, twice, ... , $(2n+1)$ times cyclically interchanges the $2n+1$ indices 1, 2, ... , $2n$, $2n+\nu$, it follows that

$$\sum^{(c)} G_1(2,3,\dots,2n,2n+\nu) = \lambda \sum^{(c)} (0,1)(2,3,\dots,2n,2n+\nu) \\ + \sum_{\kappa=1}^{m} \frac{\partial x_\kappa}{\partial t} \sum^{(c)} (\kappa,1)(2,3,\dots,2n,2n+\nu),$$

but here the right side can be written in the form

$$\lambda(0,1,2,\dots,2n,2n+\nu) + \sum_{\kappa=1}^{m} \frac{\partial x_\kappa}{\partial t},(\kappa,1,2,\dots,2n,2n+\nu)$$

and this expression vanishes identically, since by assumption all expressions (21) and consequently (see the previous page) all expressions (22) also vanish. Therefore

$$\sum^{(c)} G_1(2,3,\dots,2n,2n+\nu)$$

is identically zero, and since here the factor of $G_{2n+\nu}$ has the value $(1, 2, \dots, 2n)$ and thus is different from zero, it is herewith proved that the equations $G_{2n+1} = 0$, ... , $G_m = 0$ are all consequences of the equations $G_1 = 0, \dots, G_{2n} = 0$.

It therefore follows, at least provided $(1,2,\dots,2n) \neq 0$, that $m-2n$ of the magnitudes $\frac{\partial x_\mu}{\partial t}$ can be assumed completely arbitrarily, and in fact with GRASSMANN we set

$$\frac{\partial x_{2n+1}}{\partial t} = \dots = \frac{\partial x_m}{\partial t} = 0. \tag{25}$$

But thereby, upon dropping the superfluous equations $G_{2n+1} = 0, \dots, G_m = 0$, equations (12) are reduced to the following:

$$\lambda(0,\nu) = \sum_{\kappa=1}^{2n} (\nu,\kappa)\frac{\partial x_\kappa}{\partial t} \quad (\nu = 1,\dots,2n), \tag{12''}$$

and as with equation (14) we obtain

(26)

$$\lambda(0,1,2,\dots,\nu-1,\nu+1,\dots,2n) = (-1)^\nu(1,2,\dots,2n)\frac{\partial x_\nu}{\partial t} \quad (\nu = 1,2,\dots,2n),$$

where the factors of λ on the left side certainly do not all vanish, for otherwise (as on p. 383) $(1,2,\dots,2n)$ would also have to be zero.

Therewith it is shown that, provided $(1,2,\dots,2n) \neq 0$, equations (12) can be satisfied identically, unless λ vanishes.

It still remains to settle the case where all expressions $(\mu_1,\mu_2,\dots,\mu_{2n})$ vanish.

For this we can assume, without loss of generality, that not all of the $2n$ expressions

$$(0,1,2,\dots,\nu-1,\nu+1,\dots,2n) \quad (\nu = 1,2,\dots,2n)$$

vanish. Thus if we make the substitutions (25) in equations (12), then, as follows from (26), the resulting equations can certainly not be satisfied unless λ vanishes. On the other hand, as on p. 381, we can arrange that equations (12) are satisfied for $\lambda = 0$, and that in addition (13′) is satisfied.

Thus if we set

$$(27) \qquad \frac{\partial x_\nu}{\partial t} = \rho(-1)(0,1,\ldots,\nu-1,\nu+1,\ldots,2n) \quad (\nu = 1,\ldots,2n),$$

then for $\kappa = 0,1,2,\ldots,m$,

$$\sum_{\nu=1}^{2n}(\kappa,\nu)\frac{\partial x_\kappa}{\partial t} = \rho\sum_{\nu=1}^{2n}(\kappa,\nu)(\nu+1,\ldots,2n,0,1,\ldots,\nu-1)$$
$$= \rho(\kappa,0,1,\ldots,2n) - \rho(\kappa,0)(1,2,\ldots,2n),$$

whence under the given assumptions is equal to zero. Therefore, however, our assertion is proved.*

Now, in order to settle our problem completely, we will assume that $(0,1,2,\ldots, 2n-1) \neq 0$, as we obviously may. Then we can set $x_{2n} = t$ and under all circumstances obtain for $x_1, \ldots, x_{2n-1}$ differential equations of the form

$$(28) \qquad \frac{\partial x_\nu}{\partial t} = \xi_\nu(x_1,\ldots,x_{2n-1},t) \quad (\nu = 1,\ldots,2n-1),$$

while $x_{2n+1}, \ldots, x_m$ must satisfy equations (25). If we integrate equations (28) on the basis of the initial conditions $x_\nu = y_\nu$ for $t = 0$ and in addition set $x_{2n+1} = y_{2n+1}, \ldots, x_m = y_m$, we obtain $x_1, \ldots, x_m$ represented as functions of $y_1, \ldots, y_{2n-1}, t, y_{2n+1}, \ldots, y_m$, and it follows exactly as on p. 383 that in the new variables the equation $\sum X_\mu dx_\mu = 0$ takes the form

$$\sum_{\nu=1}^{2n-1} X_\nu(y_1,\ldots,y_{2n-1},0,y_{2n+1},\ldots,y_m)dy_\nu$$
$$+ \sum_{\kappa=1}^{m-2n} X_{2n+\kappa}(y_1,\ldots,y_{2n-1},0,y_{2n+1},\ldots,y_m)dy_{2n+\kappa} = 0.$$

This is the content of No. 519. It goes considerably beyond what was achieved by JACOBI, and shows that GRASSMANN was fully conversant with the theory of systems of equations.

In Nos. 520 and 521 is analyzed which cases the considerations thus far permit the PFAFFian equation $X_1dx_1 + \cdots + X_mdx_m = 0$ to reduce to an equation between just $m-1$ variables. The reduction considered is then always possible if all expressions of the form

$$(29) \qquad (0,\mu_1,\mu_2,\ldots,\mu_{2n'+1})$$

---

*In the previous considerations, which are only a translation of GRASSMANN's, there is hidden a theorem on the skew-symmetric determinant

$$\Delta = |(0,\mu)(1,\mu)\cdots(m,\mu)| \quad (\mu = 0,1,\ldots,m),$$

namely the theorem that the vanishing of all expressions of the form $(0,\mu_1,\ldots,\mu_{2n+1})$ entails the vanishing of all $(2n+1)$-row subdeterminants of Δ. Since every expression $(0,\mu_1,\ldots,\mu_{2n+1})$ is the square root of a $(2n+2)$-row subdeterminant of Δ (cf. Editorial Note 112), and in fact of a subdeterminant that contains exactly $2n+2$ elements of the diagonal of Δ, it is clear that we have to do here with one of the well-known FROBENIUS theorems on skew-symmetric determinants (cf. *Crelle* **82**, p. 242, Theorem V, 1877).

vanish and in addition $m \geq 2n'$, for the number* defined on p. 384 satisfies $n \leq n'$ and thus also $m \geq 2n$. For $m > 2n$ the reduction is carried out in No. 519, and for $m = 2n$ in Nos. 515–517.

In particular, the reduction is always possible if m is even, $= 2n'$, for then each of expressions (29) contain two equal indices and is therefore certainly zero.

In No. 522 is proved the important, and before GRASSMANN unknown, theorem that *the equation* $X_1 dx_1 + \cdots + X_m dx_m = 0$ *is always reducible to an equation between just* $2n' - 1$ *variables, provided all the expressions* (29) *vanish.*

Thus if $m \geq 2n'$, one can according to what has gone before always introduce such new variables $y_1, \ldots, y_{m-1}, t$ that the equation $\sum X_\mu dx_\mu = 0$ takes the form

$$\sum_{\mu=1}^{m-1} X_\nu(y_1, \ldots, y_{m-1}, 0)dy = 0. \tag{30}$$

But now by assumption all expressions (29) are equal to zero, and therefore also remain zero if in them one sets $x_1 = y_1, \ldots, x_{m-1} = y_{m-1}, x_m = 0$. Thus if $m - 1 \geq 2n'$, then the conditions are satisfied under which equation (30) can be reduced to an equation between just $m - 2$ variables. When one repeats this procedure sufficiently often one finally obtains the equation $\sum X_\mu dx_\mu = 0$ reduced to one between just $2n' - 1$ variables.

From this point, in Nos. 524 and 525 the proof that the condition derived in No. 511 is not just necessary but also sufficient is treated, and thus *the* ***Pfaffian*** *expression* $X_1 dx_1 + \cdots + X_m dx_m$ *can always be brought to the form* $U_1 du_1 + \cdots + U_{n'} du_{n'}$ *with just* n' *differentials if and only if all expressions* (29) *vanish.* This theorem too was new at GRASSMANN's time.

According to No. 511 the reduction considered is certainly only possible if all expressions (29) vanish. If this condition is satisfied, one first brings the equation $\sum X_\mu dx_\mu = 0$ to the form of an equation between just $2n' - 2$ variables, which according to the above is always possible. In this equation, according to the process of PFAFF, one of the variables, let it be u_1, is set to a constant, and then the resulting equation between $2n' - 2$ variables is reduced to one between $2n' - 3$. One of these $2n' - 3$ variables is again set to a constant, call it u_2, and so forth. If one applies this procedure r times, one has set r functions $u_1, \ldots, u_r$ of the original variables to constants and still has an equation between $2n' - 2r - 1$ variables. Thus if one makes $r = n' - 1$ and also sets the last remaining variable $u_{n'}$ to a constant, then obviously $u_1 = \text{const}, \ldots, u_{n'} = \text{const}$ is an assembly that integrates the equation $\sum X_\mu dx_\mu = 0$. According to No. 502 one must therefore have

$$\sum_{\mu=1}^{m} X_\mu dx_\mu = \sum_{\nu=1}^{n'} U_\nu du_\nu,$$

and the theorem is thereby proved. At the same time it is clear that the definition of each u_κ requires the integration of a certain ordinary differential equation that one can formulate without having previously defined $u_1, u_2, \ldots, u_{\kappa-1}$.**

(Nos. 525–529). From the theorems just proved it follows that a PFAFFian expression in $2n'$ variables is reducible to one with only n' differentials under all

*It is a defect of GRASSMANN's presentation that the letter n is used with different meanings. We therefore denote only the number defined on p. 384 by n and otherwise write n' for n.

**It is true that GRASSMANN did not say this explicitly, but it was without doubt known to him, as follows from his remarks in No. 523.

circumstances, and from this it follows again that the equation $\sum X_\mu dx_\mu = 0$ is never reducible to an equation between fewer than $2n - 1$ variables, *where by n is meant the number defined on* p. 384. Thus even if the equation could only be reduced to an equation between $2n - 2$ variables, one could reduce the expression $\sum X_\mu dx_\mu$ to an expression with just $n - 1$ differentials, and thus according to No. 511 all expressions

$$(0, \mu_1, \ldots, \mu_{2n-1})$$

would equal zero, which conflicts with the definition of the number n.

Finally, at this point the integration of the equation $\sum X_\mu dx_\mu = 0$ can be carried out, at least if one is limited to integrating assemblies of the form $u_1 = \text{const}$, $\ldots$, $u_{n'} = \text{const}$.* Thus if n is the number defined on p. 384, then there is no integrating assembly of this type, for which $n' < n$; on the other hand there is an integrating assembly for which $n' = n$, and according to No. 503 one can find all these integrating assemblies, provided one has cast the expression $\sum X_\mu dx_\mu$ to the form $U_1 du_1 + \cdots + U_n du_n$.

Now that we have become familiar with GRASSMANN's investigations into PFAFF's problem, we will briefly review what GRASSMANN has added to the contributions of his predecessors.

GRASSMANN's merit consists, first, in that—as we can express it today—he has completely developed the invariant theory of an arbitrary PFAFFian equation to a certain degree. Thus he shows that to each given PFAFFian equation $X_1 dx_1 + \cdots + X_m dx_m = 0$ there belongs a certain whole number, which is a characteristic for the equation. This whole number lies between the limits 1 and $\frac{1}{2}(m+1)$, including the limits, and can be found without integration. If it has the value n, the equation $\sum X_\mu dx_\mu = 0$ can, by the introduction of new variables, be reduced to a PFAFFian equation in $2n - 1$ variables, but not to one in less than $2n - 1$, and in addition the left side of the equation $\sum X_\mu dx_\mu = 0$, that is the PFAFFian expression $\sum X_\mu dx_\mu$, can be reduced to an expression $U_1 du_1 + \cdots + U_n du_n$ with exactly n differentials, but not to one with less than n differentials. Herewith it is actually proved that every PFAFFian equation $\sum X_\mu dx_\mu$, whose characteristic value has the value n, can take the normal form

$$y_1 dy_2 + y_3 dy_4 + \cdots + y_{2n-3} dy_{2n-2} + dy_{2n-1} = 0,$$

where $y_1, \ldots, y_{2n-1}$ denote mutually independent variables; in other words, it is proved that the number n is the only invariant of the PFAFFian equation $\sum X_\mu dx_\mu = 0$. It is true that GRASSMANN did not express his result in this form, and in addition the so-called normal form is not found in his work, but nevertheless one can justly say that he has given the criteria from which one can recognize into what normal form a given PFAFFian equation can be cast.

For the *Pfaffian expression* $\sum X_\mu dx_\mu$ the question remains to be settled whether the form $U_1 du_1 + \cdots + U_n du_n$, into which it can be cast, is capable of yet a further simplification or not, and thus whether the expression $\sum U_\nu du_\nu$ can be reduced to an expression with n differentials but only $2n - 1$ variables, or not; for the reduction to an expression with only $2n - 2$ variables is certainly impossible, since otherwise the number of differentials could be reduced to $n - 1$. For this

*That GRASSMANN considered only integrating assemblies of this form is shown in the Remark following No. 503.

question too GRASSMANN found the answer:* The reduction to an expression in $2n-1$ variables is possible if and only if, for each given expression $\sum X_\mu dx_\mu$ all expressions of the form $(\mu_1, \mu_2, \ldots, \mu_{2n})$ vanish, meaning by $\mu_1, \ldots, \mu_{2n}$ numbers from the series 1, 2, ..., m. If not all of these expressions vanish, then $\sum X_\mu dx_\mu$ can take the form

$$y_1 dy_2 + \cdots + y_{2n-1} dy_{2n},$$

while if all vanish, $\sum X_\mu dx_\mu$ can take the form

$$\rho(y_1, \ldots, y_{2n-1})(y_1 dy_2 + y_3 dy_4 + \cdots + y_{2n-3} dy_{2n-2} + dy_{2n-1}),$$

where both times the y denote mutually independent variables. There is therefore only lacking the demonstration that the last of these expressions can always be reduced to one of the same form in which ρ has the value 1; but even without this simplification, whose possibility only CLEBSCH had recognized,** what GRASSMANN had achieved for the theory of the invariants of a PFAFFian expression is still most noteworthy, for with him had basically been found the criteria by which one can recognize to which of the two possible normal forms a given PFAFFian expression is reducible, and exactly with respect to the correctness and completeness of the criteria CLEBSCH lagged considerably behind GRASSMANN.***

Whichever actual formulation of the normal form of a given PFAFFian equation may be feasible, GRASSMANN had indeed shown that it can be achieved by the integration of a series of ordinary differential equations and that all these systems can be given without integration, but one can, after the prior work of JACOBI, discover in this no special contribution. The question of whether the order of the required integrations can be reduced, GRASSMANN had not considered; in this connection one must therefore be supplied with the almost contemporaneous investigations of NATANI**** and CLEBSCH, although even these did not bring the question to a conclusion, which was reserved for the following decade.

On the other hand one cannot undervalue what GRASSMANN has achieved for systems of equations of the form (12), and in connection with this one must also mention his characteristic notation, which is exactly contemporaneous with that of JACOBI–CAYLEY, and in fact is superior, insofar as GRASSMANN's symbols always directly call to mind the PFAFFian expression for which it was formed, whereas in the symbol $(1, 2, \ldots, 2n)$, as such, no relation to PFAFF's problem can be recognized. On this account GRASSMANN's notation is applicable to systems of PFAFFian equations without further ado, which is not the case for that of JACOBI–CAYLEY.

*To be sure he had only spoken explicitly of the case $m = 2n$, see the Remark following No. 416.

See his two articles "Ueber das Pfaffsche Problem", *Crelle* **60, pp. 193–251 (1862; the article is dated 28 Sept. 1860), **61**, pp. 146–179 (1863, dated 25 Jan. 1861), as well as a short preliminary note of March 1861, **59**, pp. 190–192 (1861).

***LIE was the first to note this, and at the same time stated that GRASSMANN's main accomplishments for the theory of the PFAFFIAN problem are to be sought in the formulation of the correct criteria. Cf. the Obituary of GRASSMANN, *Math. Ann.* **14**, p. 28 (1879).

****See his article (dated January 1860) "Ueber totale und partielle Differentialgleichungen", *Crelle* **58**, pp. 301–328, which only appeared in 1861 and therefore certainly had no influence on GRASSMANN, just as little as the works of CLEBSCH.

Supplementary Notes

Part I. The Elementary Conjunctions of Extensive Magnitudes

Chapter 1. Addition, subtraction, multiples and fractions of extensive magnitudes

In his "definitions", Numbers 1 and 2, Grassmann admits the operations of a real vector space. "Magnitude" is taken as a primitive concept, as we should take "element" of a vector space, or "vector" — in the abstract, not the geometric, sense.

Definition Number 3 is a little fuzzy. Any real number is "derivable" from any other non-zero real number, so π is as good a unit as 1. One has to stretch a little beyond his context for "1". To be precise, one would say something like: "The numbers have a unit 1 that is distinguished by its multiplicative properties. I call this number the *absolute* unit." Really, any vector $\neq 0$ is a unit.

Definition Number 4 is that of a "linearly independent" set of magnitudes.

In Definition Number 5 he insists that a basis (or at least a linearly independent set) be chosen, and defines an extensive magnitude to be a linear combination of some such set. From the remarks to Number 3, every magnitude is an extensive magnitude, and vice versa.

Of course, this observation changes if one works in the *full* Grassmann algebra (exterior algebra) $\Lambda(V) = \Lambda^0(V) \oplus \Lambda^1(V) \oplus \cdots \oplus \Lambda^n(V)$. His reference to "the original units" is imprecise, for reasons cited in connection with Number 3. What he is saying here about "first order" is the extensive magnitudes of first order are to be those that are linear combinations of a fixed linearly independent set of magnitudes. Thus the concept of "first order" defines a vector space (in the abstract).

Then addition and subtraction are defined "coefficientwise" in terms of this basis, as is multiplication by numbers. He then derives the fundamental properties of these operations, summed up in Number 8. (Nowadays, we take an approximately minimal subset of these properties as axioms for a vector space, and prove the existence of bases.)

There follow, in Numbers 16–20, basic properties of linear combinations and linear (in)dependence, including the "exchange property" (Numbers 19–20).

Numbers 21–24 discuss bases and dimension. His remark after Number 24 notices the comment about Number 3 noted above. Numbers 25, 26 give what we would write $\dim(U \cap V) + \dim(U + V) = \dim U + \dim V$. Proofs are essentially those given in modern texts.

Numbers 27–32, with the exceptions of Numbers 30–32 (that really go nowhere) are just basic properties of linearly independent sets and the operations in vector spaces.

His "shadow" in Number 33 is more carefully defined than one might expect from his times. It is, in modern terms, the "projection on the subspace $\langle a_1, \ldots, a_m\rangle$ relative to the decomposition $\langle a_1, \ldots, a_n\rangle = \langle a_1, \ldots, a_m\rangle \oplus \langle a_{m+1}, \ldots, a_n\rangle$. [Here $\langle u_1, \ldots, u_t\rangle$ is used for the vector subspace of linear combinations of $u_1, \ldots, u_t$.]

Numbers 34–36 are just further properties of operations in vector spaces.

Chapter 2. The product structure in general

It is here that Grassmann launches into what is currently called "multilinear algebra". He is already breaking ground in Chapter 1, but a number of his contemporaries would probably have said "oh, yes" to that, even if they hadn't thought about it in that way before. More would have been mystified by what comes next.

In order to view Definition 37 in its proper generality one should probably speak about a bilinear pairing $U \times V \to W$, where U, V, W are vector spaces. If $\{e_1, \ldots, e_n\}$ is a basis for U, $\{f_1, \ldots, f_m\}$ is a basis for V, and if $[uv]$ denotes the image of the pair $(u, v) \in U \times V$, then all products $[uv]$ are determined by the $[e_i f_j]$ as in his formula, where he takes $V = U$. It is clear that he wants to allow (indeed, ultimately to insist) that the magnitudes in W be of different kind than those in V.

Numbers 38–42 then just express the bilinearity of such a product.

The notion of Number 43 amounts to "specialization" or "substitution" or "homomorphism". Here one notes from his remark that he is not necessarily assuming products to be associative. The way he has introduced products, really as a map $V \times V \to W$, he is not yet equipped to speak about products of more than two factors, so a good context for the remark, and for Number 45 below, has not been established. Thus to make sense of Number 45 (with conventions of Number 7), he would need something like this: "Suppose we have bilinear mappings (products) $V \times V \to W_1$, $W_1 \times V \to W_2$, $W_2 \times V \to W_3$, $\ldots$, $W_{m-1} \times V \to W_m$. Then (a) holds, where $[v_1, v_2, \ldots, v_m]$ means $[\cdots[[[v_1 v_2]v_3]v_4]\cdots v_n]$."
The rest of §2 expresses the "multilinearity" of such products.

By "unit products" in Number 48 he seems to mean products $[e_i e_j]$ (or more generally $[e_{i_1} \cdots e_{i_m}]$) where the e_i are his "original units". It seems that Definition 50 amounts to saying that the "defining equations" are bilinear, or multilinear. It's really stronger than that, and by a lot. For instance one might have a product which is bilinear with $[e_1 e_1] = 0 = [e_2 e_2] = [e_2 e_1]$, while $[e_1 e_2] \neq 0$. But this would not satisfy Number 50 as he uses it. A modern proof might go a little faster, if one gave a proper formulation to the notion of linear product structure, which might go something like this: There is a bilinear mapping $V \times V \to U$, where U is a subspace of $V \otimes V$ and our product has a factorization $V \times V \to V \otimes V \to W$, where the image of U in W is zero. If the mapping $V \times V \to U$ sends (v_1, v_2) into $v_1 \otimes v_2$, then $[v_1 v_2] = 0$. If $(v_1, v_2) \mapsto 0$, then $[v_1, v_2] \simeq v_1 \otimes v_2$, i.e. there is no defining equation. Now his definition requires something more, really that U be stable under the action of all linear substitutions from V. The symmetric and skew-symmetric elements of $V \otimes V$ form subspaces satisfying this condition, and

$V \otimes V = S^2(V) \oplus \Lambda^2(V)$. Each of these summands is irreducible for the action of the group $\mathrm{GL}(V)$ of all invertible linear transformations of V, and they are not isomorphic as $\mathrm{GL}(V)$ modules (different dimensions). It follows from general algebraic principles that $U = S^2(V)$ or $U = \Lambda^2(V)$, so that [ ] is skew-symmetric or commutative, respectively ("combinatorial" or "algebraic" in his terminology).

One may view Number 51 as providing his rationale for concentrating on products satisfying the conditions of Chapter 3. It is almost certainly his deepest theorem up to this point.

Chapter 3. Combinatorial product

In Numbers 52 and 53 he starts with a "system of units"— i.e. a basis for a vector space, and considers arbitrary products (formally) of these, always following the association $[ab \cdots z] = [[\cdots [[ab]c] \cdots y]z]$, subject to the conditions that interchanging the last two (unit) factors changes the sign. Then he extends this by multilinearity. We'd say that if V is a vector space with the original system of units as basis, then $\Lambda^j(V)$ (the jth exterior power) is the vector space linearly generated by the products involving precisely j of these units, subject to his relations. From this, he is able to speak of products of "elementary factors" and derive Number 58. Extended to multilinearity it says that if $B \in \Lambda^r(V)$, $C \in \Lambda^s(V)$, then $[BC] \in \Lambda^{r+s}(V)$ is equal to $(-1)^{rs}[CB]$. At this point he has only defined these "products" of higher order by "rearranging parentheses" in products of elementary factors. It is not until later, culminating in the associative law of Number 80, that he really justifies speaking in the terms used here.

Number 63 shows that the coordinates of the units in a product of n factors are the n-rowed minors of the matrix of coefficients. Then, up to Number 68, he goes over to the multilinear point of view and derives properties.

To §2: Number 69 is the basis for saying $\dim(\Lambda^r(V)) = \binom{n}{r}$ if $\dim V = n$, and Number 70 for saying that two "pure m-vectors" $a_1 \wedge \cdots \wedge a_m$ and $b_1 \wedge \cdots \wedge b_m$ are scalar multiples one of the other if and only if the vector subspaces $\langle a_1, \ldots, a_m \rangle$ and $\langle b_1, \ldots, b_m \rangle$ of V are equal. Thus the m-dimensional subspaces of V are a part (actually a subvariety) of the projective space of lines in $\Lambda^m(V)$; this means of geometrizing the set of such subspaces is one of the aspects of Grassmann's work that has contributed most to modern geometry. The next sections deal with the effects of linear transformation of the factors on a combinatorial product, starting with "elementary [...]" or "linear evolutions", or (after Dieudonne, *La Géométrie des Groupes Classiques*) "transvections". Number 76 is a result of the fact (and Grassmann's arguments offer a proof of this) that the "linear evolutions" generate the special linear group $V \to V$ of determinant 1).

In §3, he proceeds from exterior products of "units" in $\Lambda^m(V)$ and $\Lambda^{n-m}(V)$ to get units (or 0) in $\Lambda^n(V)$ (possibly with sign change), and by bilinearity to get the exterior products $\Lambda^m(V) \times \Lambda^{n-m}(V) \to \Lambda^n(V)$. (Here "$n$" has nothing to do with $\dim V$, although $\Lambda^n(V) = 0$ if $n > \dim V$.)

Number 84 contains the germ of the idea of giving equations that tell which elements of $\Lambda^m(V)$ are pure m-vectors, i.e. pure outer products, or alternatively, represent m-dimensional subspaces of V. Such equations define what are called "Grassmann varieties" (e.g. Hodge and Pedoe, *Methods of Algebraic Geometry*, vol. 2).

In §4, we see this: The "C" of Number 87 represents (as Grassmann notes) the intersection of the subspaces represented by A and by B.

His notion of "supplement" is defined in terms of a fixed basis for the "principal domain", taken in a fixed order: $[e_1 \cdots e_n] = 1$. This last "$= 1$" is not what we'd do today; rather, we'd say, "fix the basis $e_1 \wedge \cdots \wedge e_n$ for $\Lambda^n(V)$. Then if E is a unit of $\Lambda^j(V)$, the supplement E' of E is the unit in $\Lambda^{n-j}(V)$ such that $[EE']$ ($= E \wedge E'$, in modern notation) $= e_1 \wedge \cdots \wedge e_n$. In general if $A \in \Lambda^j(V)$, $B \in \Lambda^{n-j}(V)$, $[AB]$ is of the form $\lambda(A,B)e_1 \cdots e_n$, where $\lambda(A,B)$ is a number." Then the formula of Number 89 reads $|E = \lambda(E,E')E'$.

In Number 90, Grassmann extends the map "$|$" of units in $\Lambda^j(V)$ to units of $\Lambda^{n-j}(V)$ by linearity to a linear map $\Lambda^j(V) \to \Lambda^{n-j}(V)$.

To §5: To give the regressive product a coordinate-free setting, one really needs the notion of dual vector space to V, $V^* =$ all linear maps $V \to \mathbb{R}$, i.e. all linear functionals on V. Then $\dim V^* = n$, and one can form all $\Lambda^j(V^*)$. There is a natural identification of $\Lambda^j(V)$ with the dual space $\Lambda^j(V)^*$ of $\Lambda^j(V)$. Having chosen (as above, for example) a basic element for $\Lambda^n(V)$, one has a pairing as λ above that makes $\Lambda^{n-j}(V)$ be isomorphic to $\Lambda^j(V)^*$, so to $\Lambda^j(V^*)$. Now suppose $r+s>n$. Then $\Lambda^r(V) \simeq \Lambda^{n-r}(V^*)$, $\Lambda^s(V) \simeq \Lambda^{n-s}(V^*)$, and the outer product of elements of $\Lambda^{n-r}(V^*)$ and $\Lambda^{n-s}(V^*)$ is in $\Lambda^{2n-r-s}(V^*)$, with $2n-r-s<n$, so corresponds to an element of $\Lambda^{n-(2n-r-s)}(V)$, i.e. to an element of $\Lambda^{r+s-n}(V)$. This is the "regressive product" of the two original elements. (If $r+s=n$, Grassmann's unfortunate identification "$[e_1 \cdots e_n] = 1$" makes for some confusion.)

Normally, we nowadays have little use for the "regressive product". Grassmann does make it fit into a more general product on $\Lambda(V)$ (the "relative product") and it has a nice geometric meaning. If $A \in \Lambda^r(V)$, $B \in \Lambda^s(V)$ represent subspaces of respective dimensions r, s (say U and W), then there are subspaces $U^\perp, W^\perp \in V^*$, of respective dimensions $n-r$, $n-s$ ($U^\perp = \{f \in V^* | f(U) = 0\}$), and $U^\perp + W^\perp$ is represented by an element of $\Lambda^{2n-r-s}(V^*)$ if $U^\perp \cap W^\perp = \{0\}$. Then $U^\perp + W^\perp = (U \cap W)^\perp$, so $(U^\perp + W^\perp)^\perp = U \cap W$ back in V. This subspace corresponds to the regressive product of A and B. For example, if $A = [e_1 \cdots e_r]$, $B = [e_{n-s+1} \cdots e_n]$, $r+s>n$, then $U = \langle e_1, \ldots, e_r \rangle$, $V = \langle e_{n-s+1}, \ldots, e_r \rangle$, $U \cap V = \langle e_{n-s+1}, \ldots, e_r \rangle$, $U^\perp = \langle e^*_{r+1}, \ldots, e^*_n \rangle$, $V^\perp = \langle e^*_1, \ldots, e^*_{n-s} \rangle$, where the $\{e^*_i\}$ are the "dual basis" to $e_1, \ldots, e_n$, $U^\perp + V^\perp = \langle e^*_1, \ldots, e^*_r \rangle$, $(U^\perp + V^\perp)^\perp = \langle e_{n-s+1}, \ldots, e_r \rangle$, which is $U \cap V$, and is what the requisite product gives: $|A = [e_{r+1} \cdots e_n]$, $|B = \pm[e_1 \cdots e_{n-s}]$, $[|A, |B] = \pm[e_1 \cdots e_{n-s} e_{r+1} \cdots e_n]$, $|[|A, |B] = \pm[e_{n-s+1} \cdots e_r]$. (From this point of view, $[e_1 \cdots e_r]$ corresponds to that element of $\Lambda^{n-r}(V)^*$ that is ± 1 on $[e_{r+1} \cdots e_n]$ and 0 on other "units" if its basis, so to $\pm[e^*_{r+1} \cdots e^*_n]$, $[e_{n-s+1} \cdots e_n]$ to $\pm[e^*_1 \cdots e^*_{n-s}]$, and, in turn, $[e^*_{r+1} \cdots e^*_n] \wedge [e^*_1 \cdots e^*_{n-s}] = \pm[e^*_1 \cdots e^*_{n-s} e^*_{r+1} \cdots e^*_n]$.

Now $[e_{n-s+1} \cdots e_r] \in \Lambda^{s+r-n}(V)$ corresponds to the element of $\Lambda^{2n-s-r}(V)^*$ that is ± 1 on $[e_1 \cdots e_{n-s} e_{r+1} \cdots e_n]$ and 0 on other units, thus to $\pm[e^*_1 \cdots r^*_{n-s} e_{r+1} \cdots e^*_n]$. Thus we again have $[|A, |B] = \pm[e_{n-s+1} \cdots e_r]$. (The signs are determined in a consistent way.) The point is that it is better to think of "$|$" as a mapping $\Lambda^j(V) \to \Lambda^{n-j}(V)^*$ (or from $\Lambda^j(V^*)$ to $\Lambda^{n-j}(V)$) than without the dual spaces; of course the concept of dual space was not there for Grassmann.

In Number 110 what he is saying is that one can replace the original basis e_1, ..., e_n by any basis resulting from this by a linear transformation of determinant 1, and all his multiplication laws are unchanged.

Perhaps part of the failure of the mixture of progressive and regressive products to catch on lies in the failure of the associative law in general, as noted in Number 125.

In Number 127, he makes sure, by his exclusions, that $\{A_1, \ldots, A_u\}$ and $\{A_{u+1}, \ldots, A_v\}$ have no nontrivial linear combinations in common. Thus the map $C \mapsto C_1$ is a well-defined "projection". That it is *linear* is the content of Number 130.

In Number 129, the denominator $[BC]$ is really in $\Lambda^n(B+C)$, and is non-zero; he treats it as a number by his identification "$[e_1 \cdots e_n] = 1$".

Number 133 is just preparing ground for his presentation of "Cramer's rule" in Number 134. There he derives the rule using exterior multiplication in place of the more customary determinantal rules (in Solution 1) for the case where the matrix of coefficients is invertible. The non-invertible case can't be regarded as a "solution" until he tells how to select a maximal linearly independent set of columns from the coefficient matrix. This can of course be done by "Gaussian elimination", but that is a more efficient say of solving the original system, anyway. "Solution 2" amounts to trying to reduce to the "homogeneous" case.

In Number 136 he arrives at the resultant of two polynomials (the quantity $[u_1 u_2 \cdots u_{n_1+n_2}]$ of p. 90). It is on this discovery that he staked his priority claim against Cauchy.

Chapter 4. Inner product

This chapter is quite interesting; from a modern viewpoint what Grassmann does is to show that his identification "$[e_1 \cdots e_n] = 1$" and his "relative product" give a structure of inner product space (with a positive definite inner product and with the units of a given degree as orthonormal basis) to each $\Lambda^d(V)$, where V is the "principal domain". Once one has the inner product on V, one can reconstruct everything he does in the chapter. He does concentrate on V, the principal domain, in Numbers 188ff. He does not seem to have noticed that one must prove the Cauchy–Schwarz inequality in order for "$< AB$" of Number 195 to make sense, that is that $|[A|B]| \leq \alpha\beta$, in his notation (cf. Editorial Note 68). In that case all his inner products can then be expressed as one would contract tensors, using the inner product in V. [This is not to detract from his achievement, but rather to point out how much he has achieved.]

The concept of "circular evolution" would now be called a "plane rotation or reflection", depending on the $\pm$ sign [Number 154]. What he shows then is, in effect, that such transformations generate the orthogonal group of V (relative to its inner product), so that he can show what we'd call "independence of the choice of orthonormal basis for V" (cf. Number 168). Then "normal shadows" of §3 are orthogonal projections, in case the degrees are equal (cf. Remark after Number 164).

Chapter 5. Applications to Geometry

§1 is a summary of the notions of 3-dimensional vector geometry, with careful distinctions between points and "displacements" (or "vectors", or "translations"). In §2, "points at infinity", "lines at infinity", etc. are introduced—in a way that was probably current at the time, perhaps from Möbius—so as to complete the affine space to a projective space.

The bulk of Chapter 5 consists of expressing equations for various geometric loci in Grassmann-algebra terms. There are no remarkable derivations or impressive geometrical results here.

Part II. The Theory of Functions

Chapter 1. Functions in general

Other publications on the history of 19th century mathematics, especially analysis, have reviewed the evolution of the notion of "function"—in Grassmann's case, he leaps at once to the abstract one, although in fact all his functions are pretty classical ones.

§1 amounts to connecting systems of n real-valued functions, each of (the same) m real variables to mappings $f : \mathbb{R}^m \to \mathbb{R}^n$, with therefore vector-valued functions of vector variables.

§2 deals with symmetric functions of vector variables, as obtained by "polarization" (the "arithmetic mean" of Number 353) from arbitrary functions formed from pure products (whether "combinatorial", or "algebraic", or more general).

In §3, one would say nowadays that Grassmann is working in the "symmetric algebra" based on the underlying vector space ("principal domain", "Hauptgebiet") V. His "algebraic quotient" of Number 374 may then be viewed as an element of the "field of fractions" of the "integral domain", which is the symmetric algebra of V. This should *not be confused* with the "quotient" of Number 377, which is really a *linear transformation* $V \to V$.

§4: Here Q is the unique linear transformation sending a_j to b_j for each j, $1 \leq j \leq n$. He adds them, multiplies by scalars, and (in Number 381) shows that the so-called "matrix units" relative to a given basis $\{e_1, \ldots, e_n\}$ for V form a basis (the "$E_{r,s}$") for the linear mappings $V \to V$. The "power" $[q^n]$, with his identification of $\Lambda^n(V)$ with numbers, is just the *determinant* of the linear transformation Q. Although he does not seem ready to compose linear transformations $V \to V$, so as to obtain a connection with products of matrices, what he says in Number 384 is very close to saying that "the determinant of the composite of two linear transformations is the product of the determinants."

We'd call the vectors $a_1, \ldots, a_m$ of Number 386 a "basis for the image of Q".

Numbers 388 and 389 deal with "eigenvalues", especially the case of distinct eigenvalues.

Number 390 could be expected to culminate in a version of the "Jordan normal form" for a (complex) matrix. It seems that Grassmann almost gets there, the omission being that he apparently doesn't allow for more than one "Jordan block", i.e. matrix of the form

$$\begin{bmatrix} \lambda & 1 & 0 & \dots & 0 \\ 0 & \lambda & 1 & \dots & 0 \\ 0 & 0 & \lambda & \dots & 0 \\ \dots & \dots & \dots & \dots & \dots \\ 0 & 0 & 0 & \dots & \lambda \end{bmatrix}.$$

Number 391 has as its final conclusion what we'd call the "spectral theorem for real symmetric linear transformations" (or matrices), under the additional hypothesis $[Qa|a] \neq 0$ for $a \neq 0$ (which forces the matrix to be positive definite or negative definite, an observation he seems to miss (but Editorial Note 94 picks up)). However, his conclusions from (c) on are reached by fallacious reasoning; what he says amounts to saying that the relations (b) are preserved under every orthogonal transformation of the space spanned by $c_1, \dots, c_n$, and that is false unless Q is a scalar. In particular, the relation "$[Qb_2|a_2] = 0$" on p. 223 would force $x = 0$, so would constrain the kind of orthogonal transformation one could use. Editorial Note 93 does not pick up the error.

Nowadays one proves Number 391 either by maximizing the quadratic form $[Qx|x]$ on the unit sphere, or by appealing to the fundamental theorem of algebra after extending $[Qx|x]$ to a Hermitian form on $V + iV$. Then one gets an eigenvector for Q (necessarily one in V, with real eigenvalue); Q stabilizes the orthogonal complement, and one completes the proof by induction on dimension n. The requirement of Number 391 or that of Editorial Note 94 is not needed.

Regarding the Remark after Number 396. What can be simpler than this "proof by coordinates"? Choose the coordinate system so that center of circle is at $(0,0)$, desired point on positive axis, say at $(x,0)$. If circle has radius r, equation of circle is $x^2 + y^2 - r^2 = 0$ and "double-distance" is $(x-r)(x+r) = x^2 - r^2 = f(x,0)$. (If $(x,0)$ is inside the circle, $x^2 - r^2 < 0$, but "double distance" is product of directed distances.)

Number 402 just is the notion of defining a vector space by generators and (linear) relations. Here "linear" is not the same as "linear product"—see end of Chapter 2, Part I. The notion of "linear product" was put in modern terms in these notes there.

Number 405: The correspondence is that of projective space. That is, in each case one has coordinates $(\alpha, \beta, \gamma, \delta)$, not all 0, where for $\lambda \neq 0$, $(\lambda\alpha, \lambda\beta, \lambda\gamma, \lambda\delta)$ defines the same "point" or "circle".

Number 410: Here Grassmann comes back to the idea that he is working over a Euclidean vector space V with a distinguished ordered basis of unit vectors $e_1, \dots, e_n$. Then one has the corresponding basis (ordered monomials in $e_1, \dots, e_n$, without repetitions) for the exterior algebra $\Lambda(V)$ and that (ordered monomials, allowing repetitions) for the symmetric algebra of V.

Number 411: The earlier "matrix units" are now generalized to the corresponding ones for linear transformations of one vector space to another (in terms of fixed bases).

Number 418 is what we call the "triangle inequality".

Number 420 is the definition of "$\lim_{q\to 0+} f(q) = p$"; here of course Grassmann is getting ready for the notion of continuity, as well as that of "derivative".

Number 427: We'd write $C = \lim_{x\to a} \frac{f(x)}{f_1(x)}$.

Chapter 2. Differential calculus

In Number 431, he is working in the "symmetric algebra" of his "principal domain" or "Hauptgebiet", and A is either a monomial or a homogeneous element (of degree n).

Number 437 and some later assertions are forms of the "chain rule". All is well, because he assumes continuity of the derivatives and partial derivatives concerned.

Number 438 presents the "derivative as linear transformation", the modern way of teaching "vector calculus" (e.g. Spivak, *Calculus on Manifolds*, Theorem 2–1).

Chapter 3. Infinite series

Number 454: The notion of "true" series seems motivated by the behavior of the power series for an analytic function within its circle of convergence. It says that the series of norms is dominated by a convergent geometric series. Much of the rest of Chapter 3 is devoted to "Cauchy's integral theorem", using a discrete version of integration around a circle, with application, in Number 466, to deduce the analyticity of *twice* continuously differentiable functions of a complex variable. The last sections of Chapter 3 extend these results to multiple variables.

Chapter 4. Integral calculus

Number 505: Apparently one is thinking of $\boldsymbol{L}$ here as a multilinear function of n vector variables, with scalar values. Then the notation "$[\boldsymbol{L}a_1 \cdots a_n]$" stands for what one gets by completely skew-symmetrizing the quantity $\boldsymbol{L}(a_1, \ldots, a_n)$. The result is an alternating n-linear function $\Lambda(\boldsymbol{L})$ on the vector space V, or a linear function on $\Lambda^n(V)$. But the space of such functions is 1-dimensional, so $\Lambda(\boldsymbol{L})$ acts on $\Lambda^n(V)$ by a fixed scalar λ, and $[\boldsymbol{L}a_1 \cdots a_n] = \lambda[a_1 \cdots a_n]$ (cf. Editorial Note 111). The following Numbers, 508 and 509, seem to be a kind of "expansion by minors" in this setting. It is not clear how this connects with the last paragraph of Editorial Note 111; Cayley's "Pfaffian" is associated with a skew-symmetric matrix of $2n$ rows and columns, and is a polynomial of degree n in the entries; the square of this polynomial is the determinant (see Remark, p. 313).

The Editorial Notes on pp. 378–389 go systematically through the last sections of Grassmann, interpreting in more standard notation. One might remark that a 20th century treatment of the problem is to be found in C. Caratheodory, *Calculus of Variations and Partial Differential Equations of First Order*, 2nd English Edition, Chelsea, 1982 (Chapter 8).

Subject Index

Numbers without a preceding "p." refer to the Numbers in the text. Referenced material added by the editors of the Teubner Edition is enclosed {within braces}.